Medieval and Early Modern Philosophy and Science

VOLUME 34

The titles published in this series are listed at *brill.com/memps*

Quantifying Aristotle

Quantifying Aristotle

The Impact, Spread and Decline of the Calculatores *Tradition*

Edited by

Daniel A. Di Liscia
Edith D. Sylla

With the collaboration of

Paul J.J.M Bakker

TUTA SUB AEGIDE PALLAS · 1683 ·

BRILL

LEIDEN | BOSTON

Cover illustration: Portrait of Nicole Oresme. Ms. Paris, Bibliothèque nationale de France, fr. 565, 1r (detail).

Library of Congress Cataloging-in-Publication Data

Names: Di Liscia, Daniel A., editor. | Sylla, Edith Dudley, editor. | Bakker, Paul J. J. M.
Title: Quantifying Aristotle : the impact, spread, and decline of the Calculatores Tradition / edited by Daniel A. Di Liscia, Edith D. Sylla ; with the collaboration of Paul J.J.M Bakker.
Description: Leiden ; Boston : Brill, 2022. | Series: Medieval and early modern quantifying Aristotle philosophy and science, 2468-6808 ; volume 34 | Includes bibliographical references and index.
Identifiers: LCCN 2022011340 (print) | LCCN 2022011341 (ebook) | ISBN 9789004499829 (hardback) | ISBN 9789004512054 (ebook)
Subjects: LCSH: Aristotle. | Philosophy, Medieval. | Science, Medieval. | Science–Philosophy.
Classification: LCC B725 .Q36 2022 (print) | LCC B725 (ebook) | DDC 185–dc23/eng/20220503
LC record available at https://lccn.loc.gov/2022011340
LC ebook record available at https://lccn.loc.gov/2022011341

Typeface for the Latin, Greek, and Cyrillic scripts: “Brill”. See and download: brill.com/brill-typeface.

ISSN 2468-6808
ISBN 978-90-04-49982-9 (hardback)
ISBN 978-90-04-51205-4 (e-book)

Contents

Acknowledgements

It is my pleasant duty to thank a series of institutions and persons who, in one way or another, made possible the 2018 Munich conference, as well as the present publication. First of all, I would like to mention my 'home institute', the *Munich Center for Mathematical Philosophy* (MCMP) of the Ludwig-Maximilians-Universität München, which provided us with all the necessary assistance in organisational matters. Besides, it was a pleasure to collaborate on this occasion with the *Lehrstuhl für Wissenschafstgeschichte* of the same university. Special thanks should be addressed to the *Fritz Thyssen Stiftung* for its generous financial support of the meeting. I would also like to express my gratitude to Linda Grohmann for her valuable help with the production of the complex bibliography and a first formal verification of the chapters. I very much appreciate that Paul Bakker and Christoph Lüthy have welcomed this book to the prestigious series *Medieval and Early Modern Philosophy and Science*. Special thanks go to Paul Bakker for his accurate work as collaborator of this volume and for his engagement with the various aspects connected with the realisation of this project.

My warmest gratitude goes to my co-editor, Edith D. Sylla, for her dedicated and precious work on this volume. It is no secret that Edith Sylla more than anyone else is the person we all identify with this research topic, the *Calculatores*, a topic to which she contributed for many decades at the highest level of quality. Hence, not only is it a great honour for me that she accepted to participate in the conference and publication, setting the crown on this volume with a ground-breaking paper on Leibniz and his calculatory background. It was also a great relief for me that she was happy to assume the task of being co-editor of the present volume. She has read all the papers with her usual accuracy and provided many improvements to many of them. I am grateful for her kind corrections, suggestions, and various improvements.

Daniel A. Di Liscia
Munich / Copenhagen, October 2021.

Figures

Notes on Contributors

Harald Berger
is an Associate Professor at the University of Graz (Austria), Department of Philosophy. He teaches classic, medieval, and early modern philosophy. He specializes in late medieval philosophy, including manuscript studies and history of universities. Among his publications are the volumes: Albert von Sachsen, *Logik*. Lateinisch—Deutsch (Hamburg, 2010) and Heinrich Totting von Oyta, *Schriften zur Ars vetus* (Munich 2015) and a contribution on 'Albert of Saxony on Parts and Wholes,' in F. Amerini, I. Binini, M. Mugnai (eds.), *Mereology in Medieval Logic and Metaphysics* (Pisa 2019).

Joël Biard
is Professor Emeritus at the University of Tours (France), Centre d'Études Supérieures de la Renaissance. His research fields are the late medieval philosophy of language, theories of mind and knowledge, and natural philosophy. Focussing on John Buridan and Blasius of Parma he has published, among severals papers, *Science et nature. La théorie buridanienne du savoir*, (Paris 2012), *Les* Questions sur l'âme *de Jean Buridan* (Paris 2019) and, with Aurélien Robert, the collective volume *La philosophie de Blaise de Parme. Physique, psychologie, éthique* (Florence 2019). He is currently preparing a monograph about Blasius of Parma's theory of matter.

Daniel A. Di Liscia
is Lecturer at the Ludwig-Maximilians-Universität München (Germany), at the Munich Center for Mathematical Philosophy. He carried out several research projects on the Calculators' tradition, especially on the geometrisation of natural philosophy and metaphysics. His most recent publications are 'Biagio Pelacani da Parma's Geometrisation of Latitudes and the Problems of the Mean Degree Theorem' (in: J. Biard and A. Robert [eds.], *La philosophie de Blaise de Parme. Physique, psychologie, éthique* [Florence 2019]) and *Eine Wiener* Expositio *zum* Tractatus de latitudinibus formarum. *Edition mit Einführung und Kommentar* (Vienna, 2022). He is currently editing Jacques Legrand's *Compendium utriusque philosophie*.

Elżbieta Jung
is a historian of science, medieval philosophy and theology focusing on the mathematical physics of the Oxford Calculators. Additionally, she works on the history of late medieval universities and the history of feminism. She is Chair of

the Department of the History of Philosophy at the University of Łódź (Poland) and is Director of the Center for Philosophy of Nature. She is currently working on critical editions of Richard Kilvington's *Quaestiones super Physicam*, *Quaestiones super De generatione et corruptione* (with Robert Podkoński), and *Quaestiones super libros Sententiarum* (with Monika Michałowska). Her most recent publication (with R. Podkoński) is *Towards the Modern Theory of Motion. Oxford Calculators and the New Interpretation of Aristotle* (Łódź 2020).

Edit Anna Lukács
works at the Austrian Academy of Sciences at Vienna (Austria). She has been a Visiting Fellow at the Pontifical Institute of Mediaeval Studies in Toronto (Canada) in 2018–2019. Her research focuses on the philosophy and theology of the Oxford Calculators, and their influence at the early University of Vienna. She has published numerous papers on Thomas Bradwardine and a Latin-German edition of selected parts from Thomas Bradwardine's *De causa Dei* (Göttingen 2013). Her latest book, *Dieu est une sphère* (Aix-en-Provence 2019) explores the theological uses of a geometric metaphor in the French Middle Ages.

Richard J. Oosterhoff
teaches early modern history at the University of Edinburgh (Scotland), with specialisms in intellectual history and history of science. Previously he held posts at the University of Cambridge, and fellowships at Houghton Library, the Huntington Library, and the Warburg Institute. His publications include *Making Mathematical Culture: University and Print in the Circle of Lefèvre d'Étaples* (Oxford 2018) and *Logodaedalus: Word Histories of Ingenuity in Early Modern Europe*, co-authored with Alexander Marr, Raphaële Garrod, and José Ramón Marcaida (Pittsburgh 2018).

Robert Podkoński
is a historian of natural philosophy at the Department of the History of Philosophy, University of Łódź (Poland). His research focuses on the Oxford Calculators' philosophical ideas and their impact on the development of early modern science. He has recently published editions of three treatises from Richard Swineshead's *Liber calculationum*: *De potentia rei*, *De luminosis*, and *De motu locali*, as well as an edition of two short treatises on motion ascribed to Richard Swineshead. He is currently preparing a critical edition of Richard Kilvington's commentary on Aristotle's *De generatione et corruptione* (with Elżbieta Jung).

Stephen Read
is Professor Emeritus of History and Philosophy of Logic at the University of St. Andrews (Scotland). He is the author of *Relevant Logic* (Oxford 1988) and *Thinking about Logic* (Oxford 1995), editor of *Sophisms in Medieval Logic and Grammar* (Dordrecht 1993), editor and translator of *Thomas Bradwardine: Insolubilia* (Leuven 2010), translator of *John Buridan: Treatise on Consequences* (New York 2015), and joint editor (with Catarina Dutilh Novaes) of *The Cambridge Companion to Medieval Logic* (Cambridge 2016). In addition, he is the author of many articles on contemporary and medieval philosophy of logic and language. He led the Leverhulme-funded project 'Theories of Paradox in Fourteenth-Century Logic: Edition and Translation of Key Texts' (2017–2021).

Sabine Rommevaux-Tani
is researcher at the CNRS in France and current director of the laboratory SPHERE. Focusing on Radulphus Brito, Thomas Bradwardine, Nicole Oresme, Blasius of Parma, and Christopher Clavius, she has published extensively on the history of mathematics and natural philosophy in the Middle Ages and the Renaissance. She is the author of *Les nouvelles théories des rapports mathématiques du XIVe au XVIe siècle* (Paris 2014). She has recently published a critical edition of the anonymous treatise *De sex inconvenientibus* (Paris 2021).

Fabio Seller
is Research Fellow in the History of Medieval Philosophy at the University of Naples Federico II (Italy). He has widely published on late medieval and Renaissance philosophy and science, specifically on the history of astronomy and astrology. Among his publications are *Scientia astrorum. La fondazione espistemologica dell'astrologia in Pietro d'Abano* (Naples 2009), *La città, il sole, le stelle. Temi astrologici e astronomici in Tommaso Campanella* (Naples 2015), and *Apocalittica e tempo della fine in Tommaso Campanella* (Milan 2020).

Edith D. Sylla
is Professor Emerita at North Carolina State University (Raleigh, North Carolina, USA). She works on the history of mathematics, physics, and their interrelations from the late Middle Ages to the early eighteenth century. She published extensively on the different aspects related to the Calculators. Her most recent paper is 'Infinity and Continuity: Thomas Bradwardine and His Contemporaries,' in S. Shapiro and G. Hellman (eds.), *The History of Continua: Philosophical and Mathematical Perspectives* (Oxford 2021).

Aníbal Szapiro
is Assistant Professor of History of Science at the University of Buenos Aires (Argentina). After completing his Ph.D. in philosophy on the history of early modern Astronomy (*Changes in the Astronomical Concept of* Horizon *during Modernity*, 2016), his research has focused on the relation between optics and astronomy in late medieval and early modern thought. His contribution to the present volume was produced during his postdoc stay at the Munich Center for Mathematical Philosophy at the Ludwig-Maximilians-Universität München (Germany) (2018–2019, DAAD Scholarship) on *The Calculators' Tradition and the Geometrisation of Light.*

Mark Thakkar
is a historian of fourteenth-century scholasticism. He took his D.Phil. in Philosophy at Balliol College, Oxford (2005–2010), and worked as a lexicographer on the *Dictionary of Medieval Latin from British Sources* (2010–2014) before embarking on an edition of John Wyclif's logical works, for which he held a Leverhulme Early Career Fellowship at the University of St. Andrews (2014–2018). He is currently an Honorary Research Fellow at St. Andrews and a Research Associate at the IRHT in Paris.

Cecilia Trifogli
is Professor of Medieval Philosophy at the University of Oxford (UK), and Chairperson of the British Academy committee responsible for the series *Auctores Britannici Medii Aevi.* She has published extensively on the tradition of Aristotle's natural philosophy in the Middle Ages. She is the author of a book on the early English tradition of commentaries on the *Physics* (*Oxford Physics in the Thirteenth Century*, 2000). She also works on critical editions of medieval philosophical texts. She has recently edited (with Silvia Donati and Jennifer Ashworth) the *Quaestiones* on the *Physics* by the Oxford master Geoffrey of Aspall (Oxford 2017).

Introduction

Daniel A. Di Liscia

After the great age of the reception of Aristotle in the West—a period when the Philosopher's works were translated into Latin, variously interpreted, and gradually incorporated into the standard of the university curricula—Oxford saw the rise, during the first decades of the fourteenth century, of a new approach to natural philosophy: an approach commonly identified with a group of authors characterised, not unfairly, as the *Calculatores*. Thomas Bradwardine, Richard Kilvington, Walter Burley, Roger Swineshead, William Heytesbury, Richard Swineshead, John Dumbleton, and other thinkers more or less committed to this new 'research programme', did not abandon the Aristotelian framework, but transformed it to such an extent that it was scarcely understandable to more traditional readers. With the tools of the new logic and the development of special techniques (which John Murdoch has called 'analytical languages') they asked questions that previous generations had not asked and translated many old problems of 'qualitative physics' into problems of 'quantitative physics', primarily by formulating them using expressions which in one way or another involved the notions of infinity and continuity. '*Penes quid attenditur* …?' will henceforth be the main question through which a complex apparatus of *calculationes* and *sophismata* will support this new philosophical approach—an approach that, being internally connected to the logic and philosophy of language, began to achieve success in the field of natural philosophy, but rapidly expanded to theology, medicine, and moral philosophy. This new approach spread to almost all European universities after several decades, and its traces are observable in the works of Galileo and even Leibniz. But this success story is as naïve as it is partial. It is well documented that, simultaneously with the first generation of Calculators, its critics also arose, multiplying by the same measure. In some cases, the criticism was a question of 'internal affairs' within the calculatory tradition itself—as for instance, when rejecting Bradwardine's rule of motion in favour of another one. In other cases, however, the criticism was referred to the quantitative approach as such and its linguistic means. 'Aristotelianism' always played a decisive role in the dispute. Especially within Humanist circles, the attacks against the Calculators were sometimes disguised in a general criticism against scholastic Aristotelian philosophy. In other cases, much to the contrary, the Calculators were hardly reproved for not being 'true Aristotelians'. The fact is that the Calculators' stream not only repeated itself and conveyed its main doctrinal contents during a long period

 | DOI:10.1163/9789004512054_002

of time but also transformed their views in different ways depending on context. In both processes, in persisting on some fixed patterns and searching for innovative paths within them, the Calculators' tradition manifested a particular link to Aristotelian philosophy which deserves the best of our attention.

The following book is the final output of a meeting that took place in Munich in May 2018. Most of the presentations offered during that meeting are included in expanded and corrected versions. Our conference assembled a group of recognised scholars to discuss not only a particular point of interpretation of a single text but also, and above all, to explore how a tradition of thought as general and powerful as the *Calculatores* tradition was transmitted. Of course, we discussed single authors and the problems of interpretation that arise by reading their works, but for a good part, we tried to address some of the questions that are related to this tradition as such: How did the *Calculatores* become so widely recognised? Were they essentially 'nominalist', 'realist', or something completely different? Could they be aligned at all with a particular Aristotelian school? What is the difference between Alvarus Thomas' treatise *De triplici motu* and the works of William Heytesbury and Richard Swineshead that lie behind it? How well did the commentators on Heytesbury's edition of 1494 understand the key ideas of the master? What stopped the *Calculatores*' quantifying programme or impeded its development? What were the factors that pushed the reproduction, or at least the imitation, of the original models? What was the true aim of the geometrisation of qualities and motions that James of Naples, Nicole Oresme and others proposed? What role did logic play at the different stages in the *Calculatores* tradition and what was the link between this tradition and the standards of the university curriculum? How did the transition from a manuscript to a print tradition alter the dynamics of learning and transmission? These are but a few of the questions we addressed in our conference, trying to focus on the Calculators' tradition as a whole and in its relationship to Aristotelian philosophy. For it seems to be hardly disputable that Aristotelian philosophy offers a much more adequate background to explore these questions than the categories of the scientific revolution.

Within this context, there are problems that need to be addressed. Which were the most decisive links between the *Calculatores* tradition and the general Aristotelian background, which regulated the exchange dynamic as a whole and occasionally conditioned the relationship to other philosophical movements such as Renaissance Platonism? Can we still speak of the 'kinematic' and 'dynamic' approaches to the analysis of motion as parts of 'mechanics' after realising that this distinction was expressed in the more usual terms of

Aristotelian methodology (simply meaning that an investigation takes place according to the 'effects' and according to the 'causes')? For, by using these categories, several of the Calculators' analyses gained a new scientific status over the standards of 'logic exercises' and were integrated into the commentary tradition. One obvious result of the incorporation of this new approach into the Aristotelian commentaries is that the structure and content of such commentaries changed. Perhaps the Calculators did not provide a mathematised view of nature but of the Aristotelian text. This feature alone, however, could have been sufficient for next generations to transmit the general idea that a 'quantified Aristotle' was not impossible, whether one wanted to have it, or not. And here, of course, opinions differ.

It is true that Leibniz, especially in his relationship to the scholastic past, was in many regards an exception within a mainly mechanical Cartesian world, which programmatically turned away from *occultae qualitates* and to which, even the very definition of 'motion' in an Aristotelian sense is simply untranslatable.[1] Nevertheless, it is sufficiently known that in the midst of the 'Scientific Revolution', Leibniz initiated a wide ranging hunt for a copy of Richard Swineshead's *Liber calculationum*, from which he finally ordered a full handwritten copy of the 1520 printing, currently conserved in the Niedersächsische Landesbibliothek.

We hope that this volume can also contribute to a more adequate appreciation of some aspects of the background of seventeenth-century philosophy and science. Simply by studying the sources more attentively, previous generations of researchers have made great progress over Kästner's declaration regarding Swineshead's 'metaphysisches Spinnengewebe' and the pseudophysics of 'Spitzfindigkeiten'. At the same time, Kästner was not completely wrong, at least if one compares Swineshead's techniques of discussion with the science of motion in Galileo's *Discorsi*. There are manifest differences between the two that one cannot overlook. Still, this sort of comparison is not the best approach for an adequate understanding of the problem. Whatever might be our philosophical convictions, the truth is that we cannot help but try again

1 See R. Descartes, *Le monde ou traité de la lumière*. In C. Adam and P. Tannery (eds.), *Oeuvres de Descartes*, Paris 1996, 11: 4–118, 39: 'Ils [les philosophes] avoüent eux-mesmes que la nature du leur est fort peu connuë; & pour la rendre en queque façon intelligible, ils ne l'ont encore sceu expliquer plus clairement qu'en ces termes: motus est actus entis in potentia, prout in potentia est, lesques sont pour moy si obscurs, que je suis contraint de les laisser icy en leur langue, parce que je ne les scaurois interpreter (Et en effet, ces mots: le mouvement est l'acte d'vn Estre en puissance, entant qu'il est en puissance, ne sont pas plus clairs, pour estre François).'

and again to get a more profound understanding of 'what *Physics* was then' and reconstruct as accurately as possible the threads of this logico-metaphysical cobweb.[2]

There seems to be little doubt that our authors all stay in a common line representing a particular *Denkstil* (to use the expression of Ludwig Fleck), which often enough went beyond the support of a specific doctrinal opinion. Even the characterisation as *Calculatores*—which might be too narrow for the complete range of the tradition surveyed in this book—is well documented. Most probably, it arose in connection with Swineshead's emblematic work, but it represented quite well the general approach of our authors. John of Holland used it not only in his *De motu*.[3] That he used the same expression in a collection of *Sophismata*, a work that also belies its author's link to the *Calculatores*, shows that this characterisation was not accidental.[4]

2 In a section on 'Rechenkunst und Algebra', the famous historian of mathematics, Abraham Gotthelf Kästner, refers to Swineshead (Suisset) and his *Liber calculationum*, admitting not having read the work (*!*), but assuming for sure that it cannot contain true physics, a science first 'invented' by Galileo: 'Wegen einiger genannten Bücher, gehörte Su.⟨isset⟩ unter die arithmetischen Schriftsteller, des Calculators wegen, so weit ich aus dem Titel urtheilen kann, würde ich ihn darunter nicht zählen, sondern argwohnen, dieses Buch enthalte, nur etwa in mathematisher Einkleidung, Spitzfindigkeiten aus dem, was man damahls Physik hieß, metaphysisches Spinnengewebe. Denn, dass Suisset mathematische, eigentlich rechnende, Physik dreyhundert Jahre vor ihrem Erfinder Galileus sollte gelehrt haben, daran zweifle ich, mit aller Achtung für den berühmten Mann, dessen Werke jetzo nicht mehr zu finden sind.' See A.G. Kästner *Geschichte der Mathematik*, 1: *Geschichte der Künste und Wissenschaften seit der Widerherstellung derselben bis an das Ende des achtzehnten Jahrhunderts. Erster Zeitraum; bis zum Ende des sechszehnten Jahrhunderts*, Göttingen 1796, 52.

3 M. Clagett, *The Science of Mechanics in the Middle Ages*, Madison, WI, 1959 and 1961 (University of Wisconsin Publications in Medieval Science, 4), 249[25]: 'Motus localis uniformiter difformis describitur *a calculatoribus* sic …' (my italics).

4 John of Holland, *Sophismata*, Leipzig, Universitätsbibliothek, Ms. 1360, 207ra–282rb: *Inc.*: 'Ad utrumque respiciens non facile pronuncio, nan tacens, karissimi, non facio quod vultis. Si autem scripsero, detractorum non effugio dentes …' *Expl.*: '… et quanto minor proportio acquiritur vel deperditur, tanto etiam velocitas. Secundo supponendum est [pro regulis potentiarum quod a proportione equalitatis nec minoris inequalitatis sit motus et secundum ahoc respondetur. Et sic est finis reverendi magistri Johannis Holandrini].' The passage between brackets was written by another hand which is responsible for the next text in the manuscript, Kilvington's *Sophismata*. John of Holland's *Sophismata* seem to be well-structured in six parts, including logical questions in parts 1–4 and physical problems in the last two parts: (1) *De insolubilibus*, 215ra; (2) *De scire et dubitare*, 217vb; (3) ⟨*De terminis confundentibus*⟩: *Aliqua sophismata de terminis … vim habentibus confundendi*, 232va; (4) ⟨*De modalibus*⟩: Sophismata unum vel plura in quo vel in quibus videbitur *de causis veritatis, falsitatis, possibilitatis et necesitatis* propositionum, ubi etiam interferentur sophismata de futuris contigentibus, 244va; (5) ⟨*De maximo et minimo*⟩, 250va; (6) ⟨*De velocitate motuum*⟩, 266^{v}. In the description of his part, John still clarifies: 'In sexta autem et ultima parte tractabuntur

Agostino Nifo, taking the side of the 'Aristotelian party', used the term *Calculatores* to refer to some extremely wrong ('arithmetic') interpretation of Aristotle ... and to academic opportunism. Leibniz, being familiar also with the Humanist criticism of Vives—which he thinks is a little bit superficial—stood for Swineshead's approach, which in his correspondence with Wallis he characterised very appropriately as 'semimathematical'.

Our survey of the *Calculatores* tradition starts with Cecilia Trifogli's contribution on Thomas Wylton, an early fourteenth-century English philosopher and theologian who was Fellow of Merton College, Oxford, where he taught philosophy, and then moved to Paris in order to study theology. One of his students was Walter Burley. Trifogli's chapter offers a detailed account of the discussion about the beginning and ceasing of the instant of time as found in a *quaestio* of Wylton's commentary on Aristotle's *Physics* that asks 'Whether the instant remains one and the same throughout the whole of time.' Trifogli first explains the exegetical context of Wylton's discussion. This originates from Aristotle's puzzle about the identity and diversity of the instant of time in *Physics* IV.10. Although Aristotle leaves this puzzle without an explicit solution, his most authoritative medieval commentators (Thomas Aquinas and Giles of Rome) assume that its solution is to be found in Aristotle's treatment of the same topic in *Physics* IV.11. Wylton points out, however, that this standard interpretation is not correct and, according to Trifogli, he seems definitely right on this point. Thus, he is led to explore a way out of the puzzle independent of Aristotle's discussion in *Physics* IV.11. The crucial ingredient of Wylton's solution is the rejection of the fundamental premise from which the puzzle originates, namely, that an instant can only cease to exist at an instant. Wylton instead argues that an instant ceases to exist in time. Trifogli also gives a presentation of Wylton's sophisticated defence of his claim concerning the temporality of the ceasing of the instant against objections arising from the indivisible nature of the instant. Finally, she compares Wylton's solution of the Aristotelian puzzle to that of Walter Burley. The two solutions are *prima facie* in conflict: the essential ingredient of Burley's solution is not that the ceasing of an instant happens

sophismata *de velocitate motuum* ubi dicta *Hesbri declarabuntur et de proportionibus et de calculatoriis* atque regule utiles ad faciendum et solvendum sophismata apponentur' (207ra, my italics). Thus, for John of Holland, above all Heytesbury ('doctor reverendus Wilhelmus Hesbri', 267vb) and the proportional analyses of motion belong to the same spectrum of ideas that he associates with the *Calculatores*. However, the notion of *calculator(es)* is here not linked to 'mechanics' but, primarily, to the production and solution of *sophismata* which could partially be dedicated to the discussion of motion. He also mentioned 'Burlay' (213ra), Edward Hupton ('magister uxoniensis famosus', 212ra), Richard Billingham (211^{v}, and for instance also 267rb) and several times the opinion of 'Buridanus, magister parisiensis' (208vb).

in time but that it happens in an instant, i.e., in the instant itself that ceases to exist. Trifogli shows that the conflict is only apparent, because the difference between the two solutions can be explained by a difference in the interpretations of the Aristotelian puzzle offered by the two commentators.

The natural philosophy of Thomas Wylton is usually not considered to be an integrating part of the Calculators' tradition. However, its significance for an adequate understanding of the first phase of its development is unquestionable. Trifogli's chapter clarifies the close connection between the analytical language of the Calculators in dealing with the problem of the first and last instant of time and the interpretation of Aristotelian *Physics*. It also makes it evident that it is most probably the case that the Calculators' tradition did not start with a deafening timpani beat but rather developed gradually in a process of self-independence out of the Aristotelian commentary tradition.

Elżbieta Jung's chapter takes us to the very core of the Oxford Calculators' tradition. There is no doubt that the two thinkers she analyses, Thomas Bradwardine and Richard Kilvington, settled some of the new basic principles and methods by dealing with Aristotelian ideas in natural philosophy. It seems hardly questionable that their analysis has to do with ways of quantifying Aristotelian philosophy or, perhaps, with a search for independent quantification approaches. In this case, the focus is the question of how to understand Aristotle's claims from Books IV and VII of the *Physics* and Averroes' commentaries on these texts. In other words, following the tradition of Grosseteste, who claimed that mathematics (i.e., geometry) is the only demonstrative science, the question is how to explain Aristotle's 'rule of motion' using mathematics, and not only logic, as a tool in natural philosophy. Richard Kilvington, most likely discussing these issues with his colleagues, must have noticed the inconsistency between Aristotle's and Averroes' views and the ideal of mathematical strictness as represented by the works of Euclid, Jordanus de Nemore, and Archimedes. Evidence from Kilvington's questions on the *Physics* (written according to Jung before 1328) suggests that Bradwardine may have been using Kilvington's arguments to reinterpret Aristotle's views of motion.

The vast majority of reconstructions until now has underlined the starting and leading role of Thomas Bradwardine's *Tractatus de proportionibus velocitatum in motibus* (1328). Whereas the role of this text remains unquestioned, Jung argues that the 'first kick' for the Calculators' tradition has to be reassessed. She presents evidence that Richard Kilvington was very well aware of the mathematics of compounding ratios traditionally associated with 'Bradwardine's rule'. As Jung says: 'In an attempt to trace the impact, spread, and decline of quantifying Aristotle, we should realize that the activity of quantifying motion had a prehistory before 1328.' Thus, however might have been the personal relationship between these two central authors of our tradition, Jung's chapter sheds light on the 'collective work' of contemporary scholars dealing with the same problem and building a new conception of science brick by brick.

Speaking of the *Calculatores* undoubtedly refers to Richard Swineshead, the author of one of the most significant and complex texts in natural philosophy, the *Liber calculationum*. After having contributed to our knowledge of this text and its diffusion, Robert Podkoński now turns his attention to two short treatises (*opuscula*) on motion ascribed to Richard Swineshead, preserved in manuscripts Cambridge, Gonville and Caius College Library, 499/268, and Sevilla, Biblioteca Capitular y Colombina, 7-7-29. Podkoński presents extensive fragments of each of these treatises and compares them to one another. The mutual similarities of these texts confirm the hypothesis, first formulated by Edith Sylla and John Murdoch, that they are successive attempts to interpret and modify Aristotle's 'science of local motion' with the aid of *calculationes*.

Furthermore, these and other fragments of the *opuscula* are compared with passages from the treatise *De motu locali*, a chapter of the *Liber calculationum*, as well as with the section *De motu locali* of William Heytesbury's *Regule solvendi sophismata*, and with Thomas Bradwardine's *Tractatus de proportionibus*. The clear dependence of Swineshead's *opuscula* on Heytesbury's arguments and solutions points towards the fact that, most probably, Richard Swineshead was a student of Heytesbury, and that some of the statements Heytesbury formulated inspired Swineshead to further develop the mathematical science of local motion. In addition, the similarity of the conclusions in the *opuscula* and certain paragraphs of the treatise *De motu locali* suggests that these short treatises should be considered to be successive, intermediary steps that Richard Swineshead took on his way to construct the complete, speculative science of local motion based on the *calculationes* and inspired by Heytesbury's work. It is fairly plain that Heytesbury's claim that 'there can be formulated no rules concerning difformly difform local motions' was recognised by Swineshead as an intellectual challenge. All of Swineshead's texts on local motions focussed on a special kind of difformly difform motions, namely the uniformly accelerated/decelerated ones, eventually formulating certain rules describing such motions. What is more, some remarks found in these *opuscula* explain some salient features of the account of local motion given in the treatise *De motu locali*, like the obvious lack of *penes effectum* explanations. Podkoński's work on these short treatises on local motion opens the extraordinary opportunity for describing the successive stages of the development of the science of local motion in Richard Swineshead in particular, and in the Calculators' circle in general.

The Calculators' programme was clearly apt to be picked up and developed in natural philosophy, but not exclusively. In the preface to his famous treatise on proportions and velocities, Bradwardine suggested the project of a generalised mathematisation of knowledge. In this treatise, he proposed a new understanding of some of the most important quantitative statements connected with Aristotle's central notion in physics, i.e. motion. Nevertheless, that was by no means all of it. Theology was also a field open to *calculationes*. It is worth recalling that, as Pierre Duhem convincingly remarked, it was in the commentary tradition on the *Sentences* of Peter Lombard that emerged the first discussions on the *intensio* and *remissio* of forms, which is remarkably earlier than our authors. This is the topic of the chapter by Edit Anna Lukács, considering another work by Bradwardine that influenced European philosophy until Leibniz and far beyond the limits of the Calculators' tradition: *De causa Dei contra Pelagium et de virtute causarum* (1344). In the first book of this work, Bradwardine introduces calculatory methods and different mathematical arguments for setting the philosophical foundations of his theology. While it is not difficult to find clues about the relationship of calculatory mathematics to medicine, a central point in Lukács' chapter is the difference between Bradwardine and his fellow calculator Richard Kilvington regarding imaginary argumentation (*secundum imaginationem*).

Lukács also highlights an argument in *De causa Dei*, Book I, chapter 43, which foreshadows Blaise Pascal's Wager. From calculating sin, contrition, time, and pain, Bradwardine demonstrates that infinite grace must be posited against infinite rewards and punishments to avoid the mathematical contradictions implied in the Pelagian theory of grace. In other words, being a good mathematician makes you an anti-Pelagian and makes you assume the primacy of infinite grace. In sum, Bradwardine's theological work proves not only complementary with respect to the central work of the Calculators, the *Tractatus de proportionibus* (1328), but also plays an important role within a growingly influential trend in fourteenth-century Oxford: the rationalisation of theology through mathematics. Thus, Lukács' chapter opens a field for further research by linking the Calculators to theology.

Many of us might have experienced that teaching any other version of the 'theorem' of the middle degree and the 'infinite series' than the geometrical version formulated by Nicole Oresme raises numerous difficulties. Many of these difficulties are due to the fact that the literary, philosophical, and pedagogical context of these issues identified with the theories of Calculators was seldom mathematics or natural philosophy, but logical analysis of physical and mathematical problems. In many of her publications, Edith Sylla has shown the importance of *sophismata* for the calculator's tradition, not only for its emergence but also for its development and … its decay. At the very least, one can be certain that almost everything we identify with the Calculators must be understood within the context of logic and philosophy of language. One of the favourite fields of speculation, especially for the authors of the first generation of Calculators, was the analysis of so-called *insolubilia*, that is, of propositions that put in question their own truth by the very nature of asserting what they assert (for instance, 'this assertion is wrong').

According to Stephen Read's contribution, there were several approaches to the problem of the Liar paradox before the Oxford Calculators entered the scene (*cassatio, restrictio, mediatio, distinctio*), some of these dating back to the thirteenth century. Read's chapter focuses on the radical new approach proposed by Bradwardine and its continuation in the works written on the same subject by Richard Kilvington, William Heytesbury, Roger Swineshead, and John Dumbleton. Bradwardine's principal idea is that we have to consider 'multiple meanings' in an insoluble sentence. This approach was assumed by Kilvington when analyzing epistemic puzzles, as for instance in his sophisma 48/49 ('a *est scitum a te*'), where he maintains that such a sentence is neither absolutely false nor absolutely true, but only *secundum quid*.[5] In William Heytesbury, known for having established rules to deal with the problem, it becomes clear that *insolubilia* were to be treated in an obligational context. According to Read, Heytesbury started by distinguishing 'cases where what they usually mean is all they signify from cases where they signify more' (p. 135). The next 'calculator', Roger Swineshead—not to be confused with Richard, the author of the *Liber calculationum*—gives a subordinate place to signification in his analysis of the problem. He tries above all to modify the truth-conditions of a proposition so that those propositions that falsify themselves could be excluded. The last thinker of the series, John Dumbleton, author of a comprehensive *Summa* including logic and natural philosophy, declares himself against the positions of Burley, Bradwardine, Roger Swineshead, and Heytesbury. Dumbleton starts from the idea that terms do not have a natural meaning but that 'all signification is mediated by concepts, or simple acts of understanding' (p. 144). Thus, the first conclusion in his analysis of *insolubilia* is that 'a term signifies only while and insofar as there is actual comprehension by

5 Richard Kilvington, *Sophismata*, eds. N. Kretzmann and B.E. Kretzmann, Oxford 1990 (Auctores Britannici Medii Aevi, 12), 137–151; Richard Kilvington, *The Sophismata of Richard Kilvington*, transl. N. Kretzmann and B.E. Kretzmann, Cambridge 1990, 2nd ed., New York 1991, 134–145.

it' (p. 145). Incorporating a much broader context of discussion than Bradwardine and Heytesbury, Dumbleton argues that insolubles could only be false or true depending on the admitted 'casus'. Read's chapter refers to various other texts connected with the Oxford Calculators, that are also important for our understanding of the development of the tradition, such as the *Compendium* of Dumbleton's *Insolubles* 'approved according to usage at Oxford' (*secundum usum Oxonie*) or even the reception of these ideas in Italy by Gaetano da Thiene.

While we can link the origins of the Calculators' tradition to Oxford, it is also unquestionable that the spread of the new ideas started very early. One of the most conspicuous texts that documents the reception of the Calculators' ideas outside Oxford is the *Tractatus de sex inconvenientibus*, the object of Sabine Rommevaux-Tani's chapter. This text presents some of the hardest difficulties to assess in our field, both conceptually and historically. First of all, this treatise is not only anonymously conveyed, but it also contains references that are decisive for a reconstruction of the Calculators' tradition. For instance, the text characterises an opinion as having been accepted by 'the entire Oxford school' (*tota scola oxoniensis*). At the same time, the author of the treatise seriously considers the opinion of another scholar, Adam of Pipewell, of whom we currently do not know with certainty anything else but this quotation. Moreover, the title of the treatise is far from being obvious and its dating is uncertain. The historical puzzle becomes even more complicated when we take into consideration that a newly discovered copy (Praha, Národní Knihovna České Republiky, Ms. VIII. G. 19) plainly shows that the text must have been used at the Universities of Prague and Vienna but, at the same time, it includes a reference holding that this text was disputed at the University of Paris.

According to Rommevaux-Tani, who has recently finished a critical edition of this text, the *Tractatus de sex inconvenientibus* was written between 1335 and 1339. Its author must have been a member of the University of Oxford, and could have disputed at Paris some of the questions discussed in the treatise. After some preliminary considerations of the problems of the text and its unusual structure, Rommevaux-Tani focuses on some of the difficulties related to local motion. In the first article, the anonymous author of *De sex incovenientibus* seeks to determine the cause of the acceleration of the fall of a heavy body; in the second he asks whether the velocity of the motion of a sphere depends on a point or some space, finally, in the third article, he asks whether the velocity of a uniformly difform local motion beginning at no degree is equal to its middle degree. All three questions are central for an evaluation of the impact of late scholastic natural philosophy on the modern science of motion, since the text of *De sex inconvenientibus* was printed in Venice in 1505 and therefore could have reached many more readers than Kilvington's *Sophismata* or Dumbleton's

Summa, works that were never printed. For each article, the anonymous author proposes six difficulties that would result from a negative answer to the question asked. Concerning the first article, the author follows Adam of Pipewell according to whom the principal cause of the acceleration of the fall of a heavy body, which is the decrease in the resistance, has to be complemented by further secondary causes, such as the continuation of the motion, the proximity to its natural place, the impulse of the medium, the accidental heaviness, and the natural inclination to reach its natural place. Concerning the second article, the author supports Bradwardine's position that the velocity of the motion of a sphere depends on the point that moves most rapidly. He rejects the position of Richard de Versellys or Gerard of Brussels that the velocity of a sphere depends on the midpoint of its diameter. In the third article, the author discusses several arguments tending to prove the validity of the middle degree theorem, stated before him in the *Probationes conclusionum* attributed to William Heytesbury. The discussion of these three questions evidences the thorough knowledge that the author of the treatise had of the physics of the Oxford *Calculatores* and some of the problems that it posed. Additionally, this chapter includes a description of manuscript Praha, Národní Knihovna České Republiky, Ms. VIII. G. 19.

Mark Thakkar's contribution returns to Oxford to re-assess the classical accounts of what the very notion of 'Calculators' means. Our volume as a whole explores the spread of the 'calculatory style' in different regions of Europe from its very beginning in the fourteenth century until Leibniz, after whom the *Calculatores* became a subject more for historians than for mathematicians, physicists, and philosophers. What happened in Oxford itself after the generations of the 'core' Oxford Calculators is hardly ever considered. Some studies show that the tradition was, in fact, alive in logic and natural philosophy in scholars like Richard Feribrigge, John Chilmark, Ralph Strode, and John Wyclif, a major figure in late medieval philosophy and theology.[6] Wyclif is doubtlessly a special case, not only for his significance in England and influence on the continent but precisely for reasons linked to the content of his writings. On the one hand, he left a number of works in which he takes up the classical themes of the Calculators. On the other hand, unlike any other that is known to us, he seems to hold a kind of atomism that is not at all compatible with the continuity requirements common to his colleagues. Was he, then, a *calculator* at all? This question cannot be answered without discussing the very notion of 'cal-

6 See, for instance, the initial pages in Clagett's chapter 'The Reception and Spread of the English and French Physics, 1350–1600,' in *The Science of Mechanics*, 629–635.

culator'. The Oxford Calculators have been variously characterised and enumerated since they first emerged as a historiographical category. Thakkar finds solid reasons to criticise these characterisations and to make a fresh attempt to capture what was genuinely distinctive about their tradition. He then applies his proposed characterisation to conclude that John Wyclif should be considered an Oxford calculator. Thakkar carries out an thorough analysis at a conceptual and historical level by attempting to reconcile this view with Norman Kretzmann's portrayal of Wyclif as a sworn enemy of the Calculators. He concludes that further research is required to determine whether a rejection of atomism was essential to the calculatory tradition. Along the way, Thakkar's chapter makes various other contributions to scholarship on the Oxford Calculators, for instance, by identifying Roger Swineshead's long-lost *Sophismata logicalia* and exposing the misattribution to Gaetano da Thiene of a well-known commentary on William Heytesbury's *De tribus predicamentis*.

The gigantic figure of Galileo has guided many decades of research dedicated to finding new milestones on the way from Oxford to Florence and Padua. Supporting or rejecting the thesis of continuity, the greatest efforts focused on the Italian reception of the new trends in philosophy—including, of course, the *Calculatores*. The countereffect of this research was the disregard of the spread of this tradition in Poland and at universities of the German Empire. Whereas the impact of the *Liber calculationum* in Italy was connected to natural philosophy and medicine, it seems that in the German universities, mathematics (especially the kind of applied mathematics that was usually categorised as 'middle science'), logic, and philosophy of language played the most decisive role. The *Sophismata* collections by Heytesbury and Kilvington, and Heytesbury's *Regule*, as well as other similar texts, were eagerly copied, studied, and reworked in this region of Europe. During the fifteenth century, the most important works of the Oxford Calculators were used to compose new teaching tools for logic and philosophy of language that incorporated some of the new ideas recently developed in other centres, such as Paris, Vienna, or Heidelberg. There is still a great amount of material to be ordered, analysed, and finally, edited. Harald Berger's chapter is devoted to one outstanding scholar of the late fourteenth and early fifteenth centuries, Helmoldus de Zoltwedel (i.e., Salzwedel), who was active at the universities of Prague (from 1383 to 1409) and Leipzig (from 1409 until his death in 1441). His *magnum opus* is the *Quaestiones parvorum logicalium* (Prague ca. 1390), a monumental work divided into nine parts dealing with the treatises of the then relevant authorities, such as Thomas Maulfeld, Marsilius of Inghen, and John of Holland. The ninth and final part of this work discusses *insolubilia*, based on John of Holland's treatise, and the fifth of six questions is devoted to the most famous insoluble, the so-called Liar paradox. Besides the two known complete witnesses of Helmold's *Quaestiones*, there is also a third copy of this single question IX.5 as well as an extremely shortened version of it. It is in particular in part IX and in this very question 5, that Helmold quotes and critically discusses English authors including the *Calculatores*. Historically most interesting is the first article of Helmold's question IX.5, which discusses ten previous opinions regarding the Liar, the tenth and last of which is attributed to his *magister reverendus* Conrad of Soltau (this being the only known testimony to a work of Conrad on *insolubilia* and the Liar). These ten opinions are critically edited in the appendix. Helmold has an especially critical look at William Heytesbury's famous treatise. The second article of question IX.5 presents Helmold's own solution as the eleventh opinion. It is typically developed within the framework of the scholastic genre of *obligationes*, as does Heytesbury. Helmold becomes, thus, an important witness of the influence of logical and semantic theories of the *Calculatores* in Central Europe.

Already in the thirteenth century, there were various quantifying approaches in the field of the *intensio et remissio formarum*. This trend was further deepened and broadened from the fourteenth century onwards with the application of new 'analytical languages'. Pierre Duhem, Anneliese Maier, and Edith Sylla worked extensively on this problem, even if they have not exhausted its full potential. The various difficulties are conceptual as well as historical. Some texts deal with the pure description of the process of qualitative intensification and diminution while other writings focus on the ontological structure that underlies this process. In addition, some texts are limited to one single or a particular set of problems, and deal with *intensio* and *remissio* in that specific context while other works address the topic itself as the object of study in an independent treatise. As the questions treated are often mathematical or 'semi-mathematical', several problems of quantification find a natural place in this field. From a historical point of view, the relationship between this topic and the Aristotelian commentary tradition is one of the more complex and momentous. Emerging from the commentary tradition on the *Sentences* of Peter Lombard, the topic, which was only transiently touched upon by Aristotle himself, was later also integrated into the commentaries on his *Physica, De caelo* and *De generatione et corruptione*. Moreover, Galileo's so-called *Juvenilia* present a significant amount of material on this subject. It is clear, then, that if a process of quantification of Aristotelian thought was made possible by the calculatory tradition, questions on the *intensio* and *remissio* of forms probably played a decisive role.

With respect to this topic, Joël Biard's chapter examines the views of Blasius of Parma (ca. 1355–1416), one of the most prominent figures in the Italian reception of the calculatory ideas. Biard focuses on three texts by Blasius: the *Questiones de latitudinibus formarum*, the *Questio disputata de intensione et remissione formarum*, and question 10 on Book V of the *Physics*.

The issue of the intensification or remission of qualitative forms properly speaking is based, for Blasius, on discussions about the ontological status of qualities and their relation to their subject of inherence, placing the issue at the borderline of physics and metaphysics. Qualities can be considered either as the qualified subject itself or as a qualitative form inherent in the subject. In each case, however, we cannot completely abstract from the fact that the quality is inherent in a material subject and, hence, we have to assume that this fact has an impact on our description of the process of intensification or remission of the qualitative form. At the same time, Blasius takes sides for the calculating procedures that were on debate since the beginning of the fourteenth century. In the calculation of local motion, Blasius puts forward the theory of proportions. In the evaluation of intension and remission of forms, he evokes this theory, but the principal problems for measuring gradual qualities and their variations are (1) whether the growth of a quality is measured in connection with the weakening of the opposite quality, and (2) whether the variation is to be evaluated in relation to the zero degrees or the highest degree in a given subject. In binding the quality and its subject closely together, Blasius does not go as far as the Oxford Calculators in the mathematisation of intensive magnitudes. Nonetheless, he does attempt to integrate the new tools into his physics and his ontology.

The vast majority of the Calculators' texts, and hence of the secondary bibliography, are rarely linked to a specific natural object. In his chapter on Nicole Oresme's *De visione stellarum*, Aníbal Szapiro addresses the influence of the Calculators' tradition on late medieval discussions of a particular natural object, namely light. This topic has not yet received sufficient attention in our field. Whereas contributions on the Calculators usually focus on motion and the different problems and notions connected with it, light as such has not been perceived as an equally important subject for discussion. Historically, the reasons for this fact are evident: after the struggles surrounding the reception of the Aristotelian world view in the thirteenth century and the establishment of the *Physica* as a textbook at European learning centres, motion became the central notion in natural philosophy. But this was not always the case. A previous generation, which had contributed to a critical reception of Aristotle's views and was not yet completely committed to them, focussed on light rather than on motion as the fundamental component of the physical world. Reviving the *Hexaemeron* literature, the 'metaphysicians of light', above all Robert Grosseteste and Roger Bacon, made important steps towards a mathematisation of nature through the study of light. Optics was considered the basic and modelling discipline. The roots laid at Oxford must have made a profound impression on scholars of later generations such as Bradwardine and Kilvington. For example, in Swineshead's *Liber calculationum* and Dumbleton's *Summa*, light becomes an important object of discussion. The subject could not go unnoticed by Oresme, who connects his doctrine of configurations to Grosseteste's geometrising project and who, unlike all the other authors discussed in this volume, was not only a mathematician but also a competent theoretical astronomer. Szapiro considers specifically the theoretical and empirical aspects involved in Oresme's attempt to quantify atmospheric density to determine the existence of atmospheric refraction in the case of a uniformly difform atmosphere. Szapiro's analysis allows him to reject the accepted interpretation of Oresme's speculation as an application of the middle degree theorem and the diagram accompanying it as a configuration of qualities. His chapter contributes to our understanding of the interaction between optical, astronomical, and physical knowledge in late medieval studies of light. In particular, it opens an interesting field for future studies on the Calculators' concept of experience, especially in connection with those scientific disciplines in which empirical data play a significant role, such as astronomy and optics.

The fact that the Calculators' tradition provided a quantified view of Aristotle is well known, though not in all its breadth or depth. While the vast majority of studies conducted so far focused on the theory of motion and on different analytical languages employed in intrepreting Aristotle's *Physica, De caelo, De generatione et corruptione*, and related works, Daniel A. Di Liscia examines the new readings of the *Metaphysics* made possible by the Calculators. Starting from a theological background and first presenting a fundamental text on the subject under discussion, the *Tractatus de perfectione specierum* by James of Naples, this chapter aims to offer an overview of some of the central issues regarding the 'the great chain of being' in late medieval philosophy. It shows how, following the approach initiated by James of Naples, Nicole Oresme and others, one of the solutions consisted in a generalised application of geometry to structure all entities in the universe according to a gradual system of perfection. Geometry, a 'normal science' established on solid principles of evidence and clarity, also offered, unlike many discussions in the field of logic and the philosophy of language, the possibility of representing models of the universe according to degrees of perfection. This approach gained more and more terrain until it became a new way of discussing the old problem of the chain of being within an Aristotelian framework. Di Liscia's chapter sketches some of the developments in France, Germany, and Italy. For France, Di Liscia examines Jacques Legrand's metaphysical text on natural philosophy. For Germany, he draws the attention to the incorporation of *latitudines* and *perfectiones* in the questions on the *Metaphysics* by Nicholas of Amsterdam and John of Wesel. For the reception of these ideas in Italy, Di Liscia follows the path that goes from Blasius of Parma's questions on *De latitudinibus formarum* to Paul of Venice's section on *Metaphysics* in his *Summa naturalium* and Agostino Nifo's *Dilucidarium* of metaphysical disputations. The latter work incorporates the reaction of a 'good Peripatetic' of the Renaissance against the new developments toward a quantifying interpretation of Aristotle's *Metaphysics*.

Whereas Di Liscia's chapter on the metaphysical tradition suggests that there were many routes of transmission of the Calculators' ideas and approaches, Richard Oosterhoff's chapter explores of the critical approach championed by the Humanists. For, as the Calculators' quantitative approach grew and became widespread all over Europe, the voices against it became louder and more numerous. There are many early indications of this situation in the negative utterances by Coluccio Salutati and Leonardo Bruni, but the culmination of this process was reached at the moment when the Calculators' philosophy seemed to have the University of Paris firmly under control. John Mair and his circle installed the Calculators' quantifying approach in all its extension. Since Pierre Duhem's studies we know that in the first decades of the sixteenth century several new tracts were produced, some of the old texts commented and, above all, that *calculationes* and *sophismata* became the most usual way to comment on Aristotle's texts. However, the previous generation might have had a different attitude that is worth further consideration.

Oosterhoff's contribution deals with the question of why the 'mathematical languages of analysis' that Murdoch described as vibrant in the fourteenth century, were nevertheless not widely taught around 1500. It does so by carefully considering the presence and absence of calculatory techniques in the first generation of textbooks introducing natural philosophy that were printed at Paris in the 1490s. The usual explanation is that humanists saw scholastic university analysis—and especially mathematics—as sophistic and inelegant, and therefore entirely discouraged this type of study. But on examining the first generation of printed university textbooks, Oosterhoff finds that traditionalists like Pierre Tartaret and Thomas Bricot tended not to include any calculatory techniques. Meanwhile, Jacques Lefèvre d'Étaples, a competitor usually associated with humanism, did include a primer on such techniques in his introduction to *Physica* (albeit wrapped in qualifiers that aimed to discourage wrongheaded sophistry). Even Jacques Almain's *Embammata* (1506), neglected in William Wallace's account of sixteenth-century Calculators, on closer inspection is presented as a diversion for experts, not an introduction. With the exception of Lefèvre, it seems that calculatory techniques no longer belonged to the introductory *Physica* courses for the generation of around 1500, despite the activity of a few highly expert practitioners of calculatory techniques such as Alvarus Thomas. This was significant because this generation set in place the printed textbook culture that would define the sixteenth century—a culture that would go on to imagine the relationship between mathematics and physics quite differently. Even those educational reformers such as Giorgio Valla and Pedro Ciruelo, who were particularly keen on mathematical learning, tended to focus on Boethian or Platonic sources, which did not include calculatory analytical languages (except, obviously, the proportion theory). Therefore, mathematical practitioners who also taught physics, such as Jean Fernel and Oronce Fine, seem to have had little interest in the tradition of the Calculators.

The reception of the Calculators' tradition in Italy started very early. It covered almost all fields of academic knowledge, especially medicine, physics, and logic. Within this reception history, a particular place is to be attributed to the incorporation of William Heytesbury's writings into Italian academic life.[7] Approximately at the same time when Blasius of Parma was dealing with *intensiones* and *latitudines*, a lesser known master today, Angelo da Fossambruno, was teaching about the same subject following Heytesbury whose work was published later, in 1494, by Bonetus Locatellus in Venice. This edition, containing among others Heytesbury's major works, his *Sophismata* and his *Regule solvendi sophismata*, represented a milestone in the spread of the Calculators' ideas. One particularly interesting feature of this edition is that it also includes the commentaries on Heytesbury's analysis of motion produced by some Italian masters such as Bernardo Torni, Mesino de Codronchi, Gaetano da Thiene, and, of course, Angelo da Fossambruno.

Fabio Seller's chapter studies Angelo's analysis of motion as connected to Heytesbury's *Regule*. In all probability, the text (first published in 1482) represents a re-write of the *lectiones* Angelo delivered at Padua. Its content is particularly concerned with that part of *De tribus praedicamentis* relating to

7 Not only the manuscript tradition but also several printed books are still waiting for being studied, for instance, the *Opusculum* by Benedictus de Victoriis from Bologna on William Heytesbury (a series of eight *lectiones*, where Kilvington's *Sophismata* are also often discussed) and on Paul of Venice (a 'recollection' of the treatise on *suppositio*). Cf. Benedictus de Victoriis, *Opusculum Benedicti Victorii Bononiensis in Tisberum de sensu composito ac diviso: cum eiusdem collectaneis in suppositiones Pauli Veneti*, Bologna 1504, and the *Tractatus rationalis*, Siena 1491 (or, as established in the colophon: *Opus tripartitum*) by the master of arts and medicine Oliverius Senensis. This work, which is dedicated to Cardinal Giovanni de' Medici, the later Pope Leo x, deals with logic (terms, supposition theory, *de relativis*, *de consequentiis*), and natural philosophy, especially *maxima et minima* (called here *de potentia activa et passiva*).

the analysis of the different types of local motion, by showing the peculiarities of uniformity and difformity in these motions. After having defined motion in its most proper sense as the acquisition of parts one after the other, Angelo provides an ordered set of definitions of *motus*, and presents a series of contradictory statements to which they would lead. Once he finds a more consistent definition, his analysis continues by focusing on the examination of different types of local motion, each one supported by examples that disclose the true meaning of *intentio* and *latitudo* of motion. Angelo's work ends with a few remarks about four *sophismata*, and their solutions, involving some interesting considerations on infinity. He thus reveals an acquaintance with the ideas of the Calculators, which lies well beyond the mere commentary on Heytesbury's work.

Richard Swineshead's *Liber calculationum* was well known and appreciated by Leibniz. His complex work, above all his correspondence, presents a considerable amount of evidence in this regard that cannot be considered circumstantial. Thus, it is worth asking for the reason behind his interest in this tradition. Was it merely bibliographical (among other things Leibniz was a librarian)? Was it about the ontological foundations of Leibniz's new physico-mathematical conception, perhaps as a background for the newly developed calculus (here, for instance, the notion of velocity as an intensity of a quality could have been especially attractive to him)? Or was Leibniz moved by an epistemological and methodological interest in nature related more to the philosophy of science than to the content of science itself? Edith Sylla's chapter explores these complex questions in depth. She presents a critical and detailed analysis of a large part of the secondary bibliography and sources, many of them hitherto little known. Analysing statements of Scaliger, Cardano, and others, Sylla maintains that a large part of Leibniz's awareness of the tradition of the Calculators was linked to the Portugese Alvarus Thomas, in partitular to his work *Liber de triplici motu* (Paris 1509). Connecting some modern interpretations of Leibniz with her investigations on science and late medieval philosophy, Sylla attributes a predominant role to nominalism, not so much as a philosophy of language but in particular as a view of mathematical entities as beings of reason (*entia rationis*) which exist in the minds of mathematicians, not in the physical world.

To end this tight presentation, it might be worth adding a few words of clarification about what this book is not and does intend to be. The chapters are related to each other by a common subject matter and, on some occasions too, by years of common work in research and teaching between some of the contributors. Nevertheless, they are by no means chapters of a general book planned to cover the entire 'history of the Calculators'. This book is neither a handbook, nor a textbook, nor a 'companion' to anything. Thus, some thematical overlaps were scarcely avoidable and, for the same reasons, this book has no claim to completeness. Finally, I would like to underline the fact that as editors we have prioritised the variety of views and opinions over a 'unified' presentation of the topic.

The editors' respect for the personal choice made by each author in the use of specialised terminology has led us to accept sometimes a variety of denominations even when the idea behind it is basically the same. It will be useful for the reader to keep in mind two cases in particular which are often associated with the Calculators and that we have not tried to unify. First, the so-called 'mean speed theorem', by which a uniformly accelerated motion is reduced to a uniform motion which, during the same interval of time, maintains the 'mean degree' of velocity of the former (or also, which is certainly equivalent, which maintains 'the instantaneous velocity' corresponding to the middle instant of time interval). In the particular case of local motion, this equivalence between the two motions can be expressed in terms of an equality of the respectively covered spaces. For this assertion, there are several different denominations in this book, all referring basically to this same idea: 'mean/middle', 'speed/velocity/degree', 'rule/theorem'. Besides, the reader has surely already made the experience of discovering in the widespread bibliography on the topic, that sometimes also the local denomination of 'Merton rule' is usual.

Second, this book also contains several remarks or discussions about 'Bradwardine's function' or Bradwardine's 'rule', or even 'law' of motion. These expressions always refer to the statement according to which the *arithmetical* increase of a velocity requires the *geometrical* increase of the ratio F/R between the causal factors involved, i.e., of the ratio between the 'force' and the resistance of the motion in question. Consequently, if we wish to double the velocity, it is not enough to double the ratio F/R, but we have to square these causal factors. Similarly, for a threefold increase of the velocity we will need to cube this ratio, and so on. This idea has been usually associated, already in our sources, with the name of Thomas Bradwardine, who proposed it in a famous work called *Tractatus de proportionibus velocitatum in motibus* (mostly referred to as *De proportionibus*). After a period of discussion, exchange and, sometimes, reformulations, the editors decided to respect the denominations (and interpretation of it) used by each contributor. As a consequence, the reader may find different, sometimes even divergent approaches and opinions on the same topic.

CHAPTER 1

Thomas Wylton on the Ceasing of an Instant of Time

Cecilia Trifogli

1 Introduction

In a question of his *Quodlibet* about the instant of time[1] Thomas Wylton remarks that the issue of the ceasing (*desitio*) of the indivisible elements of change and time is a difficult one.[2] Such indivisible elements are the instantaneous phases of a change (the so-called *mutata esse* in Scholastic terminology) and the instants of time respectively. Their role in change and time is very much the same as the role of points in a line. In particular, since change and time are continua just as a line is, their indivisibles are not parts of them, just like points are not parts of a line, but they are boundaries of their parts. Any finite phase of a change or any finite period of time is bounded by two indivisibles—one at its beginning and another at its end—just like a finite line has an initial point and a final point. It is clear that for the indivisibles to play the role of boundaries of the parts of a continuum there must be more than one of them—actually very many of them—in a given continuum. Indeed, since a continuum is divisible into parts *ad infinitum*, there are infinitely many parts in a continuum and each part requires two distinct indivisibles as its boundaries. However, there is an important difference between the modes of existence of distinct indivisibles within a line and within a change or time respectively. Any two distinct points in a line exist simultaneously, that is, temporally together, whereas any two distinct indivisibles in a change or in time cannot exist simultaneously but only in

1 On Wylton's *Quodlibet* see C. Trifogli, 'The Quodlibet of Thomas Wylton,' in C. Schabel (ed.), *Theological Quodlibeta in the Middle Ages: The Fourteenth Century*, Leiden 2007 (Brill's Companions to the Christian Tradition, 7), 231–266. The question on the instant of time is edited in L.O. Nielsen and C. Trifogli, 'Thomas Wylton's Questions on Number, the Instant, and Time,' *Documenti e studi sulla tradizione filosofica medievale* 16 (2005), 88–105. Following the numbering of Wylton's quodlibetal questions given in my paper 'The Quodlibet of Thomas Wylton.' I will refer to this question as Q. 15.

2 Wylton, *Quodlibet*, Q. 15, 103–104, [4.9.1].

 | DOI:10.1163/9789004512054_003

temporal succession, that is, one after the other in time. Accordingly, when the indivisible at the beginning of a change or of a period of time exists, the indivisible at its end does not yet exist, and when the indivisible at its end exists, the indivisible at its beginning no longer exists. Thus, indivisibles in a change and time, unlike the indivisibles in a line, begin to exist and cease to exist.[3] It is difficult, however, to give a satisfactory account of the beginning and ceasing of such indivisibles. As Wylton puts it, the problem concerns the 'measure' of these events, and what he means is the temporal measure. In other words, the general question that is difficult to answer is about the 'when' of these events' occurrence: when does an indivisible in a change or in time begin to exist or cease to exist? Does this happen in time or at an instant? And if at an instant, which one? The instant at which the indivisible exists or another one?

In this paper I present how Thomas Wylton deals with these questions. I will first clarify the context of Wylton's discussion; I will then introduce and explain his reply to these questions; finally, I will compare his reply to that of Walter Burley, which I have analysed in a recent paper.[4]

2 The Aristotelian Context

When in his quodlibetal question about the instant of time Wylton remarks that there are difficulties concerning the measure of the ceasing of indivisibles of change and of time, what he has in mind is a puzzle about the 'now', that is, the present instant, raised by Aristotle in the introductory and dialectical phase of his discussion of time in *Physics* IV.10.[5] The puzzle is about the identity and diversity of the now. The question asked is whether the now is always different or always the same. And the puzzle points out that there are problems both with the alternative that the now is always different and with the alternative that it is always the same. What is relevant for our purpose is the argument that Aristotle's puzzle puts forward against the alternative that the now is always

---

3 This difference ultimately derives from the difference between the permanent being of a line and the successive being of change and time. For an informed and illuminating discussion of the distinction between permanent and successive things, see D.A. Di Liscia, 'Walter Burley, Paulus Venetus und die Tradition *De Instanti* (mit dem "*Tractatus de instanti*" des Paulus Venetus nach Hs. Florenz, Biblioteca Nazionale Centrale, II. IV. 553, foll. 68va–69va),' in A. Speer and D. Wirmer (eds.), *Das Sein der Dauer*, Berlin 2008 (Miscellanea Mediaevalia, 34), 130–135.

4 C. Trifogli, 'Walter Burley on the *Incipit* and *Desinit* of an Instant of Time,' in F. Goubier and M. Roques (eds.), *The Instant of Change in Medieval Philosophy and Beyond*, Leiden 2018 (Special issue of *Vivarium*, 55 [2017]), 85–102.

5 Aristotle, *Physics*, IV.10, 218a8–30.

different. This argument is the one Wylton refers to in his quodlibetal question and it is the focus of the present paper. The argument goes as follows: if the now is always different, it must cease to exist to give way to another now. This conclusion is not controversial, because, as has been explained above, it reflects a common assumption about the successive mode of existence of time. It gives rise, however, to a difficult question about the ceasing to exist of the now, the temporal question: given that the now ceases to exist, when does this happen, that is, when does the now cease to exist or when has it ceased to exist? Aristotle apparently considers all possible answers to this question and shows that none of them works.

I shall examine the argument of *Physics* IV.10, together with Wylton's solution to it, in more detail in the next section. In this section, in order to give a complete picture of the Aristotelian context of Wylton's discussion, I will present the other essential ingredient of it. This is Aristotle's treatment of the sameness and diversity of the now in *Physics* IV.11.[6] In this later chapter, after presenting his famous definition of time as the number of motion in respect of the before and after,[7] Aristotle formulates what looks like his considered view about the sameness and diversity of the instant. The issue at stake seems to be exactly the same or very close indeed to that which gives rise to the puzzle of *Physics* IV.10, but this time Aristotle settles it in a straightforwardly positive way without any reservations. His considered view consists in a distinction: 'The now is in a way the same, and in a way not the same,' Aristotle claims.[8] According to a common medieval formula used by Wylton to specify this distinction, the instant is the same according to its substance (*secundum substantiam*), but is not the same according to its being (*secundum esse*).[9] This formula as such is not of great help to elucidate Aristotle's distinction because it is not at all clear what the substance and the being of an instant are. It reflects well, however, the main line of argument used by Aristotle to establish his distinction concerning the instant. This is the analogy between the identity and the diversity

6 Aristotle, *Physics*, IV.11, 219b10–33.

7 Aristotle, *Physics*, IV.11, 219b1–2.

8 Aristotle, *Physics*, IV.11, 219b12–13, transl. E. Hussey, in: *Aristotle Physics Books III and IV*, Oxford 1983 (Clarendon Aristotle Series, 44).

9 E.g., Wylton, *Quodlibet*, Q. 15, 88, [1.1]: '… de nunc temporis, de quo distinguo sicut facit Philosophus 4 *Physicorum*, capitulo de tempore, ubi solvit quaestionem quam prius movebat in principio capituli, scilicet an unum et idem nunc maneat in toto tempore vel in alia et alia parte temporis sit aliud et aliud nunc, dicens quod nunc uno modo est idem, alio modo non-idem. Ipsum autem nunc idem est quod quid erat esse, *id est secundum substantiam*, esse autem alterum, *id est secundum esse* quod correspondet mobili in alia parte et alia spatii est aliud et aliud' (emphasis mine).

of the instant in time and the identity and the diversity of the thing subject to a (non-substantial) change. The example used is that of a locomotion: the thing moving in a locomotion remains the same in substance throughout its locomotion but changes with respect to its accidents corresponding to the different locations that it comes to occupy: e.g., Coriscus remains the same in substance throughout his locomotion but his being in the market place and his being in the Lyceum are different. Aristotle's analogical argument then concludes that, as a moving thing remains the same with respect to its substance throughout its locomotion but varies with respect to its accidental being, so does the instant of time: it remains the same with respect to its substance throughout time but it varies with respect to its being.

Aristotle's view about the identity and diversity of the instant in *Physics* IV.11 is very obscure. The most obscure part of it is certainly the part about the identity of the instant with respect to its substance. It is difficult to make sense of the idea that there is an indivisible element of time that persists throughout time, that is, of the idea of a persisting present, as some Aristotelian scholars describe it.[10] Many medieval commentators, however, think that this idea makes perfect sense. The widespread tendency in the medieval reception of Aristotle's physics is to accept Aristotle's account of the identity and diversity of the instant and to support it with detailed exegeses but without subjecting it to a close philosophical scrutiny, like that reserved to other aspects of Aristotle's doctrine of time (for example, the existence of time and the unity of time).

Thomas Wylton belongs to the large group of medieval commentators who accept Aristotle's view. He devotes two extensive discussions to it. One is in a question of his commentary on the *Physics*; the question asks 'Whether the instant remains one and the same throughout the whole of time' (*Utrum instans maneat unum et idem in toto tempore*).[11] The other is in the quodlibetal question I have referred to above; the question asks 'Whether the instant according to its substance is the proper measure of a thing subject to generation and corruption according to its permanent being' (*An nunc secundum substantiam sit mensura propria rei generabilis et corruptibilis secundum esse permanens eius*). Wylton's discussion in both questions is not merely exegetical.

10 See, for example, E. Hussey's comments in *Aristotle Physics Books III and IV*, 152–157; U. Coope, *Time for Aristotle. Physics IV.10–14*, Oxford 2005 (Oxford Aristotle Studies), 125–139.

11 This question, like the vast majority of Wylton's *Quaestiones libri Physicorum*, is still unedited. The quotations in this article are from manuscript Cesena, Biblioteca Malatestiana, Plut. VIII, sin. 2 (henceforth: C). I will refer to this question by its numbering within Wylton's questions on book IV of the *Physics*: IV, Q. 22.

It also contains a sustained philosophical analysis of the notions of the instant according to its substance and the instant according to its being. In particular, Wylton defends the view of the persistence of the instant against John Duns Scotus' objections and comes out with an original interpretation of this view, which goes well beyond what Aristotle's passage of *Physics* IV.11 explicitly says or suggests.[12]

The aspect of Wylton's discussion in the two questions that concerns us here, however, is not his original interpretation of Aristotle's view. Rather, it is the connection that Wylton points out between Aristotle's treatment of the question of the identity and diversity of the instant in *Physics* IV.11 and that in *Physics* IV.10. The question is left without solution in *Physics* IV.10, but is given a positive solution in *Physics* IV.11. Therefore, it is very tempting to assume that the discussion of *Physics* IV.11 also provides the solution to the puzzle of *Physics* IV.10. This is indeed the assumption of the two most important commentators on the *Physics* of the thirteenth century: Thomas Aquinas and Giles of Rome.[13] Despite such authoritative support, however, this assumption is not correct. The discussion of *Physics* IV.11 does not address at all the puzzle about the diversity of the instant of *Physics* IV.10. It does show that there is a sense in which the instant is always different but it does not show how this sense in which the instant is always different is not open to the objection raised by the puzzle of *Physics* IV.10. Unlike Aquinas and Giles, Wylton has a clear perception of this point, namely, that the argument of *Physics* IV.10 against the instant's

12 I have presented Wylton's discussion about the persisting instant in these two questions in C. Trifogli, 'Thomas Wylton on the Instant of Time,' in A. Zimmermann (ed.), *Mensch und Natur im Mittelalter*, Berlin 1991 (Miscellanea Mediaevalia 21/1), 308–318. For an outline of the content of the quodlibetal question, see also Nielsen and Trifogli, 'Thomas Wylton's Questions,' 63–68.

13 In his commentary on the passage of *Physics* IV.10 Aquinas simply gives a literal exposition of the argument; in his commentary on the passage of *Physics* IV.11 he clearly assumes that this passage contains the solution to the initial puzzle of Physics IV.10, but he does not offer any explanation of this assumption. See Thomas Aquinas, *In octo libros Physicorum Aristotelis Expositio*, ed. P.M. Maggiòlo, Turin 1954, IV, lectio XV, 274; lectio XVIII, 287–288. In his commentary on the passage of *Physics* IV.10 Giles devotes a very long discussion to the Aristotelian puzzle. Most of it is intended to explain and justify the various assumptions of the argument. At the very end of the discussion Giles anticipates that the solution to the question of whether the instant is the same in time is that based on the analogy between the instant and the thing subject to change used by Aristotle in the passage of *Physics* IV.11. In his commentary on the latter passage Giles explicitly remarks that in this passage Aristotle provides a solution to the puzzle of *Physics* IV.10 but, like Aquinas, he does not offer any explanation of this claim. See Giles of Rome, *Commentaria in octo libros Phisicorum Aristotelis*, Venice 1502, repr. Frankfurt am Main 1968, lectio XVIII, 97vb–98va; lectio XXI, 101^{rb-va}.

being always different can be repeated for the case of the instant according to its being introduced in *Physics* IV.11.

Both in Wylton's *Physics*-question and in his quodlibetal question, the argument of *Physics* IV.10 against the ceasing of the instant is incorporated within an objection against the view that the instant in its substance persists in time (the view that Wylton supports). In the formulation of the quodlibetal question the objection is the following:

> Furthermore, against the first opinion I first argue by showing that the argument of the Philosopher at the beginning of the chapter on time which he uses to prove that the instant according to its substance is not subject to corruption is not conclusive. For by the same argument I conclude that the instant according to its being is not subject to corruption but remains the same in time, because using the argument of the Philosopher but applying it to the instant according to its being I argue as follows. For I ask about the instant according to its being whether it is different now and beforehand or it is absolutely the same, just as he asks about the instant according to its substance ...[14]

This objection comes at the end of a long series of objections against the view that there is an instant that persists through time (the instant according to its substance) and has the role of measuring the permanent being of a thing subject to generation and corruption (as in the title of the quodlibetal question). All the preceding objections are attributed either to a doctor (*unus doctor*) or to his followers (*alii sequaces istius doctoris*). The doctor in question is Scotus but his followers have not been identified.[15] Thus, the fact that Wylton uses the first-person 'arguo' to introduce this objection suggests that the objection is made up by Wylton himself against his own view.

---

14 Wylton, *Quodlibet*, Q. 15, 94, [2.10.1]: 'Praeterea, contra primam opinionem arguo, primo probando quod ratio Philosophi, per quam probat in principio capituli de tempore quod nunc secundum substantiam non corrumpitur, non concludat. Nam per eandem rationem concludam quod nunc secundum esse non corrumpitur, sed manet idem in tempore, quoniam arguo sic per rationem Philosophi, sed applicando eam ad nunc secundum esse. Quaero enim de instanti secundum esse, sicut ipse facit de instanti secundum substantiam, si sit aliud in tempore nunc et prius vel omnino idem ...' The parallel passage in *Quaestiones libri Physicorum*, IV, Q. 22 is in Ms. C, 68ra, lin. 35–39.

15 Wylton, *Quodlibet*, Q. 15, 89–94, [2.1]–[2.9.3] (the references to Scotus' discussion are given in the *apparatus fontium* of this edition).

The objection is based on two main claims. (a) One is about the argument that Aristotle uses to prove that the instant according to its substance persists: the argument is that of *Physics* IV.10 against the being always different of the instant. (b) The other claim is that this argument is not conclusive. The reason given for this second claim is that the argument can be applied to the instant according to its being; therefore, if it were conclusive, it would establish that it is not the case that the instant according to its being is always different, a conclusion that all parties involved in this dispute about the instant reject.

Wylton's reply to this objection consists in denying claim (a). He correctly points out that the argument Aristotle uses in *Physics* IV.11 to conclude the existence of a permanent instant is not that of *Physics* IV.10 against the instant's being always different but the argument based on the analogy between the instant and the thing subject to change.[16] Wylton, however, accepts claim (b) together with the reason given for it. The argument of *Physics* IV.10 does indeed apply to the case of the instant according to its being, but the argument is not a valid one:

> Therefore, I say that that argument does not conclude either about the instant according to its substance or about the instant according to its being that it persists throughout the whole of time, but it is a probable argument that reaches a false conclusion, and yet it is difficult to find a solution to it because it is difficult to give the measure of an indivisible either in a change or in time ...[17]

16 Wylton, *Quodlibet*, Q. 15, 103–104, [4.9]–[4.9.1]: 'Ad alias rationes quas adduxi contra primam opinionem quas non reputo minus difficiles aliis respondeo secundum ordinem. Ad primam dico quod Philosophus ibi non adduxit illam rationem tanquam demonstrativam nec propter eam tenet conclusionem quam tenet de nunc secundum substantiam, sed propter aliam rationem quam ponit ubi solvit quaestionem de unitate instantis in toto tempore. Et ratio propter quam tenet illam conclusionem [istam] quod nunc est mensura mobilis [motu] sicut tempus est mensura motus, et per consequens ex unitate et alietate ex parte mobilis contingit concludere unitatem et alietatem ex parte ipsius nunc, ut superius deductum est.' The parallel passage in *Quaestiones libri Physicorum*, IV, Q. 22 is in Ms. C, 68rb, lin. 3–9.

17 Wylton, *Quodlibet*, Q. 15, 104, [4.9.1]: 'Unde dico quod illa ratio non concludit nec de nunc secundum substantiam nec de nunc secundum esse quod maneat in toto tempore, sed est ratio probabilis concludens falsum, tamen bene difficilis ad solvendum quia difficile est [videre] dare mensuram desitionis alicuius indivisibilis vel in motu vel in tempore, quia tamen non est ad propositum, ideo transeo.'

In the quodlibetal question Wylton does not offer a solution to the argument of *Physics* IV.10,[18] but he does offer it in the *Physics*-question, to which we now turn.

3 The Argument against the Ceasing of the Instant and Wylton's Solution to It

The following passage contains Wylton's formulation of Aristotle's argument of *Physics* IV.10:

> Since there cannot be two things belonging to time that exist simultaneously except in the case that one of them contains the other, but it cannot be said about the instant that it contains something, given that the instant is indivisible, therefore, when one instant exists, the other instant is corrupted. Therefore, either (i) the instant is corrupted in itself, which is not true, because then it would simultaneously exist and not exist; or it is corrupted in another instant. (ii) Not in a prior instant, because then it would be corrupted before it existed. (iii) Not in a posterior instant, because then the instant in which it is corrupted will either be the one immediate to the instant in which it existed or one separated from it by a medium. Not in an immediate instant, because an instant is not immediate to an instant. Nor is it corrupted in an instant separated from it by a medium, because between those two instants an intermediate time intervenes, and in any time there are infinitely many instants, so that one instant will exist simultaneously with infinitely many instants, which is impossible.[19]

18 See the very end of the passage quoted in the previous note.

19 Wylton, *Quaestiones libri Physicorum*, IV, Q. 22, Ms. C, 68ra, lin. 39–68rb, lin. 3: 'Cum non possint esse aliqua simul de tempore nisi unum contineat aliud (ad C), quod non contingit dicere de instanti, cum sit indivisibile, quod aliquid contineat, ergo, cum unum est, aliud corruptum est. Aut ergo corrumpitur in se ipso, quod non est verum, quia tunc simul esset et non esset; aut corrumpitur in alio nunc. Non in nunc priori, quia tunc corruptum esset antequam esset; nec in posteriori, quia tunc illud instans in quo corrumpitur vel erit immediatum instanti in quo erat vel mediatum. Non immediatum, quia instans instanti non est immediatum. Nec corrumpitur in instanti mediato, quoniam, cum inter illa duo instantia cadat tempus medium, et in quolibet tempore sunt infinita instantia, ergo unum instans erit simul cum infinitis instantibus, quod est impossibile.'

Let us focus on the logic of this argument. The first step is to move from the hypothesis that the instant is always different to the conclusion that the instant is subject to corruption, that is, it ceases to exist. The question about the temporal measure of its ceasing to exist is then asked: when does an instant *t* cease to exist? An implicit assumption here is that an instant can only cease to exist at an instant, so that there are only three possible answers to this question: instant *t* ceases to exist either (i) at *t* or (ii) at an instant *s* before *t* or (iii) at an instant *w* after *t*. The first alternative (i) is ruled out because it would lead to the conclusion that instant *t* would simultaneously exist and not exist. The second alternative (ii) is obviously unsound: instant *t* cannot cease to exist before it exists. The third alternative (iii) is further articulated: the later instant *w* at which *t* ceases to exist is either immediate to *t* or separated from *t* by an intervening period of time. Because of the continuity of time, the first case is not possible: instants are not immediate to one another but between any two instants there is a period of time and hence infinitely many instants. But then the second case is not possible either because instant *t* would persist throughout the whole period of time between itself and the instant *w* of its corruption, which is against the assumption that it lasts for exactly one instant or, more literally, against the assumption that distinct instants cannot exist simultaneously but only one after the other.

Following the context of Aristotle's discussion, Wylton explicitly formulates the argument for the case of the temporal instant. He clearly assumes, however, that it applies to all instantaneous things, things that last for only one instant. The other relevant things of this kind in Aristotle's physics are the instantaneous phases of a change, the so-called *mutata esse*. Accordingly, in Wylton's view, an appropriate solution to this argument is one that is valid for all instantaneous things. Thus, he considers and rejects a solution to the argument that appeals to the alternative property of the instant of being a measure because this solution would leave the problem open for the case of the *mutatum esse*, which is not a measure.[20]

The alternative solution that Wylton supports is introduced in a very condensed way as follows:

---

20 Wylton, *Quaestiones libri Physicorum*, IV, Q. 22, Ms. C, 68[rb], lin. 9–17: 'Ad rationem istam tamquam sophisticam vel probabilem solum respondetur multis modis. Uno modo dicitur quod instans secundum esse non corrumpitur in aliquo instanti nec in aliqua mensura. Est enim mensura per essentiam nec eius est mensura altera eiusdem generis. Contra: haec responsio tollitur, si fiat ratio de mutato esse in motu, cuius mensura est instans secundum esse, in quo instanti corrumpitur. Et est eadem difficultas hic, sicut in instanti temporis, et cum hoc hic oportet dare mensuram extrinsecam in qua corrumpitur vel deficit.'

> Therefore, if the argument is made about the instantaneous phase of a change (*mutatum esse*), it is replied in another way that it is not corrupted in an instant, as is assumed in the argument, but in time. For an instant ceases to exist in virtue of the ceasing of a time, the ceasing of which is divisible.[21]

The crucial aspect of this solution is the rejection of the implicit premise of the argument of *Physics* IV.10 that the corruption of an instant *t* can only occur at an instant (either at *t* itself or at a prior or at a posterior instant). While this premise is supposed to exhaust all possible alternatives about when the corruption of instant *t* occurs, it neglects the alternative that the corruption of the instant could occur in time, that is, in an extended period of time after *t*. In the solution approved by Wylton instead this alternative is the one which does not encounter any of the difficulties of the others and which therefore is the right one.

As to what it means for an instant (or an indivisible phase of a change) to cease to exist in time, the short explanation given in the last sentence of the passage just quoted, namely, that an instant ceases to exist in virtue of the ceasing to exist of a time, is not very illuminating. It seems to rely on some implicit assumptions about the ontology of time and of the instant. One natural suggestion about what these assumptions could be is that what is at work here is a kind of reductionist account of the instant according to which the ceasing to exist of an instant is reducible to the ceasing to exist of a period of time. This reductionist account could then be used to give an easy solution to the problem of the ceasing of an instant along the following lines: there is nothing more to the ceasing of an instant apart from the ceasing of a period of time to which that instant belongs. However, this natural reductionist reading of the claim that an instant ceases to exist in virtue of the ceasing of a period of time is not that adopted by Wylton.

Note that in the passage above Wylton is reporting the opinion of others. He clearly agrees with this opinion that what is wrong with Aristotle's argument is the premise that an instant ceases to exist in an instant rather than in time. In rejecting this premise, however, he does not simply appeal to the reductionist assumption as this opinion does. When later on in his discussion he presents his own assessment of this opinion, he provides an explanation of the ceasing

21 Wylton, *Quaestiones libri Physicorum*, IV, Q. 22, Ms. C, 68[rb], lin. 17–20: 'Ideo aliter dicitur, si fiat argumentum de mutato esse, quod ipsum non corrumpitur in aliquo instanti, ut accipitur in ratione, sed in tempore. Per hoc enim quod tempus aliquod cedit, cuius cessio est divisibilis, cedit instans.'

of an instant according to which such a ceasing is dependent on the ceasing of a period of time but is not reducible to it.

This is how Wylton's explanation goes:

> For an indivisible ceases to exist only because of the ceasing of a divisible. Accordingly, the instant that is now present ceases to exist only because of the fact that it becomes past; but it does not become past because of the mere succession of an indivisible. The proof of this claim is evident. For, once the instant that is now present has ceased to exist, I ask whether there is something intermediate between the instant that is then present and the instant that has just become past or there is not. If there is something intermediate, then something divisible has ceased to exist, and this is the intended conclusion. If there is not something intermediate, therefore the instant that is now present and the instant that has just become past are simultaneous. And it follows from this 'therefore, they are totally simultaneous', because they are indivisibles. Therefore, since one of them is present, the other is present too, and thus it has not ceased to exist. Therefore, the proposition that states that an indivisible ceases to exist in an indivisible way should always be denied. And the reason why an indivisible ceases to exist in a divisible way is because it ceases to exist as the result of the ceasing of something else that is divisible.[22]

One preliminary move of Wylton's explanation is to understand the claim that an instant ceases to exist in time in the sense that the ceasing of an instant is such that it is not the case that that instant only ceases to exist but a whole period of time ceases to exist too, that is, a period of time starting from that instant. The intuition here is that one cannot just cut off an instant from time; on the contrary, one must always cut off a slice of time together with an instant. The proof of the claim thus interpreted is then the following. Suppose that the

---

22 Wylton, *Quaestiones libri Physicorum*, IV, Q. 22, Ms. C, 69[ra], lin. 7–17: 'Indivisibile enim non cedit (cadit C) nisi per successionem alicuius divisibilis (*corr. ex* indivisibilis C). Unde instans nunc praesens non cedit nisi per hoc quod fit praeteritum, non autem fit praeteritum per solam successionem indivisibilis, cuius probatio evidens est. Nam ipso cesso, quaero an inter instans quod tunc est praesentialiter et nunc quod iam factum est praeteritum sit medium vel non. Si sic, ergo aliquod divisibile cessit, et hoc est propositum. Si non sit medium, ergo sunt simul, et sequitur "ergo sunt totaliter simul", cum sint indivisibilia. Ergo ex hoc quod unum est praesens, et aliud praesens, et per consequens non est cessum. Igitur semper est haec propositio neganda, quod indivisibile cedit indivisibiliter. Causa autem quare cedit divisibiliter est quia cedit ad cessionem alterius quod est divisibile.'

present instant *t* ceases to exist, that is, becomes past; and suppose that the new present instant is *w*. Because of the continuity of time, instants are dense so that between *t* and *w* there is a period of time. But this period of time is past too, given that it is prior to the new present instant *w*. Hence, it is not only instant *t* that has become past, but the whole period of time between *t* and the actual present instant *w* has become past too. With the concise formulation used by Wylton at the end of the passage just quoted: an indivisible ceases to exist in a divisible way (*divisibiliter*), that is, together with the ceasing to exist of something divisible, that is, in the present case, a period of time.

Much light on Wylton's solution is shed by his reply to an objection to it. The objection says that the claim that an instant of time ceases to exist in a divisible way is incompatible with the indivisible existence of the instant:

> But against this there is the following objection: from the mere fact that the instant according to its being is indivisible it follows that it ceases to exist in an indivisible way. For, suppose that the opposite is the case, that is, that its corruption is divisible in time. Since everything that is corrupted in a divisible way always has being until the terminus of its corruption, it follows that it exists in a divisible way, so that it will have a being that lasts for some time.[23]

The general idea on which this objection relies is that there must be an isomorphism between the way in which something exists and the way in which it ceases to exist, so that things that exist in a divisible way cease to exist in a divisible way whereas things that exist in an indivisible way cease to exist in an indivisible way. Therefore, since an instant exists in an indivisible way, that is, just for one instant, it must also cease to exist in an instant, and not in time, which is contrary to what Wylton's solution assumes.

Wylton replies to this objection as follows:

> ... about the assumption made that 'if the instant ceases to exist in a divisible way, therefore it exists in a divisible way,' I draw a distinction about 'ceasing', because it can either express (i) the way to having ceased to exist or (ii) the having ceased to exist itself. If it is taken in the first way (i), then I

23 Wylton, *Quaestiones libri Physicorum*, IV, q. 22, Ms. C, 68rb, lin. 20–25: 'Sed contra hoc videtur esse: eo ipso quod instans secundum esse est indivisibile sequitur quod cedit indivisibiliter. Nam detur oppositum, quod eius corruptio sit divisibilis in tempore. Cum omne quod corrumpitur divisibiliter semper ante terminum corruptionis habeat esse, sequitur quod divisibiliter erit, ita quod habebit esse durativum.'

> say that, as the instant ceases to exist, it ceases to exist when it exists. And it does not follow from this that 'therefore it at the same time exists and does not exist', because what is on its way to ceasing to exist before it has ceased to exist exists, and for then it should not be conceded that it does not exist. If by 'ceasing' the having ceased to exist is understood (ii), then I say that it does not follow 'it has ceased to exist in a divisible way, that is, because of the ceasing of something divisible, therefore it exists in a divisible way', because when it has ceased to exist it does not exist. But I do concede that its having ceased to exist is divisible and that it is not possible to give a first instant or a first part of time in which it has ceased to exist, just as it is not possible to give the first part of a change or a first instant of a change after its initial terminus or a first part of time, as has been proved in *Physics* VI.[24]

Thus, Wylton thinks that the objection arises from an ambiguity about ceasing (*cedere*): when we talk about something's ceasing to exist we can either mean that (i) it is on its way to having ceased to exist but has not yet ceased to exist or that (ii) it has already ceased to exist. The having ceased (*cessum esse*) to exist is here understood as the terminus to which the ceasing to exist as a way leads and as temporally posterior to it. In terms more intelligible to us, let us distinguish the two cases using the present tense of the verb 'to cease' for the first case and the past tense for the second case, so the distinction becomes that between 'to cease to exist' and 'to have ceased to exist'.

It is important to point out that to this distinction about ceasing there corresponds a distinction between two ways of understanding the original Aristotelian question about the temporal aspects of the ceasing of the instant. Understood in the first way, the question asks when an instant ceases to exist (that is, it is on its way to having ceased to exist). Understood in the second way, it asks when an instant has ceased to exist. For Wylton, both questions can be

---

24 Wylton, *Quaestiones libri Physicorum*, IV, Q. 22, Ms. C, 69ra, lin. 18–29: 'Cum accipitur "si cedit divisibiliter, igitur divisibiliter est", distinguo de hoc quod est cedere, ex hoc quod potest dicere viam ad cessum esse vel ipsum cessum esse. Si primo modo, tunc dico quod, sicut cedit, cedit quando est. Nec sequitur "ergo simul est et non est", quia quod est in cedendo ante cessum esse est, et pro tunc non est concedendum quod non est. Si intelligatur per cedere cessum esse, dico quod non sequitur "cessum est divisibiliter, quia scilicet ad cessionem alicuius divisibilis, ergo divisibiliter est", quia quando cessum est non est. Sed bene concedo quod suum cessum esse est divisibile, nec contingit dare primum instans, nec primam partem temporis in quo cessum est, sicut nec contingit dare primam partem motus, nec primum instans post terminum a quo nec temporis, ut probatur sexto *huius*.'

answered in a non-aporetic way, but the respective answers are different. The answer to the first (present-tense) question is that an instant *t* ceases to exist at *t*, that is, when it exists. The answer to the second (past-tense) question is that instant *t* has ceased to exist in time, that is, in the period of time starting from *t*.

I will return to the answer to the first (present-tense) question in the next section. It is the answer to the second (past-tense) question that is relevant for the solution of the argument of *Physics* IV.10 that Wylton supports. He points out that this answer is not open to the objection against his solution. If instant *t* has ceased to exist in a period of time, it does not follow that it exists in a divisible way, because it does not exist at all in that period of time. The having ceased to exist is a state in which something does not exist. When it has ceased to exist, instant *t* does not exist. Therefore, the divisibility of the having ceased to exist of an instant is associated to its non-being and not to its being. And the non-being of an instant is indeed divisible, that is, lasts for a time, so that the association between the divisibility and the non-being of an instant is not problematic (is not in contrast with any isomorphism).

Wylton admits, however, that the divisibility associated to the non-being of the instant prevents an answer to the more specific question of when instant *t* has first ceased to exist. Instant *t* has ceased to exist in the period of time starting at *t*, but there is no instant within that period (or also a part of that period) in which it has first ceased to exist. Take any instant *w* in that time. Instant *t* was already a non-existent in the period of time between *t* itself and instant *w*, so that *w* is not the first instant of the non-existence of *t*.

4 Comparison between Wylton's and Burley's Solutions

Wylton and Burley have very similar views about the two general doctrinal areas to which the particular issue of the ceasing to exist of an instant of time belongs. These are Aristotle's theory of time and the so-called *incipit* and *desinit* problem, that is, the problem of assigning temporal limits to the beginning and ceasing of things. As to time, both commentators hold a strongly realist view about its ontological status, according to which time is an extra-mental thing distinct from change, and defend this view with very similar arguments.[25] As to

25 See C. Trifogli, 'Il problema dello statuto ontologico del tempo nelle *Quaestiones super Physicam* di Thomas Wylton e di Giovanni di Jandun', *Documenti e studi sulla tradizione filosofica medievale* 1 (1990), 491–548; C. Trifogli, 'Motion and Time', in A.D. Conti (ed.), *A companion to Walter Burley. Late Medieval Logician and Metaphysician*, Leiden 2013 (Brill's

the *incipit* and *desinit* problem, they both devote an extensive discussion to the case of permanent things. Burley's treatise *De primo et ultimo instanti* is one of the most important medieval works on this problem,[26] and the question on the same topic discussed by Wylton in his *Physics*-Commentary is a major source of it.[27] Finally, both commentators offer an explicit solution to the argument of *Physics* IV.10 about the ceasing to exist of the instant. Therefore, because the two commentators deal with this argument within the same theoretical background, it makes good sense to compare their respective solutions; more than that, one may even expect their solutions to be the same. Surprisingly, however, this is not the case: the two solutions are different and *prima facie* in conflict.

As we have seen, the crucial ingredient of Wylton's solution consists in rejecting the assumption of the Aristotelian argument that the ceasing to exist of an instant *t* occurs at an instant, and arguing for the claim that it occurs in time. On the contrary, Burley's solution does not challenge the Aristotelian assumption. Rather, its crucial ingredient consists in defending one of the three alternatives in which that assumption is articulated, namely, that the ceasing to exist of instant *t* occurs at instant *t* itself. Burley then argues that, contrary to what the argument assumes, it does not follow from this that instant *t* simultaneously exists and does not exist when it is subject to corruption. Instant *t* both exists and is subject to corruption at instant *t*, but it is not also the case that it does not exist at *t*.[28]

In short, for Wylton the ceasing to exist of instant *t* occurs in the time after instant *t*, whereas for Burley it occurs at instant *t* itself. Isn't there a sharp conflict between these two replies? I do not think that this is the case. And the reason for this is that they are not replies to the same question. The relevant distinction here is that drawn by Wylton in the passage quoted above, namely, the

---

Companions to the Christian Tradition, 41), 267–299; C. Trifogli, 'The Reality of Time in the Commentary Tradition on the *Physics*: the case of Wylton and Burley', in G. Fioravanti, C. Leonardi and S. Perfetti (eds.), *Il commento filosofico nell'Occidente Latino (secoli XIII–XV)*, Turnhout 2002 (Rencontres de Philosophie Médiévale, 10), 233–251.

26 It is edited in C. Shapiro and H. Shapiro, 'De primo et ultimo instanti des Walter Burley', *Archiv für Geschichte der Philosophie* 47 (1965), 159–173. On this treatise and Burley's application of the logic of first and last instants presented in it to some issues about the Eucharist, see the recent paper by E.D. Sylla, 'Mathematics and Physics of First and Last Instants: Walter Burley and William of Ockham,' in F. Goubier and M. Roques (eds.), *The Instant of Change in Medieval Philosophy and Beyond*, Leiden 2018 (Special issue of *Vivarium*, 55 [2017]), 103–129.

27 For the edition of Wylton's question and its close relationship with Burley's treatise see C. Trifogli, 'Thomas Wylton's Question *An contingit dare ultimum rei permanentis in esse*,' *Medieval Philosophy and Theology* 4 (1994), 91–141.

28 Trifogli, 'Walter Burley on the *Incipit* and *Desinit* of an Instant of Time,' 90–91.

distinction between the present-tense question and the past-tense question. As I pointed out in my paper on Burley's view about the ceasing of the instant,[29] Burley's reply addresses the present-tense question 'when does instant *t* cease to exist?', whereas Wylton's reply answers the past-tense question 'when has instant *t* ceased to exist?' Thus, the contrast between the two commentators resides in their interpretations of the Aristotelian argument about the ceasing of the instant. In Wylton's interpretation, the argument asks the question about the having ceased to exist of the instant (the past-tense question), whereas in Burley's interpretation it asks the question about its ceasing to exist (the present-tense question).

Wylton's interpretation is the correct one from the exegetical point of view: Aristotle uses the past-tense 'has ceased to exist' in his formulation of the question.[30] On the other hand, Burley's unorthodox interpretation of the Aristotelian argument gives rise to a very sophisticated discussion of the notion of ceasing to exist and its applicability to the case of indivisibles like instants.[31] A proper assessment of the respective merits of the two interpretations is not so important in this context. What is more relevant is to compare the replies of the two commentators to each of the two questions. As to the present-tense question, it is clear from the passage quoted above that Wylton's reply is the same as that of Burley, that is, instant *t* ceases to exist at instant *t* itself, so that it ceases to exist when it exists. What about Burley's reply to the past-tense question? Like Wylton, Burley too is led to distinguish between the two questions in dealing with an objection against his solution to the Aristotelian argument. And his reply to the past-tense question is conceptually (although not linguistically) the same as that of Wylton, namely, that instant *t* has ceased to exist in the period of time starting at *t* but there is not an instant within this period at which *t* has first ceased to exist.[32]

Conclusion

The puzzle about the ceasing of the instant of time in *Physics* IV.10 is very challenging for Aristotelian commentators. The puzzle itself is a difficult one, because it seems to rule out all the possible ways in which an instant can cease to exist. But it does need to be solved, given that the assumption that there are

29 Trifogli, 'Walter Burley on the *Incipit* and *Desinit* of an Instant of Time,' 91.
30 See, for example, Coope, *Time for Aristotle*, 26–27.
31 Trifogli, 'Walter Burley on the *Incipit* and *Desinit* of an Instant of Time,' 92–96.
32 Trifogli, 'Walter Burley on the *Incipit* and *Desinit* of an Instant of Time,' 100–102.

many instants existing one after the other plays an essential role in Aristotle's account of time. Aristotle, however, does not provide an explicit solution to it. What is more, other parts of Aristotle's discussion of the instant—most notably his treatment of the sameness and diversity of the instant in *Physics* IV.11—complicate the task of finding an appropriate solution to the puzzle because they may induce commentators to look for such a solution in the wrong direction. In this paper I hope to have shown that Thomas Wylton takes up this challenge and deals with it in a brilliant way. The most fundamental merit of Wylton's approach is that it sets the puzzle of *Physics* IV.10 in its correct theoretical framework. The puzzle is about the temporal account of the ceasing of the instant, that is, about the questions of 'when' an instant ceases to exist or of 'when' it has ceased to exist. Questions of this kind naturally belong the domain of the *incipit-desinit* theory. And it is to this theory that Wylton turns to find a solution to the puzzle. The solution he proposes involves counter-intuitive claims like the one that an indivisible thing has ceased to exist in a divisible thing, or that an indivisible thing ceases to exist when it exists. Wylton takes great care to show that these claims make perfect sense when interpreted within the logic of *incipit-desinit.* His discussion of the logic of the ceasing of the instant is extremely clear, compelling, and subtle. Although strictly speaking Wylton himself does not belong to the *Calculatores*, the *incipit-desinit* theory is indeed a crucial part of the *Calculatores* theoretical framework. Thus, Wylton's discussion shows how powerful the approach of the *Calculatores* can be when applied to traditional problems of Aristotle's natural philosophy.

CHAPTER 2

The New Interpretation of Aristotle: Richard Kilvington, Thomas Bradwardine, and the New Rule of Motion

Elżbieta Jung

1 Introduction

The main purpose of this paper is to show that one of the most famous theories of motion, formulated in 1328 by Thomas Bradwardine as most of the historians of medieval science believe, was not his original achievement. As I have argued in my paper *Works by Richard Kilvington*, it was Richard Kilvington who offered weighty arguments, later used by Bradwardine to reinterpret Aristotle's theory.[1] That article was the result of my extensive research on the Oxford Calculators in general and on Richard Kilvington in particular.[2] I began my study with Kilvington's questions on motion and at the outset I noticed their dependence, as I thought, on Thomas Bradwardine's *Tractatus de proportionibus velocitatum in motibus*. Thus, my first step was a careful study of Norman and Barbara Kretzmann's works, especially a critical edition, English translations, and detailed study of Kilvington's *Sophismata*.[3] The Kretzmanns accentuate the obvious need for further study of the historical and philosophical relationships between these two men.[4] Therefore I transcribed and attentively read all of Kilvington's works, most of them still available only in manuscripts.[5] My in-depth studies of

1 E. Jung-Palczewska, 'Works by Richard Kilvington,' *Archives d'histoire doctrinale et littéraire du Moyen Âge* 67 (2000), 181–223.

2 I am grateful to Edith Sylla, who first drew my attention to Richard Kilvington's questions on motion found in Venezia, Biblioteca Nazionale Marciana, Ms. lat. VI. 72 (2810).

3 Richard Kilvington, *Sophismata*, eds. N. Kretzmann and B.E. Kretzmann, Oxford 1990 (Auctores Britannici Medii Aevi, 12) and [Richard Kilvington], *The Sophismata of Richard Kilvington*, transl. N. Kretzmann and B.E. Kretzmann, New York 1991.

4 Kretzmann and Kretzmann, 'Introduction,' in [Richard Kilvington], *The Sophismata*, XXVIII.

5 A critical edition by Monika Michałowska of Kilvington's questions on Aristotle's *Ethics* has been published recently: Richard Kilvington, *Quaestiones super libros Ethicorum*, ed. M. Michałowska, Leiden 2016 (Studien und Texte zur Geistesgeschichte des Mittelalters, 121). Robert Podkoński and I have been working on Kilvington's questions on the *De generatione et corruptione*. I am also working on a critical edition of the *Questiones super Physicam*, which, most likely, will be published next year.

 | DOI:10.1163/9789004512054_004

all Kilvington's works confirmed Kilvington's probable biography and the dates of his works, as suggested by the Kretzmanns.[6] Still, ongoing historical and doctrinal research of the 1320s brought into light new discoveries and revealed unknown facts which are significant with regard to the main subject of this paper. Thus it seems worth rediscussing the new interpretation of Aristotle's theory of motion traditionally ascribed to Thomas Bradwardine in the light of those findings.

2 Richard Kilvington and Thomas Bradwardine: The Founders of the Calculators' School

Before discussing Kilvington's and Bradwardine's new theory of motion I will briefly outline their relationship as seen through the eyes of the historians of medieval science. Anneliese Maier described Kilvington as not only Bradwardine's friend but also his student.[7] Norman Kretzmann and William Courtenay rejected the teacher-student relationship between Bradwardine and Kilvington.[8]

There is no doubt that the roads of Bradwardine and Kilvington continued to cross. Thomas Bradwardine (ca. 1295–1349) was a Bachelor of Arts at Balliol College, Oxford, in 1321. In 1323 he became a fellow of Merton College, Oxford, where he probably remained for the next twelve years.[9] In the same year, he became a Master of Arts, in 1340 a Doctor of theology.[10] Richard Kilvington (ca. 1302–1361) began his study in the Arts at Oxford around 1317. Most likely, he first entered Balliol College, where he met Bradwardine. He studied theo-

6 Kretzmann and Kretzmann, 'Introduction,' in [Richard Kilvington], *The Sophismata*, XXIII–XXVII.

7 See A. Maier, *Die Vorläufer Galileis im 14. Jahrhundert* (Studien zur Naturphilosophie der Spätscholastik, 1), 2nd ed., Rome 1966 (Storia e letteratura, 22), 174; Eadem, *An der Grenze von Scholastik und Naturwissenschaft* (Studien zur Naturphilosophie der Spätscholastik, 3), 2nd ed., Rome 1952 (Storia e letteratura, 52), 266–267; Eadem, *Ausgehendes Mittelalter. Gesammelte Aufsätze zur Geistesgeschichte des 14. Jahrhunderts*, 1, Rome 1964 (Storia e letteratura, 97), 75.

8 N. Kretzmann, 'Richard Kilvington and the Logic of Instantaneous Speed,' in A. Maierù and A. Paravicini Bagliani (eds.), *Studi sul XIV secolo in memoria di Anneliese Maier*, Rome 1981 (Storia e letteratura, 151), 143–178, at 144–145; W.J. Courtenay, *Schools and Scholars in Fourteenth-Century England*, Princeton, NJ, 1987, 244.

9 See E.B. Fryde and J.R.L. Highfield, 'An Oxfordshire Deed of Balliol College', *Oxoniensia* 20 (1955), 40–45.

10 See J.A. Weisheipl, 'The Place of John Dumbleton in the Merton School,' *Isis* 50 (1959), 439–454, at 446, n. 45.

logy at Oriel College, Oxford.[11] In 1321, he became a Bachelor of Arts, in 1324 or 1325 a Master of Arts, then a Doctor of Theology in ca. 1335.[12]

Both Bradwardine and Kilvington belonged to the circle of friends and courtiers of Richard de Bury, Bishop of Durham, who, starting in the year 1334, helped them to advance in an ecclesiastical and political career. They both were members of de Bury household in 1334–1335.[13] Richard de Bury also introduced them to the royal court of Edward III. Bradwardine and Kilvington were members of a diplomatic mission in 1338–1340.[14] Their academic career was followed by an ecclesiastical one. In 1333 Bradwardine was made Canon at Lincoln Cathedral, and his career was crowned with his election, in 1349, as Archbishop of Canterbury. As chancellor of St. Paul's Cathedral in London, he was made royal chaplain in 1337 and, probably, the king's confessor. He accompanied Edward III on his travels to Flanders and France during the campaign of 1346. Immediately after his episcopal consecration, which was held in Avignon, Bradwardine returned to England to assume his position, yet he died one month later, on the 26th of August 1349, as a victim of the first wave of the Black Death.[15] Kilvington's ecclesiastical career culminated in his appointment as Dean of St. Paul's Cathedral in 1354. Along with Richard FitzRalph, Kilvington was involved in the battle against the mendicant friars, a dispute that continued almost until his death in 1361.[16]

Thomas Bradwardine authored many significant works, which cover a number of scholarly domains. His insight and intellectual inquisitiveness earned him the title of *Doctor profundus* and a mention in Chaucer's *Canterbury Tales*. Works by Bradwardine that have been preserved to our time are the following: two treatises in mathematics, *Arithmetica speculativa* and *Geometria specu-*

11 See Kretzmann and Kretzmann, 'Introduction,' in [Richard Kilvington], *The Sophismata*, XXIV.

12 See Kretzmann and Kretzmann, 'Introduction,' in [Richard Kilvington], *The Sophismata*, XXVII; Jung-Palczewska, 'Works,' 222.

13 See G. Leff, *Bradwardine and the Pelagians*, Cambridge 1957 (Cambridge Studies in Medieval Life and Thought, Series 2, 5), 3.

14 See Kretzmann and Kretzmann, 'Introduction,' in [Richard Kilvington], *The Sophismata*, XXVII.

15 See E. Wilks Dolnikowski, *Thomas Bradwardine. A View of Time and a Vision of Eternity in Fourteenth-Century Thought*, Leiden 1995 (Studies in the History of Christian Thought, 65), 5.

16 See E. Jung, 'Richard Kilvington,' in E.N. Zalta (ed.), *The Stanford Encyclopedia of Philosophy* (Winter 2016 Edition), URL = https://plato.stanford.edu/archives/win2016/entries/kilvington/. See also E. Jung, 'Richard Kilvington,' in E. Jung and R. Podkoński, *Towards the Modern Theory of Motion. Oxford Calculators and the New Interpretation of Aristotle*, Łódź 2020 (Research on Science and Natural Philosophy, 4), 13–17.

lativa;[17] a number of logical treatises (all written before 1328); a famous work on motion, *Tractatus de proportionibus velocitatum in motibus*, written in 1328;[18] a *Tractatus de continuo*;[19] two theological works, a commentary on the *Sentences*, which includes a question on future contingents, edited by Jean-François Genest as a separate treatise *De futuris contingentibus*,[20] and *De causa Dei contra Pelagium et de virtute causarum ad suos Mertonenses*,[21] composed in 1344. Bradwardine also authored *De memoria artificiali adquirenda*.[22]

While most of Bradwardine's philosophical works seem to have been composed as student textbooks or 'guides' for his colleagues (Mertonians), all of Kilvington's known works, except for a few sermons, stem from his teaching at Oxford and often reflect lively discussions in the classroom.[23] None of his works is written in the usual commentary format, following the order of books

---

17 Thomas Bradwardine, *Geometria speculativa*, ed. and transl. A.G. Molland, Stuttgart 1989 (Boethius. Texte und Abhandlungen zur Geschichte der exakten Wissenschaften, 18).

18 Thomas Bradwardine, *Tractatus de proportionibus velocitatum in motibus*, ed. H.L. Crosby, Jr., *Thomas of Bradwardine. His* Tractatus de proportionibus. *Its Significance for the Development of Mathematical Physics*, Madison, WI, 1955 (University of Wisconsin Publications in Medieval Science, 2). The colophon of one of the copies of Bradwardine's treatise mentions the year 1328: 'Explicit tractatus de proportionibus editus a magistro Thoma de Bradelbardin. Anno Domini MCCC28' (ibid., 140).

19 Thomas Bradwardine, *Tractatus de continuo*, ed. J.E. Murdoch, *Geometry and the Continuum in the Fourteenth Century: A Philosophical Analysis of Thomas Bradwardine's* Tractatus de Continuo, unpublished Ph.D. dissertation, University of Wisconsin, 1957.

20 On Bradwardine's questions on the *Sentences* see, e.g., J.-F. Genest and K.H. Tachau, 'La lecture de Thomas Bradwardine sur les Sentences,' *Archives d'histoire doctrinale et littéraire du Moyen Âge*, 57 (1990), 301–306; J.-F. Genest, 'Les premiers écrits théologiques de Bradwardine: textes inédits et découvertes récentes,' in G.R. Evans (ed.), *Mediaeval Commentaries on the* Sentences *of Peter Lombard. 1: Current Research*, Leiden 2002, 395–421. For a critical edition of the question *Utrum Deus habeat prescienciam omnium futurorum contingencium ad utrumlibet*, see J.-F. Genest, 'Le *De futuris contingentibus* de Thomas Bradwardine,' *Recherches Augustiniennes et Patristiques*, 14 (1979), 249–336. Chris Schabel and Severin Kitanov prepared a critical edition of questions 7–9 from the set of nine (for the titles, see Genest, 'Les premiers écrits théologiques,' 397). See S. Kitanov and C. Schabel, 'Thomas Bradwardine's Questions on Grace and Merit from His *Lectura* on the *Sentences* at Oxford, 1332–1333,' *Archives d'histoire doctrinale et littéraire du Moyen Âge* 89 (2022) (forthcoming).

21 Thomas Bradwardine, *De causa Dei contra Pelagium et de virtute causarum ad suos Mertonenses libri tres*, ed. H. Savile, London 1618.

22 Edited by M. Carruthers, 'Thomas Bradwardine, *De memoria artificiali adquirenda*,' *The Journal of Medieval Latin* 2 (1992), 25–43. For an English translation, see M. Carruthers, *The Book of Memory: A Study of Memory in Medieval Culture*, New York 1990, 281–228. Additionally, see also M. Carruthers and J.M. Ziolkowski (eds.), *The Medieval Craft of Memory. An Anthology of Texts and Pictures*, Philadelphia, PA, 2002, 205–214.

23 See E. Jung, *Arystoteles na nowo odczytany. Ryszarda Kilvingtona 'Kwestie o ruchu'*, Łódź 2014, 107–316.

in the respective works of Aristotle. Kilvington reduced the number of topics discussed to certain central issues, which were fully developed with no more than ten questions constituting a commentary. The reduction in the range of topics is counterbalanced by a deeper analysis in the questions chosen for treatment.[24] Some of Kilvington's questions cover twenty folios, which in a modern edition yield about 150 pages. Only his logical treatise was not written as a commentary, but rather as 'a guide' for students showing how to solve sophisms. In the preface to his *Sophismata* Kilvington writes:

> When we are able to call both sides into question, we will readily discern what is true and what is false, as Aristotle says in Book One of his *Topics*. Therefore, in order that we may more readily discern what is true and what is false, in the present work, which consists of *sophismata* to be thoroughly investigated, I intend, to the best of my ability, both to demolish the two sides of the contradiction and also to support them by means of clear reasoning. I am led to do this by the request of certain young men who have been pressing their case very hard. And so, wishing to give them something I have often heard them ask for, I have undertaken to make an attempt in that direction.[25]

The chronology of Kilvington's works is well established. The *Sophismata* and *Quaestiones super De generatione et corruptione*, composed before 1325, came from his lectures as a bachelor of arts; the *Quaestiones super Physicam* were composed at the latest in 1326; the *Quaestiones super libros Ethicorum* (before 1332) date from his time as a Master of Arts; after he advanced to the Faculty of Theology, he produced eight questions on Peter Lombard's *Sentences* (1333 or 1334).[26] As there have been some recent discoveries, I will briefly outline these findings to reassert the view that the commentary on the *Sentences* was composed in these years. In a recent paper, Chris Schabel convincingly argued that in Oxford, like in Paris, it was common practice to lecture on the *Sentences* in one year only.[27] William Courtenay suggests Kilvington lectured on the *Sentences* between 1331 and 1335, perhaps around 1333–1334, since he is familiar

24 See for example Kilvington, *Quaestiones super libros Ethicorum*, 63–336.

25 [Richard Kilvington], *The Sophismata*, 1.

26 See Jung, 'Works by Richard Kilvington,' 222.

27 See C. Schabel, 'Ockham, the *Principia* of Holcot and Wodeham, and the Myth of the Two-Year *Sentences* Lecture at Oxford,' *Recherches de Théologie et Philosophie Médiévales*, 87 (2020), 59–102. See also W.O. Duba and C. Schabel, 'Remigio, Scotus, Auriol and the Myth of the Two-Year *Sentences* Lecture at Paris,' *Recherches de Théologie et Philosophie Médiévales* 84 (2017), 143–179.

with Wodeham's and Bradwardine's theological opinions.[28] Bradwardine lectured in the year 1331–1332 or 1332–1333.[29] Rega Wood is of the opinion that Wodeham lectured at Oxford after the *Norwich Lecture* (*Lectura secunda*), i.e., after 1332.[30] Paul Streveler and Katherine Tachau provide us with the date 1332–1333 for Wodeham's first year as a bachelor lecturing on book I of the *Sentences*, and 1333–1334 for his second year lecturing on books 2–4.[31] According to Schabel, Wodeham's Oxford lectures took place in 1331–1332.[32] Thus, the *terminus ante quem* for Kilvington's *Sentences* commentary is 1333, that is after Wodeham's and Bradwardine's lectures. We do not have any evidence that Kilvington lectured on the Bible, so we can safely assume that he lectured only on the *Sentences*, and following Schabel, we can safely place his lectures on the *Sentences* for one academic year only in 1333–1334, and less likely in 1334–1335.

Kilvington's *Sentences* commentary consists of a set of eight questions,[33] his *Ethics* form a set of eight questions,[34] his *Physics* commentary contains

---

28 See Courtenay, *Schools and Scholars*, 272.

29 See C. Schabel, 'Richard FitzRalph vs. William Skelton, 1331–1332: The Attribution of the *Determinationes* in a Florence Manuscript,' in M. Dunne and I. Nolan (eds.), *A Companion to Richard FitzRalph* (forthcoming). I am grateful to Chris Schabel for sharing his texts with me.

30 See Adam Wodeham, *Lectura secunda in librum primum Sententiarum*, eds. R. Wood and G. Gál, New York 1990, 1: 32*–37*.

31 Robert Holcot, *Questions on Future Contingents*, eds. P.A. Streveler and K.H. Tachau, *Robert Holcot. Seeing the Future Clearly. Questions on Future Contingents*, Toronto, ON, 1995 (Studies and Texts, 119), 27.

32 See Schabel, 'Ockham, the *Principia* of Holcot and Wodeham,' 71.

33 The commentary on the *Sentences* is found in the following manuscripts: (1) Bologna, Biblioteca comunale dell'Archiginnasio, Ms. 985, 1ra–52va; (2) Brugge, Stedelijke Openbare Bibliotheek, Ms. 188, 3ra–54vb; (3) Brugge, Stedelijke Openbare Bibliotheek, Ms. 503, 80ra–105rb; (4) Erfurt, Universitätsbibliothek, Dep. Erf., CA F. 105, 122rb–182rb; (5) London, British Library, Harley, Ms. 3243, 111rb–131rb; (6) Paris, Bibliothèque Nationale de France, Ms. lat. 14.576, 117ra–199vb; (7) Paris, Bibliothèque Nationale de France, Ms. lat. 15.561, 198ra–226vb; (8) Praha, Národní Knihovna České Republiky, Ms. III. B. 10, 130ra–139vb, 140va–212vb; (9) Wrocław, Biblioteka Uniwersytecka, Ms. IV. F. 198, 15ra–45rb; (10) Città del Vaticano, Biblioteca Apostolica Vaticana, Ms. Vat. lat. 4353, 1^{r}–58^{r}; (11) Tortosa, Biblioteca de la Catedral y del Cabildo de la Sanctísima Iglesia Catedral, Ms. 186, 35ra–66ra. For a description of the manuscripts see M. Michałowska (ed.), *Richard Kilvington on the Capacity of Created Beings, Infinity, and Being Simultaneously in Rome and Paris*. Critical Edition of Question 3 from *Quaestiones super libros Sententiarum*, Leiden 2021 (Studien und Texte zur Geistesgeschichte des Mittelalters, 130), 35–49. The eight questions, from the Bologna manuscript are titled as follows: (1) *Utrum Deus sit super omnia diligendus*; (2) *Utrum per opera meritoria augeatur habitus caritatis quo Deus est super omnia diligendus*; (3) *Utrum omnis creatura sit suae naturae certis limitibus circumscripta*; (4) *Utrum quilibet actus voluntatis per se malus sit per se aliquid*; (5) *Utrum peccans mortaliter per instans solum mereatur puniri per infinita instantia interpolata*; (6) *Utrum aliquis nisi forte in poena peccati possit esse perplexus in his quae pertinent ad salutem*; (7) *Utrum omnis actus factus extra gratiam*

eight questions,[35] his commentary on the *De generatione et corruptione* consists of nine questions.[36] As it appears, Kilvington's works are of an intricate manuscript tradition. Most problematic, with regard to both the manuscript tradition and the structure, and at the same time most interesting in terms of originality, are his questions on the *Physics*. The complete set consisting of eight

---

sit peccatum; (8) *Utrum aliquis possit simul peccare venialiter et mereri vitam aeternam*. A critical edition by E. Jung and M. Michałowska of Kilvington's question 4 will be submitted to Brill in 2022.

34 Richard Kilvington, *Quaestiones super libros Ethicorum*, 11–26.

35 The questions are the following: *Expositio super primum librum Physicorum* (Città del Vaticano, Biblioteca Apostolica Vaticana, Ms. Vat. lat. 4353, 125^{r}–129^{v}); (1) *Utrum omne scitum sciatur per causam* (Città del Vaticano, Biblioteca Apostolica Vaticana, Ms. Vat. lat. 4353, 130^{r}–143^{v}); (2) *Utrum omne quod generetur ex contrariis generetur* (Città del Vaticano, Biblioteca Apostolica Vaticana, Ms. Vat. lat. 4353, 124^{v}; Sevilla, Biblioteca Capitular y Colombina, Ms. 7-7-13, 40ra–50vb; Erfurt, Universitätsbibliothek, Dep. Erf., CA O. 74, 70ra–86va); (3) *Utrum in omni generatione tria principia requirantur* (Sevilla, Biblioteca Capitular y Colombina, Ms. 7-7-13, 27rb–37ra; Paris, Bibliothèque Nationale de France, Ms. lat. 6559, 71rb–88rb; Brugge, Stedelijke Openbare Bibliotheek, Ms. 503, 41va–50vb); (4) *Utrum omnis natura sit principium motus et quietis* (Sevilla, Biblioteca Capitular y Colombina, Ms. 7-7-13, 37ra–40vb); (5) *Utrum potentia motoris excedit potentiam rei motae* (Venezia, Biblioteca Nazionale Marciana, Ms. lat. VI. 72 [2810], 81ra–89rb; Città del Vaticano, Biblioteca Apostolica Vaticana, Ms. Vat. lat. 2148, 76^{ra-vb}); (6) *Utrum qualitas suscipit magis et minus* (Venezia, Biblioteca Nazionale Marciana, Ms. lat. VI. 72 [2810], 89rb–101ra; Paris, Bibliothèque Nationale de France, Ms. lat. 16.401, 149^{v}–166^{v}; Città del Vaticano, Biblioteca Apostolica Vaticana, Ms. Vat. lat. 2148, 71ra–75ra; Città del Vaticano, Biblioteca Apostolica Vaticana, Ms. Vat. lat. 4429, 64^{r}–70^{v}; Paris, Bibliothèque Nationale de France, Ms. lat. 6559, 121ra–131ra; Oxford, Bodleian Library, Ms. Canon. Misc. 226, 61^{v}–65^{r}; Praha, Národní Knihovna České Republiky, Ms. III. B. 10, 140va–152vb ; Cambridge, Peterhouse Library, Ms. 195, 70rb–72ra); (7) *Utrum aliquod motus simplex possit moveri aeque velociter in vacuo et in pleno* (Venezia, Biblioteca Nazionale Marciana, Ms. lat. VI. 72 [2810], 101ra–107vb); (8) *Utrum omne transmutatum in transmutationis initio sit in eo ad quod primitus transmutatur* (Venezia, Biblioteca Nazionale Marciana, Ms. lat. VI. 72 [2810], 107vb–112rb).

36 The questions are the following: (1) *Utrum augmentatio sit motus ad quantitatem*; (2) *Utrum numerus elementorum sit aequalis numero qualitatum primarum*; (3) *Utrum ex omnibus duobus elementis possit tertium generari*; (4) *Utrum continuum sit divisibile in infinitum*; (5) *Utrum omnis actio sit ratione contrarietatis*; (6) *Utrum omnia elementa sint adinvicem transmutabilia*; (7) *Utrum mixtio sit miscibilium alteratorum unio*; (8) *Utrum omnia contraria sint activa et passiva adinvicem*; (9) *Utrum generatio sit transmutatio distincta ab alteratione*. They are to be found, as a complete or incomplete set, in the following manuscripts: (1) Brugge, Stedelijke Openbare Bibliotheek, Ms. 503, 20vb–50vb; (2) Cambridge, Peterhouse Library, Ms. 195, 60ra–81rb; (3) Erfurt, Universitätsbibliothek, Dep. Erf., CA O. 74, 35ra–86va; (4) Kraków, Biblioteka Jagiellońska, Ms. 648, 40ra–53rb; (5) Paris, Bibliothèque Nationale de France, Ms. lat. 6559, 61ra–132vb; (6) Sevilla, Biblioteca Capitular y Colombina, Ms. 7-7-13, 9ra–27ra; (7) Firenze, Biblioteca Nationale Centrale, Cod. Conv. Soppr. B. VI. 1681, 37ra–77vb.

questions is not preserved in any codex: we find an *expositio* of the *Physics* and one question in a Vatican manuscript only, two questions are to be found only in a Sevilla manuscript, the other three only in Venice, and the next two in different libraries in Europe.

With regard to the main subject of this paper, the most interesting is the question titled *Utrum in omni motu potentia motoris excedit potentiam rei motae*, because here Kilvington offers a new interpretation of Aristotle's rules of motion. This question was well known by the anonymous author of the commentary on the *Sentences*, to be found in ms Vatican library Vat. Lat. 968, where he discusses the new rules of motion. In her *Verschollene Aristoteleskommentare des 14. Jahrhunderts*, Maier gives the following argument to justify the date of Kilvington's commentary on the *Physics*:

> Richard Killington is the author of a now-lost commentary on the *Physics*. We learn this from the anonymous *Sentences* commentary ... which is in Vat. lat. 986, the author of which appears to be a student of Bradwardine. Killington is quoted here several times, once with these words: *hoc probat Kilmiton in questionibus suis super librum Physicorum*. The proof quoted is based on a famous rule that Bradwardine had established in his *Tractatus proportionum*: Killington's commentary on the *Physics* must therefore follow this, i.e., must have been after 1328.[37]

I verified Maier's statement and came to the conclusion that the anonymous author of the *Sentences* commentary, when debating the problem of a proportion between an active power and resistance, quoted the opinions of both Kilvington and Bradwardine and made a clear distinction between them. He also argued separately against both claims.[38] Maier rightly recognized that

37 Maier, *Ausgehendes Mittelalter*, 1: 253: 'Richard Killington ist der Verfasser einer heute verschollenen Erklärung der Physik. Wir erfahren dies aus dem in philosophischer und naturphilosophischer Beziehung hochinteressanten Anonymen Sentenzenkommentar, der sich im Vat. Lat. 986 findet und dessen Verfasser ein Schüler Bradwardines zu sein scheint. Killington wird hier mehrfach zitiert, einmal mit diesen Worten: *hoc probat Kilmiton in questionibus suis super librum Physicorum*. Der zitierte Beweis bezieht sich aus einer berühmten Regel, die Bradwardine in seinem *Tractatus proportionum* aufgestellt hatte: Killingtons *Physikkommentar* muss also nach diesem, d. h. nach 1328 entstanden sein.'

38 Città del Vaticano, Biblioteca Apostolica Vaticana, Ms. Vat. lat. 986, 56rb: 'Primus articulus est utrum proportio actionum sequatur proportionem potentiae activae ad passivam seu potentiam resistivam ... *Primo hoc probat Kilmington in quaestionibus suis super* Physicorum, nam si sic, igitur subduplus motor non movebit in eodem tempore aeque velocitatis medietatem mobilis, quod movet duplus motor. Quod videtur falsum: cum universaliter enim videtur, qualis est proportio totius ad totum, talis est proportio partis ad

Kilvington's proof is related to Bradwardine's function. She, however, misintepreted that Kilvington used Bradwardine's rule for his own purpose.

Accordingly to Norman Kretzmann:

> … Kilvington, unlike Bradwardine and Swineshead, for instance, did not contribute significantly to the mathematical or strictly calculatorial side of the Calculator's work. Even when he is dealing with material we would think of as belonging to physics or mathematics. Kilvington, at least in his *Sophismata*, employs only the detailed argumentation and the analysis of concepts and of language that have part of the method of philosophy in almost every period of its history.[39]

My research confirmed Kretzmann's opinion that Kilvington, as a logician, was not interested in calculations. He, however, made a broad use of mathematics in all philosophical inquires as well as in theology.[40]

My claim that Kilvington's use of the new calculus of proportions in the description of motion enabled Bradwardine to formulate the New Rule of Motion, and that Kilvington was the inventor of the new theory of motion is, nevertheless, still disputed by Edith Sylla. On many occasions she has repeated her claim, formulated in her paper in 2008:

> Jung comes to the conclusion that Kilvington's arts lectures likely occurred before 1327, thus before Bradwardine's *On the ratios of velocities in motions* of 1328. As evidence against Weisheipl's opinion it is only necessary to note, however, that, as Jung herself reports, Thomas Bardwardine

---

partem. Probatur consequentia nam totus motor excedit totum mobile per certum excessum et medietas motoris excedit medietatem motoris per medietatem excessus et per consequens, si velocitatis sequitur excessum potentiae non erit tanta velocitas etc. *Secundo ad hoc arguit Bradwadinus in* Proportionibus *suis*, quia si sic, igitur sequitur quod si aliquis motor excedet suam resistentiam per minorem excessum quam alius suam, tardius moveret illam. Consequens falsum; igitur. Consequentia clara … *Ad primam dico* quod intelligit quod omnis motus provenit a proportione potentiae maioris aequalitatis motoris ad motum … *Ad secundam* iste est intellectus quod secundum quod proportio potentiae motoris ad potentiam rei motae est maior secundum hoc maior velocitas est, sed hoc non est contra me.' See, Jung-Palczewska, 'Works,' 215.

39 N. Kretzmann, '*Tu scis hoc esse omne quod est hoc*: Richard Kilvington and the Logic of Knowledge,' in N. Kretzmann (ed.), *Meaning and Inference in Medieval Philosophy. Studies in Memory of Jan Pinborg*, Dordrecht 1988 (Synthese Historical Library, 32), 225–245, at 225–226.

40 See, e.g., Jung-Palczewska, 'Works,' 199–203, 218–220; E. Jung and R. Podkoński, 'Richard Kilvington on Proportions,' in J. Biard and S. Rommevaux (eds.), *Mathématiques et théorie du mouvement, XIVe–XVIe siècles*, Villeneuve d'Ascq 2008 (Histoire des sciences), 81–101.

> is supposed to have read the *Sentences* at Oxford around 1332–1333, while we know that his *On the ratios* is dated 1328, less than seven years earlier. It seems more reasonable to assume at most that Bradwardine and Kilvington may have been working in parallel and that each knew the work of the other before completing the edited versions of his own work.[41]

I would readily accept Sylla's conclusion were it not for the fact that most of the arguments justifying the 'New Rule of Motion', known also as 'Bradwardine's Rule', are *verbatim* quotes from Kilvington's works. This very fact, in my opinion, cannot be explained by claiming either that Kilvington and Bradwardine had been working in parallel or that Kilvington's question was the result of that 'he heard some ideas aurally within the Oxford context in 1320s.'[42] It rather testifies that one of them had the work of the other put on his desk.

At the beginning of his *De proportionibus*, Bradwardine declares that his work is the first dealing with the problem of motion in a strictly mathematical way. He writes:

> Since each successive motion is proportionable to another with respect to speed, natural philosophy, which studies motion, ought not to ignore the proportion of motions and their speeds, and, because an understanding of this is both necessary and extremely difficult, nor has as yet been treated fully in any branch of philosophy, we have accordingly composed the following work on the subject.[43]

This paragraph can be read either as a declaration that Bradwardine considers himself to be the inventor of the new theory, or that he considers himself to be the first who organized the material on this subject in a coherent, systematic

41 See E.D. Sylla, 'The Origin and Fate of Thomas Bradwardine's *De proportionibus velocitatum in motibus* in Relation to the History of Mathematics,' in W.R. Laird and S. Roux (eds.), *Mechanics and Natural Philosophy before the Scientific Revolution*, Dordrecht 2008 (Boston Studies in the Philosophy of Science, 254), 67–119, at 95. See also E.D. Sylla, 'Guide to the Text,' in John Buridan, *Quaestiones super octo libros Physicorum Aristotelis (secundum ultimam lecturam)*, Libri III–IV, eds. M. Streijger and P.J.J.M. Bakker, Guide to the Text by E.D. Sylla, Leiden 2016 (Medieval and Early Modern Philosophy and Science, 27), XX–CCXVIII, at XXXIV, LXXXI, LXXXII, CXXIX, CLXXXI.

42 E.D. Sylla's comment to my paper.

43 Thomas Bradwardine, *De proportionibus*, 65^{1-5}: 'Omne motum successivum alteri in velocitate proportionari contingit; quapropter philosophia naturalis, quae de motu considerat, proportionem motuum et velocitatum in motibus ignorare non debet. Et quia cognitio illius est necessaria et multum difficilis, nec in aliqua parte philosophiae tradita est ad plenum, ideo de proportione velocitatum in motibus fecimus istud opus' (transl. Crosby, 65).

way with an introduction presenting the mathematical calculus of proportions, and the next three chapters using these calculi. If the first reading were correct, Bradwardine's claim would not be true, as will be shown below, since there were others who reinterpreted Aristotle's theory of motion by using the new calculus of proportions; following the second reading, Bradwardine should have mentioned the others who had already argued against the false theories of motion.

The information found in the anonymous treatise *De sex inconvenientibus*, at the beginning of the fourth question on local motion (*Utrum in motu locali sit certa servanda velocitas*) is also significant.[44] The anonymous author writes:

> Firstly I argue that this is not possible, since, as the first theory states, it would follow that such speed is determined by the excess of acting potencies over resistances, or, as the second theory states, by proportions of excesses of acting potencies to resistances … The first two theories are demonstratively disproved by many, and most precisely by two: magister Thomas Bradwardine in his treatise *De proportionibus* and by Adam of Pippewell, who argues subtly.[45]

The treatise was written by someone who was associated with the Oxford Calculators. Although the exact date of its composition is unknown, it can be narrowed down considerably to 1335–1339.[46] We only know that Adam of Pipewell, mentioned in the anonymous treatise, was a fellow of Balliol College in 1321, then a fellow at Merton College by 1325, and still present there in 1327.[47] We

44 Anonymous, *Tractatus de sex inconvenientibus*, ed. S. Rommevaux-Tani, *De sex inconvenientibus, un traité anonyme de philosophie naturelle au XIVe siècle*, édition, introduction et étude doctrinale, Paris 2022, 297–364. For a semi-critical edition of question IV see Anonymous, *Tractatus De sex incovenientibus*, q. IV, ed. J. Papiernik, in Jung and Podkoński, *Towards the Modern Theory of Motion*, 297–390.

45 The corresponding passage of the Latin text according to Rommevaux-Tani's edition runs as follows: 'Et arguo primo quod non, quia ex isto tunc sequitur quod talis velocitas attenderetur penes excessum potentiarum moventium ad potentias resistentes, sicut ponit una positio; aut penes proportionem excessuum potentiarum moventium ad potentias resistentes, sicut ponit secunda positio … Prime due positiones demonstrative a pluribus improbantur, precise a duobus famosis, a magistro Thoma de Bradwardyn in tractatu suo *De proportionibus* et a magistro Adam Pippewell, qui subtiliter hoc demonstravit' (ed. Rommevaux-Tani, 295). All English translations are mine unless noted otherwise.

46 For the problems of dating this text see Rommevaux-Tani's contribution in this book and her introduction in the edition previously mentioned. Additionally, see also Jung and Podkoński, *Towards the Modern Theory of Motion*, 22–29.

47 See G.C. Brodrick, *Memorials of Merton College with Biographical Notices of the Wardens*

can assume that Adam of Pipewell produced the work the anonymous author refers to, which was well known at Merton, before Bradwardine's treatise. However, we have no evidence (neither the work nor any mention in the other works, except the *De sex inconvenientibus*) confirming this hypothesis, but we have at our disposal Kilvington's commentary on the *Physics* (ca. 550 pages), composed before 1326.

Although, in my opinion, Richard Kilvington was the real inventor of the new theory of motion, it seems that he was unaware of the significance of his discovery. It was Thomas Bradwardine, the greatest beneficiary of Kilvington's work, who used his arguments with great skill in formulating the New Rule of Motion that continued to be supported by Aristotelians until the early sixteenth century.[48] In the following sections of this article, I will provide arguments to support my opinion.

3 Richard Kilvington and Thomas Bradwardine on Local Motion

Before discussing Kilvington's and Bradwardine's arguments for the new rule of motion, and their new interpretation of Aristotle's 'laws of dynamics', I would like to present the most important topics raised by Kilvington in his question *Utrum in omni motu potentia motoris excedit potentiam rei motae*,[49] composed no later than 1326, to show the parallels between this work and Bradwardine's *De proportionibus*, written in 1328.

Kilvington's question is divided into two main parts between which one finds a counterargument (*contra*) based on the authorities of Aristotle and Averroes, followed by a *determinatio quaestionis*. In Part I, to justify his point of view, Kilvington presents *pro* and *contra* arguments for the discussed opinions, while in Part II, he concentrates on some particularly relevant objections to the arguments previously presented. For that reason, the reader is forced to seek Kilvington's opinions in every single argument in both parts of the question.

The four articles contained in Part I concentrate on various problems concerning the conditions of motion, respectively: an excess of an acting potency over a passive one; the limits of the acting potency of the mover; the limits

---

and Fellows, Oxford 1885, 195; A.B. Emden, *A Biographical Register of the University of Oxford to A.D. 1500*, 3: P to Z, Oxford 1959, 1484.

48 See, e.g., Sylla, 'The Origin and Fate,' 67–95; S. Rommevaux, 'A Treatise on Proportion in the Tradition of Thomas Bradwardine: The *De proportionibus libri duo* (1528) of Jean Fernel,' *Historia Mathematica* 40 (2013), 164–182.

49 For Jung's critical edition of this question, see Jung and Podkoński, *Towards the Modern Theory of Motion*, 215–266.

of the passive potency of a body moved; and finally, when the conditions are established, the results of their interaction, i.e., the speed of motion. In the *determinatio quaestionis* Kilvington declares that, in accord with the definitions of terms presented in Part I, the answer to the question whether in every motion the acting potency (i.e., the power of a mover [F]) exceeds the passive potency (i.e., the resistance [R]) is affirmative.

Bradwardine's *De proportionibus* consists of four chapters. The first recapitulates the theories about proportionality found in Boethius' *Arithmetic* and Campanus of Novara's *Commentarium super quartum librum Elementorum Euclidis*. In the second chapter, Bradwardine criticizes four theories interpreting Aristotle's theory that speed is proportional to the acting and passive potencies involved. In the third chapter, Bradwardine introduces his own solution to the problem and 'he commences his exegesis by quoting Aristotle and Averroes in general support of his view, after which he launches directly into his twelve theorems concerning velocity.'[50] Finally, chapter four deals with circular motion.

Both Kilvington and Bradwardine declare that Aristotle and Averroes maintain that the potency of the mover exceeds the potency of the thing moved, and that motion must necessarily take place according to some proportion.[51] For both 'Calculators' Aristotle and Averroes are the most important authorities. They refer to Aristotle's *Physica*, *De caelo*, *De anima*, *Metaphysica*, *De generatione et corruptione* and *De motu animalium*, and to Averroes' commentaries on these works. But the Oxford way of teaching they adopted while studying at the university is also evident in their frequent and constructive application of the treatises *De ponderibus* by Euclid, Archimedes, and Jordanus de Nemore.[52]

50 Crosby (ed.), *Thomas of Bradwardine*, 38.

51 See Richard Kilvington, *Utrum in omni motu*, 251–252, § 91. See also Thomas Bradwardine *De proportionibus*, 108, 470–500. Quotations from Kilvington's text come from Jung's critical edition, mentioned above in n. 49. Quotations and English translations from Bradwardine's text come from Crosby's edition and translation mentioned above in n. 18.

52 For the Oxford *curriculum* see, e.g., Courtenay, *Schools and Scholars*, 219–250; W.J. Courtenay, 'The Role of English Thought in the Transformation of University Education in the Late Middle Ages', in J.M. Kittelson and P.J. Transue (eds.), *Rebirth, Reform and Resilience. Universities in Transition, 1300–1700*, Columbus, OH, 1984, 108–115; J.E. Murdoch, '*Mathesis in philosophiam scholasticam introducta*. The Rise and Developement of the Application of Mathematics in Fourteenth-Century Philosophy and Theology,' in *Arts libéraux et philosophie au Moyen Âge. Actes du quatrième Congrès international de philosophie médiévale, Montréal 1967*, Montreal 1969, 216–227; J.E. Murdoch, 'Mathematics and Sophisms in Late Medieval Natural Philosophy,' in *Les genres littéraires dans les sources théologiques et philosophiques médiévales: définition, critique et exploitation. Actes du colloque international de Louvain-la-Neuve, U.C.L., Institut d'études médiévales*, Louvain-la-

4 Condition *sine qua non* for Motion: An Excess of the Acting Potency of the Mover over the Passive Potency of the Thing Moved

At the beginning of the first article of his question, Kilvington quotes four different opinions: (1) as some say a minimum excess for motion may be assigned (*minimum quod sic*); (2) others claim that a maximum excess which is not sufficient for motion may be assigned (*maximum quod non*); (3) still others argue that there is an excess which is sufficient to continue motion but not to initiate it; (4) finally others maintain that the same excess which is sufficient for the continuation of motion is sufficient to initiate it.[53] While the first two opinions state that it is possible to assign an intrinsic or an extrinsic limit of the sequences of the force [F] to resistance [R] proportions, the last two views differ in deciding whether it is necessary to initiate a motion with a greater excess of an active power over resistance than the one that sustains the motion.

Kilvington refutes the first three opinions on the basis of imaginable, theoretical cases and by means of an observation of the facts.[54] He grounds his criticism on Averroes' statement according to which 'any element, in its natural motion caused by its essential form, reaches its natural place and at the same time its essential form.'[55] Kilvington first describes the following situation: 'Suppose that a fire initiates its acting on a piece of earth for which

---

Neuve 1982 (Textes, Études, Congrès, 5), 85–100; E.D. Sylla, *The Oxford Calculators and the Mathematics of Motion, 1320–1350. Physics and Measurement by Latitudes*, dissertation, Harvard University, 1970 (repr. with preface and errata, New York 1991), 182–304; M. Clagett, 'The Impact of Archimedes on Medieval Science,' *Isis* 50 (1959), 419–429, n. 162; M. Clagett and E.A. Moody, *The Medieval Science of Weights* (*Scientia de ponderibus*)*: Treatises Ascribed to Euclid, Archimedes, Thabat Ibn Qurra, Jordanus de Nemore and Blasius of Parma*, Madison, WI, 1952 (University of Wisconsin Publications in Medieval Science, 1), 3–20.

53 Richard Kilvington, *Utrum in omni motu*, 215, §1: 'Et probo quod non, quia tunc vel esset dare: (1) minimum excessum sufficientem ad motum, ut ponunt quidam, (2) vel maximum excessum non sufficientem ad motum, ut ponunt alii; (3) vel quicumque excessus sufficeret ad continuandum motum non tamen ad incohandum, ut ponunt tertii, (4) vel quicumque excessus sufficeret ad incohandum motum sicut ad continuandum, ut ponunt quarti.'

54 Richard Kilvington, *Utrum in omni motu*, 215–219, §§2–11.

55 Averroes, *Commentum magnum super libro De celo et mundo Aristotelis* IV, comm. 22, eds. F.J. Carmody and R. Arntzen, Leuven 2003 (2 vols.) (Recherches de Théologie et Philosophie médiévales. Bibliotheca, 4), 694^{98-101}: '… illud quod movetur ad suum locum non movetur nisi secundum quod aliquod movetur ad suam formam; quod enim movetur ad formam est illud quod est in potentia forma.'

the natural place is the centre of the Earth. Acting on this piece of earth, the fire adds levity to it and, as a consequence, causes its desire for a superior place.'[56]

One of the arguments based on this scenario is particularly interesting here. Kilvington maintains that in the above presented case a pure heavy element, such as a piece of earth, having been warmed by fire, when it will be equally heavy and light, would be fixed half way between the inner surface of the Moon and the centre of the Earth, i.e., much above the outer surface of the air and much above a large part of the fire. This Kilvington proves as follows. Take four elements as sign: earth (E) by 1, water (W) by 10, air (A) by 100, and fire (F) by 1000.[57] Thus, according to Aristotle, Kilvington argues, the sphere of fire does not exceed all the other elements 1000 times but more than a proportion of 9 to 1.[58] In fact, his proof is limited to the assumption and conclusion while the most interesting is the calculation which leads to the conclusion. Since all the four elements are continuously proportional ($W : E = A : W = F : A = 10 : 1$), the proportion between the last and the first ($F : E$) is equal to $(10 : 1)^3$, which is why Kilvington claims that the fire exceeds the earth 1000 times, i.e., in a proportion $(10 : 1)^3$. However, this is not a proper calculation, because it takes into

56 Richard Kilvington, *Utrum in omni motu*, 215, § 4^2–216, § 4^6: 'Et consequentiam probo, quia pono quod ignis incipiat agere in terram puram, cuius centrum est centrum mundi. Tunc quacumque levitate inducta in illa terra erit terra mixta, ergo appetit locum superiorem quam prius quando fuit pura, quia illud quod minus grave est, intensive appetit locum superiorem.' A similar reasoning is found in Kilvington's commentary on *De generatione et corruptione*, written before his *Physics* commentary, and in the questions on the *Physics*: *Utrum in omni generatione tria principia requirantur* and *Utrum natura sit principium motus et quietis*.

57 This assumption is based on the text of Aristotle's *De generatione et corruptione*, II, 333a21–27: 'If it is meant that they are comparable in their amount, all the "comparables" must possess an identical something whereby they are measured. If, e.g. one pint of Water yield ten of Air, both are measured by the same unit; and therefore both were from the first and identical something. On the other hand, suppose they are not "comparable in their amount" in the sense that so-much of the one yields so much of the other, but comparable in "power of action" (a pint of Water, e.g. having a power of cooling equal to that of ten pints of Air); even so, they are "comparable in their amount" though not qua "amount" but qua "so-much power".' See Aristotle, *De generatione et corruptione* (*On Generation and Corruption*), transl. H.H. Joachim, in *The Basic Works of Aristotle*, ed. R. McKeon, New York 2001, 518.

58 Richard Kilvington, *Utrum in omni motu*, 216–217, § 7^{3-8}: 'Signetur ergo terra per unum, aqua per 10, aer ⟨per⟩ 100, et ignis per 1000. Tunc sequitur quod ex dicto Philosophi, quod ignis in sua sphaera excedit alia omnia tria elementa residua in maiori proportione quam in millecupla qualis est proportio 9 ad 1; ergo punctus medius inter centrum et concavum orbis lunae est supra valde magnam partem ignis.'

account only two elements (the first and the last), while Kilvington talks about the distance of the sphere of the three other elements to the sphere of fire. The sum of the three other elements is 111 = 100(E) + 10(W) + 1(F) and it is equal to the sphere composed of these three elements. Now, if the fourth element (F) is divided by this sum, the quotient is 9+1/111. The fourth element, therefore, contains the other elements taken together 9 times and a fraction, and this is what should be demonstrated.

A similar calculation, but much more elaborated and much better proven, is found in chapter IV of Bradwardine's *De proportionibus*. Bradwardine begins his argumentation as follows:

> Since, on the basis of the foregoing, and by assuming only a few more propositions to be true, it is possible to find with ease the proportion of the elements to each other, and because the discovery of this is most appropriate to natural philosophy, and has, up to this day, remained unknown, therefore (although this is admittedly not too pertinent to the present undertaking) we will, nevertheless, uncover the secret of it.[59]

This artful declaration opens a detailed mathematical proof based on the works of Alpharganus, Ṯābit ibn Qurra, and Ptolemy. Bradwardine's calculus goes as follows:

> The proportion between the sphere of fire and that composed of the three remaining elements is greater than the proportion of 31 to 1; for each element contains the one immediately lesser more than 32 times. Proof: Corresponding, therefore, to the four elements, take these four terms, continuously proportional in the aforementioned proportion: 1, 32, 1024, 32768, the sum of whose first three terms is 1057 and represents the sphere composed of the lower elements. If the fourth term (which represents the sphere of fire) is divided by this sum, the quotient is 31+1/1057. The fourth term, therefore, contains the above-mentioned sum 31 times and a fraction, and this is what we were seeking for demonstrate.[60]

59 Thomas Bradwardine, *De proportionibus*, $132^{196-200}$: 'Quia per quaedam praedictorum, paucis aliis vero coassumptis, proportio elementorum adinvicem faciliter sciri potest, et eius scientia multum philosophiae congruit naturali, et hucusque latuit cooperta (licet praesenti negotio non multum pertineat) eius latentiam detegemus' (transl. Crosby, 133).

60 Thomas Bradwardine, *De proportionibus*, $136^{270-281}$: 'Proportio spherae ignis ad spheram compositam ex tribus elementis residuis est maior proportione 31 ad 1. Nam (per proximam) quodlibet elementum maius trigesies et bis continet proximam sibi minus. Igitur,

The comparison of Kilvington's and Bradwardine's way of reasoning allows one to formulate the following conclusions: (1) both authors are thoroughly familiar with the new calculus of ratios; (2) Kilvington relies on the authority of Aristotle and Averroes; Bradwardine uses the theories of Alpharganus, Ṯābit ibn Qurra, and Ptolemy; (3) Bradwardine guides his readers by showing each step of his proceeding and makes a broad use of his previously proved (in Part I) axioms and theorems; Kilvington, on the other hand, forces us to reconstruct his calculations. However, it is well known that Bradwardine wrote a treatise that was used as a textbook of mathematical physics. Kilvington, on the other hand, was the author of the question *Utrum in omni motu potentia motoris excedit potentiam rei motae* that was the result of his lecturing on the *Physics*, which allowed him to ignore certain levels of reasoning visible to his students and fellow medieval readers.

Article I of Kilvington's question ends with the acceptance of the fourth opinion above mentioned: 'the same excess which is sufficient for the continuation of motion is sufficient to initiate it' (*quicumque excessus sufficeret ad incohandum motum sicut ad continuandum*). Kilvington's arguments in favour of this opinion are based on the authorities of Aristotle, Averroes, Euclid, Archimedes, and Jordanus de Nemore. In justifying his position, Kilvington first gives an example that was later repeated by Bradwardine but used for a different purpose. Bradwardine uses this argument in section three of chapter II where he disproves 'the third erroneous theory, which claims that: (with the moving power remaining constant) the proportion of the speeds of motions varies in accordance with the proportion of resistances, and (with the resistance remaining constant) that it varies in accordance with the proportion of moving powers.'[61]

---

ad istas quattuor elementorum disponantur isti quattuor termini continue proportionales proportione praedicta (1, 32, 1024, 32768) quorum, si res primi congregantur in unum, 1057 constituent: quod spheram ex inferioribus elementis compositam representat. Per quod congregatum, si quartus terminus (qui ignem significat) dividatur, 31 exhibunt et remanet unitas dividenda. Quartus igitur terminus congregatus praedictum, trigesies et semel amplius continebit. Et hoc cuius demonstrationem quaesivimus' (transl. Crosby, 137). Bradwardine also operates with the series (1, 10, 100, 1000) of four terms but in order to prove that the distance of the outer surface of the air from the center of the Earth is more than 10 and less than 11 times the radius of the Earth. See Thomas Bradwardine, *De proportionibus*, $138^{298-306}$.

61 Thomas Bradwardine, *De proportionibus*, $94^{183-185}$: 'Sequitur de tertia opinio erronea, quae ponit proportionem velocitatum in motibus (manente eodem motore vel aequali) sequi proportionem passorum, et (manente eodem passo vel aequali) sequi proportionem motoris' (transl. Crosby, 95).

Richard Kilvington, *Utrum in omni motu*, 220, § 12$^{9-30}$	**Thomas Bradwardine, *De proportionibus*, 98$^{265-285}$**
… nec potest dici, quod in aliquo casu aliquis sit maximus excessus non sufficiens ad motum, quia sit excessus iste A, adhuc minor excessus A sufficit ad motum, ergo A excessus sufficit ad motum. Argumentum probo, quia in circumvolutione spherae vel rotae provenit motus alicuius partis propinquioris centro ex minori excessu quam sit A excessus, quia quanto partes spherae vel rotae centro fuerint propinquiores tanto tardius moventur, ut patet VI *Physicorum* commento 41. Capiatur ergo aliqua pars talis in rota, cuius motus provenit ex minore excessu quam sit A, ad quam figatur aliqua corda per clavum, ad cuius cordae extremitatem sit aliquis lapis colligatus. Tunc sic: volvatur rota ut prius et ita tarde movebitur corda fixa ad praedictam rotam, sicut et ista pars et ita tarde, ergo movebitur lapis colligatus ad extremum cordae sicut corda et pars rotae praedictae. Ergo si motus partis praedictae proveniat ex minori excessu quam sit A excessus, sequitur quod motus lapidis colligati, qui est motus per se et in actu, provenit ex minori excessu, qui sit, ergo A sufficit ad motum, quia quicumque excessus in motu rotae sufficit ad motum, quia aliqua pars rotae movetur in duplo tardius quam suprema superficies et aliqua pars in quadruplo tardius et sic in infinitum, ut apparet capitulo de vacuo commento 71; ergo quicumque excessus sufficit ad motum.	Nec potest dici quod tarditas motus non potest in infinitum duplari: quia si sic, sit A tarditas mobilis quae duplari non potest. Volvatur igitur sphera, seu corpus columnare, super axem quiescentem. Tunc in aliqua parte iuxta polum spherae, seu axem corporis columnaris, est tarditas dupla ad A ut est satis notum et facile demonstrare. Tunc cum ista parte colligetur corda fortis et longa, in cuius extremo alligetur aliquod ponderosum, quod sit B. Tunc tarditas motus B est dupla ad A tarditatem; et hoc est quod volumus demonstrare. Nec potest dicere cavillator quod motus B est motus per accidens, et in potentia tantum, ideo non facit ad propositum, quia iste motus habet motorem in actu et motum seu mobile in actu, terminus a quo et terminus ad quem in actu, tempus in actu, et spatium seu locum pertransitum in actu. Igitur est motor in actu.

According to Kilvington, the above presented argument proves that any excess is sufficient for motion on the *sine qua non* condition that $F : R > 1$; if $F = R$, the excess is zero and motion stops. Bradwardine uses this example to undercut Aristotle's theory:

> The theory is, on the one hand, to be refuted on the ground of falsity, for the reason that a given motive power can move a given mobile with a given degree of slowness and can also cause a motion of twice that slowness. According to this theory, therefore, it can move double the mobile. And, since it can move with four times the slowness, it can move four times the mobile, and so on *ad inifnitum*. Therefore, any motive power would be of infinite capacity.[62]

Kilvington puts forward the following three arguments in support of this opinion. First, it is impossible to observe the exact moment when motion begins and when it lasts continually, and hence, it is impossible to assign a minimum excess *quod sic* or a maximum excess *quod non*. Second, an observation of different weights placed on a balance with equal arms shows that even a minimally heavier body lifts the other one. Third, in accordance with Aristotle and Averroes, motion is initiated and sustained in every case when the active power is greater than the resistance.[63] As a consequence, motion occurs whenever the ratio of F to R is a 'ratio of *maioris inaequalitatis*', i.e., when F : R is greater than 1. Thus, Kilvington affirms that any proportion of force to resistance greater than 1 can produce motion.[64]

This theory has an unexpected consequence: the Earth can move rectilinearly. The motion of the Earth is caused by its inclination to bring the geometric centre of the universe into coincidence with its centre of gravity, and because the Earth as a whole is unequally heavy and dense, and geological changes perpetually alter it, its centre of gravity shifts. If the world were eternal, this slight rectilinear motion would be infinite.[65]

---

62 Thomas Bradwardine, *De proportionibus*, $98^{254-259}$: 'Est autem ista ex mendacio arguenda, quia aliqua potentia motiva localiter potest movere aliquod mobile aliqua tarditate, et potest movere dupla tarditate. Ergo (per istam positionem) potest movere duplum mobile. Et potest movere quadrupla tarditate: igitur quadruplum mobile, et sic in infinitum. Igitur quaelibet potentia motiva localiter esset infinita.' (transl. Crosby, 99).

63 Richard Kilvington, *Utrum in omni motu*, 252–258, §§ 93–105.

64 Richard Kilvington, *Utrum in omni motu*, 253, § 93^{12-13}: 'Pro primo articulo dico, quod quilibet excessus sufficit ad motum.'

65 Richard Kilvington, *Utrum in omni motu*, 253, § 95^{3-21}: 'Sed dico: omnibus aliis paribus concedi potest conclusio et iuxta sententiam Philosophi, quod terra est centrum in continuo motu. Et quando dicitur quod Philosophus vult in *De motu animalium* quod omne motum in suo motu necessario indiget fixo, ideo centrum indiget terra quiescente, dico quod intelligit quod terra quiescit a tali motu circulari quali movetur coelum. Non enim volo ponere quod terra moveatur motu circulari aeternaliter circa suum centrum, sicut posuit Plato ... Sed dico, quod ceteris ⟨paribus⟩ terra movetur aeternaliter dicendo alterius motibus secundum partem graviorem, verum tamen motu valde tardo et insensibili.

5 The Limits for Active and Passive Potencies

Having established a major condition for motion, Kilvington concentrates, in the next two articles, on setting boundaries to the range of causes of motion, i.e., active and passive potencies. An active potency is bounded by the patient, e.g., the weight that can be lifted or the distance that can be traversed. A passive potency is bounded by the agent by which it can be affected, as sight is bounded by the smallest object that can be seen. Commenting on Aristotle and Averroes, Kilvington raises some important queries, which open a new perspective on the solution of the problem. He is interested in answering the following questions: how is the power to be bounded if it is active or passive; if it is subject to weakening or not; if it is mutable or immutable? How to assign the boundaries of active potencies if a body moves in a medium that is not uniformly resistant? Most of the cases considered are posed *secundum imaginationem* which, however, does not make empirical verification irrelevant. Kilvington's mathematical interest is to be observed, at the outset, in his classification of all potencies as active or passive ones, even in cases in which it is hardly possible to define them as such in accordance with Aristotelian terms. In the Oxford Calculators' School this tradition was followed by William Heytesbury.[66]

While discussing the problem of the limits for active potency, Kilvington presents four opinions popular at that time (he uses the wording *dicitur a quibusdam*), in the case of passive potency he mentions five opinions. His view is that an active potency is determined by an upper limit, a *maximum quod sic*, and by a lower limit, a *minimum quod non*.[67] In order to determine Socrates'

Et quiescit terra a motu sensibili, non obstante quod insensibiliter moveatur ... Dico etiam ultra quod, si terra naturaliter debet esse sphaerica, est tam vallosa et in quibusdam partibus plana, propter convenientiorem habitudinem animalium, sicut aqua congregata in certis locis non obstante quodam terram undique circumdare, ita quod ista sunt quodammodo violenta et aliqualiter propter naturam, non tamen violenta propter peius sed propter melius, et talia violenta sunt sive possunt esse aeterna.' See also Jung-Palczewska, 'Works,' 216–217; E. Grant, 'Cosmology,' in D.C. Lindberg (ed.), *Science in the Middle Ages*, Chicago 1978 (Chicago History of Science and Medicine), 265–302 esp. 290–291. John Buridan also presents the same conclusions. See John Buridan, *Quaestiones super libris quattuor De caelo et mundo*, ed. E.A. Moody, Cambridge, MA, 1942, 231.

66 See, e.g., E. Jung, 'Richard Kilvington,' in E.N. Zalta (ed.), *The Stanford Encyclopedia of Philosophy* (Winter 2016 Edition), URL = https://plato.stanford.edu/archives/win2016/entries/kilvington/, and M. Hanke and E. Jung, 'William Heytesbury,' in E.N. Zalta (ed.), *The Stanford Encyclopedia of Philosophy* (Spring 2018 Edition), URL = https://plato.stanford.edu/archives/spr2018/entries/heytesbury/.

67 For a detailed discussion of the limit-decision problem see, e.g., C. Wilson, *William Heytes-*

power we can observe that he lifts 5 pounds (and not sensibly or notably more) or traverses 10 miles (and not notably more). The most proper way, however, to describe a capacity of an active potency is to determine a *minimum quod non*-limit, since every excess of an active potency over a passive one is sufficient for motion. With this conclusion Kilvington can be placed among all those masters, who, according to the anonymous author of the *Treatise on maxima et minima*,

> claiming that every excess does suffice to motion, must grant and would grant, that an active capacity is bounded by a minimum upon which it cannot act. And this is the resistance equal to the active capacity, because the active capacity cannot act upon that resistance since action does not occur through a proportion of equality.[68]

Kilvington then goes on to discuss the limits of passive potencies. According to the definition Heytesbury later adopted, 'A passive potency is one which, inasmuch as it is susceptible to less or can be affected by less, is susceptible to a greater, or can be affected by a greater, and not vice versa.'[69] Kilvington accepts the *minimum quod sic*-limit with respect to circumstances for a passive potency, since—as he puts it—it happens that we cannot see not only a small thing, but also a big one, for instance a cathedral, if we stay too close to it. When we ask, however, about Socrates' capacity of good vision, we point out the smallest thing he can scarcely see. And this is also Aristotle's point of view. What happens, however, is that a weak active power, such as a drop of water, overcomes the resistance of a rock. Also a small amount of fire can overcome the resistance of a large amount of water. This happens because every passive potency can be affected in part.[70]

To sum up, Kilvington indicated most of the issues concerning the problem of setting boundaries to potencies involved in an action-passion process.

---

bury: Medieval Logic and the Rise of Mathematical Physics, Madison, WI, 2nd ed., 1960 (University of Wisconsin Publications in Medieval Science, 3), and J. Longeway, *William Heytesbury: On Maxima and Minima. Chapter 5 of* Rules for Solving Sophismata *with an Anonymous Fourteenth-Century Discussion*, Dordrecht 1984 (Synthese Historical Library, 26).

68 See Longeway, *William Heytesbury*, 135.

69 Richard Kilvington, *Utrum in omni motu*, 259, §114$^{27-29}$: 'Et ideo verum est quod, sicut potentia passiva terminatur per minimum cum certis circumstantiis, ita potest terminari per maius cum circumstantiis proportionabilibus.'

70 Richard Kilvington, *Utrum in omni motu*, 259, §114.

Although he did not explicitly formulate general rules concerning different types of division, his debates reveal that he approved the following conditions for the existence of limits. First, there must be a range in which the potency can act or be acted upon, and another range in which it cannot act or be acted upon. Second, the potency should be capable of taking on a continuous range of values between zero potency and the value that serves as a boundary, and no other values.[71]

6 The Measurement of the Speed of Motion

The fourth and final article of Kilvington's question contains an extensive debate on how to correctly 'measure' the speed of motion (v). This time, Kilvington does not mention any other opinions of his contemporaries. He begins the debate with the following statement:

> If the question is true, the speed of motion varies either as the difference whereby the power of the mover exceeds the resistance offered by the thing moved or it varies with the proportion of the active power of the mover to the passive power of the thing moved.[72]

The two theories mentioned in this passage state respectively that the speed of motion is proportional to the arithmetical difference between an active potency and a passive one ($v \sim F - R$) or that it is proportional to the proportion of F to R ($v \sim F : R$).

Presented and reviewed are only these two theories. Bradwardine refutes four theories: the first one claims that the speed of motion is proportional to the arithmetical difference between an active potency and a passive one; the second one states that speed is proportional to $F - R/R$ (as it seems, this is originally Bradwardine's idea); the third theory states that the speed is proportional to the proportion of F to R; 'the fourth erroneous theory takes … a non-mathematical view of the relation of velocity to forces, claiming that velo-

71 For a detailed discussion, see E. Jung, 'Richard Kilvington on Local Motion,' in P.J.J.M. Bakker (ed.) with the collaboration of E. Faye and C. Grellard, *Chemins de la pensée médiévale. Études offertes à Zénon Kaluza*, Turnhout 2002 (Textes et études du Moyen Âge, 20), 116–128.

72 Richard Kilvington, *Utrum in omni motu*, 233, § 48[29–31]: 'Item, si quaestio esset vera, tunc velocitas motus vel sequeretur excessum potentiae moventis super potentiam rei motae vel proportionem potentiae moventis super potentiam rei motae.'

cities vary neither as a proportion of forces nor as an arithmetical excess of motive force over resistive force,' but vary, instead, by a 'natural relation of mover to moved.'[73] Since I am of the opinion that Kilvington composed his questions on the *Physics* in 1326, at the latest, and Bradwardine composed his treatise on proportions in 1328, I am presenting Kilvington's arguments first and then Bradwardine's ones.

In refuting the first opinion ($v \sim F - R$), both Kilvington and Bradwardine begin with quotes from Averroes' commentary on *Physics* IV, commentum 71 (*omnis motus est secundum excessum potentiae moventis super rem motam*)[74] and *Physics* VII, commentum 39 (*secundum excessum potentiae alterantis super potentiam alterati erit velocitas motus alterationis in quantitate temporis*).[75] Bradwardine adds a quote from Aristotle's *De caelo et mundo* I and from *Physics* IV, commentum 35.[76] Then they both make a similar statement. Kilvington: 'But this theory may be refuted in many ways.' Bradwardine: 'The present theory, may, however, be torn down in several ways.'[77] Kilvington presents the following arguments against this theory.

(1) Undoubtedly, this theory challenges Aristotle's and Averroes' rule of motion, which holds that two movers would move two *mobilia* taken together with the speed equal to the speed with which one of them would move one *mobile*. Suppose that $F_1 = F_2 = 2$ and $R_1 = R_2 = 1$, then $F_1 + F_2 = 4$, $R_1 + R_2 = 2$ and, in accordance with the rule: $v_2 \sim (F_1 + F_2) - (R_1 + R_2) = 4 - 2 = 2$ while in the case of the motion caused by a separate acting mover $v_1 \sim F_1 - R_1 = 2 - 1 = 1$, thus $v_1 \neq v_2$.

73 Crosby (ed.), *Thomas of Bradwardine*, 38. 37.

74 Richard Kilvington, *Utrum in omni motu*, 234, § 49; Thomas Bradwardine, *De proportionibus*, 86^{9-11}. Averroes, *In Physicam* IV, comm. 71, 161^{rb-va}: 'Proportio tarditatis ad tarditatem est sicut proportio impedimentis ad impediens ut dicit Avempace … Et si concesserimus quod proportio motuum, que fuerit in medio, adinvicem est sicut proportio impedimentis ad impedientem, quando idem motum movetur in vacuo, movetur motu indivisibili, et in instanti, sed sequitur necessario ut moveatur in tempore, cuius proportio ad tempus in quo movetur in medio est sicut proportio excessus potentie motoris super rem motam.'

75 Richard Kilvington, *Utrum in omni motu*, 234, § 49; Thomas Bradwardine, *De proportionibus*, 86^{14-15}. Averroes, *In Physicam* VII, comm. 39, 337va.

76 Thomas Bradwardine, *De proportionibus*, 86^{5-16}.

77 Richard Kilvington, *Utrum in omni motu*, 234, § 50^{8}: 'Sed contra istam opinionem potest sic argui multipliciter.' Thomas Bradwardine, *De proportionibus*, 86^{17}: 'Haec autem opinio destrui poterit multis modis.'

Richard Kilvington, *Utrum in omni motu*, 234, § 50⁹–235, § 50⁴	Thomas Bradwardine, *De proportionibus*, 86¹⁸–88⁴³
… si haec esset vera, tunc falsificaretur regula Aristotelis et Commentatoris VII *Physicorum* ⟨textu⟩ commenti 38, et isto commento 38, quod moventia aequalia aggregata aeque velociter praecise movent sicut unum movens praedictorum separatorum moveret unum mobilium separatorum. Nam ex ista opinione sequitur, quod duo aequalia coniuncta mobilia in duplo velocius ⟨moventur⟩ quam unum movens divisum moveret unum mobilium praedictorum per se. Quia per duplum excessum excedunt duo moventia aequalia duo mota aequalia quando unum movens de numero istorum mobilium excedit unum illorum mobilium, ⟨ut⟩ manifeste patet per exemplum in numero. Nam unus binarius excedit unam unitatem per excessum qui est unitas, et alius binarius separatus excedit aliam unitatem per aequalem excessum. Si congregantur illi duo binarii, excedunt duas unitates congregatas per duplum excessum qui per binarium est duplus ad unitatem, qui fuit primus excessus in binariis separatis.	Secundum istum modum sequitur quod, aliquo motore movente aliquod mobile per aliquod spatium in aliquo tempore, medietas motoris [movet] medietatem moti per medietatem excessus; sicut quaternarius excedit binarium per binarium, et medietas eius (scilicet binarius) excedit medietatem eius (scilicet unitatem) per unitatem tantum, que est medietas prioris excessus. Et falsitas consequentis patet per Aristotelem, septimo *Physicorum* … Consequentia patet, quia excessus istorum duorum motorum coniunctorum ad ista duo mobilia coniuncta est duplus ad excessum unius istorum motorum super suum mobile; sicut quilibet binarius excedit unitatem per unitatem, duo autem binarii (qui quaternarium faciunt) excedunt duas unitates (qua dualitatem constituunt) per dualitatem, quae est dupla ad excessum binarii super unitatem … Falsitas consequentis patet per Aristotelem septimo *Physicorum*, ut prius …

(2) This theory is also contrary to the following rule of Aristotle and Averroes: 'half of a given mobile is moved by a half of a given mover in the same time and on the same distance as a whole mobile moved by a whole mover.'[78] The given

78 Cf. Aristotle, *Physica* VII, 250a25–28; Averroes, *In Physicam* VII, comm. 36, 335va. Averroes writes: 'Ut declaverunt Geometrae.' Kilvington writes: 'Sic declarant geometrae et dicit Commentator VII *Physicorum* commento 36.' Bradwardine states: 'Sicut universaliter demonstrant geometri,' and supplements this statement with a mathematical argument referring to the introductory chapter of his treatise.

numerical example is the same: suppose that $F = 4$, $R = 2$, then $F - R = 2$ and the whole mover exceeds the whole mobile by 2, while a half of $F = 2$ exceeds half of a mobile $R = 1$ by 1 ($\frac{1}{2}F - \frac{1}{2}R = 2 - 1 = 1$).

Richard Kilvington, *Utrum in omni motu*, 235, § 51[5–18]	Thomas Bradwardine, *De proportionibus*, 88[60–77]
Item, ex ista opinione sequitur quod alia regula Aristotelis sit falsa, quae ponitur in textu commenti 36 in VII *Physicorum*, quae est: Si aliquis motor moveat aliquod mobile per aliquod spatium in aliquo tempore, subduplus motor movebit medietatem mobilis tanti per tantum spatium in aequali tempore et aeque velociter, quia unitas est eadem proportio medietatis ad medietatem sicut totius ad totum. **Sic declarant geometrae et dicit Commentator VII *Physicorum* commento 36.** Sed illud foret falsum, si velocitas motus sequitur excessum, nam totus motor excedit totum mobile per certum excessum, et medietas motoris excedit medietatem moti per medietatem illius excessus tantum, ut patet in exemplo priori, sicut quaternarius excedit binarium ⟨per binarium⟩, ita medietas eius, scilicet binarius, excedit medietatem binarii per unitatem, quae est medietas prioris excessus.	Nec potest dici quod Aristoteles et Averroes intelligent, in locis praedictis, per proportionem seu analogiam, proportionalitatem arithmeticam seu aequalitatem excessuum, ut dicunt quidam, quia septimo *Physicorum* probat Aristoteles istam conclusionem: Si aliqua potentia moveat aliquod mobile per aliquod spatium in aliquo tempore, medietas motoris movebit medietatem moti per aequale spatium in aequali tempore, quoniam similiter secundum analogiam sicut se habet medietas motoris ad medietatem moti, et totus motor ad totum motum. Quod tamen de proportionalitate arithmetica, quae significat aequalitatem excessuum, dignoscitur esse falsum (ut primo argumento contra hanc opinionem sufficienter est ostensum). Et Averroes ibidem dicit quod sic erit eadem proportio, **sicut universaliter demonstrant geometri.**

(3) Now to demonstrate the flaw in Averroes' theory, Kilvington gives a series of examples from sense experience:

(a) 'If to a single man who is moving some weight which he can scarcely manage with a very slow motion, a second man joins himself, the two together can move it much more than twice as fast.'[79]

79 Transl. Crosby, 99.

Richard Kilvington, *Utrum in omni motu*, 235, § 52^{19-26}	Thomas Bradwardine, *De proportionibus*, $98^{287-291}$
Item, probo quod velocitas motus non semper sequitur excessum, quia videmus quod, si aliquis unus homo moveat aliquod ponderosum motu valde tardo, quod ponderosum potest vix ille homo movere, si alius homo sibi coniungitur, illi duo homines movebunt illud ponderosum multo plus quam in duplo velocius; et tamen excessus non duplicatur nisi ⟨duplo⟩ praecise; ergo velocitas non sequitur excessum.	Tertio ex ista positio super mendacio arguenda, quoniam experimentum sensibile docet huius positionis contrarium. Videmus enim quod, uno homine movente aliquod ponderosum (quod potest vix solus movere motu valde tardo), si aliquid sibi adiungatur, illi duo movent illud multo plus quam in duplo velocius.

(b) The same principle is quite manifest in the case of weight suspended from a revolving axle, which it moves insensibly during the course of its own downward motion (as is the case with clocks): 'If an equal clock weight is added to the first, the whole descends and the axle, or wheel, turns much more than twice as rapidlym (as is sufficiently evident to sight).'[80]

Richard Kilvington, *Utrum in omni motu*, 235, § 53^{27}–236, § 53^{3}	Thomas Bradwardine, *De proportionibus*, $98^{292-297}$
Istud confirmo: si in horilogio per certum tempus moveretur rota certa velocitate, movente addito, licet non dupletur pondus, dupla velocitate movebitur, ut experimento notum est. Illud apparet manifeste de pondere suspenso ad aliquod circumvolubile, quod per suum descensum movet insensibiliter et movet⟨ur⟩ rota quodam motu insensibili. Sic accidit in horilogio: ad pondus si suspendatur iterum tantum pondus, totum descendet cum impetu et circumvolvet rotam multo plus quam in duplo velocius, ut patet manifeste sensationi.	Et patet manifeste de pondere suspenso ad axem circumvolubilem, quod per suum descensum movet insensibiliter et volvit axem seu rotam motu insensibili (sicut accidit in horologio) ad quod si suspendatur tantum pondus, totum descendet et circumvolvet axem seu rotam plus quam in duplo velocius (ut sensui sufficienter constat).

80 Transl. Crosby, 99.

Bradwardine uses these two examples to disprove the second false theory: 'with the moving power remaining constant, the proportion of the speeds of motions varies in accordance with the proportion of resistances. And, with the resistance remaining constant, that it varies in accordance with the proportion of moving powers'.[81]

(c) The next example Kilvington gives is as follows: 'suppose that one man carries a weight on some distance, running as fast as he can. Then if another man, who runs as fast as the first one, joins him, they both would carry the weight and move with the same speed as before.'[82] This means that, despite the active power being doubled, the speed of motion would remain the same. As a consequence, speed does not vary in accordance with the arithmetical proportion between an active power and a mobile.

(d) Sense experience shows that 'if three men haul a ship with ineffective power, they can only turn it around and not move it forward; however, if a fourth joins them, they can tow the ship in a straight line over a great distance.'[83]

(e) 'A man, carrying a load with which he can move very slowly, being burdened with an additional, much smaller load, he will move twice as slowly and with great difficulty.'[84] Bradwardine gives a similar example of a fly, a boy, and a strong man who, when burdened with an extra load, do not move much slower than without that additional load.[85]

---

81 Thomas Bradwardine, *De proportionibus*, $94^{183–186}$. For Latin quote see above, n. 62 (transl. Crosby, 95).

82 Richard Kilvington, *Utrum in omni motu*, 236, § $54^{4–9}$: 'Item, posito quod aliquis homo trahat unam fabam per unam cordam currendo ita velociter sicut potest, tunc si alius homo tantae potentiae ad currendum sibi iniungatur ad trahendum illam fabam praedictam, illi duo homines non trahent velociter quam unus illorum per se.'

83 Richard Kilvington, *Utrum in omni motu*, 236, § $55^{9–18}$: 'Posito quod tres homines forte nitantur trahere navem quam non possunt movere nisi girando huc illuc, tunc isti tres homines unam actionem in navem faciunt. Si addatur quartus homo aequalis potentiae alicuius praedictorum, illi quattuor homines movebunt navem directe per magnum spatium, ut experimento notum est. Ex quo sequitur, quod minus quam agens duplum facit plus quam duplam actionem, quia si primi tres homines agant in navem prius directe per medietatem tanti spatii, per quantum movent isti quattuor homines, fecissent medietatem actionis praecise quam faciant isti quattuor homines.' See also Averroes, *In Physicam* VIII, comm. 23, 359ra; Aristotle, *Physica* VIII, 253b.

84 Richard Kilvington, *Utrum in omni motu*, 236, § $56^{21–24}$: 'Item, exemplum potest adduci de homine portante aliquod pondus cum quo vix potest ire lento passu. Tunc minori addito quam aequali pondere movebitur in duplo tardius et vix aliquo modo vadit.'

85 Thomas Bradwardine, *De proportionibus*, $92^{145–150}$: 'Videmus enim quod musca portando aliquod modicum velociter multum volat, et puer aliquod modicum velociter satis movet, et homo fortis unum magnum mobile (quod vix potest movere) movet valde tarde. Et

(4) Finally, the presented opinion has to be rejected because a geometric proportion (i.e., a similarity of proportions) of movers to the *mobilia* would not produce equal speeds, since it does not represent an equality of excesses. For, although the proportion of 4 to 2 and 2 to 1 are the same, the excess of one term over the other is 2 in the first case, and 1 in the second. The conclusion is contrary to Aristotle's and Averroes' statement that equal proportions between movers and mobiles result in equal speed.

Richard Kilvington, *Utrum in omni motu*, 236, § 57^{25}–257, § 57^{4}	**Thomas Bradwardine, *De proportionibus*, $88^{50–55}$**
Item, si velocitas motus sequeretur excessum, tunc ex similitudine motus ad mota et ex proportione istorum geometrica non sequitur velocitas aequalis, quia non sequitur ex aequalitate proportionum aequalitas excessuum, ut manifeste patet. Nam aequalis est proportio quattuor ad duo et duorum ad unum, et tamen excessus sunt inaequales sicut binarius et unitas. Et per consequens ad quod deducitur est contra Aristotelem et Commentatorem VII *Physicorum* commento 31, 36 et illis commentis capitulo de vacuo commento 71 et multis aliis locis, ubi semper ex aequalitate proportionum moventium ad mota arguunt aequalem velocitatem motuum illorum mobilium.	Tertio sic: Tunc ex proportione geometrica, scilicet similitudine proportionum motorum ad sua mota, non sequitur aequalis velocitas motuum, quia nec excessuum; quoniam eadem est proportio duorum ad unum et sex ad tres; excessus tamen unius est unitas, alius autem ternarius. Consequens autem ad quod deducitur est falsum et contra Aristotelem septimo *Physicorum* in fine et multis locis, ubi semper ex aequalitate proportionum motorum ad sua mota arguit aequalitatem velocitatum in motibus. Idem vult Averroes super loca praedicta, et similiter quarto *Physicorum* commento 71.

Some of these arguments, as it was shown above, are to be found in the first part of Bradwardine's chapter II, where he criticizes the first false theory: 'the proportion between the speeds with which motions take place varies as the difference whereby the power of the mover exceeds the resistance offered by the thing moved.'[86]

licet illi mobile apponatur maius quam Musca vel puer posset movere, movet totum non multum tardius tunc quam prius.'

86 Thomas Bradwardine, *De proportionibus*, $86^{5–6}$: 'Prima ponit proportionem velocitatum in motibus sequi excessum potentiae motoris ad potentiam rei motae' (transl. Crosby, 87).

According to Kilvington, the above mentioned arguments *contra* justify the need to adopt the second theory, that speed of motion is proportional to the proportion of F : R. In support of his opinion, he gives the last argument quoted below, which Bradwardine also uses to close the discussion at the end of the first part of Chapter II. To reinforce his view, Kilvington refers to authorities. Bradwardine uses the expression *ex prioribus*, i.e., the arguments taken from chapter I. So again he leaves the impression that he is the first one who criticizes this false theory.

Richard Kilvington, *Utrum in omni motu*, 237, § 58[5]–238, § 58[6]	**Thomas Bradwardine, *De proportionibus*, 92[126–132]**
Pro istis et consimilibus dicitur secunda opinio quod velocitas motuum sequitur proportionem moventis ad motum et non excessum, quia si sequeretur excessum, tunc sequeretur quod, si aliquod mobile simplex quod in dupla proportione excedit, non posset moveri in aliquo medio subtiliori mundi (*sic*) quam in duplo velocius, quia nullum medium potest excedere per plus quam per duplum excessum. Tunc per subtiliationem medii in infinitum non potest motus istius mobilis velocitari in infinitum; **quod est contra Aristotelem et Commentatorem capitulo de vacuo, commento 71.**	Tunc sequeretur quod, si terra pura movetur in aliquo medio quod in dupla proportione excederet vel maiori, non potest moveri in duplo velocius in aliquo medio alio. Non enim posset excedere aliquod medium per duplum excessum, quoniam tunc totum esset excessus. Et tunc, manente eodem motore, non in infinitum per subtiliationem medii posset velocitas motus generari; **quod ex prioribus** constat esse falsum.

Chapter III, in which Bradwardine presents his own theory, begins with the following grandiose 'opening':

> Now that these fogs of ignorance, these winds of demonstration, have been put to flight, it remains for the light of knowledge and of truth to shine forth. For true knowledge proposes a fifth theory which states that the proportion of the speeds of motions varies in accordance with the proportion of the power of the mover to the power of the thing moved.[87]

87 Thomas Bradwardine, *De proportionibus*, 110[1–5]: 'His igitur ignorantiae nebulis demonstrationum flatibus effungatis, superest ut lumen scientiae resplendeat veritatis. Scientia autem veritatis ponit quintam opinionem, dicentem quod proportio velocitatum in motibus sequitur proportionem potentiae motoris ad potentiam rei motae' (transl. Crosby, 111).

Kilvington begins article iv of his question, where he presents the new calculus of proportions and the new interpretation of Aristotle's rules of motion, saying: 'from what has been said the second theory follows,'[88] i.e., speed is proportional to the proportion of F to R (v ~ F : R). To support their views, both Kilvington and Bradwardine quote the same paragraphs from Averroes.[89]

Richard Kilvington, ***Utrum in omni motu,*** **237, § 58^{16}–238, § 58^{6}**	**Thomas Bradwardine,** ***De proportionibus,*** **110$^{6-26}$**
… dicit Commentator sic uniformiter: 'cum ergo sic fuerint duo motores et duo mota, et proportio unius motoris ad suum motum fuerat sicut proportio reliqui motoris ad suum reliquum motum, tunc illi motus erunt aeque veloces, et cum diversabitur proportio diversabitur et motus.' Et in fine in eodem dicit sic: 'diversitas motuum in velocitate et tarditate est secundum proportionem hanc, quae est inter duas potentias, motivam scilicet et resistivam.' Et ii *De coelo* commento 36 dicit sic Commentator: 'velocitas motus et tarditas non fiunt nisi secundum proportionem potentiae motoris ad potentiam rei motae; quanto fuerit potentia motoris ad motum maior, tanto motus erit velocior, et quanto minor, tanto motus erit tardior.' Et vii *Physicorum* commento 35 ex duplicatione proportionis motoris ad motum arguit Commentator duplam velocitatem motus et dicit sic: 'Cum diviserimus motum, contingit necessario quod proportio illius motoris ad medietatem moti sit dupla ad proportionem quae praefuit inter motorem et totum motum, et ideo movebit idem motor medietatem moti in duplo velocior quam totum motum istum vel mobile.'	Et hoc est quod vult Averroes, super quarto *Physicorum*, commento 71 sic dicens 'Universaliter manifestum est quod causa diversitatis et aequalitatis motuum est aequalitas et diversitas proportionis motoris ad rem motam. Cum igitur fuerint duo motores et duo mota, et proportio alterius motoris ad alterum motum fuerit sicut proportio reliqui motoris ad reliquum motum, tunc duo motus erunt aequales in velocitate; et cum diversatur proportio, diversabitur motus secundum istam proportionem.' Et infra in eodem dicit: 'diversitas motuum in velocitate et tarditate est secundum proportionem hanc quae est inter duas potentias, scilicet motivas et resistivas.' Et septimo *Physicorum*, commento 35 ex duplicatione proportionis potentiae motoris ad motum arguit duplicationem velocitatis in motu, sic dicens: 'Cum diviserimus motum, contingit necessario ut proportio potentiae motoris ad motum sit dupla istius proportionis, et sic velocitas dupla ad istam velocitatem.'

88 Richard Kilvington, *Utrum in omni motu*, 237, § 59^{7}: 'Ex quibus concluditur haec opinio secunda.'

In the first part of his questions, Kilvington formulates seven theorems (*conclusiones*)—as Crosby translates this term—against this theory,[90] and in part II he addresses these arguments, some of which he totally accepts, others he modifies.[91] Some of these arguments are also found in chapters II and III of Bradwardine's treatise. Since Kilvington's discussion spreads out over 16 pages, I will present only the main theorems.

Theorem I: 'A body twice as heavy as another body would not move in the same medium with a double speed.'[92] The following two numerical examples confirm this claim:

(1) 'Suppose that a body moves downward with its natural motion and that the resistance of a medium is exceeded by the moving potency (gravity) like 3 to 1; then if we double the gravity of the body (i.e., if we double its weight), the potency would exceed the resistance as 6 to 1, though it does not mean that the speed of motion would be doubled.'[93] In accordance with Euclid's definition of a double proportion, a double proportion is a proportion multiplied by itself. Thus, a 'correct' double proportion of 3 to 1 is 9 to 1, and not a proportion of 6 to 1, since $9 : 1 = (9 : 3)(3 : 1) = (3 : 1)(3 : 1)$.[94] The proportion of

89 Cf. Averroes, *In Physicam*, IV, comm. 71, 160vb; VII, comm. 35, 335ra; *In De coelo* II, comm. 36.

90 Richard Kilvington, *Utrum in omni motu*, 238–251, §§ 60–89.

91 Richard Kilvington, *Utrum in omni motu*, 262–266, §§ 123–135.

92 Richard Kilvington, *Utrum in omni motu*, 238, § 61^{9}: 'Duplum grave praecise non movebitur praecise in duplo velocius in eodem medio quam grave subduplum.'

93 Richard Kilvington, *Utrum in omni motu*, 238, § 61$^{11-14}$: 'Posito quod grave simplex excedat resistentiam medii sicut tria excedunt unum, tunc duplata sua gravitate se habebit ad eandem resistentiam sicut sex ad unum, quae proportio non est dupla ad primam proportionem.'

94 On the medieval theory of proportion, see Sylla, 'The Origin and Fate.' As J.E. Murdoch and E.D. Sylla, 'The Science of Motion,' in D.C. Lindberg (ed.), *Science in the Middle Ages*, Chicago 1978 (Chicago History of Science and Medicine), 206–264, at 225–226, explain: 'The standard medieval mathematics of proportions in general terms tells us that the only way to increase or decrease proportions is to do so by (in *our* terms) multiplying or dividing proportion by proportions. For example, to increase the proportion 3 to 1 to its double would be to multiply it by itself (yielding 9 to 1 as its double); to decrease the proportion 16 to 1 to its half would be to divide it by proportion 4 to 1 (yielding 4 to 1 as its half) or to increase the proportion 3 to 2 to the proportion of 2 to 1 would be to multiply it by the proportion 4 to 3 (yielding 12 to 6, which equals 2 to 1). One should note a difference between the first two of these examples and the third. In the first two, we have a multiplication or a division which amounts, respectively, to taking powers or roots. Thus, for the first example we have (in modern notation): $9/3=3/1$, and thus multiplying $3/1$ by itself gives $9/3 \cdot 3/1=9/1$. For the second example, we have: $16/4=4/1$ and thus dividing $16/1$ by $4/1$ gives $16/1=16/4 \cdot 4/1$. However, in the third example, we do not have equal proportions, and, therefore, we do not have a problem of powers or roots.

6 to 1 is less than 9 to 1, and if we take for granted the theorem of Averroes and Aristotle that speed is proportional to the proportion of F to R, it turns out that the speed would be less than twice the previous one; so a body twice as heavy in the same medium would not move twice as fast: $v_2 < 2v_1$ and not $v_2 = 2v_1$.[95]

Bradwardine's reasoning is the same. However, he does not provide any numerical example, but instead he formulates the following general conclusion (his Theorem IV): 'If the proportion of the power of the mover to that of its mobile is greater than two to one, when the motive power is doubled the motion will never attain twice the speed.'[96]

(2) We can also prove, Kilvington argues, that in the same medium, the speed of a body twice as heavy might be greater than twice the previous one. 'Suppose that an active power exceeds a resistance in a proportion of 6 to 4; hence, in accordance with Averroes' theorem, a doubling of the active power gives

---

But the multiplication of 3/2 by the *unequal* proportion 4/3 is still what is involved in the "increase" of 3/2. Thus: 3/2·4/3=12/6.' For more information about the standard medieval calculus of proportions, see also J.A. Weisheipl, 'Ockham and some Mertonians,' *Mediaeval Studies* 30 (1968), 163–213, repr. in J.I. Catto and T.A.R. Evans (eds.), *The History of the University of Oxford, 2: Late Medieval Oxford*, Oxford 1992, 607–658; M. Clagett, *The Science of Mechanics in the Middle Ages*, Madison, WI, 1959 and 1961 (University of Wisconsin Publications in Medieval Science, 4), 440–444, 465–503; A.G. Molland, 'The Geometrical Background to the "Merton School",' *British Journal for the History of Science* 4 (1968–1969), 108–125; J.E. Murdoch, 'The Medieval Language of Proportions: Elements of the Interaction with Greek Foundations and the Development of New Mathematical Techniques,' in A.C. Crombie, *Scientific Change. Historical Studies in the Intellectual, Social and Technical Conditions for Scientific Discovery and Technical Invention, from Antiquity to the Present. Symposium on the History of Science, University of Oxford 9–15 July, 1961*, London 1963, 237–271.

95 Richard Kilvington, *Utrum in omni motu*, 238, § 61[14]–239, § 61[4]: 'Nam proportio nonupla est praecise dupla ad proportionem triplam, ut potest demonstrari per definitionem proportionis duplicatae positam v Euclidis, nam sicut unum ad tria, ita tria ad novem. Ergo proportio novem ad unum est praecise dupla ad proportionem ⟨tres⟩ ad unum, per definitionem proportionis duplicatae quae est quod, si fuerint tria continue proportionabilia proportione inaequalitatis, tunc proportio tertii ad primum est proportio secundi ad primum duplicata. Ex quo sequitur, quod proportio nonupla est praecise dupla ad proportionem triplam. Et per consequens proportio sextupla est minor quam dupla ad triplam. Et sequitur quod, si grave excedit resistentiam medii in quo movetur in tripla proportione praecise, quod duplum grave non movebitur in illo medio in duplo velocius, immo minus quam in duplo velocius, cum velocitas motus sequatur proportionem, ut ponit haec opinio.'

96 Thomas Bradwardine, *De proportionibus*, 112[64–66]: 'Si potentiae moventis ad potentiam sui moti sit maior quam dupla proportio, potentia motiva geminate motus eiusdem duplam velocitatem nequaquam attinget' (transl. Crosby, 113).

the proportion of 12 to 4, which is greater than the proportion of 6 to 4.'[97] The correct calculus, however, gives a proportion of 9 to 4, since: $(9:4) = (9:6)(6:4) = (3:2)(3:2)$ which is a proportion double of 6 to 4, while 12 to 4 is not.[98] Since the proportion of 12 to 4 is greater than 9 to 4, the speed of motion would be greater than the double of the previous one, so that a body twice as heavy in the same medium would not move twice as fast: $v_2 > 2v_1$ and not $v_2 = 2v_1$.

97 Richard Kilvington, *Utrum in omni motu*, 239, § 62[5–11]: 'Posito quod aliquod grave excedat resistentiam medii in quo movetur sicut sex excedunt quattuor. Tunc duplata sua gravitate se habebit ad suam resistentiam eandem sicut duodecim ad quattuor, quae est maior quam dupla proportio ad proportionem sex ad quattuor. Nam sicut sex ad quattuor, ita novem ad sex, quia utraque proportio est sexquialtera. Ergo proportio novem ad quattuor est dupla ad proportionem sex ad quattuor, quoniam sexquialtera proportio duplicata est dupla ad unam sexquialteram.'

98 In the fourteenth century, various formulations of this definition were known. See Euclid (Campanus), *Elementa*, ed. H.L.L. Busard, *Campanus of Novara and Euclid's Elements*, Wiesbaden 2005 (Boethius. Texte und Abhandlungen zur Geschichte der Mathematik und der Naturwissenschaft, 51/1), book v, def. x, 168: 'Cum fuerint tres quantitates continue proportionales, dicetur proportio prime ad tertiam proportio prime ad secundam duplicata'; Euclid, *Elementa*, eds. H.L.L. Busard, *The Mediaeval Latin Translation of Euclid's Elements Made Directly from the Greek*, Wiesbaden 1987 (Boethius. Texte und Abhandlungen zur Geschichte der Mathematik und der Naturwissenschaft, 15), book v, def IX, 109: 'Quando vero tres quantitates proportionales fuerint, prima ad tertiam duplicem proportionem habere dicitur quam ad secundam'; Euclid (Adelard of Bath), *Elementa*, ed. H.L.L. Busard, *The First Latin Translation of Euclid's* Elements *Commonly Ascribed to Adelard of Bath*, Toronto, ON, 1983 (Studies and Texts, 64), 146: 'Cum fuerint tres quantitates proportionales, erit proportio prime ad tertiam sicut proportio prime ad secundam repetita'; Euclid (Robert of Chester), *Elementa*, eds. H.L.L. Busard and M. Folkerts, *Robert of Chester's (?) Redaction of Euclid's* Elements, *the so-called Adelard II Version*, Basel 1992 (Science Networks-Historical Studies, 9), 1: 161: 'Cum fuerint tres quantitates proportionales, dicetur proportio prime ad terciam proportio prime ad secundam duplicata'; Euclid (Johannes de Tinemue), *Elementa*, eds. H.L.L. Busard, *Johannes de Tinemue's redaction of Euclid's* Elements *the so-called Adelard III version*, Stuttgart 2001 (Boethius. Texte und Abhandlungen zur Geschichte der Mathematik und der Naturwissenschaft, 45), 129: 'Cum fuerint tres quantitates proportionales, erit proportio prime ad tertiam, prime ad secundam geminata. Verbi gratia: Proportio 9 ad 4 est quasi proportio 9 ad 6 dupla quia constat quasi totaliter ex ea que est inter 9 ad 6 et 6 ad 4 cuius equaliter sit, utrobique sit sexquialtera.' It seems that Kilvington follows Johannes de Tinemue while Bradwardine follows Axiom 7 from Book v of Euclid's *Elements*: 'Si fuerit proportio maioris inequalitatis primi ad secundum ut secundi ad tertium, erit proportio primi ad tertium dupla ad proportionem primi ad secundum et secundi ad tertium' (Thomas Bradwardine, *De proportionibus*, 78[297–302]).

Richard Kilvington, *Utrum in omni motu*, 239, § 61$^{16–22}$	Thomas Bradwardine, *De proportionibus*, 78$^{297–302}$
Ut potest demonstrari per definitionem proportionis duplicate positam v Euclidis … si fuerint tria continue proportionabilia proportione inequalitatis, tunc proportio tertii ad primum est proportio secundi ad primum duplicata.	Si fuerit proportio maioris inequalitatis primi ad secundum ut secundi ad tertium, erit proportio primi ad tertium dupla ad proportionem primi ad secundum et secundi ad tertium.

However, it can be noticed, at first glance, that such calculation of proportions forces changes in the interpretation of Aristotle and Averroes, Archimedes, Euclid, and Jordanus de Nemore, who state that speed is proportional to the proportion of the moving power to the resistance of the body moved.[99] While noticing the contradiction, Kilvington confirms that the value of speed depends on the proportion of active power to resistance, and only in a case when F : R is 2 : 1 would the doubling of F result in double the speed. Bradwardine says the same in his Theorem III.

Richard Kilvington, *Utrum in omni motu*, 362, § 124$^{13–16}$	Thomas Bradwardine, *De proportionibus*, 112$^{60–63}$
Et numquam movebitur duplum grave praecise in duplo velocius in eodem medio nisi quando grave subduplum excedat resistentiam medii in proportione dupla.	Si potentiae moventis ad potentiam sui moti sit dupla proportio, eadem potentia movebit medietatem eiusdem moti velocitate praecise duplata.

Theorem II: A mobile, e.g., a piece of earth, which moves in water will not move twice as fast in air twice as rare. 'This conclusion follows from what has been said above,' Kilvington writes, since if the proportion of the heaviness of a simple body in downward motion (F) to the resistance of a medium (R) is

99 Richard Kilvington, *Utrum in omni motu*, 240, § 62. Euclid, *Liber de ponderoso et levi et comparatione corporum ad invicem*, ed. M. Clagett, in M. Clagett and E.A. Moody, *The Medieval Science of Weights (Scientia de ponderibus): Treatises Ascribed to Euclid, Archimedes, Thabat Ibn Qurra, Jordanus de Nemore and Blasius of Parma*, Madison, WI, 1952 (University of

greater than 2 : 1, then in a medium doubly rare, e.g. in air, the body would move slower than twice. If $F : R_w > 2 : 1$, and $R_a = 2R_w$, then $F : R_a < 2v$.[100] The same point is made by Bradwardine in his Theorem IV quoted above.[101]

Richard Kilvington, *Utrum in omni motu*, 241, § 63$^{12-19}$	**Thomas Bradwardine, *De proportionibus*, Theorem IV, 112$^{64-66}$**
Et si illud grave simplex excedat resistentiam aquae in proportione maiori quam in dupla, tunc excedet resistentiam duplae subtilitatis in proportione maiori quam dupla ad proportionem primam, sicut demonstrari potest ut prius per definitionem proportionis duplicatae. Ergo tunc movebitur tale grave in aere duplae subtilitatis praecise plus quam in duplo velocius quam movebatur in aqua.	Si potentiae moventis ad potentiam sui moti sit maior quam dupla proportio, potentia motiva geminata motus eiusdem duplam velocitatem nequaquam attinget.

Wisconsin Publications in Medieval Science, 1), 23–32, at 28; Jordanus de Nemore, *Liber de ponderibus*, ed. E.A. Moody, in Clagett and Moody, *The Medieval Science of Weights*, P.01, 154: 'Inter quelibet gravia est virtutis et ponderis eodem ordine sumpta proportio'; R1.01, 174: 'Inter quelibet gravia est virtutis et ponderis eodem ordine sumpta proportio'; Jordanus de Nemore, *Elementa super demonstrationem ponderum*, ed. E.A. Moody, in Clagett and Moody, *The Medieval Science of Weight*, E.1, 128: 'Inter quelibet gravia est velocitatis in descendendo eodem ordine sumpta proportio, descensus autem et contrarii motus proportio eadem sed permutata'; Thomas Bradwardine, *De proportionibus*, 96$^{236-238}$; 100^{368}–102^{343}; 102^{364}–104^{385}.

100 Richard Kilvington, *Utrum in omni motu*, 241, § 63$^{4-18}$: 'Secundo ex ista opinione sequitur ista conclusio, quod idem grave simplex motum in aqua alicuius densitatis non movebitur praecise in duplo velocius in aere duplae subtilitatis. Haec autem conclusio sequitur per similes determinationes prioribus. Nam si illud grave excedat resistentiam aquae in proportione maiori quam dupla, tunc excedit aerem duplae subtilitatis in proportione minori quam dupla ad illam proportionem, ut patet per deductionem priorem; ergo illud grave movebitur in aere duplae subtilitatis minus quam in duplo velocius quam movebitur in ista aqua. Et si illud grave simplex excedat resistentiam aquae in proportione maiori quam in dupla, tunc excedet resistentiam duplae subtilitatis in proportione maiori quam dupla ad proportionem primam, sicut demonstrari potest ut prius per definitionem proportionis duplicatae. Ergo tunc movebitur tale grave in aere duplae subtilitatis praecise plus quam in duplo velocius quam movebatur in aqua.'

101 See below, p. 74. The same situation occurs when the resistance is halved. See Crosby (ed.), *Thomas of Bradwardine*, 39.

Theorem III: The following rules given by Aristotle and Averroes are false: (1) If a given power moves a given mobile through a given distance in a given time, the same power will move twice the same mobile through half the distance in an equal time, and through the same distance in twice the time. (2) If a given power moves a given mobile through a given distance in a given time, double the power will move that mobile through double the distance in an equal time.[102] If $F_1 = F_2$ and $R_1 = 2R_2$, then $t_1 = t_2$ (t-time) and $s_1 = 2s_2$ (s-distance), or then $s_1 = s_2$ and $t_1 = 2t_2$. If $F_1 = 2F_2$ and $R_1 = R_2$, then $t_1 = t_2$, and $s_1 = 2s_2$. Bradwardine quotes these rules in Chapter II, part 3, in support of, as he says, the third wrong theory: '(with the moving power remaining constant) the proportion of the speeds of motions varies in accordance with the proportion of resistances, and (with the resistance remaining constant) that it varies in accordance with the proportion of moving powers.'[103] In his opinion, 'the theory is, however, refutable on two grounds: first, on that of insufficiency, second, because it yields false consequences.'[104]

In Kilvington's opinion, the above mentioned rules are false, since when the proportion of $F_1 : R_1 = 3:1$, and when F would be doubled, such a moving power would not move a body with a speed $v_2 = 2v_1$, but with $v_2 < 2v_1$. If $F_1 : R_1 = 3:1$, then double F_1 (read multiplied by 2) results in the proportion $F_2 : R_2 = 6:1$, if $R_1 = R_2$, and $6:1$, as was said above, does not double the proportion of $3:1$, but $9:1$ doubles $3:1$, and $6:1 < 9:1$. On the other hand, when $F : R = 3:2$, then: $v_2 > 2v_1$, because if $F_2 = 2F_1$, then $F_2 : R_2 = 6:2$, which is not a correct calculus, since $9:4$ doubles $3:2$, and $9:4 < 6:2$, so $v_2 > 2v_1$. $3:2$, and $9:4 < 6:2$, so $v_2 > 2v_1$.

Kilvington's calculation appears to be confusing. What he seems to be doing here is applying the new calculus of proportions for Aristotle's old theory: speed is proportional to a proportion of $F : R$. Thus in both cases, either when $F_1 : R_1 > 2:1$ or when $F_1 : R_1 < 2:1$, v_2 is not equal to $2v_1$. In the first case, speed is pro-

102 Richard Kilvington, *Utrum in omni motu*, 242, § 64^{1-11}: 'Tertio ex ista opinione sequitur ista conclusio, quod istae regulae Philosophi in VII *Physicorum* sint falsae. Quarum una ponitur in textu commento 37 et isto commento, quae est illa: "si aliquis motor moveat mobile per aliquod spatium in aliquo tempore, eadem potentia motoris movebit medietatem illius mobilis per duplum spatium in aequali tempore et per aequale spatium in medietate temporis," ut dicitur commento 36. Et alia regula ponitur textu commenti ultimi, quae est quod "si aliqua potentia moveat aliquod mobile per aliquod spatium in aliquo tempore, dupla potentia movebit idem mobile per duplum spatium in aequali tempore," quod est ex praecedentibus notum esse falsum.' See Aristotle, *Physica* VII, 250a1–4; Averroes, *In Physicam* VII, comm. 36, 335va; Thomas Bradwardine, *De proportionibus*, $96^{215-219}$.

103 For the Latin, see above, n. 61.

104 Thomas Bradwardine, *De proportionibus*, $96^{247-249}$: 'Ista tamen positio est dupliciter arguenda; primo super insufficientia, secundo super mendacio consequentiae' (transl. Crosby, 97).

portional to a proportion $v_2 \sim 2F_1 : R_1 = 6:1$ and this does not double $3:1$, and a proportion $6:1$ is less than $9:1$. So that is the reason why he states that speed $v_2 < 2v_1$. In the second case, when speed–v_1 is proportional to $3:2$, double the proportion would result in the speed $v_2 > 2v_1$.

In his comment to the above-mentioned theorem III, Kilvington, followed later on by Bradwardine, explains how the double proportion should be understood:

Richard Kilvington, ***Utrum in omni motu,*** **262, § 126**[21]**–263, § 126**[4]	**Thomas Bradwardine,** ***De proportionibus,*** **100**[323–338]
Et per idem ad tertiam conclusionem dico quod non oportet quod, 'si aliquis motor possit movere aliquod mobile per aliquod spatium in aliquo tempore, quod possit movere medietatem illius mobilis per duplum spatium in equali tempore.' Immo dico quod Philosophus intelligit per 'medietatem mobilis' talem partem mobilis quae habet se ad motorem in proportione subdupla ad proportionem totius mobilis ad eundem motorem. Et in aliis regulis intelligit per 'duplum agens' habens proportionem ⟨duplam⟩ ad eandem resistentiam. Et sic intelligendo verificentur omnes istae regulae, si cetera fuerunt paria. Et sic loquitur Philosophus in locis preallegatis 'quod grave duplum movebitur praecise in duplo velocius in eodem medio,' id est, grave habens praecise proportionem duplam ad resistentiam illius medii; et ⟨haec est⟩ regula proportionis inter grave subduplum et illud medium.	Conclusio autem allegata ex septimo *Physicorum*, quae dicit 'si aliqua potentia moveat aliquod mobile per aliquod spatium in aliquo tempore, eadem potentia movebit medietatem illius per duplum spatium in aequali tempore,' intelligit per 'medietatem mobilis' partem mobilis habentem ad illam potentiam motivam medietatem proportionis totius mobilis ad eandem ... Et per ista auctoritatum sequentium glosa patet.

Kilvington believes that the above definition of a double proportion allows explaining Archimedes', Jordanus de Nemore's and Euclid's theorems, which were quoted above to support the contrary theory. Likewise Bradwardine claims that Jordanus' theorem should be read in this way.[105]

105 Richard Kilvington, *Utrum in omni motu*, 263, § 126[8–10]: 'Et per hanc glossam dissolvi poterunt conclusions Archimedis et Jordani *De ponderibus* et Euclidis, quae sunt in contrarium

Theorem IV: Two heavy simple bodies, one of which moves in water one foot deep and the other in subdouble dense air two feet deep, would not move downward with the same speed in the same time, since even though in both cases the proportion of the active power to the resistance is the same, the resistances of air and water are not the same with regard to their intensity, and thus the speed would not be the same.[106] In Kilvington's opinion this is so, because there are two different resistances: one is intrinsic (*intensive*), and depends on the density of the medium, the other is extrinsic (*extensive*) and depends on the distance to be traversed.[107]

Theorem V: 'An object may fall in the same dense medium faster than another.'[108] To prove this conclusion, Kilvington presents two arguments based on the new calculus of proportions and on Aristotle's and Averroes' statement that a medium can be rarefied *ad infinitum*. He also gives evidence for contradictory statements: (1) 'a heavy mixed body may not fall in another medium twice as fast as it moves now,' (2) 'a heavy mixed body may fall in another medium twice as fast as it moves now.'[109] Bradwardine presents the above conclusions in a shorter version in his Theorem XI: 'An object may fall in the same medium both faster, slower, and equally with some other object that is lighter than itself'.[110] As Crosby explains:

> An important idea embodied in Bradwardine's development of this theorem is that, in motions involving mixed bodies, the internal resistance is to be added to any external resistance which is present, i.e., if a mixed

---

allegatae.' Thomas Bradwardine, *De proportionibus*, 100[338]: 'Conclusio autem allegata *De ponderibus* similiter debet intelligi ...'

106 Richard Kilvington, *Utrum in omni motu*, 263, § 127[13–21]: 'Dico quod duo motus duorum gravium aequalium, quorum unus descendit in aqua pedalis quantitatis et aliud in aere bipedalis quantitatis duplae tamen subtilitatis, non sunt aeque veloces nec aequalis est proportio motoris ad suam resistentiam utrobique. Et quando dicitur quod resistentia aeris et resistentia illius aquae sunt aequales, dico quod non, quia non sunt aequales extensive (notum est) nec intensive, quia iste aer propter eius subtilitatem non potest tantum resistere suo motori ad causandum motum retardari, sicut potest ista aqua propter eius densitatem, propter hoc quod partes eius propinquius iacent et melius poterunt applicari ad causandum motum tardum.'

107 For Kilvington's discussion of this distinction see *Utrum in omni motu*, 243, § 67; 263, §§ 127–128. For the parallel in Bradwardine, see *De proportionibus*, 116–120 (transl. Crosby, 117–121).

108 Richard Kilvington, *Utrum in omni motu*, 244, § 686[6–7]: 'Aliquod corpus movetur velocius alio et hoc in medio denso.'

109 Richard Kilvington, *Utrum in omni motu*, 244–246, §§ 68–75.

110 Thomas Bradwardine, *De proportionibus*, 114[98–99]: 'Quantumcumque gravius in eodem medio tardius et velocius illo et aequali velocitate potest descendere' (transl. Crosby, 115).

> body may be represented by f/r, then the velocity of such a body through a medium possessing the resistance R, is to be represented as $v = f/R+r$, and not $v = (f/r)/R$. Therefore, though velocity remains the function of a proportion of forces, coacting forces are to be added to each other rather than made proportionate; the resultant motive force is proportionate to the resultant resistive force.[111]

The meaning of Kilvington's proofs can be described in the same way.

Theorem VI: 'A body will move at equal speed in a medium and in a vacuum.'[112] Kilvington offers a proof of this statement by using the new calculus of proportions.[113] His final conclusion, presented in Part II of his question, is the same as Bradwardine's Theorem XII: 'All mixed bodies of similar composition will move at equal speeds in vacuum'.[114]

Richard Kilvington, *Utrum in omni motu*, 264, § 130[13–14]	**Thomas Bradwardine, *De proportionibus*, 116[127–128]**
Omnia talia mixta proportionalis compositionis aeque velociter moverentur in vacuo.	Omnia mixta compositionis consimilis aequali velocitate in vacuo movebuntur.

Theorem VII: 'If the speed varies in accordance with the proportion of an active power to resistance, a heavy pure body may move infinitely fast.'[115] To support this claim, Kilvington shows that in a downward motion the speed of motion of a pure heavy body increases infinitely, since at the beginning of the second part of a distance traversed, a body has to overcome half as great a resistance of the medium, because if at the beginning $F : R = 2 : 1$, in half the distance $F : R = F : \frac{1}{2}R = 4 : 1 = (2 : 1)^2$, and thus a body moves twice as fast; at the beginning of the fourth part $F : R = (2 : 1)^3$, and so on, thus before it attains its place, it will move with an infinite speed.[116] In Part II of his question, Kilvington affirms

111 Crosby (ed.), *Thomas of Bradwardine*, 41.

112 Richard Kilvington, *Utrum in omni motu*, 246, § 74[11–12]: 'Aliquid movetur in pleno et in vacuo aeque velociter.'

113 Richard Kilvington, *Utrum in omni motu*, 246–247, §§ 74–75.

114 Thomas Bradwardine, *De proportionibus*, 116[127–128] (transl. Crosby, 117).

115 Richard Kilvington, *Utrum in omni motu*, 247, § 76[17–18]: 'Aliquid movebitur infinita velocitate, si velocitas motus sequatur proportionem.'

116 Richard Kilvington, *Utrum in omni motu*, 247–248, §§ 76–80.

that we can state that a body might move with an infinite speed only if the term 'infinite' is understood syncategorematically, which means that a body would move twice as fast, three times, four times, and so one *in infinitum*, since there is no *maximum quod sic*-limit for the speed of motion. Moreover, since there is no proportion between the speed of accelerated motion and a uniform motion, one can say, syncategorematically, that the speed of accelerated motion is infinite in comparison to uniform motion.[117]

7 Conclusion

To conclude, in his questions on the *Physics*, Kilvington made it clear that he was aware that Aristotle's and Averroes' rules of motion, given in *Physics* VII, were not universally applicable. He was convinced that the correct calculus of proportions was formulated by Euclid in his *Elements*. As a consequence, he noticed that the proper understanding of Euclid's definition of operations on proportions necessitates a new interpretation of Aristotle's and Averroes' theory of motion. On the one hand, Euclid's and Archimedes' theory of operations on proportions concludes that doubling a ratio corresponds to squaring the fraction which we form from the ratio. On the other hand, Aristotle's and Averroes' statements clearly indicate that speed is proportional to a proportion of an active power to resistance, which is not squared but simply multiplied by two. Having noticed the contradiction of these two views, Kilvington first presented two main arguments against the Aristotelian proposition and finally concluded that, when talking about a power moving one half of a mobile, Aristotle means precisely a double ratio between F and R; when talking about a power moving a mobile twice as heavy, he means taking the square root of the ratio of F : R. The new mathematical rule describing the speed of motion is in accord with those of Aristotle only in one case: if the ratio of the power of the mover to that of its mobile is 2 : 1, the same power will move half the mobile with exactly twice the speed. It has already been shown that Kilvington's reinterpretation of Aristotle's and Averroes' rules of motion reveals that the speed cannot be described by sole multiplication. Thus, as it was shown above, on the new interpretation, 'double' the ratio of 3 : 2 (or the ratio 3 : 2 *duplicata*) is the ratio 9 : 4, not 6 : 2, and the ratio of 3 : 1 *dupla* or *duplicata* is equal to 9 : 1, not to 6 : 1. Kilvington's calculus provides values of the ratio of F to R greater than 1 : 1 for any speed

117 Richard Kilvington, *Utrum in omni motu*, 265–266, §§ 133–135. Cf. Thomas Bradwardine, *De proportionibus*, $124^{293–296}$.

down to zero, since any root of a ratio greater than 1 : 1 is always a ratio greater than 1 : 1. And, with his additional assumption that any excess, however small, of an active power over resistance is sufficient to initiate motion and to continue it, he may have described a very slow motion with a speed greater than 0 and less than 1 ($0 > v < 1$). Hence, he avoids a serious weakness of Aristotle's theory, which cannot explain the mathematical relationship of F and R in very slow motions, when speed is lesser than 1.

Murdoch and Sylla emphasized that with Bradwardine's treatise *On the proportions of velocities* 'the situation changed rather dramatically in 1328.' In their opinion, Bradwardine 'removed the whole problem of relating velocities, forces, and resistances from the context of an exposition of Aristotle's words and investigating it in its own right.'[118] Although Sylla identified four questions on motion, found in a Venice manuscript (Biblioteca Nazionale Marciana, Ms. lat. VI. 72 [2810]), as authored by Richard Kilvington, she did not notice the strong similarities between Kilvington's and Bradwardine's works, and she repeated Maier's views.[119] Maier, as it was said at the beginning of this paper, noticed the contextual relations between Kilvington's and Bradwardine's works, but she was convinced that Kilvington used 'Bradwardine's rule'. My research shows that it was in fact Kilvington who initiated the study of the mathematics of motion and used the 'correct' technique of calculus of proportions, which allowed him to resolve the inconsistences of Aristotle's theory, although he did not further develop the new theory in such a consistent and precise way as Bradwardine did. Kilvington seems to have taken the first step towards reinterpreting Aristotle and Averroes. He, and most likely some others, opened the new chapter in the history of medieval physics.

As Sylla explains with respect to Bradwardine:

> Bradwardine's function can be represented as a correlation between the latitude of proportion and of velocity, such that the proportion of equality corresponds to zero velocity and greater proportions correspond to greater velocities.[120]

Precisely the same conclusion applies to Kilvington.[121] It seems that Kilvington invented the new theory of motion and applied the new calculus of proportions

118 Murdoch and Sylla, 'The Science of Motion,' 224.

119 See Sylla, *The Oxford Calculators*, 436.

120 Sylla, *The Oxford Calculators*, 403.

121 Since both Kilvington's question and Bradwardine's treatise were put together by the authors themselves, and were not *reportationes* of classroom disputations, it seems that

to Aristotle's 'laws of motion', while Bradwardine employed many of Kilvington's arguments and examples to substantiate the theory. Kilvington, however, was not the only one. Also 'a third author', as S. Rommevaux-Tani notices, 'plays a role in this story, Adam of Pipewell, with whom the author of the *Tractatus de sex inconvenientibus* says he agrees. The precise connection between Kilvington and Pipewell is unknown, but Adam of Pipewell was a fellow at Oxford when Kilvington probably wrote his *Questiones*.'[122] Regrettably, none of Pipewell's works have survived to our time. It is worth noting that not only Kilvington's *Quaestiones super Physicam* but also his *Quaestiones super de generatione et corruptione*, written ca. 1322/1323, testify to a lively discussion of the new calculus of ratios.[123] At that time Kilvington was appointed at Balliol College, where both Bradwardine and Pipewell were most likely residing. Thus, it is very well possible that they, and maybe some others, were debating the mathematics of motion. It is also possible that Bradwardine lectured in arithmetic and geometry related to his work on motion at that time.

In his question *Utrum potentia motoris excedit potentiam rei motae*, which belongs to his commentary on the *Physics* written in 1326 at the latest, Richard Kilvington clearly explains the mathematical conception of compounding proportions implicit in what is generally called 'Bradwardine's Rule'. So here we seem to have a case of 'Stigler's law of eponymy', which states that no scientific discovery is named after its original discoverer.[124] By studying Kilvington's work, we can trace what was going on with the Oxford Calculators before Bradwardine's *De proportionibus*, which has previously been considered the founding document of Oxford calculatory work. In an attempt to trace the impact, spread, and decline of quantifying Aristotle, we should realize that the activity of quantifying motion had a prehistory before 1328.

Bradwardine, like Ockham before, did not assume that his readers would be using his treatise in conjunction with Kilvington's work. See E.D. Sylla, 'Walter Burley's *Tractatus primus*: Evidence Concerning the Relations of Disputations and Written Works,' *Franciscan Studies* 44 (1984) (*William of Ockham [1285–1347] Commemorative Issue*), 1: 257–274, at 258.

122 See Rommevaux-Tani's contribution to the present volume.

123 See Jung and Podkoński, 'Richard Kilvington on Proportions.'

124 See S.M. Stigler, 'Stigler's Law of Eponymy', *Transactions of the New York Academy of Sciences* 39 (1980), 147–158.

Acknowledgements

I gratefully acknowledge that this chapter is a result of project nr 2018/31/B/HS1/00472 funded by the National Science Centre, Poland. I thank Edith Sylla for her comments on my paper.

CHAPTER 3

The *Opuscula de motu* Ascribed to Richard Swineshead: The Testimony of the Ongoing Development of the Oxford Calculators' Science of Motion

Robert Podkoński

Introduction*

The *Liber calculationum* (*Book of Calculations*) is arguably the most sophisticated work originating from the Oxford Calculators' School, presenting the last stage in the development of a mathematically reformulated natural philosophy.[1] On the basis of colophons and notes that are to be found in some of the thirteen extant manuscript copies of this work it can be safely assumed that the author of the *Liber calculationum* was Richard Swineshead, about whom we know almost nothing else.[2]

The complete handwritten copies of the *Liber calculationum* as well as its fifteenth- and sixteenth-century printed editions contain sixteen separate treatises (*tractatus*) or chapters. Interestingly, in almost every manuscript of Swineshead's work these parts are transcribed in a different order, and in some codices certain treatises are just absent—most often these variations concern the chapters *De motu locali* (*On local motion*), *De medio non resistente* (*On a non-resisting medium*), and *De inductione gradus summi* (*On the induction of*

* The present study is the result of the National Science Centre project *Miniatura*, 2017/01/X/HS1/01937.

1 E.D. Sylla, 'The Oxford Calculators,' in N. Kretzmann, A. Kenny, J. Pinborg and E. Stump (eds.), *The Cambridge History of Later Medieval Philosophy. From the Rediscovery of Aristotle to the Disintegration of Scholasticism, 1100–1600*, Cambridge 1982, 540–563, at 560–562.

2 Città del Vaticano, Biblioteca Apostolica Vaticana, Ms. Chigi E. IV. 120, 112ᵛ: 'Explicit tractatus datus a magistro Ricardo de Svvyinshede'; Perugia, Biblioteca Comunale Augusta, Ms. 1062, Iᵛ: 'Joannis Suiset tabula Calculationum que in hoc libro continentur'; Cambridge, Gonville and Caius College Library, Ms. 499/268, 203ᵛᵇ: 'Explicit tractatus dᶜᵒ Swynyshed.' Cf. J.E. Murdoch and E.D. Sylla, 'Swineshead (Swyneshed, Suicet, etc.) Richard,' in C.C. Gillispie (ed.), *Dictionary of Scientific Biography*, 13, New York 1976, 184–185.

DOI:10.1163/9789004512054_005

the highest degree). All printed editions of the *Liber calculationum* include all sixteen treatises constituting this work and present them in the same order.[3]

I have discussed in detail the problem of the composition of the *Liber calculationum* elsewhere, but in the present study let me just point out that in the two oldest, fourteenth-century manuscript copies of Richard Swineshead's work (Cambridge, Gonville and Caius College Library, Ms. 499/268, and Paris, Bibliothèque Nationale de France, Ms. lat. 6558) we find only fourteen chapters in the following order:[4]

1. *De intensione et remissione*;
2. *De difformibus*;
3. *De intensione elementi*;
4. *De intensione et remissione mixtorum*;
5. *De raritate et densitate*;
6. *De velocitate motus augmentationis*;
7. *De reactione*;
8. *De potentia rei*;
9. *De difficultate actionis*;
10. *De maximo et minimo*;
11. *De medio non resistente*;
12. *De loco elementi*;
13. *De luminosis*;
14. *De actione luminosi*.[5]

3 R. Podkoński, 'Richard Swineshead's *Liber calculationum* in Italy: Some Remarks on Manuscripts, Editions and Dissemination,' *Recherches de Théologie et Philosophie Médiévales* 80 (2013), 307–361, at 311–334, and 337–338; R. Podkoński, 'Richard Swineshead's *Liber calculationum* in Italy: the Codex Bibl. Naz. San Marco lat. VI.226 and its significance,' *Recherches de Théologie et Philosophie Médiévales* 84 (2017), 407–426, at 411–414. The order of chapters of the *Liber calculationum* established in printed editions of this work is as follows: 1. *De intensione et remissione* (*On intension and remission*); 2. *De difformibus* (*On difformly qualified subjects*); 3. *De intensione elementi* (*On the intension of an element*); 4. *De intensione et remissione mixtorum* (*On the intension and remission of mixed subjects*); 5. *De raritate et densitate* (*On rarity and density*); 6. *De velocitate motus augmentationis* (*On the velocity of augmentation*); 7. *De reactione* (*On reaction*); 8. *De potentia rei* (*On the power of a thing*); 9. *De difficultate actionis* (*On the difficulty of action*); 10. *De maximo et minimo* (*On maximum and minimum*); 11. *De loco elementi* (*On the place of an element*); 12. *De luminosis* (*On luminous bodies*); 13. *De actione luminosi* (*On the action of a luminous body*); 14. *De motu locali* (*On local motion*); 15. *De medio non resistente* (*On the non-resisting medium*); 16. *De inductione gradus summi* (*On the induction of the highest degree*).

4 Cf. Podkoński, 'Richard Swineshead's *Liber calculationum* in Italy: Some Remarks,' 311–318, 331–334; Podkoński, 'Richard Swineshead's *Liber calculationum* in Italy: the Codex,' 414–421.

5 Cf. Podkoński, 'Richard Swineshead's *Liber calculationum* in Italy: Some Remarks,' 350.

It is easy to notice that the treatise *De motu locali* is absent here. Interestingly, in the Cambridge codex after the *Liber calculationum* there are also three shorter texts transcribed, the colophons of two of them mentioning the name of Richard Swineshead too.[6] The first of these texts is described in the secondary literature as a fragment of a commentary on Aristotle's *De caelo*.[7] The two remaining texts are short treatises (*opuscula*) on motion that interest us here.[8] It is worth noting that only the *explicit* of the second of these short treatises refers to Richard Swineshead.[9] Still, the similarities between these two texts cannot be taken as a mere coincidence and suggest that these treatises should be regarded as Swineshead's successive accounts of the same topic. Both *opuscula* concern the most specific problem for all Oxford Calculators, namely establishing the 'proper rules' of local motion.[10] A quick reading leaves one with the impression that these short treatises should be seen as intermediate 'steps' leading straight from William Heytesbury's *Regulae solvendi sophismata* (the section *De motu locali*) to *De motu locali*, a chapter of Richard Swineshead's *Liber calculationum*. A similar hypothesis has been already formulated by John Murdoch and Edith Sylla in their presentation of Swineshead's works in the *Dictionary of Scientific Biography*, so I just pay honour to them both here, presenting *in extenso* fragments of the above-mentioned texts as well as some other fourteenth-century Oxford natural philosophical works that clearly confirm this hypothesis.[11] Also, I point out the specific statements from Swineshead's *opuscula* that help explain—at least partly—his motivations for composing

6 Cambridge, Gonville and Caius College Library, Ms. 499/268, 211vb: 'Explicit tractatus de Swynyshed'; 215rb: 'Explicit tractatus de Swynyshed de motu locali.'

7 Cf. Murdoch and Sylla, 'Swineshead,' 185, 206; Cambridge, Gonville and Caius College Library, Ms. 499/268, 204ra: 'In primo de celo philosophus commento 35 arguit corpus infinitum circumvolvi non posse …'

8 Cambridge, Gonville and Caius College Library, Ms. 499/268, 212ra: 'Cum omnis effectus naturalis per motum ad esse producitur inquisitioni naturalium volenti insistere precognitioni eius necesse …'; 213rb: 'Multe possunt elici conclusiones sive regule super variationem proportionum et motus ex variatione potentie motive ad suam resistentiam et econtra …'

9 Cf. note 6 above.

10 Cf. Murdoch and Sylla, 'Swineshead,' 206–207.

11 Cf. Murdoch and Sylla, 'Swineshead,' 206: '[The] similarity [of the *Opusculum de motu*] to the last short work of the manuscript [i.e., to the *Opusculum de motu locali*], which is ascribed to Swineshead is so great that there is every reason to ascribe it also to Swineshead. In the Seville manuscript Colombina 7-7-29 the works (the *De motu* and the following *De motu locali*) appear as one, but this seems hardly plausible since there is a great deal of overlap and repetition between them. It is more likely that they are successive drafts of the same work than that they are both sections of a single larger work.'

the whole treatise *De motu locali* included in his *Liber calculationum*, as well as the goal of this text.

In order to avoid confusion, I accept here the titles given by Murdoch and Sylla for these short treatises. Hence the first and shorter of these texts is referred to as the *Opusculum de motu*, and the second and longer one as the *Opusculum de motu locali*, even though the final sections of the latter text actually concern the problem of the proper description of qualitative changes (*alterationes*).[12]

At the moment we know only two manuscript copies of the *Opusculum de motu locali*. One of them is included in the already mentioned Cambridge manuscript (Gonville and Caius College Library, Ms. 499/268), the other is found in Sevilla, Biblioteca Capitular y Colombina, Ms. 7-7-29 (30^v–34^r). In both manuscripts this text is preceded directly by the *Opusculum de motu* (Sevilla, Biblioteca Capitular y Colombina, Ms. 7-7-29, 28^v–30^v). There is also a third copy of the latter *opusculum*, contained in Oxford, Bodleian Library, Ms. Digby 154 (42^r–44^v) which, interestingly enough, is probably the only manuscript copy of any of Richard Swineshead's works preserved in an Oxford library.[13]

As it has been announced above, in what follows I will first present the similarities between both *opuscula* on motion ascribed to Richard Swineshead. These texts will also be compared to his treatise *De motu locali* from the *Liber calculationum* in order to confirm the hypothesis that they were all composed by the same author. Next I will focus on the dependence of the contents of both *opuscula* on William Heytesbury's statements from *De motu locali*, a part of his *Regulae solvendi sophismata*. The borrowings and parallel passages presented there confirm the above-mentioned hypothesis even more strongly. Further on

12 Murdoch and Sylla, 'Swineshead,' 206–207.

13 For a detailed description of the manuscripts and for the mutual relations between the copies of Richard Swineshead's works, see R. Podkoński, *Richard Swineshead's Theory of Motion*, Łódź 2019 (Research on Science and Natural Philosophy, 2), 115–132. All references to Swineshead's *Opuscula de motu* are taken from the edition included in this book (137–189). It is a mystery of a kind why so few copies of some of the Oxford Calculators' works have survived in England. With regard to the *Liber calculationum*, for example, there remains only one manuscript copy of this treatise, namely the above-mentioned Cambridge, Gonville and Caius College Library, Ms. 499/268, and a student's notebook in Worcester, Cathedral and Chapter Library, Ms. F. 35, that contains few selected passages from Richard Swineshead's greatest work. Cf. Murdoch and Sylla, 'Swineshead,' 211. It is possible that, being recognised as mathematical treatises, they were deliberately destroyed in the period 1550–1650 as were many other such works, considered at the time as dangerous and deceitful. See, J.P. Zetterberg, 'The Mistaking of "the Mathematics" for Magic in Tudor and Stuart England,' *The Sixteenth Century Journal* 11 (1980), 83–97, at 85. We have no arguments, however, in support of this hypothesis at the moment.

I will point out the differences between all texts on local motion ascribed to Richard Swineshead. These attest to a certain progress in his attempts to construct the science of local motion. I shall propose the most probable temporal order of the composition of Swineshead's *opuscula* and the treatise *De motu locali* from the *Liber calculationum*, as well as their relation to some contemporary texts on the same topic. Finally, I will return to William Heytesbury's *De motu locali*, specifically to one of the authoritative statements regarding the (im)possibility of formulating rules that would describe difformly difform motion. On the basis of conclusions that can be found in Richard Swineshead's texts on local motion, I will argue that Swineshead took Heytesbury's statements as a kind of intellectual challenge that inspired him to construct the complete set of 'rules' of local motion that ended up in Swineshead's treatise *De motu locali* from the *Liber calculationum*. Since the editions of Swineshead's text on local motion are still not easily accessible I will cite extensive fragments of these works.

1 Richard Swineshead's Successive Accounts of Local Motion: Mutual Similarities

The most obvious similarity between the *Opusculum de motu* and the *Opusculum de motu locali* that suggests that they constitute Swineshead's successive accounts of the same topic is their internal structure. In both texts the author first focuses on the description of motion *penes causam* ('with regard to the cause'), taking into account changes of velocity relative to changes of the intensity of agents (i.e., motive and resistive powers), and later proceeds to the description of motion *penes effectum* ('with regard to the effect'), concerning the relations between such factors as velocities, times, and distances covered. In each text this change of the point of view is explicitly announced by the author.[14]

Both *opuscula* also contain almost identical argumentations that at first sight look counterintuitive, but in fact are logical consequences of earlier

---

14 Richard Swineshead, *Opusculum de motu*, § 61, 151: 'Dicto generaliter penes quid habet attendi velocitas in quocumque motu quo ad causam, videndum est penes quid attenditur ⟨velocitas⟩ in motu quo ad effectum.' Richard Swineshead, *Opusculum de motu locali*, § 54, 174: 'Quia prius dictum est diffuse de velocitate motus quo ad eius causam, videlicet quod sequitur proportionem potentie motive ad potentiam resistivam, et causam sufficientem universaliter sequitur suus effectus correspondens, videtur igitur modo penes quid attendi debeat velocitas motus quo ad eius effectum.'

accepted or proven statements. Swineshead argues here that the velocity of a body moved by a greater motive power in a given uniformly difformly resistant medium will diminish faster than the velocity of a body moved by a lesser motive power.

Superius enim dictum est quod *maior potentia pertranseundo aliquam resistentiam difformem velocius remittet motum suum versus extremum intensius quam potentia minor* … (Richard Swineshead, *Opusculum de motu*).[15]	Et ex isto sequitur quod, si ab eodem puncto alicuius resistentie difformis incipiant duo motores inequales moveri eandem resistentiam dividendo, uterque continue secundum proportionem sui ad gradum correspondentem puncto contactus, cum ⟨ceteris⟩ paribus, velocius incipit maior remittere vel intendere motum suum; ⟨et etiam⟩ *quod si versus extremum intensius illius resistentie incipiant moveri, tunc maior incipit velocius remittere motum suum ipso minori.* Et si ab eodem puncto versus extremum remissius incipiant moveri, tunc maior cum ⟨ceteris⟩ paribus intenderet motum suum velocius ipso minori (Richard Swineshead, *Opusculum de motu locali*).[16]

Interestingly, there is a similar conclusion in the treatise *De motu locali* in the *Liber calculationum.*

> Si aliqua potentia in medio difformi remittat motum suum uniformiter ad non gradum, *omnis maior infinite velociter remittet motum ad extremum intensius deveniendo* (Richard Swineshead, *Liber calculationum, De motu locali,* Conclusio 39).[17]

The opening fragments of both *opuscula* contain a series of six consecutive 'rules' that are aimed at determining the changes of velocities relative to

15 Richard Swineshead, *Opusculum de motu*, § 44, 146. All emphases in the quotations are mine.

16 Richard Swineshead, *Opusculum de motu*, § 20, 166.

17 Richard Swineshead, *Liber calculationum, Tractatus de motu locali*, § 98, in: R. Podkoński, *Suisetica inania. Ryszarda Swinesheada spekulatywna nauka o ruchu lokalnym*, Wydawnictwo Uniwersytetu Łódzkiego, Łódź 2018, 310. All references to the *Tractatus de motu locali* of the *Liber calculationum* are taken from the edition included in this work (271–341). For other passages from Swineshead's *Liber calculationum* I shall use the edition Venice 1520.

uniform changes (i.e., increase or decrease) of motive power or resistance, or both agents concurrently. In the early *Opusculum de motu*, these 'rules' run as follows:

(1) Pro quo generaliter tenetur illa regula, quod universaliter ex uniformi intensione potentie motive, manente resistentia eadem, sequitur difformis intensio motus, quia tardius et tardius. Probatur sic: quia tardius et tardius adquiritur proportio, et consequenter motus sicut proportio, igitur tardius et tardius adquiritur motus.[18]

(2) Tenetur etiam hec regula pro generali, quod universaliter ex uniformi remissione potentie, stante resistentia, difformiter remittetur motus, quia velocius et velocius.[19]

(3) Teneantur et ille regule generaliter, quod, signata quacumque potentia motiva se habente ad suam resistentiam in proportione maioris inequalitatis, ex uniformi intensione potentie cum equali intensione resistentie sequitur difformis remissio motus.[20]

(4) Hec est alia, quod, signata quacumque resistentia se habente ad certam potentiam in proportione minoris inequalitatis, ex uniformi remissione illius potentie, cum equali remissione resistentie, sequitur motum intendi velocius et velocius.[21]

(5) Alia regula est item quod universaliter ex uniformi remissione resistentie, stante potentia, difformiter intenditur motus, quia continue velocius et velocius.[22]

(6) Alia regula est hec: ex uniformi intensione resistentie, potentia manente non variata, difformiter remittetur motus, quia continue tardius et tardius.[23]

In the later *Opusculum de motu locali*, the 'rules' are the following:

(1) Super quo tenetur illa regula tanquam firma, quod si uniformiter crescat potentia motoris, totali resistentia sua manente eadem vel equali, difformiter, quia continue tardius et tardius, intenditur, ceteris paribus, motus suus.[24]

---

18 Richard Swineshead, *Opusculum de motu*, §§20–21, 140. Here and in the two consecutive lists I have introduced the numbering of 'rules' (in brackets) in order to facilitate their mutual comparison.

19 Richard Swineshead, *Opusculum de motu*, §22, 141.

20 Richard Swineshead, *Opusculum de motu*, §24, 141.

21 Richard Swineshead, *Opusculum de motu*, §26, 142.

22 Richard Swineshead, *Opusculum de motu*, §28, 142.

23 Richard Swineshead, *Opusculum de motu*, §30, 142.

24 Richard Swineshead, *Opusculum de motu locali*, §4, 162.

(2) Item, si uniformiter decrescat potentia motoris, eadem continue manente totali potentia resistiva, difformiter, quia velocius et velocius, ceteris paribus, remittetur motus eius.[25]

(6) Econverso modo se habet respectu decrementi resistentie, non variata potentia motoris. Unde, si uniformiter crescat potentia resistiva respectu eiusdem motoris non mutati, difformiter, quia continue tardius et tardius, remittetur motus suus.[26]

(5) Item, si uniformiter decrescat potentia resistiva respectu eiusdem motoris omnino non variati, difformiter, quia continue velocius et velocius, ceteris paribus, intendetur motus.[27]

(3) Iuxta illud primum, scilicet quod idem excessus additus vel ablatus etc., sequitur quod, si uniformiter crescat potentia motoris et equaliter etiam crescat sua resistentia, continue tardius et tardius remittetur eius motus cum ⟨ceteris⟩ paribus.[28]

(4) Item, si uniformiter decrescat potentia motoris et continue eque velociter decrescat eius resistentia, continue velocius intenditur motus illius.[29]

Four of these 'rules' are equivalent to four consecutive *conclusiones* from the treatise *De motu locali* in Richard Swineshead's *Liber calculationum*:

(1) Si potentia crescat uniformiter respectu resistentie non variate, tardius et tardius intendetur motus.[30]

(2) Et si potentia uniformiter decrescat respectu resistentie non variate, velocius et velocius remittetur motus.[31]

(5) Et si decrescat resistentia uniformiter respectu potentie non variate, illa potentia velocius et velocius intendet motum suum.[32]

(6) Si resistentia uniformiter crescat respectu alicuius potentie, illa potentia continue tardius et tardius remittet motum suum.[33]

In the treatise *De medio non resistente* there is an argumentation similar to one of the remaining 'rules', the conclusion of which is, however, different:

(3) Si ab extremo medii non resistentis generaretur latitudo resistentie, quiescente extremo remissiori, cuius quilibet gradus uniformiter a non gradu intendat motum suum, manente illa latitudine continue uniformiter

25 Richard Swineshead, *Opusculum de motu locali*, § 5, 162.
26 Richard Swineshead, *Opusculum de motu locali*, § 6, 162.
27 Richard Swineshead, *Opusculum de motu locali*, § 7, 163.
28 Richard Swineshead, *Opusculum de motu locali*, § 11, 164.
29 Richard Swineshead, *Opusculum de motu locali*, § 12, 164.
30 Richard Swineshead, *Liber calculationum*, *Tractatus de motu locali*, Conclusio 18, § 27, 277.
31 Richard Swineshead, *Liber calculationum*, *Tractatus de motu locali*, Conclusio 19, § 28, 277.
32 Richard Swineshead, *Liber calculationum*, *Tractatus de motu locali*, Conclusio 21, § 30, 277.
33 Richard Swineshead, *Liber calculationum*, *Tractatus de motu locali*, Conclusio 20, § 29, 277.

> extensa; et in eodem extremo incipiat potentia motiva crescere a non gradu, que etiam uniformiter crescat, illa tardius et tardius intendit motum suum, deductis aliis impedimentis et iuvamentis extrinsecis.[34]

It is worth noting that the 'rules' common to both *opuscula* and the treatise *De motu locali* in Richard Swineshead's *Liber calculationum* are presented in the latter text as quite advanced ones—namely as rules 18, 19, 20, and 21—and derived as logical consequences of a few earlier proven conclusions, with no further justification.[35] In both *opuscula* these 'rules' are accompanied with more or less extensive arguments based on some more or less advanced *calculationes*.

The treatise *De motu locali* in the *Liber calculationum* should be taken as (an almost successfully accomplished) attempt to construct the 'science of local motion' in a strict 'geometrical' axiomatic form, following the pattern of Euclid's *Elements* and of the third part of Thomas Bradwardine's *Tractatus de proportionibus*. Thus, from the above-mentioned four *conclusiones* Richard Swineshead inferred no less than five other *regulae*, all concerning increases or decreases of velocity that occur 'faster and faster' or 'more and more slowly'.[36] Four of these consecutive 'rules' are logical consequences of the above-mentioned ones, proven simply with a '*per locum a maiori*' generalisation. The axiomatic style of the treatise *De motu locali* can serve as an explanation why there are no two remaining 'rules' that are present in both *opuscula*. It could be that either it was too difficult, or even impossible, to infer them from other *conclusiones*, or Richard Swineshead found these 'rules' just superfluous and purposeless in the course of the reasonings included in this treatise.

Interestingly, there is one *conclusio* in the treatise *De motu locali* that is formulated almost explicitly in each of the *opuscula*. In the *Liber calculationum*, this *conclusio* runs as follows:

> Si aliqua potentia non variata in medio difformi remittat motum uniformiter ad non gradum vel ad gradum, nulla potentia maior nec minor non variata potest uniformiter remittere motum suum idem medium transeundo.[37]

In the *Opusculum de motu*, the corresponding passage is the following:

34 Richard Swineshead, *Liber calculationum*, *De medio non resistente*, 51[ra].

35 See notes 30–33 above.

36 Cf. Richard Swineshead, *Liber calculationum*, *Tractatus de motu locali*, Conclusiones 22–25, §§ 31–34, 277–278; Conclusio 32, §§ 47–48, 282.

37 Richard Swineshead, *Liber calculationum*, *Tractatus de motu locali*, Conclusio 38, § 80, 301.

> Ex istis patet, quod si aliquam resistentiam pertranseundo aliqua potentia uniformiter intendet vel remittet motum suum, nulla potentia maior vel minor illam eandem resistentiam pertranseundo uniformiter intendet vel remittet motum suum.[38]

Finally, the parallel passage in the *Opusculum de motu locali*:

> Item, quod in nulla resistentia difformi in qua remitteret aliquis motor motum suum uniformiter, remitteret aliquis alius, scilicet inequalis, secundum ultimum suum movendo uniformiter motum suum, nulla facta variatione in ipsa resistentia ... Sequitur igitur propositum principale, videlicet quod in nulla resistentia difformi, in qua remittet aliquis motor uniformiter motum suum, remittet aliquis alius inequalis illi, cum ⟨ceteris⟩ paribus, uniformiter motum suum.[39]

All these statements concern the phenomenon of 'multiple continua', aptly described by Stillman Drake, occurring inevitably when the Eudoxean-Euclidean calculus of ratios is applied to the Aristotelian rules of local motion. As Drake explained it, on the basis of the Calculators' 'new rule of motion', velocities associated with a particular integral ratio of motive to resistive power through repeated squaring or square-root extraction of this ratio are entirely incommensurable with the velocities similarly associated with any other integral ratio of motive to resistive power, prime to the first ratio.[40] This phenomenon was first discovered by Thomas Bradwardine, but Richard Swineshead presented it in a more 'imaginable' fashion, referring simply to a motion in a medium where resistance increases uniformly relative to distance. It is important to note here, that in both *opuscula* the above statements are proven with the aid of the so-called mean-speed theorem,[41] while in the treatise *De*

---

38 Richard Swineshead, *Opusculum de motu*, § 78, 155.

39 Richard Swineshead, *Opusculum de motu locali*, § 39, 170, and § 48, 172.

40 S. Drake, 'Bradwardine's Function, Mediate Denomination and Multiple Continua,' *Physis* 12 (1970), 51–68, at 55.

41 Richard Swineshead, *Opusculum de motu*, § 79, 155–156: 'Hoc probatur sic: capiatur certa resistentia quam pertranseundo certa ⟨potentia⟩ uniformiter remittet motum suum ad non gradum in hora. Tunc sic: prius dictum est ⟨quod⟩ triplum precise pertransibit potentia illa in prima medietate temporis ad illud quod pertransibit in secunda; igitur quecumque potentia, sive minor ⟨sive maior⟩, que eandem resistentiam remittet uniformiter pertranseundo motum suum, proportionaliter de resistentia in parte correspondente sui temporis pertransiret. Si sic, tunc sequitur quod aliqua potentia a certo gradu ad aliquem gradum certum uniformiter remittendo motum suum plus quam in triplo plus

motu locali from the *Liber calculationum* the 38th *conclusio* is directly followed by four additional assumptions (*suppositiones*), one of which is a reformulated mean-speed theorem too, and proved in due course with no less than four different arguments.[42] Later on, in the treatise *De motu locali*, Swineshead discussed the possibility of adjusting any motive power different from (i.e., either greater or lesser than) the initial one in order to effect the uniformly difform motion in the very same medium. Eventually, he presented detailed descriptions of how and when such or such a power should be changed for this purpose.[43]

---

pertransiret in prima medietate temporis quam in secunda, quod est impossibile. Et etiam sequitur quod alicuius latitudinis terminate ad duos gradus certos gradus medius est subduplus ad gradum terminantem illam latitudinem secundum extremum intensius. Hoc patet ex hoc, quod cuiuslibet latitudinis terminate ad non gradum gradus medius est precise subduplus ad gradum terminantem in extremo intensiori.' Richard Swineshead, *Opusculum de motu locali*, §§ 45–48, 171–172: 'Quibus intellectis, sequitur quod nulli motores inequales super eandem partem alicuius resistentie uniformiter remitterent motus suos. Quia si sic, nulla esset ratio quin esset assignari aliquam resistentiam difformem, in qua motor aliquis uniformiter remitteret motum suum ad non gradum ipsam complete pertranseundo, in qua etiam motor subduplus uniformiter remitteret motum suum etiam usque ad non gradum transeundo illam partem illius, in qua primus motor duplus deperdet medietatem sui motus, nulla facta remissione in illa resistentia. Sit igitur primus motor A ⟨et⟩ secundus, qui est subduplus, B. Sit quoddam punctus terminans tres quartas primas C, et punctus terminans tres quartas primas illarum trium quartarum totius E. Tunc arguitur sic: A motor in remittendo motum suum ad non gradum uniformiter in illa resistentia transibit in illa prima medietate temporis usque ad C punctum, per illam regulam prius dictam, videlicet quod in omni uniformi remissione alicuius motus ab aliquo gradu ad non gradum triplum precise pertransiretur in prima medietate temporis ad illud quod pertransiretur in secunda. Etiam per idem B motor in prima medietate sui temporis transibit usque ad E punctum, eo quod tantam latitudinem motus precise deperdet A motor in prima quarta totius sui temporis sicut B motor in prima medietate sui temporis; igitur correspondenter sicut B motor transibit triplum in prima medietate totius sui temporis ad illud quod pertransibit idem B in secunda medietate eiusdem temporis, ita A motor in prima medietate prime medietatis sui temporis transiret triplum ad illud quod transibit in secunda medietate eiusdem prime medietatis sui temporis, quod est omnino impossibile. Et consequentia satis patet, quia A motor in deperdendo primam quartam totius sui motus transiret usque ad E punctum, et hoc in prima quarta totius temporis que est prima medietas prime medietatis, eo quod uniformiter deperdet totum suum motum in toto tempore. Et in deperdendo secundam quartam totius sui motus transiret precise ab ipso E puncto usque ad C, que quidem pars est subtripla ad priorem. Et illam pertransiret precise in secunda quarta totius temporis per idem argumentum, sicut iam dictum est de priori quarta. Sequitur idem impossibile, quod universaliter motor uniformiter remittens motum suum ad non gradum triplum transiret in prima medietate prime medietatis temporis ad illud quod transiret in secunda medietate eiusdem prime medietatis.'

42 Cf. Richard Swineshead, *Liber calculationum*, *Tractatus de motu locali*, §§ 81–97, 301–310.

43 Cf. Richard Swineshead, *Liber calculationum*, *Tractatus de motu locali*, §§ 102–148, 312–332.

2 William Heytesbury's *De motu locali* and Its Impact on Richard Swineshead's Science of Local Motion

The so-called mean-speed theorem, also called the 'Merton rule', is generally recognised as the most important achievement of the Oxford Calculators, and it is William Heytesbury who is rightly credited with the authorship of this theorem.[44] In fact, in the section *De motu locali* of Heytesbury's *Regulae solvendi sophismata* we encounter two versions of this rule. In the first formula Heytesbury determines the mean degree of motion, that is the mean velocity, in a uniformly accelerated or decelerated motion as a degree (or velocity) equal to the arithmetical mean between initial and final degrees (velocities) of this motion. In the case of a uniformly changing motion, when one of its limits is the state of rest (*non-gradus*) the mean velocity is simply calculated as a half of the final, or initial, velocity. Heytesbury proves the rule stating simply that the same effect will be achieved, that is the same distance traversed in the same period of time, with a uniformly changing motion and with a uniform motion characterized by the mean degree (velocity) calculated in the described way.[45] In the second formula of the mean-speed theorem provided by Heytesbury in the same treatise he determines the ratio of the distance traversed in one half of the duration of such a uniformly changing motion to the distance traversed in the other half of its duration. He proves that this ratio will be as three to one, that is three times greater distance will be traversed in half a given time when the velocity is 'more intense' compared to the distance traversed in half a given time when the velocity is 'less intense'. The 'less intense' motion occurs in that

44 Cf. for example A.C. Crombie, *Medieval and Early Modern Science*, 2: *Science in the Later Middle Ages and Early Modern Times: XIII–XVII Centuries*, New York 1959, 93–95; E.A. Moody, 'Laws of Motion in Medieval Physics,' in E.A. Moody. *Studies in Medieval Philosophy, Science and Logic. Collected Papers 1933–1969*, Berkeley, CA, 1975, 189–201, at 195.

45 Cf. William Heytesbury, *Regulae solvendi sophismata, De motu locali*, in E. Jung and R. Podkoński, *Towards the Modern Theory of Motion. Oxford Calculators and the New Interpretation of Aristotle*, Łódź 2020 (Research on Science and Natural Philosophy, 4) § 26, 276–277: 'Omnis enim latitudo sive a non gradu incipiat, sive a gradu aliquo, dum tamen ad gradum aliquem terminetur finitum, et uniformiter acquiratur seu deperdatur, correspondebit aequaliter gradui medio sui ipsius, sic scilicet quod mobile illud ipsam uniformiter acquirens vel deperdens in aliquo tempore dato aequalem omnino magnitudinem pertransibit, sicut si ipsum per aequale tempus moveretur medio gradu illius. Cuiuslibet etiam latitudinis incipientis a quiete et terminatae ad aliquem gradum est gradus suus medius subduplus ad gradum eandem latitudinem terminantem.' For Heytesbury's *De motu locali* (part of his *Regulae*), I will refer to the text included in the above-mentioned work. For other references to the *Regulae* I will use the edition Venice 1494.

half of a given period of time when a given body either begins to accelerate its motion from the state of rest or when it ceases to move finally.[46]

Richard Swineshead referred to both versions of William Heytesbury's mean-speed theorem in each of his texts on motion discussed in the present article (the *Opusculum de motu*, the *Opusculum de motu locali* and the treatise *De motu locali* from the *Liber calculationum*). The formulas corresponding with the first version of Heytesbury's theorem are found in the following passages:

> Tamen motus localis uniformiter difformis correspondet quoad effectum suo medio gradui, sic videlicet quod tantum per idem tempus possit pertransiri per illum gradum sicut per illum motum localem uniformiter difformem.[47]

> Unde universaliter omni uniformi adquisitioni motus a non gradu ad aliquem gradum finitum et omni uniformi remissioni motus ab aliquo gradu ad non gradum correspondet gradus subduplus ad ipsum gradum terminantem ipsam latitudinem motus sic uniformiter deperditam vel adquisitam, ita videlicet quod equalis magnitudo transiretur gradu subduplo ad gradum terminantem illam latitudinem in tanto tempore vel in tanto, quanto scilicet adquireretur uniformiter seu mediante latitudine tali.[48]

> Omnis latitudo motus uniformiter acquisita vel deperdita suo gradui medio correspondet. Hoc demonstratur primo sic: cum voco latitudinem,

46 William Heytesbury, *Regulae solvendi sophismata, De motu locali*, §§35–36, 281–282: 'Cum enim a non gradu ad aliquem gradum uniformis fiat alicuius motus intensio, subtriplum praecise pertransibit in prima medietate temporis ad illud quod pertransibit in secunda. Et si alias ab eodem gradu aut ab aliquo alio quocumque ad non gradum uniformis fiat remissio, triplum precise pertransiretur in prima medietate temporis ad illud quod pertransiretur in secunda. Totus enim motus factus in toto tempore correspondebit medio gradui suo, illi scilicet quem habebit in instanti medio illius temporis; et secunda medietas illius motus correspondebit gradui medio secundae medietatis eiusdem motus, qui est subquadruplus ad gradum illum terminantem illam latitudinem. Ideo, cum ipsa secunda medietas durabit solum per subduplum tempus, in quadruplo minus pertransiretur per illam secundam medietatem quam per totum motum. Igitur totius magnitudinis pertranseundae toto motu pertransirentur tres quartae per primam medietatem totius motus et ultima quarta pertransiretur per secundam medietatem eiusdem. Sequitur igitur quod in huiusmodi uniformi remissione vel intensione alicuius motus ab aliquo gradu usque ad non gradum, vel a non gradu in aliquem gradum, in triplo plus praecise pertransiretur per medietatem illius latitudinis intensiorem quam remissiorem.'

47 Richard Swineshead, *Opusculum de motu*, §72, 153.

48 Richard Swineshead, *Opusculum de motu locali*, §70, 177–178.

> ut acquiretur, correspondere gradui medio sic quod tantum pertransietur precise mediante illa latitudine sic acquisita sicut mediante gradu medio eiusdem, si per totum tempus illo gradu medio moveretur.[49]

The following passages echo the second version of Heytesbury's theorem:

> Ex istis patet universaliter, quando a non gradu ad certum gradum uniformiter intenditur motus vel ab aliquo gradu ad non gradum uniformiter remittitur motus, triplum pertransiretur precise per medietatem talis motus intensiorem quam remissiorem, quia talis motus correspondet gradui medio equaliter totius motus, et quelibet medietatis talis motus ⟨correspondet⟩ medio gradui talis medietatis. Et cuiuslibet latitudinis talis terminate ad non gradum medius gradus medietatis intensioris est triplus ad medium gradum medietatis remissioris, quia quilibet gradus in quacumque ⟨qualitate⟩ in tanto est alio intensior, quanto sub se continet maiorem latitudinem. Sed gradus medius medietatis intensioris triplam precise continet latitudinem sub se ad illam latitudinem quam continet gradus medius medietatis remissioris, et per consequens triplum precise pertransiretur ab illo gradu, et tantum precise illa medietate sicut illo gradu, igitur etc.[50]

> Et iuxta illud satis faciliter poteris elicere illam regulam prius dictam, quod universaliter in omni intensione motus uniformi a non gradu ad aliquem gradum et similiter in omni uniformi remissione motus ab aliquo gradu ad non gradum, triplum precise pertransiretur per medietatem illius intensiorem quam per medietatem remissiorem eiusdem latitudinis. Et utrumque istorum satis diffuse est alibi determinatum.[51]

> ⟨Ubi⟩ quecumque potentia a non gradu uniformiter intendit motum suum, in triplo plus pertransibit in secunda medietate temporis quam in prima ... Ergo, cum prima pars proportionalis temporis sit dupla ad secundam, patet quod in quadruplo plus pertransibit in prima medietate temporis quam in secunda, et in quadruplo plus in secunda quam in tertia, et sic in infinitum ... Sequitur ergo quod in toto tempore pertransibit in quadruplo plus quam in secunda medietate. Et per consequens in

49 Richard Swineshead, *Liber calculationum, Tractatus de motu locali*, Conclusio 38, Suppositio 2, §§82–83, 302.

50 Richard Swineshead, *Opusculum de motu*, §76, 153–154.

51 Richard Swineshead, *Opusculum de motu locali*, §72, 178.

> prima medietate in triplo plus pertransibit quam in secunda, ubi scilicet est uniformis remissio motus ad non gradum. Ergo econverso intendendo in triplo plus pertransiretur in secunda medietate temporis quam in prima.[52]

In both *opuscula* on motion ascribed to Richard Swineshead there are many more explicit examples of their dependence on William Heytesbury's *De motu locali*. For example, the two *opuscula* mention the 'procedure' for determining the uniform velocity of a given body 'with regard to the effect' (*penes effectum*) defined by Heytesbury,[53] and his reduction of a notion of 'retardation' (*tarditas*), that he seems to recognise as superfluous:

> *Tarditas enim est quasi privatio velocitatis, non obstante quod* de virtute sermonis *omnis tarditas sit velocitas, et econverso*, sicut quaelibet parvitas est magnitudo. (William Heytesbury, *Regulae solvendi sophismata: De motu locali*).[54]

> *Tarditas se habet per modum privationis respectu velocitatis, non obstante quod omnis velocitas sit tarditas et econtra*. (Richard Swineshead, *Opusculum de motu*).[55]

This dependence is further confirmed by more or less verbatim quotations from Heytesbury's text. For example, in the final section of the *Opusculum de motu locali* that discusses the problem of the velocity of alteration we find statements taken almost word for word from Heytesbury:

---

52 Richard Swineshead, *Liber calculationum, Tractatus de motu locali*, Tertia probatio suppositionis 2, § 88, 304–305.

53 Cf. William Heytesbury, *Regulae solvendi sophismata, De motu locali*, § 6, 271: '*In uniformi itaque penes lineam a puncto velocissime moto descriptam, si quis huiusmodi fuerit, quanta sit totius magnitudinis motae velocitas universaliter metietur*'; Richard Swineshead, *Opusculum de motu*, § 63, 151: 'Unde pro motu locali uniformi est illud sciendum, quod *velocitas in motu locali uniformi attenditur simpliciter penes lineam discriptam a puncto velocissime moto* in tanto tempore vel in tanto'; Richard Swineshead, *Opusculum de motu locali*, § 58, 175: 'Loquendo igitur *de motu locali penes quid attenditur eius velocitas, ⟨dicitur quod attenditur⟩ penes maximam lineam quam discriberet ⟨punctus⟩ velocissime motus* uniformiter contenta tali velocitate per tempus, *ubi videlicet est dare punctum velocissime motum*.'

54 William Heytesbury, *Regulae solvendi sophismata, De motu locali*, § 15, 272–273. I did not translate this and the following fragments from Latin into English in order to better show the dependence of Richard Swineshead's text on Heytesbury's handbook.

55 Richard Swineshead, *Opusculum de motu*, § 7, 137.

> *Universaliter omnis magnitudo mota localiter ita velociter accedit ad aliquem terminum fixum sicut ad eundem accedit aliqua pars eiusdem magnitudinis propinquissima eidem termino.* Non enim citius deveniet pars ad talem terminum vel ad talem quam faciet totum cathegorematice, cuius est illa pars. *Sed in alteratione si est ita quod non ita cito erit tota qualitas intendenda sub gradu ita intenso sicut erit aliqua eius pars, nec ita velociter calefit vel frigefit tota sicut aliqua eius pars.*[56]

> Si autem queratur ratio diversitatis, quare non omne quod alteratur, ⟨alteratur⟩ sicut aliqua sui pars, sicut omne quod movetur localiter ita velociter movetur sicut aliqua sui pars, ex quo universaliter ita velox est quilibet motus alterationis sicut aliqua sui pars—huic est dicendum, quod ratio diversitatis potest assignari, quia *universaliter quicquid movetur localiter accedit ad aliquem terminum, vel saltem accedet ad aliquem terminum imaginatum distantem in loco ita velociter per talem motum vel talem sicut aliqua sui pars ad eundem terminum accederet que in principio illius motus equaliter cum ipso moto toto distaret.* Sed sic non est respectu motus alterationis. Quando enim calefit aqua, non ita velociter calefit ipsa tota caliditate sicut aliqua sui pars propinqua ipsi agenti, *nec ita velociter per consequens accedit ipsa tota aqua ad hunc gradum caliditatis vel ad illum gradum sicut aliqua pars illius per alterationem eandem quantitatis,* enim nulla pars illius alterationis sit velocior quam ipsa tota alteratio.[57]

Much earlier in the same *Opusculum de motu locali* Swineshead employed an argument that was obviously copied from Heytesbury's *Regulae* too:

> Ad hoc quod aliqua duo puncta seu aliqua alia mobilia aeque velociter moveantur non requiritur quod in aequali tempore aequales pertranseant magnitudines, sed stat quod inaequales pertranseant in quacumque proportione volueris imaginari. *Posito enim quod* A *punctus continue uniformiter moveatur* C *gradu velocitatis per horam pertranseundo pedalem quantitatem, et* B *punctus a quiete incipiat uniformiter intendere motum suum in medietate illius horae acquirendo illum* C *gradum, et in secunda ⟨medietate⟩ uniformiter remittat ille ab eodem gradu ad quietem. Tunc notum est quod in medio instanti totius horae movebitur* B *punctus* C *gradu velocitatis et aeque velociter omnino cum ipso* A *puncto. Et tamen in instanti*

56 William Heytesbury, *Regule solvendi sophismata, De velocitate motus alterationis,* Venice 1494, 51vb–52ra.

57 Richard Swineshead, *Opusculum de motu locali,* §§ 97–99, 185–186.

medio illius horae non erit tanta linea in pertransiri per B *punctum sicut per* A, *ceteris paribus.*[58]

Non sequitur ⟨quod⟩ licet aliqua duo puncta equaliter moveantur pro nunc, quod equale sit spatium in pertransiri ab uno sicut a reliquo in equali parte temporis. Stat enim quod aliqua duo, scilicet A et B, iam moveantur eque velociter, et tamen in qualibet parte temporis plus pertransibit A quam in eadem parte pertransibit B. Sicut *posito, quod* ⟨A⟩ *movetur continue uniformiter* C *gradu velocitatis, et quod* B *a principio eiusdem temporis intendet uniformiter motum suum ab aliquo gradu remissiori ipso* C *gradu, et sic intenderet ita quod in instanti iam presenti movetur primo* C *gradu, et iam omnino incipiat remittere motum suum sicut prius intendebat. Tunc sequitur, quod* A B *puncta moverentur equaliter, quia eodem* C *gradu velocitatis omnino. Et tamen in omni parte temporis equali maius spatium est in pertransiri ab ipso* A *quam ab ipso* B, *quia continue ante hoc instans movebatur* B *tardius ipso* A, *et continue post hoc tardius movebitur ipso eodem.*[59]

Similarly, in the *Opusculum de motu* we find the following argumentation that clearly was inspired by one of the examples proposed by Heytesbury:

Item, posito quod A *et* B *motus uniformiter intendantur ab illo* C *gradu aut ab alio aliquo gradu, et* A *in duplo velocius ipso* B *continue, arguitur quod* A *motus aliquando erit duplus ad* B, quia quocumque gradu intensiori isto C signato, in duplo citius acquiret A gradum illum quam faciet B; igitur gradum quadruplum ad C in duplo citius acquiret A quam faciet B … *Sequitur quod* A *in primo casu numquam erit in duplo velocior seu intensior ipso* B, *quod erat probandum* … ⟨Sed⟩ *posito quod* A *et* B *motus incipiant uniformiter intendi uterque a quiete,* A *continue in duplo velocius* B *praecise … sequitur quod* A *continue erit motus in duplo praecise velocior ipso* B, quod prius fuit assumptum et probandum.[60]

Similiter *de intensione motus est sciendum quod, si ab eodem gradu incipiant intendi duo motus quorum unus continue alio in duplo velocius intenditur, numquam erit unus motus duplus ad alium, licet in infinitum sic intendentur* … Et ex hoc patet quod, *si a non gradu incipiant aliqua duo*

58 William Heytesbury, *Regulae solvendi sophismata, De motu locali*, § 18, 273–274.

59 Richard Swineshead, *Opusculum de motu locali*, §§ 60–61, 175–176.

60 William Heytesbury, *Regulae solvendi sophismata: De motu locali*, §§ 45–47, 286–287.

intendere motus suos in duplo velocius unum quam aliud continue, ceteris paribus, in duplo velocius movebitur unum quam aliud, quia illud quod tardius intenditur continue erit sub gradu medio totius latitudinis adquisite ab illo quod velocius intenditur.[61]

3 Differences between Richard Swineshead's Texts on Local Motion

On the basis of the above-presented fragments one might be tempted to conclude that both *opuscula* on local motion ascribed to Richard Swineshead are virtually identical. In the present section I focus on the mutual differences between these texts and also relative to the treatise *De motu locali* included in the *Liber calculationum*. This will allow me to determine the most plausible temporal order of the composition of these texts.

First, it is important to notice that only in the shorter *Opusculum de motu* we find the introductory explanations of the new rule of motion, where Swineshead presents a quite clear paraphrase of *conclusiones* 2 to 7 of the third part of Thomas Bradwardine's *Tractatus de proportionibus*.[62] In this part of

---

61 Richard Swineshead, *Opusculum de motu*, § 80, 156; § 82, 157.

62 Thomas Bradwardine, *Tractatus de proportionibus velocitatum in motibus*, ed. H.L. Crosby, Jr., *Thomas of Bradwardine. His* Tractatus de proportionibus. *Its Significance for the Development of Mathematical Physics*, Madison, WI, 1955 (University of Wisconsin Publications in Medieval Science, 2), 110: 'Secunda conclusio: Si potentiae moventis ad potentiam sui moti sit dupla proportio, potentia motiva geminata movebit idem motum praecise in duplo velocius … Tertia conclusio: Si potentiae moventis ad potentiam sui moti sit dupla proportio, eadem potentia movebit medietatem eiusdem moti velocitate praecise duplata … Quarta conclusio: Si potentiae moventis ad potentiam sui moti sit maior quam dupla proportio, potentia motiva geminata motus eiusdem duplam velocitatem nequaquam attinget … Quinta conclusio: Si potentiae moventis ad potentiam sui moti sit maior quam dupla proportio, eadem potentia movente medietatem eiusdem moti velocitas motus nullatenus fiet dupla … Sexta conclusio: Si potentiae moventis ad potentiam sui moti sit minor quam dupla proportio, dupla potentia movente idem motum, motus ultra duplum velocitatem excrescet … Septima conclusio: Si fuerit potentiae moventis ad potentiam sui moti minor quam dupla proportio eodem potentia movente medietatem eiusdem moti, motus ultra duplam velocitatem transibit.' Cf. Richard Swineshead, *Opusculum de motu*, § 12, 139: 'Et ex hoc sequitur quod universaliter ad duplationem motus requiritur duplatio proportionis, quia cum causa motus sit proportio, ad duplationem effectus requiritur duplatio cause.' § 14: 'Et causa est quia non universaliter ex duplatione potentie vel subduplatione resistentie sequitur duplatio proportionis, et non sequitur motus ⟨duplus⟩ nisi ex duplatione proportionis, igitur etc. Antecedens patet per multas conclusiones de proportionibus satis notas.' § 15: 'Ideo tenetur pro regula illud, quod universaliter ad hoc quod velocitas alicuius motus dupletur solum per intentionem potentie ⟨motive⟩ requiritur, quod proportionaliter tanto intendatur supra gradum suum, quanto ipsamet iam est

his *Tractatus* Bradwardine introduces in a systematic manner the new interpretation of Aristotle's 'laws' of local motion presented in the final chapter of Book VII of his *Physics*. Bradwardine's interpretation, based in general on Euclidean theory of proportionality, and in particular on the notion of 'geometrical proportionality', made the Aristotelian science of motion logically and mathematically consistent.[63] In the longer *Opusculum de motu locali*, Swineshead merely offers the following brief remark: 'Universaliter equaliter varietur motus sicut proportio, ceteris paribus.'[64]

The purpose of this latter statement was probably to assure readers that the author accepted the new rule of motion here. Similarly, Swineshead begins the treatise *De motu locali* in his *Liber calculationum* with the declaration that is later referred to in the text as the 'first assumption' (*suppositio prima*): 'Hic incipiunt quedam regule de motu locali supponendo motum attendi penes proportionem geometricam.'[65]

Presumably, when Richard Swineshead was composing the *Opusculum de motu locali* and the treatise *De motu locali* included in his *Liber calculationum*, he was convinced that the new theory of motion based on the notion of 'geometrical proportionality' was so well known among his contemporaries that there was no need to explain its details any more. Therefore, we may assume that the *Opusculum de motu* had been written before the *Opusculum de motu locali*, most probably when Richard Swineshead was a regent master at the Faculty of Arts, or even when he himself was a student there, and was still either acquainting himself or his students with the new rule of motion. The latter scenario, namely that the *Opusculum de motu* was composed by a student or a young master of Arts, could also provide an explanation for the philosophically unusual introduction to this *Opusculum*, where Swineshead defines motion in terms of the Aristotelian four causes:

---

maior quam sua resistentia. Si sit in duplo maior, sic oportet, quod dupletur; si in triplo, sic oportet quod tripletur; et sic in infinitum. Et etiam ⟨si sit⟩ in sesquialtero ⟨maior⟩, oportet quod sesquialteretur, et sic in infinitum.' § 16, 140: 'Et correspondenter de remissione resistentie oportet, quod proportionaliter tanto fiet minor quam iam est, quanto ipsamet est minor quam sua potentia. Sicut, si iam sit resistentia subdupla, oportet quod subdupletur; et si subtripla, oportet, quod subtripletur; et sic in infinitum.'

63 Cf. J.A. Weisheipl, 'The Interpretation of Aristotle's Physics and the Science of Motion,' in N. Kretzmann, A. Kenny, J. Pinborg and E. Stump (eds.), *The Cambridge History of Later Medieval Philosophy. From the Rediscovery of Aristotle to the Disintegration of Scholasticism, 1100–1600*, Cambridge 1982, 519–536, at 535–536.

64 Richard Swineshead, *Opusculum de motu locali*, § 9, 163.

65 Richard Swineshead, *Liber calculationum*, *Tractatus de motu locali*, § 1, 271. Cf. Murdoch and Sylla, 'Swineshead,' 201.

> Ideo motus cognoscere volenti ipsius causas principales expediens est investigari, que sunt: materialis, formalis, efficiens et finalis. Unde causa materialis motus, sive materia motus, quod idem est, est ipsum adquisitum per motum. Causa formalis est transmutatio quedam coniuncta cum tempore. Causa efficiens est proportio maioris inequalitatis potentie motive supra potentiam resistivam. Causa finalis est terminus intentus per motum.[66]

Even though we are able to identify the most probable sources of each of the above-presented individual statements, namely certain of Averroes' commentaries on Aristotle's *Physics*,[67] it is must be observed, that in general Swineshead's explanation is somewhat unorthodox. According to the standard reading of Aristotle, the four causes serve primarily to define substances, or—secondarily—qualities, not to define phenomena, processes, or passions, under any of which categories motion seemingly should be included. Interestingly, in the *Opusculum de motu locali* Swineshead adopted William of Ockham's concept of motion, defining motion simply as a change and continuous succession of places, quantitative parts, or degrees of quality, respectively.[68]

66 Richard Swineshead, *Opusculum de motu*, §§2–6, 137. Cf. Weisheipl, 'The Interpretation of Aristotle's Physics,' 536.

67 Averroes, *In Physicam* II, comm. 70, Venice 1562, repr. Frankfurt am Main 1962 (Aristotelis opera cum Averrois commentariis, 4), 73^{vb}M: 'Quoniam homo generatur ab homine ideo generans et generatum sunt unum secundum formam et forma generata est finis motus materiae ab agente'; V, comm. 6, 209^{vb}M: 'Omnis motus constituitur per quinque, scilicet motorem et motum et illud ex quo est et illud ad quod est et illud in quo est'; V, comm. 9, 215^{ra}B: 'Motus igitur habet duplicem considerationem, quoniam secundum suam materiam est in genere eius ad quod est motus; secundum autem formam, id est secundum quod est transmutatio coniuncta cum tempore, est in praedicamento passionis'; V, comm. 48, 237^{ra}B: 'Substantia motus est de natura eius ad quod est, non eius ex quo est, ideo motus denominatur ab illo ad quod est.' Cf. J.E. Murdoch and E.D. Sylla, 'The Science of Motion,' in D.C. Lindberg (ed.), *Science in the Middle Ages*, Chicago 1978 (Chicago History of Science and Medicine), 206–264, at 214.

68 Richard Swineshead, *Opusculum de motu locali*, §§56–57, 174–175: '*Motus enim proprie dictus est unus motus successivus et continuus transitus ab uno termino ad alium terminum*, sive una talis successiva augmentatio. Quia *alie mutationes subite non sunt proprie motus* sed improprie tantum, quia, sicut iam dictum est, *loquendo proprie de motu duo importantur per hoc nomen "motus", puta mutatio et continua successio*. Ideo, cum tales mutationes sicut inductio forme substantialis et illuminatio, generatio et corruptio substantie solum habeant unum illorum duorum, puta mutationem, et continuam successionem non, huiusmodi mutationes proprie non sunt motus.' Cf. William of Ockham, *Expositio in libros Physicorum Aristotelis*, III, cap. 2, §5, eds. R. Wood, R. Green, G. Gál, J. Giermek, F. Kelley, G. Leibold and G. Etzkorn, St. Bonaventure, NY, 1985 (Opera philosophica, 5), 421: 'Est autem primo sciendum quod *"motus" aliquando accipitur stricte secundum quod*

On the basis of the differences I pointed out here (i.e., the entirely different definitions of motion included in both *opuscula* and the presence of the short introduction to the new [mathematical] rule of motion in the *Opusculum de motu* and the lack of such explanations in the *Opusculum de motu locali*), we can assume that with regard to all known writings on motion by Richard Swineshead, there is an observable progress in his treatments of the science of local motion. Admittedly, there is a possibility that both *opuscula* are just students' notes from Swineshead's lectures, or even his own notes for this purpose. Still, the *Opusculum de motu locali* reveals the more methodologically and philosophically conscious attitude of its author than the earlier *Opusculum de motu*. Therefore, we face the rare situation here of being able to reconstruct the successive stages of the development of Richard Swineshead's science of local motion from the early *Opusculum de motu*, via the *Opusculum de motu locali*, to the treatise *On local motion* included in the *Liber calculationum*. The latter had been composed from the outset following a pattern imitating Euclid's *Elements*, a work that in the Middle Ages was considered to be the exemplary way of constructing a theoretical science. Even though Richard Swineshead did not completely reach his goal in the latter text, it still presents the most developed stage of the Oxford Calculators' theory of local motion in terms of the types of local motion covered by the rules or conclusions conceived on the basis of the new rule of motion.[69]

There is one more important feature that distinguishes Richard Swineshead's *opuscula* on motion from his treatise *De motu locali*. In the beginning of this article, I have pointed out that in both *opuscula* Swineshead first focused on the description of motion *penes causam* and later on he introduced the *penes effectum* explanations, most probably following the pattern of William Heytesbury's *De motu locali*. Surprisingly enough, there is no such differentiation in the treatise *De motu locali* included in the *Liber calculationum* since in the latter text consecutive conclusions are inferred only 'with regard to the cause'. Interestingly, there are two passages in Swineshead's *Opusculum de motu* that can serve as an explanation why in the treatise *De motu locali* he limited himself to the *penes causam* conclusions.

In the very beginning of his early *Opusculum de motu* Richard Swineshead, quoting Aristotle's *Metaphysics*, stated: 'Perfecte non cognoscitur res aliqua nisi

includit successionem, aliquando generaliter pro omni mutatione, sive sit subita sive successiva.' See also William of Ockham, *Quaestiones in libros Physicorum Aristotelis*, q. 22, ed. S. Brown, St. Bonaventure, NY, 1984 (Opera philosophica, 6), 452: 'Motus localis est spatium partibiliter et continue adquisitum mobili.'

69 Cf. Podkoński, *Suisetica inania*, 199–204.

per suas causas propinquas et remotas.'[70] In my opinion, this statement should be taken as a suggestion that for Richard Swineshead, as for almost all other medieval natural philosophers, the *penes causam* description is the right or more adequate one within the scope of the theoretical science of local motion in the strict Aristotelian sense.

What is more, in the same *Opusculum de motu*, Swineshead established the equivalence between the *penes causam* and *penes effectum* descriptions:

> Cuicumque gradui in motu locali correspondet certa distantia linearis, que in tanto tempore vel in tanto cum ⟨ceteris⟩ paribus tali gradu discriberetur. Consimiliter in toto, sicut certo gradui proportionis agentis ad passum correspondet certus gradus motus, sic certo gradui motus correspondet certa distantia, que in tanto tempore vel in tanto tali gradu pertransitur.[71]

Even though he did not allude explicitly to this equivalence in any other text, most probably when composing his treatise *De motu locali* Richard Swineshead kept in mind the above assumption and thus could unhesitatingly skip the considerations on motion *penes effectum* and limit his arguments and conclusions to the description *penes causam*, as perfectly equivalent ways of describing local motion. We must not forget here, with regard to the *penes effectum* description of motion, that for all philosophers before Descartes it was absurd and methodologically forbidden to determine a ratio of intensity or extension or degrees of two items belonging to different categories, as for example the well-known Newtonian ratio of a distance over time that let modern scientists easily calculate a mean velocity of any local motion. As a consequence, the *penes effectum* description of motion proposed by the Oxford Calculators was comparatively equally complicated as the *penes causam* description, but the latter kind of description was philosophically more well-founded, since it fits better in an Aristotelian framework.

4 The Question of the Description of Difformly Difform Motions

Richard Swineshead's predecessor (possibly his teacher) William Heytesbury stated authoritatively in the *Regulae solvendi sophismata* that it is impossible

70 Richard Swineshead, *Opusculum de motu*, § 1, 137.

71 Richard Swineshead, *Opusculum de motu*, § 65, 152.

to establish any rules covering difformly difform motions. Heytesbury's statement is quoted, almost verbatim, in Richard Swineshead's *Opusculum de motu locali*:

> Unde ista discriptio in motibus, puta quod motus ille sunt eque veloces quibus equalia spatia erunt pertransita in equali parte temporis, maxime verificetur de motibus uniformibus, quia in motibus difformibus stat quod infinitum maius spatium erit pertransitum per unum motum quam per alium, et tamen eque velox erit uterque sicut alter eorundem, quia, cum motuum difformium aliqui sint uniformiter difformes, aliqui difformiter difformes, et *ipsorum difformiter difformium nulla poterit esse regula, cui gradui uniformi correspondent* ...[72]

> De difformi autem intensione vel remissione, sive ab aliquo gradu usque ad non gradum, vel econverso, seu ab aliquo gradu usque ad aliquem alium, *nulla potest esse regula* quantum in tanto tempore vel in tanto pertransiretur, aut *cui gradui intrinseco illius latitudinis correspondebit talis latitudo motus difformiter deperdita seu acquisita*. Quia sicut infinitis modis contingit talem remissionem seu intensionem difformem variari, ita etiam infinitis gradibus intrinsecis eiusdem latitudinis, immo cuilibet gradui intrinseco illius latitudinis sic acquisitae vel deperditae poterit totus ille motus correspondere.[73]

Among the introductory explanations of the same *Opusculum de motu locali*, however, Richard Swineshead, referring to the difform changes of the causes of motion, observed that it is impossible to formulate any 'rules' covering the changes of this kind 'unless they concerned anything that is uniform.'[74] In his early *Opusculum de motu* Swineshead only superficially touched upon difform motions, stating that concerning any such motion it is possible that its mean degree or mean velocity corresponds to any of its intrinsic degrees.[75] It

72 Richard Swineshead, *Opusculum de motu locali*, § 63, 176.

73 William Heytesbury, *Regulae solvendi sophismata, De motu locali*, § 40, 284.

74 Richard Swineshead, *Opusculum de motu locali*, § 2, 162: 'Omnis variatio potentie motive seu resistive vel est uniformis vel difformis, et *de difformi nulla potest esse regula nisi quatenus refertur ad illud quod est uniforme*, quia infinitis modis contingit diversitas difformitatis.'

75 Richard Swineshead, *Opusculum de motu*, §§ 73–75, 153: 'Unde quiscumque motus difformis et etiam quecumque alia qualitas, alicui gradui ⟨uniformi⟩ correspondet. Et sic possit dici quod motus et quecumque alia qualitas est difformis quod cuicumque gradui intrinseco poterit correspondere. Unde, capto quocumque motu difformi, seu qualitate,

is worth noting here that six of the final *conclusiones* of the treatise *De motu locali* included in the *Liber calculationum* concern difform motions of such a kind that an acceleration (or deceleration) is uniform.[76]

The last of these *conclusiones* is reconstructed on the basis of a quotation found in another chapter of the *Liber calculationum*, namely in the chapter *De inductione gradus summi*.[77] Nevertheless, the *conclusiones* present in the preserved (manuscript or printed) copies clearly show that Swineshead managed to establish some specific rules concerning difformly difform motions, that are uniformly accelerated or decelerated. According to the rules that are transcribed in the preserved copies of this text, the middle degree (i.e., the mean velocity) of a motion that is characterised by uniformly increasing acceleration is greater than its velocity in the middle instant of a given time, but smaller than the arithmetically calculated middle velocity of this motion (i.e., an arithmetical mean between its initial and final velocity in a given time). The same description concerns the mean velocity of a motion characterised by uniformly decreasing deceleration. With regard to a motion characterised by uniformly increasing deceleration as well as with regard to a motion characterised by uniformly decreasing acceleration there remained (or were formulated) only rules determining the relation of the mean velocity of any such motion and the arithmetically calculated middle velocity.[78]

What is more, Richard Swineshead also formulated the following general *conclusio* (the 56th in order): 'Si aliqua latitudo motus consimiliter acquiritur sicut deperditur, eidem gradui ⟨medio⟩ correspondet.'[79] The context as well

---

⟨gradus⟩ ad quem terminatur in extremo intensiori est remissimus citra aliud extremum, cui non possit correspondere ⟨motus⟩ talis seu qualitas; etiam gradus ad quem terminatur in extremo remissiori, si ad aliquem gradum terminaretur, est intensissimus circa aliud extremum, cui non possit talis motus correspondere.'

76 Richard Swineshead, *Liber calculationum*, *Tractatus de motu locali*, Conclusio 52, § 154, 335: 'Omnis motus velocius et velocius deperditus quantum ad pertransitionem spatii gradui intensiori medio correspondet.' Conclusio 53, § 157, 338: 'Omnis motus tardius et tardius deperditus, gradui remissiori medio correspondet.' Conclusio 54, § 159, 339: 'Ubicumque velocius et velocius motus intenditur, correspondet idem gradui remissiori medio.' Conclusio 55, § 161, 339: 'Ubicumque motus tardius et tardius intenditur, correspondebit gradui intensiori medio.' Conclusio 57, § 166, 340: 'Omnis motus velocius et velocius acquisitus, gradui intensiori quam sit gradus quem habebit in instanti medio quo ad pertransitionem spatii correspondet.' Conclusio 58, § 168, 341: 'Omnis motus tardius et tardius deperditus, gradui intensiori quam sit gradus habitus in instanti medio correspondet.'

77 Cf. Podkoński, *Suisetica inania*, 120 (n. 317).

78 Cf. the quotations provided in note 76 above.

79 Richard Swineshead, *Liber calculationum*, *Tractatus de motu locali*, § 163, 339.

as the logical argumentation that follows this conclusion assure us that it is a kind of generalisation of William Heytesbury's mean-speed theorem concerning not only uniformly difform but also difformly difform motions. As it seems, all the presently described seven final *conclusiones* of the treatise *De motu locali* constitute the most important achievement of his science of local motion. It even looks as if Swineshead took Heytesbury's statement that it is impossible to establish any 'rules' concerning difformly difform motions as a kind of intellectual challenge, even though in his own *opuscula* he repeatedly quoted this statement. Still, we must not forget that in both these short treatises Swineshead first discussed the conclusions referring to motions the velocities of which are changing either 'faster and faster' (*velocius et velocius*) or 'more and more slowly' (*tardius et tardius*), therefore referring to motions that are difformly difform.[80]

Most probably, according to Swineshead, a motion the velocity of which changes 'faster and faster' is a motion that can be characterised by uniformly increasing acceleration or deceleration, and a motion that changes 'slower and slower' can be characterised by uniformly decreasing acceleration or deceleration. As we have seen, the final *conclusiones* of the treatise *De motu locali* concern precisely these kinds of difformly difform motions.[81]

Conclusion

With all the preserved texts on local motion ascribed to Richard Swineshead at hand we have a rare opportunity to reconstruct the consecutive stages of the development of his speculative science of local motion. In both his *opuscula* we find more or less explicit references to the texts of his predecessors: William of Ockham, Thomas Bradwardine, and most importantly William Heytesbury. On the other hand, both *opuscula* on local motion reveal so many obvious similarities with Richard Swineshead's *opus magnum* devoted to establish the complete set of 'rules' of local motion that the most plausible scenario is that he repeatedly tried to construct the ultimate version of the speculative science of local motion embracing all kinds of local motion that can be described with the *calculationes*, not only the uniform and uniformly difform ones described by Heytesbury. It must be said here that, as far as we know, no one else—at least at Oxford—managed to supplement or even tried to correct Swineshead's 'rules'

80 Cf. notes 18–23 and 24–29 above.
81 Cf. note 76 above.

of local motion.[82] What is more, some statements he included in these *opuscula* allow us to understand certain specific features of his treatise *On local motion*, most notably the lack of *penes effectum* descriptions. Therefore, these two *opuscula* on local motion ascribed to Richard Swineshead undoubtedly should be seen as a testimony of the gradual development of the Oxford Calculators' science of motion.

82 The only thinker that developed Swineshead's 'science of local motion' further was the sixteenth-century Portuguese, Alvarus Thomas, who studied at Paris, not at Oxford. See, Podkoński, *Suisetica inania*, 216–219, and E.D. Sylla, 'Alvarus Thomas and the Role of Logic and Calculations in Sixteenth-Century Natural Philosophy,' in S. Caroti (ed.), *Studies in Medieval Natural Philosophy*, Florence 1989 (Biblioteca di Nuncius, 1), 257–298. See also Edith Sylla's contribution to the present volume.

CHAPTER 4

Calculations in Thomas Bradwardine's *De causa Dei*, Book I

Edit Anna Lukács

Introduction

In 1328, Thomas Bradwardine, who may be considered the founder of the Oxford Calculators, in the treatise called *De proportionibus velocitatum in motibus* formulated an influential revision of Aristotle's mathematics of motion. According to this treatise, velocity varies as the ratio of force causing motion to resistance. This sounds like what Aristotle had said, but Bradwardine understood ratios to vary not simply as magnitudes but such that the ratios would vary geometrically as the velocities varied arithmetically. Soon after the publication of the treatise's modern edition in 1955,[1] scholars started to reflect on the origin of Bradwardine's rule. Marshall Clagett suggested that the English mathematician may have drawn his inspiration from a 'rule concerning the potencies of compound medicines, traceable to Al-Kindi.' This inspiration Clagett supposed was doubly external: it came from a different tradition, namely an Arabic source, and from a field different from the one to which it was applied. Clagett's thesis was supported by Michael McVaugh, who traced a similar mathematics in Montpellier medical theories. Soon thereafter, Stillman Drake argued, against McVaugh's evidence, that Bradwardine's theorem of proportions in motion comprised 'a piece of creative mathematics not in and not clearly implied by the medical rule.'[2] In a 1984 paper and in later work, Edith

1 For the manuscript transmission and early editions of the treatise, see Thomas Bradwardine, *Tractatus de proportionibus velocitatum in motibus*, ed. H.L. Crosby, Jr., *Thomas of Bradwardine. His* Tractatus de proportionibus. *Its Significance for the Development of Mathematical Physics*, Madison, WI, 1955 (University of Wisconsin Publications in Medieval Science, 2), 8–9, 59–61, and J.A. Weisheipl, 'Repertorium Mertonense,' *Medieval Studies* 31 (1969), 174–274, here 179–180. For Bradwardine's rule as formulated here, see Thomas Bradwardine, *De causa Dei contra Pelagium*, ed. and transl. (partial) E.A. Lukács, Göttingen 2013, 10.

2 M. Clagett, *Studies in Medieval Physics and Mathematics*, London 1979 (Collected Studies Series, 103), 284–285; M.R. McVaugh, 'Arnald of Villanova and Bradwardine's Law,' *Isis* 58 (1967), 56–64; E.D. Sylla, 'Medieval Concepts of the Latitude of Forms: The Oxford Calculators,' *Archives d'histoire doctrinale et littéraire du Moyen Âge* 40 (1973), 223–283, here 226–228;

 | DOI:10.1163/9789004512054_006

Sylla partially supported Drake in arguing that Bradwardine's theorem relied not upon an analogy to medical theories, but upon a mathematical understanding of compounding ratios that, in modern terms, understood the double or triple of a ratio to be the square or cube of that ratio, rather than that ratio multiplied by two or three.[3] Since these contributions by Clagett, McVaugh, Drake and Sylla, hypotheses about the origins of Bradwardine's theorem of the proportions of velocities in motions have not been further challenged, excepting those instances where priority is transferred to Richard Kilvington.[4] What has been established is that Bradwardine's understanding of compounding ratios was shared by others at Oxford some years before it appeared in Bradwardine's book in 1328, as is shown by its appearance in Richard Kilvington's questions on the works of Aristotle in the preceding ten years.

Bradwardine, however, was influential not only for his rule, but also for his more general programme of introducing mathematics into physics and other disciplines. He took a mathematical approach, for example, in his *De causa Dei* (1344) by writing in a sort of Euclidean axiomatic format, as well as by using quantification to solve problematic issues.[5] Bradwardine's use of mathematics in so doing is distinctive for his attempt to apply ratios and proportions where infinities are involved. He applies proportions, for instance, to questions concerning the eternity of the world, space outside the cosmos, and the remission of eternal punishments for mortal sins.

De causa Dei and especially Book I indeed abounds with applications of the mathematics of ratios or proportions as developed in the *Tractatus de proportionibus*. But according to the *Tractatus de proportionibus*, infinite values would

---

S. Drake, 'Medieval Ratio Theory vs. Compound Medicines in the Origins of Bradwardine's Rule,' *Isis* 64 (1973), 67–77, here 77.

3 E.D. Sylla, 'Compounding Ratios: Bradwardine, Oresme, and the First Edition of Newton's *Principia*,' in E. Mendelsohn (ed.), *Transformation and Tradition in the Sciences. Essays in Honour of I. Bernard Cohen*, Cambridge 1984, 11–43; E.D. Sylla, 'The Origin and Fate of Thomas Bradwardine's *De proportionibus velocitatum in motibus* in Relation to the History of Mathematics,' in W.R. Laird and S. Roux (eds.), *Mechanics and Natural Philosophy before the Scientific Revolution*, Dordrecht 2008 (Boston Studies in the Philosophy of Science, 254), 67–119.

4 See E. Jung, 'Works by Richard Kilvington,' *Archives d'histoire doctrinale et littéraire du Moyen Âge* 67 (2000), 207–214. See also Jung's contribution to the present volume. This thesis did not receive general assent: cf. S. Rommevaux, *Les nouvelles théories des rapports mathématiques du XIVᵉ au XVIᵉ siècle*, Turnhout 2014, 5, n. 5.

5 M. Sbrozi, 'Metodo matematico e pensiero teologico nel *De causa Dei*,' *Studi medievali* 31 (1990–1991), 143–191. I will be quoting *De causa Dei* in Savile's edition: Thomas Bradwardine, *De causa Dei contra Pelagium et de virtute causarum ad suos Mertonenses libri tres*, ed. H. Savile, London 1618. I have changed the punctuation and orthography in some places.

not have been admitted into ratios, whereas in *De causa Dei* they are common. Systematically collecting these elements, this chapter seeks to survey the application of mathematics to theology. It will do so while dividing into three unequal parts an inquiry into four texts: first, introducing some complementary, but hitherto unknown elements for the development of the calculations from *De causa Dei*, Book I, chapter I, corollary, part 32; secondly, investigating the concept of local motion in Book I, chapter 1, corollary, part 40, and Book I, chapter 5; and thirdly, presenting the theological use of proportional calculation in Book I, chapter 43.

1 The Mathematics of Knowledge of God

The corollary to Book I, chapter 1 provides evidence of calculatory mathematics. Part 32 of the corollary criticizes philosophers who have excessive confidence in their ability to know God. In the context of arguments against the demonstrability of all divine operations, Bradwardine introduces concepts that may be connected to the Calculators. In passing he uses the word 'latitude'.[6] His starting point is the definition of God as a rational potency endowed with free will, which allows God to do or not to do something. In a second step, Bradwardine rejects the principle that God's actions, understood as the best of all possible actions, must be circumscribed by his will:

> Therefore, God cannot make something … that is the best He can make. If He could [make something better], let this something be *a*, whether *a* is a god or not. If it is a god, then it is a newly made god, and there are several gods,[7] contrary to supposition I and corollary, part 17. If it is not

6 Further uses of latitude include: (1) *De causa Dei*, Book I, chapter 1, corollary, part 35, 74D, where Bradwardine quotes Ptolemy's *Almagest* on the principle that longitude must always exceed latitude; (2) *De causa Dei*, Book I, chapter 39, 338E, where latitude is briefly mentioned with regard to grace *de condigno* in beatific life; (3) *De causa Dei*, Book I, chapter 46, 433A, with a metaphorical mention of the heart's latitude, namely charity, from Augustine; (4) *De causa Dei*, Book I, chapter 51, 829D, where Bradwardine quotes Gregory the Great's *Moralities* on time and eternity. For Bradwardine's references to the mathematical latitude in the *De continuo*, see Sylla, 'Medieval Concepts,' 282; for the range of meanings the word had in the Middle Ages, see R.E. Latham et al. (eds.), *Dictionary of Medieval Latin from British Sources*, London 1986 and 1997, V, 1566, especially 7 ('latitudo').

7 The intended meaning in the first sentence is as translated (literally, the text says *melius scilicet optimum*). The translation omits to render 'melius, scilicet' (see below). The capitalization in the original sentence suggests that the reference is to 'many Gods', but the intended meaning is 'divinities'.

> a god, then it is less perfect and good than God, and not indivisibly less, but divisibly according to infinite latitude, as it is easy for philosophers to prove. Since, according to corollary, parts 4 and 7, God has infinite power and is plainly omnipotent, He can produce something better and better and better, without end.[8]

Here Bradwardine makes a casual use of the word 'latitude' to indicate something with extension as opposed to something that is indivisible. But the whole paragraph actually does something much more significant, in that it makes use of the distinction between categorematic and syncategorematic infinites. Bradwardine says that God cannot make a single most perfect thing (that would be a categorematic infinite), but he can make something greater than any given finite thing and he can do that again and again (syncategorematic infinite). This is the meaning of the last sentence: 'Since God ... has infinite power and is plainly omnipotent, He can produce something better and better and better, without end.'

Later in the same part of the corollary, Bradwardine combines philosophical with mathematical argumentation concerning the eclipse of the sun that may have occurred at the time of Jesus' death.[9] By way of explanation for why Calvary during the day suddenly turned dark, whereas the same phenomenon was not observed at the same time in Greece, one might explain that solar eclipses naturally are visible from limited positions only, unlike lunar eclipses which can be seen from any facing vantage point. The argumentation is mathematical and not only philosophical simply for being related to astronomical calculation.

---

8 Bradwardine, *De causa Dei*, 29B–C: 'Deus enim non potest facere melius, scilicet optimum quod potest facere. Si enim possit, ponatur quod et sit *A*: vel ergo *A* est Deus vel non. Si est Deus, est aliquis Deus factus et novus, et dii multi, contra primam suppositionem et 17. partem huius. Si non est Deus, est minus perfectum et bonum quam Deus; et non indivisibiliter, sed divisibiliter per latitudinem infinitam, sicut facile est philosopho demonstrare. Cum ergo per quartam et septimam partes huius, Deus sit potentiae infinitae et plene omnipotens, potest facere aliquid melius et melius, et melius sine fine.'

9 Thomas Bradwardine, *De causa Dei*, 38C: 'Posses quoque contra alia signa coelestia memorata superius similiter cavillare. Verum transcende, quaeso, modicum statum sophistae, et disce eclipsim Solis etiam universalem, nequaquam ubique et omnibus apparere, sicut eclipsis Lunae apparet, et hoc est propter diversitatem aspectus, sicut non sophistica, sed philosophica, scilicet mathematica theoremata clare monstrant, quare et posset eclipsis Solis in aliqua terra contingere, etsi ipsi in Graecia non viderent etiam claro die.' On this argument against pseudo-Dionysius' opponent Apollophanes the Sophist, see E.A. Lukács, '"Our Stories" and "Your Stories": Narrating the Passion and the Resurrection in Thomas Bradwardine's *De causa Dei*,' *Essays in Medieval Studies* 31 (2015), 183–200, here 187–188, 193.

In the same part, Bradwardine also deals with Christian miracles as such. He draws a distinction between two groups among those who recognize the reality of miracles: those who identify their true cause, and those others who ascribe miracles to an untrue cause. The representatives of the latter group vary, but Bradwardine first attacks the physicians (*medici*) who attribute the effective cause of the miracles brought about by Christ to his goodness and the optimal property of his complexion—that is, the medical balance of humours—tempered to his justice.[10] This approach, Bradwardine claims, cannot account for the miracles Christ effectuated at a distance. The healing processes doctors describe when talking about Christian miracles, in reality, function physically:

> The philosophers and doctors knew that the mover and the moved are next to each other in every real and spiritual action and motion; as the great philosopher, Aristotle, testifies in *Physics* 7,9 and *On the Soul* 2,74. That complexion[11] and its whole property was still of finite strength. This is the reason why cleaning the lepers, solidifying the paralysed, and healing other infirmities requires a certain time, a certain change and motion according to the proportion of the healing power related to the resistance in the infirmity of the person to be healed.[12] In the person in whom the weakness is more resistant and powerful, more time will necessarily be needed, as philosophers and doctors well know. There will necessarily be some middle state between the infirmity and the perfectly healed state that doctors are accustomed to calling the convalescence.[13]

---

10 The second group argues as follows (Thomas Bradwardine, *De causa Dei*, 44C–D): 'Alii autem videntes nullum corpus posse sufficere ad miracula facienda, ponunt spiritum causam horum; et hii ulterius variantur. Aliqui namque putant quod spiritus hominis, anima rationalis humana sit sufficiens causa horum, non autem quaelibet (quilibet *ed.*) anima quocunque modo disposita.'

11 *Complexio* is initially a medical term which stands for the balance of humours. Bradwardine describes it as an irrational power (Thomas Bradwardine, *De causa Dei*, 42B): 'Constat quoque complexionem esse potentiam irrationalem, quare et secundum philosophos ad alterum oppositorum tantum valere; potentiam vero factivam miraculorum ad utrumque, secundum quod exigent miracula facienda. Nonne etiam complexio est potentia irrationalis, quare et secundum philosophos non agit ex electione et libere, sed de necessitate naturae sicut ignis calefacit?'

12 Cf. Thomas Bradwardine, *Tractatus de proportionibus*, 113: 'Proportio velocitatum in motibus sequitur proportionem potentiarum moventium ad potentias resistivas, et etiam econtrario.'

13 Thomas Bradwardine, *De causa Dei*, 41D–E: 'Norunt autem philosophi et medici, quod in omni actione et motione reali et spirituali, movens et motum sunt simul; testis est ille magnus philosophus Aristoteles 7. *Physicorum* 9 et 2 *De anima* 74. Illa quoque complexio,

This section contains both the confirmation of a basic assumption of the Aristotelian physics of motion and a considerable step away from it through the reformulation of natural healing in terms of the calculation of proportions in motion.[14] Later in the same corollary, Bradwardine emphasizes that he is arguing 'against philosophers and doctors', according to experience, adding a taste of lived reality to his reflections.[15]

Finally, there remain some authors lurking behind the concepts: Avicenna's variable *complexio* as well as Galen's triple division of healthy bodies appear in the above-quoted section. While none of the medical works of these authors elsewhere credited with the origins of the concept of latitude and proportional calculation is quoted,[16] the application of proportional calculation to healing demonstrates that medicine interacted with mathematics.

2 Aristotelian Motion in *De causa Dei*, Book I

The last, fortieth part of the corollary to *De causa Dei*, Book I, chapter 1 could be considered a virtual manifesto of proportional calculation. Noteworthy here is Bradwardine's introduction of quotations from Alhazen's *De aspectibus*, Book II, wherein Alhazen enumerates different causes of beauty, Bradwardine choosing to quote only the sections concerned with place and number, and the characteristics that produce beauty. From this selection, proportion emerges as something that makes composite things better suited to order and beauty than any other possible characteristic: 'And when you examine the beautiful

---

et tota eius proprietas fuit tantum finitae virtutis, quare et mundando leprosum, solidando paralyticum, et caeteras infirmitates sanando, requireret certum tempus, certam alterationem et motum, secundum proportionem illius potentiae sanativae ad resistentiam infirmitatis sanandae; et ubi esset resistentior et maior infirmitas, ibi requireretur necessario maius tempus, sicut et philosophi et medici bene norunt; essetque necessario aliquod medium inter infirmitatem et sanitatem perfectam, quod medici solent convalescentiam appellare.' Note the distinction between real and spiritual motion.

14 S.J. Livesey, 'The Oxford Calculatores, Quantification of Qualities, and Aristotle's Prohibition of Metabasis,' *Vivarium* 24 (1986), 50–69, here 67.

15 Thomas Bradwardine, *De causa Dei*, 57D; see also 42E.

16 Sylla, 'Medieval Concepts,' 226–228. In corollary 32, Bradwardine quotes mainly al-Ghazali's *Physics* and Avicenna's *On the Soul* 4 and *Metaphysics* 10: Thomas Bradwardine, *De causa Dei*, 32B, 46C, 59A. On al-Ghazali's arguments being quotations from Avicenna, see E.D. Sylla, 'Thomas Bradwardine's *De continuo* and the Structure of Fourteenth-Century Learning,' in E.D. Sylla and M.R. McVaugh (eds.), *Texts and Contexts in Ancient and Medieval Science. Studies on the Occasion of John E. Murdoch's Seventieth Birthday*, Leiden 1997 (Brill's Studies in Intellectual History, 78), 148–186, here 181, n. 74.

forms of every kind of visible object, you will find that proportionality creates beauty more than any other characteristic on its own, or, for that matter, any conjunction [of characteristics] on its own.'[17] This initially Greek aesthetic principle of proportion as harmony receives repeated application in the mathematical arguments Bradwardine formulates against the eternity of the world in Book I, chapter 1, corollary, part 40. Since infinity is shown to be unproportional, and thus non-harmonious, the argumentation appears from the very outset to oppose a series of very similar paradoxes.[18]

Part 40 consists of a fictional dialogue between Bradwardine and the principal upholder of the thesis of an eternal world, Aristotle. In order to prove that time cannot be infinite—in terms of an infinite past rather than an infinite future[19]—Bradwardine presents a series of four arguments. Some of these involve a fictional or unknown opponent called *falsigraphus*:

---

17 Thomas Bradwardine, *De causa Dei*, 120E–121B, here 121A: '… et cum consideraveris formas pulchras de omnibus modis visibilium, invenies quod proportionalitas facit pulchritudinem magis quam aliqua alia intensio vel etiam aliquae aliae coniunctae per se.' Before quoting Alhazen, Bradwardine mentions animals, buildings, pictures, handwriting, and 'further examples' of heterogeneous things that demonstrate the principle of beauty through proportionality. Note the mixture of natural beings and artefacts in this enumeration: proportion is both in the physical world and constructed. On proportions in architecture, painting and related fields, see S. Rommevaux, P. Vendrix and V. Zara (eds.), *Proportions. Science, musique, peinture et architecture*. Actes du LIe Colloque International d'Études Humanistes, Turnhout 2011 (Études Renaissantes, 6). For Alhazen, see A.M. Smith, *Alhacen's Theory of Visual Perception. A Critical Edition, with English Translation and Commentary, of the first three Books of Alhacen's* De aspectibus, *the Medieval Latin Version of Ibn al-Haytham's Kitāb al-Manāẓir*, Philadelphia, PA, 2001 (Transactions of the American Philosophical Society, 91), 1: 206–207, 213; 2: 509 and 555 for the notes on Greek origins of the idea. The mathematical approach related to the aesthetical one is shown in the following (Thomas Bradwardine, *De causa Dei*, 120D): 'Boni autem maxime species et ordo, commensuratio et determinatum, quod maxime ostendunt mathematicae scientiae' (see also 254A). The quotation stems from Aristotle, *Metaphysics* XIII, 3, 1078a33–b2; it enjoyed some popularity in Bradwardine's time. Bradwardine also contrasts mathematical demonstration with principles known *per se* to philosophy of nature, metaphysics and ethics: Thomas Bradwardine, *De causa Dei*, 59B–C. Nonetheless, there are principles upon which mathematicians and natural philosophers agree, such as the totality exceeding its quantitative parts according to Euclid, *Elements* I. See Thomas Bradwardine, *De causa Dei*, 132D. On Euclid, see n. 41.

18 Sylla, 'Thomas Bradwardine's *De continuo*,' 183. For the paradoxes, see E. Jung and R. Podkoński, 'Richard Kilvington on Continuity,' in C. Grellard and A. Robert (eds.), *Atomism in Late Medieval Philosophy and Theology*, Leiden 2009 (Medieval and Early Modern Philosophy and Science, 9), 65–84, here 82.

19 Sylla, 'Thomas Bradwardine's *De continuo*,' 154: 'For infinity, among the most frequently repeated arguments are paradoxes of the infinite that are said to arise if past time is infinite.'

I. *per viam quantitatis discretae*—contra falsigraphum[20]

II. *per viam quantitatis continuae*

III. *per viam fortitudinis, potentiae seu vigoris harum innumerabilium animarum*

IV. *per viam actionis et passionis*

  IV.1. *per viam motus localis*

  IV.2. *per viam alterationis*

  IV.3. *per viam locationis*

  IV.4. *per viam solius intensionis simpliciter infinitae* [contra falsigraphum][21]

Compared to earlier arguments against the world's infinite temporal existence, this series seems like a showcase for a method of pairing elements of infinite sets. This is indeed what Bradwardine presents in a first section. Arguments I to III compare infinite sets, but argument IV questions whether there can be ratios between infinite and finite entities. While the first procedure produces disorder, and is therefore impossible, comparing the infinite with the finite produces a similarity, and is therefore also impossible.[22] Within each of the four arguments, Bradwardine recalls its enabling condition, which is 'by the omnipotence of God or by imagination,' or 'really by the omnipotence of God or fictionally by imagination.'[23] But this seemingly general feature is not considered in each and every case. Before turning to the exceptions, I

20 R. Podkoński, 'Thomas Bradwardine's Critique of the *Falsigraphus*'s Concept of Actual Infinity,' *Studia Antyczne i Mediewistyczne* 36 (2003), 147–153, has identified Bradwardine's *falsigraphus* as Richard Kilvington. According to Latham, et al. (eds.), *Dictionary*, III, 901, 'falsigraphus' denotes in geometry a 'drawer of false diagrams', or is simply a writer of falsehoods. Roger Bacon's use of words (quoted in the *Dictionary*) most closely resembles Bradwardine's in one place. See Roger Bacon, *Liber secundus Communium naturalium Fratris Rogeri De celestibus*, ed. R. Steele, Oxford 1913 (Opera hactenus inedita, 4), 351: 'Et quod modo sine figuracione demonstratum est, necesse est quod appareat in figura propter *cavillacionem* quam *falsigraphus* induceret.' Cf. Thomas Bradwardine, *De causa Dei*, 123C: 'Et si *falsigraphus* forsitan de maiori voluerit *cavillare*, dicatur ...' For more on *falsigraphus*, see section 3.

21 Thomas Bradwardine, *De causa Dei*, 133B: 'Nec potest quis dicere *D* non esse medietatem *C*; est enim pars *C*: constat, et falsigraphus confitetur;' 134D: 'Praeter haec autem ista falsigraphia, etsi impossibili isto dato, aliquas rationes praemissas posset utcunque desinire; ad multas tamen nec vere nec apparenter sufficit respondere, sicut considerare volentibus facile est videre.'

22 Podkoński, 'Thomas Bradwardine's Critique,' 145–146; Sylla, 'Thomas Bradwardine's *De continuo*,' 181–185. Henry of Harclay argued against the future eternity of the world from the proportionality of finites and infinites. See Sylla, 'Thomas Bradwardine's *De continuo*,' 166–177.

23 Sylla, 'Thomas Bradwardine's *De continuo*,' 183, and E.D. Sylla, 'Infinity and Continuity:

shall devote some remarks to the argumentation with divine omnipotence and *secundum imaginationem*. First of all, these two methods explicitly distinguished in Bradwardine's text, are to be considered analogous.[24] They appear together as providing alternative access to the consideration of hypothetical possibilities: either God's omnipotence or the imagination, either reality or fiction, allowing Bradwardine to ponder the world's eternity which he will reject. At the same time, the two methods appear to be bestowed with different, if not opposite qualities (in reality—fictionally).

Further to these arguments against the Aristotelian temporal infinity of the world, we find in Book I, chapter 5 of *De causa Dei* the same division into divine power and human imagination, which Bradwardine uses against the Aristotelian spatial finitude of the cosmos and the Aristotelian refusal of an extramundane void. The analogical structure of the argumentation in the two chapters draws on the deeply anti-Aristotelian conclusions that follow: the world's creation in time, and the existence of an extramundane void. Both chapters discuss theological contexts which challenge two basic assumptions of the Aristotelian cosmos, whereas in the latter chapter—Book I, chapter 5—this orientation is even more articulate.[25] The case of the extramundane void is further paralleled in the *Tractatus de proportionibus*, where Bradwardine also evokes the division between the real and the imaginary, albeit by carefully avoiding a concept present in *De causa Dei*, namely God's power.[26] Hence, neither the void nor any possible entities, only infinites are connected to the divine.

Thomas Bradwardine and His Contemporaries,' in S. Shapiro and G. Hellman (eds.), *The History of Continua: Philosophical and Mathematical Perspectives*, Oxford 2021, 49–81.

24 E. Jung, 'Mathematics and the *secundum imaginationem* Procedure in Richard Kilvington,' *Przegląd Tomistyczny* 22 (2016), 109–120, here 118–119.

25 Having left the framework of the first chapter's anti-philosophical discussions, Bradwardine transfers the Earth's rectilinear motion, a hypothetical possibility according to the Parisian condemnation of 1277, to a theological reality. On *De causa Dei*, Book I, chapter 5, see E. Grant, *Much Ado About Nothing: Theories of Space and Vacuum from the Middle Ages to the Scientific Revolution*, Cambridge 1981, 135–144. E.A. Lukács, 'Bradwardine and Buckingham on the Extramundane Void,' *Bochumer Philosophisches Jahrbuch für Antike und Mittelalter* 17 (2014), 123–149, here 131–132. The real-imaginary division appears a third time in *De causa Dei*'s argumentation concerning God's infinite knowledge, framed again by the Aristotelian eternity of the world: Thomas Bradwardine, *De causa Dei*, 184D–E.

26 Thomas Bradwardine, *Tractatus de proportionibus*, 130. Rommevaux notices that here Bradwardine is possibly alluding to motion in the void: Thomas Bradwardine, *Traité des rapports entre les rapidités dans les mouvements*, transl. S. Rommevaux, Paris 2010 (Sagesses médiévales), XLIX; Lukács, 'Bradwardine and Buckingham,' 129–136.

Picking up on the idea of exception to the imaginary argumentation, the fourth and last argument of the above-mentioned series contains a number of sub-arguments involving calculation of proportions applied to (1) local motion, (2) alteration, (3) location, (4) infinite intension,[27] and finally (5) God's omnipotence and highest eminence. In this sub-set, all aspects of motion in the Aristotelian sense appear condensed: local motion, alteration, and augmentation. Significantly, it is not until he engages with the fourth sub-argument that Bradwardine evokes God's omnipotence or even fictional argumentation. He therefore makes a different, wider use of divine power or human imagination than does, for instance, his fellow Mertonian, Richard Kilvington, as described by Elżbieta Jung. According to Jung, Kilvington distinguishes between four levels of cases: on the first level we find real cases, observable in nature; on the second, not observable, but naturally possible cases (the rectilinear motion of the Earth); on the third, not observable, but theoretically possible cases (infinite speed in an instant); and finally, only theoretically possible cases (motion in the void).[28] In Kilvington's fourfold vision, there are two natural and two theoretically possible cases. They are constructed as the progressive separation of nature and of its observer. If we apply Bradwardine's analysis to Kilvington's scheme, imagination would function on all three levels from the second to the fourth, equating imagination with the notion of the 'not observable'. Thus, in Bradwardine's approach, even only theoretically possible cases become imaginary ones.[29]

As already stated, sub-arguments IV.1–4 follow from the incommensurability of finite and infinite qualities with various intensities. For instance, an infinitely heavy body is supposed to move with local motion; infinitely warm fire is approached by its opposite or seeks its natural place; intensively infinite forms and the infinitely numerous gods seek existence. The first case is worth examining at length, since it completes Bradwardine's previous discussion of

27 Intension is also used in Book III, chapter 30, in an argument about divine volition; the demonstration is mathematical (*mathematice*). See Thomas Bradwardine, *De causa Dei*, 752C–E.

28 Jung, 'Mathematics and the *secundum imaginationem*,' 117.

29 Compare to this Nicole Oresme's very different use of divine power and imagination that still allows him to arrive at the same tight connection between mathematics and natural philosophy as we find in Bradwardine: H. Hugonnard-Roche, 'Analyse sémantique et analyse *secundum imaginationem* dans la physique parisienne au XIVe siècle,' in S. Caroti (ed.), *Studies in Medieval Natural Philosophy*, Florence 1989 (Biblioteca di Nuncius, 1), 133–153, and J. Celeyrette, 'L'argumentation mathématique dans la physique d'Oresme,' in C. Grellard (ed.), *Méthodes et statut des sciences à la fin du Moyen-Âge*, Villeneuve d'Ascq 2004 (Histoire des sciences), 201–215.

proportions in several ways. The case concerns the local motion of a body with a finite quantity, but an infinite quality ('simply intensive infinite weight'):[30] let body *a*, bestowed with an infinite weight, be posited on a high level, from which it descends towards its natural place, the centre. Its motion will be either instantaneous or successive; and in both cases the body will experience resistance from the medium it traverses.[31] Yet both motions prove impossible, argues Bradwardine. For the sake of arguing against instantaneous motion, instantaneous motion is presented as impossible. Successive motion was already prohibited from Avempace's theory of motion in the *Tractatus de proportionibus*, where his name was not mentioned: the Arabic philosopher's calculation of velocity as the excess of the mover's power over the resistance of the thing moved would induce body *a* to move with infinite speed.[32] The case of an infinitely heavy body proves once more that Avempace is not right in establishing an arithmetic proportion from Aristotle's calculation of motion. In *De causa Dei*, Bradwardine deals with the sources of the controversy over the Aristotelian view of motion with more clarity than he did in the *Tractatus*.

---

30 Typically, quantities remain finite, but for the argument's sake, qualities become infinite: J. Biard and J. Celeyrette (eds.), *De la théologie aux mathématiques. L'infini au XIVe siècle*, Paris 2005 (Sagesses médiévales), 'Introduction,' 15.

31 Thomas Bradwardine, *De causa Dei*, 129E–130A: 'Et primo per viam motus localis hoc modo: Sit *A* unum corpus grave finitae quantitatis simpliciter circumquaque, et infinitae gravitatis simpliciter intensive. Quod si ponatur sursum, in medio descendet, cum sit extra locum suum naturalem minime impeditum; sed vel descendet subito vel successive. Non subito, cum habeat resistentiam aliqualem, et quia tunc in omni medio et cum omni resistentia maiori et minori finita, sic faceret. Omne insuper tale corpus maius et minus in eodem medio, et cum eadem resistentia descenderet aeque cito. Quomodo insuper corpus naturale et rude per potentiam naturalem transibit ab extremo ad extremum medio praetermisso, exibitque locum clausum fortissimum circumquaque sine ruptura, apertione, coextensione, aut compenetratione quacunque? Nec successive, quia quacunque velocitate descenderet, tanta descenderet aliquod corpus gravitatis finitae simpliciter omnimodo, aliquodque maiori. Nec potest Avempace contra sententiam Aristotelis respondere dicendo corpora gravia et levia habere ex se inclinationem determinatam et tantum finitam, ad velocitatem determinatam atque finitam. Licet enim de elementis simplicibus secundum dispositionem quam nunc habent, posset sic dici utcunque, quia habent formas finitae intensionis tantummodo; de corpore tamen infinitae gravitatis simpliciter intensive sic dici non potest. Quanto namque secundum istam responsionem habet formam et gravitatem intensiorem, tanto et appetitum et inclinationem ad motum, et velocitatem maiorem, tantoque velocius naturaliter moveretur.' Except for light, everything was supposed to have resistance and thus move with successive motion: E.D. Sylla, 'Medieval Quantifications of Qualities: The "Merton School",' *Archive for History of Exact Sciences* 8 (1971–1972), 9–39, here 13, n. 13.

32 Thomas Bradwardine, *Tractatus de proportionibus*, 104–111; Cf. the introduction by S. Rommevaux in Thomas Bradwardine, *Traité des rapports*, X, XXXVII.

A further case study on the incommensurability of an infinite and a finite entity is offered in argument IV.5, which compares forms. Framed by imagination or God's omnipotence, the argument claims that subtracting finite parts from the infinite leaves the remainder infinite, and adding two finite parts together does not produce an infinite. Conversely, adding finite parts to an infinite form increases it only finitely. With these statements, Aristotle's concept of an infinite time both *a parte ante* and *a parte post* is jeopardized: since forms are finite, time is finite too. The section's last sentences summarize all the possibilities reviewed so far in the sub-sets of argument IV. In focusing on heat, Bradwardine writes:

> It is ended further up intrinsically at some finite degree, and further down extrinsically at nonbeing of heat, just as any form that can be intensified is limited. It also has an absolutely certain and finite proportion to every other absolutely finite heat in intension, just as in the velocity of motion and action, just as such weight would have to every other weight in intension and in the velocity of local motion. Wherefore, as such, even if only a little diminished, warmth is altogether and absolutely finite, as the mathematical theorems of proportions make obvious.[33]

The closing sentence is a condensed statement. It repeats that the mathematics of proportions prove that the quantification of qualities is necessarily finite: it is according to the theorem of proportions that the absolute finitude of warmth must be conceded. It also contains an important indication on the status of the calculation with proportions: in this sentence, as nowhere else in *De causa Dei* or any other scientific writing by Bradwardine, the word 'theorem' is appended to the approach of the physical world and the quantification of qualities.[34]

---

33 Thomas Bradwardine, *De causa Dei*, 130C–D: 'Finitur enim superius intrinsece ad certum gradum finitum, et inferius extrinsece ad non esse caliditatis, sicut et quaelibet forma intensibilis terminatur. Habet quoque proportionem certam et finitam simpliciter omniquaque ad omnem aliam caliditatem finitam simpliciter in intensione, sicut et in velocitate motionis et actionis, sicut et gravitas talis haberet ad omnem aliam gravitatem in intensione et in velocitate motus localis. Quare et omnis talis, quantumlibet parum remissa caliditas, omniquaque et simpliciter est finita, sicut mathematica theoremata de proportionibus manifestant.'

34 Cf. E.D. Sylla, Review of S. Rommevaux, *Thomas Bradwardine: Traité des rapports entre les rapidités dans les mouvements*, *Aestimatio* 8 (2011), 23–29, here 25. Bradwardine did not create a law in a medieval, modern, or any other sense of the word; he and his fellow Calculators only invented a new mathematical calculation of motion. Not insignificantly, Bradwardine notes in *De causa Dei* Book II, chapter 18 that every science handles action and motion, except for arithmetic and geometry, which also prove to have some parts con-

Unambiguously, it is not conceived as reflecting any repetition in nature but merely as an implication of the mathematical abstraction applied to it.

3 The Mathematics of Justification

Theories of proportions receive an unexpected application in a later chapter, in which Bradwardine proves once again the propensity of mathematics to generate arguments, and this time, not against philosophers, but against his fellow theologians: he opposes the Pelagian doctrine of justification in handling proportions of varying penances and punishments. Book I, chapter 43 opens a debate in which Bradwardine intends to contradict the theologians who posit 'natural' penance first, and supernatural grace and justification second and/or third. To these, Bradwardine opposes the primacy of supernatural grace, after which the sinner is led to penance and justification, as the Anti-Pelagian relationship between God and man. While his case against the Pelagians could be a banal transposition of the theory of proportions to spiritual change and motion, it is most interesting for its repetition of arguments modelled on those in Book I, chapter 1, corollary, part 40.

The case comes unexpectedly. After a long section of authoritative quotations mainly from the Bible and Augustine, Bradwardine suddenly introduces the following mathematical contradiction he considers inherent in the Pelagian doctrine of justification:

> If someone, as they say, deserves the absolution of his mortal fault through contrition, he similarly deserves the absolution of his eternal punishment for that fault, either entirely, or according to one part of the infinite duration, since at least the transference of that eternal punishment into a temporal punishment is deserved. If he deserves it in one instant or a brief moment for his penance over one fault, then, if similarly exercising penance for another fault in another instant or time, he would deserve complete remission of his temporal punishment, since any temporal punishment would be incomparably less than the eternal.[35]

---

nected to motion, as shown in Euclid, *Elements* XI and Jordanus de Nemore, *De triangulis* (*Liber phylotegni de triangulis*): Thomas Bradwardine, *De causa Dei*, 259B–C. For another chapter on the sciences, see the corollary to Book II, chapter 3, 467–472, esp. 468B–C.

35 Thomas Bradwardine, *De causa Dei*, 380C–D: 'Item si quis, ut dicunt, per contritionem meretur remissionem peccatis mortalis, meretur similiter remissionem poenae aeternae debitae pro peccato, vel secundum totum, vel secundum partem eius infinitam duratione,

Later in the section, Bradwardine succinctly redefines the problem again: 'It also follows against them that contrition, how little soever, would erase every mortal fault plainly as a punishment.'[36] In Bradwardine's eyes, the contradiction is primarily inherent in the incommensurability of finites (contrition) and infinites (eternal punishment). His answer is composed of at least ten different points, all relying on the mathematical approach through proportions to the given mathematico-theological contradiction. I will point only to three of them: first, to Bradwardine's thesis and its quantification of punishment through fire as a case study of a quantified quality; secondly, to the argument with multiple proportions; and finally, I will present a hitherto unacknowledged allusion to the already mentioned *falsigraphus*.

Bradwardine divides penance into two types: the first purges original sin in baptism; the second compensates for sins committed. The second is further subdivided into interior and exterior penance. Interior penance is the mind's virtue (*virtus mentis*), while exterior penance is the sacrament of confession. Both are necessary constitutive elements of the sinner's justification and salvation, but only one, the interior penance, constitutes a quantifiable quality. Departing from the case of two sinners, Peter and Paul, Bradwardine argues against the Pelagian position:

> Let Peter have fault *a*, to which the eternal punishment *b* belongs, and that by contrition *c* he merits the remission of *a* as far as punishment and as far as *d* compensation, which is part of the infinite punishment *b*, and not as far as the whole *b*. Let Paul also have two or more similar faults, and similarly have contrition concerning his multiple faults, like Peter for his only one fault. Then Paul merits via his contrition the remission of one infinite part of an eternal compensation, which is equal to the weak *b*, for one of his faults, and another infinite part of the other eternal punishment, equal to the light *b*, for his other fault. These two parts, joined together at the same time, exceed, even infinitely exceed, the whole of *b*. If then Paul merits via his contrition such a remission, and Peter, via his own contri-

quia saltem meretur translationem illius poenae aeternae in poenam aliquam temporalem; et si hoc meretur in uno instanti vel brevi momento, una vice scilicet poenitendo; si alio instanti vel tempore, scilicet alia vice simili modo poeniteat, merebitur plenam remissionem poenae cuiuslibet temporalis, cum poena quaelibet temporalis sit incomparabiliter minor aeterna.' For a condensed presentation of this chapter, see Thomas Bradwardine, *De causa Dei* (ed. and transl. Lukács), 27.

36 Thomas Bradwardine, *De causa Dei*, 382A [actually 381A]: 'Amplius autem et sequitur contra istos quod quantulacunque contritio deleat omne peccatum mortale plenarie quoad poenam.'

> tion, merits an equal remission, the latter would merit the remission of the whole of *b* much more rationally and honourably, especially because Peter is more innocent and disposed to merit.[37]

This section makes use of another, widely accepted mathematical truth, namely that the parts of an infinite cannot exceed the totality of the infinite.[38] Bradwardine supports his argument with one 'evident truth' (*veritas nota*): 'Of each of two uniform compensations of the same species, which have a beginning and no end and are unequal in intension and remission, the more intense, however less extended or briefer it should be, will simply be bigger.'[39] This evident truth, a mathematical proposition, corresponds to a rational proportion between unequal quantities in accordance with Bradwardine's former definition in the *Tractatus de proportionibus*.[40]

Bradwardine proves his position as follows:

> Let *k* be a uniformly intense fire punishment which starts in *l* future instant, and will last interminably; and *m* another, similar, uniformly less intense [fire punishment], which starts in present instant *n*, and will last eternally. Therefore, *k* is not equal to *m*; nor smaller; therefore greater. If

37 Thomas Bradwardine, *De causa Dei*, 382A [actually 381A]: 'Ponatur enim Petrum habere *A* peccatum, cui debeatur *B* poena aeterna, et per *C* contritionem mereri remissionem *A* quoad culpam et quoad *D* poenam, partem *B* infinitam, et non quoad totum *B*. Ponaturque Paulum habere duo similia peccata vel plura, et similiter conteri pro suis multis, sicut Petrum pro suo uno; tunc Paulus per suam contritionem meretur sibi remitti unam partem infinitam unius poenae aeternae aequalis *B* debitae pro uno suo peccato, et aliam partem infinitam alterius poenae aeternae aequalis *B* debitae pro alio suo peccato, quae duae partes simul iunctae excedunt, imo et quodammodo infinite excedunt totum *B*. Si ergo Paulus per suam contritionem tantam remissionem sibi meretur, et Petrus per suam aequalem, multo rationabilius et dignius meretur sibi remissionem totius *B* longe minorem, maxime cum Petrus sit innocentior et dispositior ad merendum.'

38 See Podkoński, 'Thomas Bradwardine's Critique,' 148 (for its acceptance by Richard Kilvington) and 152 (for its uses by Kilvington and Bradwardine).

39 Thomas Bradwardine, *De causa Dei*, 382B: 'Omnium duarum poenarum uniformium et similium specie habentium principium et non finem, inaequalium in intensione et duratione, intensior quantumlibet quantumcunque duratione sit minor seu brevior, est simpliciter tamen maior.'

40 Thomas Bradwardine, *Tractatus de proportionibus*, 97: 'Proportio autem quae proprie est accepta, in solis quantitatibus reperitur. Quae definitur hoc modo: *Proportio est duarum quantitatum eiusdem generis unius ad alteram habitudo*. Et haec est duplex; nam rationalis, et in primo gradu proportionalitatis, est illa quae immediate denominatur ab aliquo numero: sicut proportio dupla, et tripla, et sic de aliis.'

> the adversary [a Pelagian] were to say that *k* is equal to *m*, let *o* be a part of *m*, which starts in *l* and is then to last eternally; and *p* its remaining part. Let *q* be the intensive, imaginary or true, part of *k*, as some assume, equal to *o*, and let *r* be the remaining part of *k*. Since therefore *k* and *m* are equal, if equal parts are removed from it, namely *o* and *q*, then the remaining parts are equal, as appears evident to everyone, and Euclid in *Elements* Book 1 posits it as manifest. Hence *p* and *r* are equal, which no one ignores is false, since *p* is finite in every respect, while *r* is infinite in duration and entirely uniform, as it follows from the premises, which are false, as could be easily demonstrated.[41]

The mathematical procedure Bradwardine uses in this example merely reproduces the one he introduced at the beginning of the section on the case of Peter and Paul.

In an example which makes use of multiple proportions applied to continuous quantities, Bradwardine alludes once again to a *falsigraphus*:

> If the deceptive writer does not cease to be bold in saying that *p* does not exceed *r* in any multiple proportion, but in some other, let us extend *p* or diminish *r*, until there is a multiple proportion between them, etc. as before. If we say that in the first ones, *k* is smaller than *m*, and *m* is bigger than *k*, we extend *k* or diminish *m* uniformly, unless they are equal punishments in adding the quantity of the prior duration, etc., as before. There

---

41 Thomas Bradwardine, *De causa Dei*, 382B–C: 'Sit *K* una poena ignis uniformis intensa incipiens in *L* instanti futuro interminabiliter duratura; et *M* alia poena similis uniformis remissior, incipiens in *N* instanti praesenti aternaliter permansura. *K* igitur non est aequalis *M*, nec minor, ergo maior. Si namque adversarius dixerit *K* esse aequalem *M*, sit *O* pars *M*, incipiens in *L* et ex tunc aeternaliter duratura, et *P* reliqua pars illius, *Q* vero pars intensiva *K* imaginaria vel vera, ut aliqui opinantur, aequalis *O* et *R* residua pars *K*. Cum ergo *K* et *M* sint aequales, si ab eis demantur aequalia, puta *O* et *Q*, partes residuae sunt aequales, sicut cuilibet per se patet, et Euclides 1 *Elementorum* suorum hoc pro manifesto supponit. *P* ergo et *R* sunt aequales, quod esse falsissimum nullus nescit, cum *P* sit omni modo finita, *R* vero infinita secundum durationem et uniformis per totum, sicut lucide sequitur ex praemissis, quod et potest ostendi faciliter esse falsum.' Euclid is referred to on many other occasions: 126B–127E, 129B, 130B, 132D, 133, 351B, 529C, 610A; indeed, Book V, central for the theory of proportions, is omnipresent. On Book V as distorted in the Latin translation and its impact on the theorems of proportions, see S. Rommevaux, *Les nouvelles théories des rapports mathématiques du XIVe au XVIe siècle*, Turnhout 2014 (De diversis artibus, 92), 8, especially n. 14. Bradwardine also mentions the commensurability of the square's diameters and sides: Thomas Bradwardine, *De causa Dei*, 187A and 210B. For the relevance of this mathematical puzzle for the theorems of proportions, see Rommevaux, *Les nouvelles théories*, 69–74.

> remains no doubt that by sole intensification *k* could be a simply greater punishment than *m*, and that *m* by sole remission would be smaller than *k*.[42]

In the arguments under review, Bradwardine proposed the prevention of the problem of having infinite punishments redeemed by finite entities such as a short amount of pain or of time. His goal was to emphasize grace as the only element capable of preventing eternal punishment and leading to penance, justification, and finally salvation.[43] In doing so, however, the punishments for mortal sins, which are infinite, end up unbalancing the calculation of punishments for venial sins, the severity of which is multiplied by time, not infinity. Infinity, the key element, is in reality the one that weakens the theological argument and divine justice as just. Broadly considered, hardly any Pelagian, whether modern or older, could refute such arguments without appropriate knowledge of calculation: with this chapter in *De causa Dei*, its author underlined the validity of mathematical argumentation in theological works.

42 Thomas Bradwardine, *De causa Dei*, 382A: 'Si autem falsigraphus non desinat protervire dicendo *P* non excedere *R* in aliqua proportione multiplici, sed alia qualicunque; intendatur *P* vel remittatur *R*, donec fiat proportio multiplex inter illas, etc. sicut supra. Si vero dicatur in primis *K* esse minorem *M*, et *M* maiorem *K*, intendatur *K* vel remittatur *M* uniformiter, donec sint poenae aequales, quantitate durationis priori servata, etc. sicut prius. Nulli enim est dubium, quin *K* possit per solam intensionem fieri maior poena simpliciter quam sit *M* et *M* per solam remissionem minor quam sit *K*.' In order to complete the review of Bradwardine's references to the *falsigraphus*, see also *De causa Dei*, 12E; recorded in Latham, et al. (eds.), *Dictionary*, III, 901.

43 The solution ultimately concerns the habitual disposition of the person (Thomas Bradwardine, *De causa Dei*, 384E): 'Peccatum proprie non est aliquid positivum sed tantummodo privativum; privatio autem tolli non potest nisi per positionem habitus praecedentem saltem natura; nec tanta vel tanta privatione plenarie potest tolli sine positione tanti vel tanti habitus praecedente. Si enim ponatur minor habitus, et non tantus, tota illa privatio nequaquam plenarie tollitur, sed aliqua portio eius manet, sicut patet in naturalibus, in artificialibus, et in moralibus evidenter. Non potest ergo Deus, ut videtur, dimittere et tollere plenarie peccatum quoad culpam, nisi prius natura ponendo plenarie habitum quem privavit seu privat.'

Conclusion: Mathematics in *De causa Dei*

With this limited and schematic system, could the meaningful introduction of mathematical treatises into theological calculations be a realistic aim? It seems that in *De causa Dei*, Book I, it was at least a programmatic system. With respect to some other sciences, Bradwardine wrote, 'It would be highly fitting and would resolve many problems if theologians and perfect Christians did not ignore the mathematical sciences, but practised them either for their clearest benefits, or for better refuting their misusers.'[44] This statement perfectly reflects the ambiguity of the relationship between theology and mathematics in Bradwardine's mind: the former taking advantage of the latter and dominating it—paralleling the mathematics of apology with an apology of mathematics.

This ambiguity constitutes a constant feature in *De causa Dei*. The connection of theology and infinites in Book I, underlines this ambiguity as something underpinning the simple refusal to see devoted mathematicians working on a secularisation of the divine attributes.[45] As we have seen, the theological corrective uses of calculating proportions about infinites include non-theological corrective uses too. These rare, neutral uses of mathematical calculation refrain from resorting to concepts such as God's power or even human imagination. Through them, the physical world and its Aristotelian motions—and only these—appear as proper objects of the theorems of proportions. In this limited way, Bradwardine's *De causa Dei* posits a counterpart and completion to the *Tractatus de proportionibus*.

De causa Dei, Book I also offers an exceptional retrospective of the calculation of proportions: it highlights its Greco-Arabic origins; it appends the term 'theorem' to it, and provides the clearest formulation of physical healing in terms of varying proportions. Throughout Book I, it becomes increasingly evid-

---

44 Thomas Bradwardine, *De causa Dei*, 468B: 'Nonne et diligenter consideratis et debite ponderatis praemissis, et aliis similibus praetermissis videretur multum decere, et plurimum expedire Reipublicae Christianae, quod theologi et perfecti Catholici astrologiam, et alias doctrinales seu mathematicales scientias non nescirent, vel propter utilitatem earum praeclaras, si oportuerit, exercendas, vel saltem propter abusores earum [et falso nomine sapientium gloriantes] efficacius confutandos.' Bradwardine makes a similar statement at the beginning of his *Tractatus de proportionibus* in relation to the role of mathematics in philosophy (Thomas Bradwardine, *Tractatus de proportionibus*, 64): 'Et quia, testante Boethio, primo *Arithmeticae* suae: Quisquis scientias mathematicales praetermiserit, constat eum omnem philosophiae perdidisse doctrinam ...' On this relationship, see also above, n. 17.

45 Cf. Sylla, 'Thomas Bradwardine's *De continuo*,' 186.

ent that in Bradwardine's theology, anything that is in motion or change—a stone, a sick person, or a sinner—is equally subject to calculation. The corrective, moreover, negative use of mathematics about the infinite is paralleled by the extension of that with which motion and change are concerned and about which calculations are made. Among these, medicine appears as a privileged domain which on the evidence of *De causa Dei* clearly interacted with the mathematics of proportions. Whether it constitutes an influential application of mathematics or simply one of its many domains of application, the presence of medicine demonstrates the depth of vision Bradwardine applied in his approach to physical reality, but also, more importantly, it shows that he remains the Calculators' most creative mathematician.[46]

Finally, when enlarging the historical perspective on Bradwardine's mathematics of justification, one encounters a significantly similar proof.[47] Centuries later, another mathematician, anti-Pelagian, and apologist for the Christian faith, Blaise Pascal created the argument known as the Wager. Its starting point is an existential alternative: 'Either God is or he is not'; its conclusion is concerned with virtuous life: 'But here there is an infinity of infinitely happy life to be won, one chance of winning against a finite number of chances of losing, and what you are staking is finite.'[48] While the calculations of probabilities, infinitesimals and decision-making theories of the Wager were unknown to Bradwardine, several points emerge as presuppositions or theses common to the two authors. In the Wager, Pascal is committed to the idea of grace as irresistible and necessary for both salvation and justification—the justification Bradwardine tries to prove in his argument—as well as to an absolute predestination to salvation. Structurally, infinity plays a key and much commented

46 See Drake, 'Medieval Ratio Theory,' 77.

47 In the contemporary reception of Bradwardine's application of mathematics to theology in *De causa Dei*, we can point to two figures: Richard Kilvington responded to some of Bradwardine's theological theses, but the most extensive discussion is to be found in Bradwardine's pupil, Thomas Buckingham. See J.-F. Genest, *Prédétermination et liberté créée à Oxford au XIVe siècle. Buckingham contre Bradwardine*, Paris 1992 (Études de philosophie médiévale, 70), 88–89. The 1505 edition of Thomas Buckingham's commentary on the *Sentences* is unreliable. For his *Quaestiones theologicae*, I consulted Oxford, New College Library, Ms. 134, where on fols. 363r–366v, Buckingham deals with Bradwardine's argumentation in Book I, chapter I, corollary, part 40, then with the refutation of the Pelagian doctrine of justification. On fol. 404r, he restarts the debate on grace and penance. His treatment of Bradwardine's arguments deserves fuller treatment and could form the basis of a separate article.

48 See P. Bartha and L. Pasternack (eds.), *Pascal's Wager*, Cambridge 2018 (Classic Philosophical Arguments, 4), 'Introduction,' 13. The argument is formalised on 8, 14, 279, 294. See also W. Wood, 'The Wager and Pascal's Theology,' ibid., 45–63.

role in both arguments.[49] The divergence of the audiences—an audience of many for Pascal, a restricted readership for Bradwardine—ultimately shows how the sciences at stake in this chapter were affected by a fundamental change at the time of the Wager's formulation: namely, theology becoming no less secularized than mathematics.

49 See the chapter by A. Hájek, who tries to invalidate the Wager by replacing infinite utilities with 'arbitrarily large' finite values: A. Hájek, 'The (In)validity of Pascal's Wager,' ibid., 123–147.

CHAPTER 5

The Calculators on the Insolubles: Bradwardine, Kilvington, Heytesbury, Swyneshed, and Dumbleton

Stephen Read

1 Introduction

This paper considers the work on logic of five of the Oxford Calculators, in particular, their novel approaches to the insolubles, that is, the logical paradoxes. In the thirteenth century, right up until 1320, among the various solutions on the market, two dominated, those of the *cassantes* and the *restringentes*.[1] Nonetheless, although these are often the only solutions discussed in treatises on insolubles at this time, only two treatises are extant in which *cassatio* is defended, both early;[2] in the others, it is dismissed in favour of *restrictio*, of which several variants were developed. Then in the early 1320s, possibly when Bradwardine was still at Balliol, or in his first year at Merton in 1323–1324,[3] he argued powerfully against four versions of *restrictio*, then dismissed two versions of *cassatio* and two further proposals (those of the *mediantes* and the *distinguentes*) before presenting, defending and developing a radically new approach.

The problem should be familiar, often encapsulated in the famous Liar paradox, in the form 'This proposition (sentence, statement, assertion) is not true' or 'What I am saying is false'.[4] The most common form in which it was given

1 See, e.g., P.V. Spade, 'Five Early Theories in the Mediaeval Insolubilia-Literature,' *Vivarium* 25 (1987), 24–46, §§VI–VII.

2 See P.V. Spade, *The Mediaeval Liar: A Catalogue of Insolubilia-Literature*, Toronto, ON, 1975 (Subsidia Mediaevalia, 5), items IX and XXI.

3 B. Fleming, *Thomas de Bradwardine, Oxford Scholar, Royal Servant, and Archbishop of Canterbury*, unpublished Ph.D. dissertation, Leuven University, 1964, 78, dates Bradwardine's treatise to the years 1324–1327, but the argument in footnote 10 below suggests it must have been written earlier in the 1320s. The explicit of the Madrid manuscript of Bradwardine's treatise describes its author as 'regens Oxonii': see Thomas Bradwardine, *Insolubilia*, ed. and transl. S. Read, Leuven 2010 (Dallas Medieval Texts and Translations, 10), 2.

4 Note that throughout this paper, I will use 'proposition' to refer to what the medievals referred to as 'propositiones', that is, concrete sentences, whether spoken, written or mental. See, e.g.,

 | DOI:10.1163/9789004512054_007

by the Calculators was a scenario in which Socrates (a standard term in this context for an awkward customer, without necessarily any historical connotation) says just one thing, namely, 'Socrates says something false'. Did Socrates say something false? Suppose he did: then the only thing he said was false, so it was false that he said something false, so he must have said something true. But if what he said was true, then it was true that what he said was false, so it was false. So what he said was false if and only if it was true, which is paradoxical, and seems to entail that it was both, which is impossible.

The *cassantes* looked to solve the problem by declaring that, despite appearances, Socrates had said nothing. If he had failed to say anything (coherent), then there was nothing to be true or false and certainly not both. Bradwardine dismissed this with the observation that clearly Socrates had uttered letters, syllables and so on in a coherent way. The suggestion that this was an illusion, as was our reasoning about what he said, is itself paradoxical and not to be taken seriously. The proponent of *cassatio* can reply, however, as we shall see in § 6 below, that Socrates may well have made an intelligible sound, but not one coherent enough to constitute a proposition with a truth-value.

The *mediantes* proposed that, though Socrates did say something, it was neither true nor false. We shall consider a version of this view later. The *restringentes* proposed a restriction on the reference (or as they called it, the supposition[5]) of the terms in a proposition, that such terms could not refer to a whole of which they themselves were a proper part. (Note that a term can supposit for itself, in material supposition, just not, they said, for a greater whole of which it is part.) Of the various versions, one banned all such reference to the whole by the part, another only such reference in the presence of privative semantic expressions like 'false' and 'not true'. The former restriction seems implausible in the light of innocuous examples like 'This sentence contains five words'. The latter seems ad hoc unless some independent reason can be given to suppose that such reference is unacceptable, which Bradwardine claimed had not been done.

The *distinguentes* made a distinction, saying that Socrates' utterance is true in one sense and false in another. Bradwardine refuted this proposal by the

---

N. Kretzmann, 'Medieval logicians on the meaning of the propositio,' *The Journal of Philosophy* 67 (1970), 767–787, and L. Cesalli, 'Propositions: their Meaning and Truth,' in C. Dutilh Novaes and S. Read (eds.), *The Cambridge Companion to Medieval Logic*, Cambridge 2016, 245–264, § 10.1.

5 On the medieval theory of supposition of terms, and how it is both like and unlike the modern notion of reference, see, e.g., S. Read, 'Medieval Theories: Properties of Terms,' in E.N. Zalta (ed.), *The Stanford Encyclopedia of Philosophy* (Spring 2019 Edition), URL = https://plato.stanford.edu/archives/spr2015/entries/medieval-terms/, § 3.

usual method of 'revenge':[6] take the sense in which it is said to be true, call it ϕ, and consider the strengthened paradox 'This proposition is not true in sense ϕ'. If it is true in sense ϕ, it isn't, and if it isn't true in sense ϕ, then surely it is, for that is what it says. Paradox has returned.

2 Bradwardine's Radical Solution

Bradwardine was born in Hartfield, Sussex, around 1300. He was regent master at Balliol in 1321 and joined Merton College in 1323.[7] He was the first of the Oxford Calculators. He wrote his *De proportionibus* in 1328 and *De continuo* in the early 1330s. He became a member of the circle of Richard de Bury, the Bishop of Durham, in 1335. He was Chancellor of St. Paul's Cathedral, London, from 1337, and in the 1330s started work on his *De causa Dei contra Pelagium*, published in 1344. Consecrated Archbishop of Canterbury at Avignon in July 1349, he died at Lambeth from the Black Death within days of his return to England, in August 1349.

Bradwardine's radical idea for a solution to the insolubles can be thought of like this: if Socrates' utterance simply means that what Socrates said was false, we seem to land in contradiction. It turns out that it is true if and only if it is false, so if it is either, it is both. It is implicitly contradictory, so perhaps it really is actually contradictory, signifying not only that it is false, but also, somehow implicitly, that it is true. In that case it must simply be false, since it cannot be both. This is his 'multiple-meanings' solution: insolubles must mean more than at first appears. Bradwardine, whose later career shows how smart, subtle and deep a thinker he was, came up with a smart, subtle and deep argument to establish this.[8]

Take any proposition, α, he said, which signifies or means that it itself is false—like Socrates' utterance, for example. (Recall that Bradwardine has already dismissed the *restringentes*'s suggested ban on self-reference.) Either that is all α signifies, or not. Suppose it is all α signifies, and suppose α is false. As later in his mathematical and theological works, Bradwardine sets out his defini-

6 See Thomas Bradwardine, *Insolubilia*, § 5.8. On revenge paradoxes and Bradwardine's use in particular, see S. Read, 'Bradwardine's Revenge,' in J.C. Beall (ed.), *Revenge of the Liar*, Oxford 2007, 250–261.

7 According to J.A. Weisheipl, 'The Place of John Dumbleton in the Merton School,' *Isis* 50 (1959), 439–454, at 441, Balliol College was founded exclusively for the study of arts, so to proceed to one of the higher faculties, young scholars, such as Bradwardine and later John Wyclif, had to transfer to another college.

8 See Thomas Bradwardine, *Insolubilia*, ch. 6.

tions and postulates explicitly and systematically. First, a proposition is true if and only if it signifies wholly as it is, otherwise false, so every proposition is either true or false. This is his first postulate—bivalence. Everything a proposition signifies must obtain for it to be true; if anything it signifies fails to hold, that is enough to render it false. His second postulate, again consonant with the thought that a proposition might signify many things (and more than might at first appear), is that propositions signify anything that follows from what they signify. If α is false, and assuming it only signifies that α is false, then α signifies wholly as it is, so α is true. But α signifies that α is false, and we have shown that if it is false, it is true, so α must signify that α is true, by the second postulate. So it does not signify only that it is false, for we have shown that if all it signifies is that it is false then it also signifies that it is true. It does not yet follow that it does signify that α is true, for that only follows from the counterfactual assumption that being false is all that α signifies.

So α signifies more than that α is false—call the extra signification ϕ. So α signifies both that α is false and ϕ. Again, suppose that α is false. Then things are not wholly as it signifies, so something it signifies must fail, i.e., either α is not false or not-ϕ. But α signifies that α is false, so by the second postulate, α signifies that either it is not false or not-ϕ. But it also signifies that ϕ, and it follows from ϕ and 'α is not false or not-ϕ', by logical rules encapsulated in his fourth, fifth, and sixth postulates, that α is not false, and so by bivalence, that α is true. So again by the second postulate, it follows that α signifies that α is true. So α signifies both that it is false and that it is true. Thus α is indeed contradictory, and it cannot be wholly as α signifies, so α is false. But α was any proposition which signified (*inter alia*) that it itself is false. So we have proved Bradwardine's second thesis, that any proposition which signifies itself to be false, also signifies itself to be true and is false. Similarly, any proposition which signifies that it's not true, signifies that it's false, by the first and second postulates, and so signifies that it is both true and false, and thus is false too. Bradwardine continued in the rest of his treatise to deal with the problem of revenge, and to show how his solution deals with other paradoxes and insolubles.

Two of the manuscripts of his treatise contain what one of them calls 'an incidental chapter'. It is unclear whether it is by Bradwardine himself, but it is certainly by an adherent of his solution. It opens with the words:

> After these insolubles were first written ... besides the insolubles presented in the fourth chapter, which the solution of the *restringentes* did not solve, an insoluble was found that is forever incapable of solution in that way. For let *A* be (one of) 'God exists' (call it [*D*]) and 'Nothing granted by

> Socrates is known by you' (call it C), where you do not know if A is [D] or C, and let [B] be 'A is known by you', where only [B] is granted by Socrates.[9]

This is the last of Richard Kilvington's 48 sophisms from his *Sophismata*, written in Oxford shortly after Bradwardine's treatise.[10]

3 Richard Kilvington

Kilvington was the son of a priest from the diocese of York (probably from Kilvington, near Thirsk). He studied at Oriel College in the 1320s. He wrote his *Quaestiones super de Generatione et Corruptione* before 1325, *Quaestiones super Physicam* in 1325–1326, *Quaestiones super Libros Ethicorum* between 1326 and 1332, and *Quaestiones super libros Sententiarum Petri Lombardi* in 1334. He became Master of Theology around 1335. He was another member of the intellectual circle round Richard de Bury, the Bishop of Durham, from 1335, and took part in diplomatic missions in the service of Edward III. He was Dean of St. Paul's Cathedral by 1354, and died in a second phase of the Black Death in 1361.

Much of Kilvington's *Sophismata* concerns logical puzzles about physical matters, e.g., sophisms about beginning and ceasing. But the last four of the 48 sophisms concern epistemic puzzles, culminating in the fiendish Sophism 48. Note the reference to granting in proposition C. This is an allusion to the practice of obligations, and the literature on insolubles and sophisms (and more widely) is thoroughly ridden with the language of obligations.

There is still disagreement about what (logical) obligations were for and why they played such a large role in fourteenth-century logic, but at least we now understand how they worked. There is always an Opponent and a Respondent.

---

9 Thomas Bradwardine, *Insolubilia*, §A.0, 176: 'Postquam ista insolubilia fuerunt primo scripta, … preter insolubilia quarto capitulo posita, que solutio restringentium non dissolvit, inventum fuit unum insolubile per istam in perpetuum non dissolvendum. Sit enim a: deus est, que sit b, et nullum concessum a Sorte scitur a te, que sit c, et nescias an a sit b vel c, et sit d ista: a scitur a te, que sola conceditur a Sorte.' I have interchanged the labels '*B*' and '*D*' to fit in with Kilvington's presentation below.

10 Richard Kilvington, *Sophismata*, eds. N. Kretzmann and B.E. Kretzmann, Oxford 1990 (Auctores Britannici Medii Aevi, 12). N. Kretzmann and B.E. Kretzmann date the *Sophismata* between 1321 and 1326 (Richard Kilvington, *The Sophismata of Richard Kilvington*, transl. N. Kretzmann and B.E. Kretzmann, New York 1991, xxvii); and E. Jung before 1325 (E. Jung, 'Richard Kilvington,' in E.N. Zalta [ed.], *The Stanford Encyclopedia of Philosophy* Winter 2016 Edition, URL = https://plato.stanford.edu/archives/win2016/entries/kilvington/, §1). But the passage cited dates the discovery of Sophism 48 after Bradwardine's *Insolubles*. They must both belong to the early to mid-1320s.

The Opponent proposes a background scenario (*casus*), which may or may not be counterfactual, but at least establishes the context. Then, in the primary type of obligation, *positio*, a proposition is proposed (the *positum*), usually one that is false in the scenario. The Respondent was expected to admit the obligation provided the *positum* could be true in the scenario, otherwise to reject it. The obligation on the Respondent was then to respond, in accordance with strict rules and without contradicting himself, to a series of propositions put forward by the Opponent. His response could be to grant, to deny, or to doubt each of them in turn. Propositions which follow from or are inconsistent with the *positum* and any propositions previously granted or the contradictories of those denied, were said to be 'relevant' (*pertinens*), and should be granted if they follow (*pertinens sequens*) and denied if inconsistent (*pertinens repugnans*). The rest were said to be 'irrelevant' (*impertinens*), and should be granted if known to be true in the scenario, denied if known to be false in the scenario, and otherwise doubted. This particular scheme of rules was known later as the *responsio antiqua*. A revised scheme was proposed by Roger Swyneshed (whom we will meet later), known as the *responsio nova*,[11] and an idiosyncratic revision to the criterion for relevance was proposed by Kilvington in his discussion of Sophism 47, of which more later.

What is the puzzle with Sophism 48? Recall the constituent propositions:

B '*A* is known by you'
A either 'God exists'–*D*, for short
or 'Nothing granted by Socrates is known by you'–*C*, for short[12]

where the only proposition granted by Socrates is *B*. So *C* is equivalent to '*B* is not known by you', that is, 'That *A* is known by you is not known by you'. *C* is true, because *B* is not known by you since you don't know if *A* is known by you, for you don't know what *A* is. Then if *A* is *C*, *A* is true. Moreover, if *A* is *D*, then *A* is true and known by you to be true. So either way, *A* is true, so you do know *A* (or at least, that *A* is true), so *B* is true. But then *C* is false, for we have shown, and so we know, that *A* is known by you.[13] But we showed that *C* was true. Paradox.

---

11 See, e.g., C. Dutilh Novaes and S. Uckelman, 'Obligationes,' in C. Dutilh Novaes and S. Read (eds.), *The Cambridge Companion to Medieval Logic*, Cambridge 2016, 389–390.

12 *A* is one of *C* or *D*, not their disjunction, as F. Bottin, 'L'*Opinio de insolubilibus* di Richard Kilmyngton,' *Rivista critica di storia della filosofia* 28 (1973), 408, repeated in F. Bottin, *Le antinomie semantiche nella logica medievale*, Padua 1976, 87, seems to think.

13 This form of inference is sometimes known as Gödel's Rule, from its role, in the form $\vdash A \Rightarrow \vdash \text{Prov}[A]$ (that is, having proved *A*, we have *ipso facto* proved that *A* is provable),

Kilvington uses this sophism, among other things, as a springboard to develop his own solution to the insolubles. He writes:

> I say, then, that no insoluble that is presently under discussion is absolutely true or absolutely false; instead each is true in a certain respect and false in a certain respect.[14]

He applies this to the Socratic liar, 'Socrates says what is false' (*Sortes dicit falsum*):

> Suppose, for example, that Socrates says this: 'Socrates says what is false' and nothing else. In that case I say that the term 'what is false' can be taken in one way for what is false in a certain respect and in another way for what is false absolutely. If for what is false in a certain respect, it is to be granted. If for what is absolutely false, then it must be denied that Socrates says what is false.[15]

He likens the situation to one Aristotle describes in Book III of his *Nicomachean Ethics* (1110a4–12). To save his ship, the captain may choose to throw his goods into the sea, but he does not do this voluntarily without qualification (or absolutely–*simpliciter*), but only in a certain respect (that is, to save his life and that of his passengers). So the action is voluntary in a certain respect and at the same time not voluntary without that qualification: 'I say that one must not grant or deny absolutely that [he] wants to throw his goods into the sea.'[16] Similarly, a proposition may be neither true nor false without qualification, but

---

in the proof of Gödel's theorem. See, e.g., M. Löb, 'On a Solution of a Problem of Leon Henkin,' *Journal of Symbolic Logic* 20 (1955) 115–118, Rule IV. In epistemic logic, the rule means that if we have proved something, we can infer that we know it. In alethic modal logic, it is known as Necessitation.

14 Transl. from Richard Kilvington, *The Sophismata*, 142: 'Dico, igitur, quod nullum insolubile de quo praesens est locutio est simpliciter verum vel simpliciter falsum; sed quodlibet est verum secundum quid et falsum secundum quid' (Latin text from Richard Kilvington, *Sophismata*, 146).

15 Richard Kilvington, *The Sophismata*, 142: 'Ut, verbi gratia, posito quod Socrates dicat istam: "Socrates dicit falsum" et nullam aliam. Tunc dico quod iste terminus "falsum" sumi potest uno modo pro falso secundum quid et alio modo pro falso simpliciter. Si pro falso secundum quid, concedenda est. Si pro falso simpliciter, tunc est negandum quod Socrates dicit falsum' (Richard Kilvington, *Sophismata*, 146).

16 Richard Kilvington, *The Sophismata*, 142: 'Et dico quod non debet concedi vel negari simpliciter quod … vult proicere merces in mari' (Richard Kilvington, *Sophismata*, 146).

both true in a certain respect and false in a certain respect. In particular, Kilvington says, an insoluble like 'Socrates says what is false'

> is made true in a certain respect on this account, that the part of the insoluble supposits for the whole of which it is a part.

But in addition

> it is false in a certain respect because from that insoluble together with a possible hypothesis [viz. that that is all Socrates says] its opposite follows ... The part, however, does not supposit for its whole in such a way that the insoluble is true absolutely (*simpliciter*) in virtue of the supposition of the part for the whole.[17]

Applied to Sophism 48, Kilvington distinguishes two ways of taking the phrase 'known by you'. *D* ('God exists') is known by you absolutely, so if *A* is *D*, so too is *A*:[18]

> But if *A* is *C*, then *A* is true in a certain respect and not absolutely. And because I doubt whether *A* is 'God exists' or *C*, I doubt both senses of the sophism. And the reason why I say that if *A* is *C*, *A* is true, is this: because if *A* is *C*, then it is an insoluble, and each insoluble is true in a certain respect and false in a certain respect, and neither true nor false absolutely.[19]

17 Richard Kilvington, *The Sophismata*, 142: 'verificatur secundum quid propter hoc, quod pars illius insolubilis supponit pro toto cuius est pars ... et est falsum secundum quid eo quod ex ipso insolubili cum casu possibili sequitur suum oppositum ... Non tamen sic supponit pars pro suo toto ut insolubile sit verum simpliciter propter suppositionem partis pro toto, licet possit esse verum secundum quid propter hoc, quod pars supponit pro suo toto' (Richard Kilvington, *Sophismata*, 147).

18 Bradwardine's solution is different, and depends on a further thesis that he argues for in ch. 9, that if a proposition signifies that it itself is unknown (to someone), then it also signifies that it is unknown to them that it is unknown to them. Thomas Bradwardine, *Insolubilia* § 9.3. The solution to Sophism 48 is then given in § ad A1: 'For [*B*] should be doubted, because if *A* is [*D*], [*B*] is true, and if *A* is *C*, [*B*] is false, because it signifies itself to be unknown to you and consequently it signifies that it is unknown to you that [*B*] is unknown to you, as is clear in chapter 9. If ⟨*C*⟩ is proposed, it should be conceded; and if it is further argued: *C* is known to you and [*D*] is known to you, so *A* is known to you, the conclusion should be conceded. And if it is objected: therefore [*B*] should be conceded, which earlier was doubted, it must be said that it does not follow, but what you concede is a proposition similar to [*B*].'

19 Richard Kilvington, *Sophismata*, 139. 'Sed si *A* sit *C*, tunc *A* est verum secundum quid et non

4 William Heytesbury

Writing ten years later, William Heytesbury published a solution very similar to Bradwardine's, using an argument made famous in modern times by Arthur Prior.[20] Heytesbury was a Fellow of Merton College by 1330 and according to a note in one manuscript, wrote his *Rules for Solving Sophisms* (*Regulae*) in 1335.[21] He also wrote the treatises *On Compounded and Divided Senses*, *Sophismata*, *Sophismata asinina* and *Iuxta hunc textum* during the 1330s. He was a foundation Fellow of The Queen's College at its foundation in 1341, but shortly afterwards returned to Merton. He was Doctor of Theology by 1348, Chancellor of University of Oxford in 1352–1354(?) and 1370–1372 and died in 1372 or 1373.

Suppose Epimenides really did say 'All Cretans are liars', as Clement of Alexandria claimed,[22] and suppose he was justified in his pessimism about his compatriots—all other Cretan utterances were false. Then if Epimenides' utterance was true, it was false (for it is a Cretan utterance); and conversely, if it was false, it must be true (for then all Cretan utterances would be false). That is impossible, so not all other Cretan utterances were false. Simply in virtue of Epimenides saying what he did, something said by a Cretan must be true, but it cannot, it seems, be what Epimenides himself said. That is in itself paradoxical.

Bradwardine and Heytesbury drew a slightly different conclusion from Prior's. Prior infers that some other Cretan utterance was true. Bradwardine and Heytesbury infer that Epimenides' utterance must mean something more than at first appears. When Epimenides uttered 'All Cretans are liars' he must have said more than just that all Cretans are liars. As we saw, Bradwardine has

simpliciter. Et quia dubito utrum *A* sit ista "Deus est" vel *C*, ideo utrumque sensum sophismatis dubito. Et causa quare dico quod si *A* sit *C*, *A* est verum, est ista: quia si *A* sit *C*, tunc est unum insolubile, et quodlibet insolubile est verum secundum quid et falsum secundum quid et neque verum neque falsum simpliciter' (Richard Kilvington, *The Sophismata*, 142–143).

20 A.N. Prior, 'Epimenides the Cretan,' *Journal of Symbolic Logic* 23 (1958), 261–266.

21 See William Heytesbury, *On 'Insoluble' Sentences: Chapter One of his Rules for Solving Sophisms*, transl. P.V. Spade, Toronto, ON, 1979 (Mediaeval sources in translation, 21), 2.

22 Clement of Alexandria, *Stromateis*, Books 1–3, transl. J. Ferguson, Washington, DC, 1991 (The Fathers of the Church, 85), I 14, 66, writing in the second century CE, seems to have been the first to identify Epimenides the Cretan as the author of the famous remark of St. Paul's in his *Epistle to Titus* that 'Cretans always lie'. However, the earliest occurrence of the passage from Epimenides' (lost) panegyric to Minos, written around 600 BCE, is in Callimachus' *Hymn to Zeus*, written in the third century BCE, where Crete is said to possess Zeus' tomb, as well as his birthplace—but gods never die. See Callimachus, *Hymn to Zeus*, ed. and comm. G.R. McLennan, Rome 1977 (Testi e commenti, 2) 15 and 35 ff. Epimenides himself was drawing on Hesiod and possibly even Homer.

a subtle argument depending on his second postulate to tell us what this further hidden meaning is. Heytesbury draws back, and exploits the apparatus of obligations to keep silent. Taking the now familiar obligational scenario where Socrates says 'Socrates says something false' and nothing else, he observes:

> But if [the Opponent] asks what the proposition uttered by Socrates signifies under the scenario other than that Socrates says something false, I say to him that the Respondent does not have to settle this or that question, because it follows from this scenario that this proposition signifies other than that Socrates says something false, but that scenario does not specify what ⟨that other signification⟩ is, and so the Respondent does not have to settle that question any further.[23]

Bottin attributes to Bradwardine a distinction between the principal signification of a proposition and its consequential signification.[24] This is tendentious, for the idea appears in Bradwardine in the mouth of an opponent, and arguably Bradwardine rejects the distinction.[25] Recall from § 2 that while for Bradwardine every proposition that signifies that it itself is false also signifies that it is true, that implicit signification that it is true is not consequential (only) on its signifying that it is false.[26] However, such a distinction became popular in later authors, notably in Heytesbury, who introduces the idea of what the words usually mean (*sicut verba communiter pretendunt*), and distinguishes cases where what they usually mean is all they signify from cases where they signify more.[27]

---

23 L. Pozzi, *Il Mentitore e il Medioevo. Il dibattito sui paradossi dell'autoriferimento. Scelta di testi, commento, traduzione*, Parma 1987 (Quaderni di filosofia, 10), 240: 'Si autem queratur in casu illo quid significabit illa propositio dicta a Sorte aliter quam quod Sortes dicit falsum, huic dicitur quod respondens non habebit illud seu illam quaestionem determinare, quia ex casu isto sequitur quod ista propositio aliter significet quam quod Sortes dicit falsum, sed casus ille non certificat quid illud sit et ideo non habet respondens quaesitum illud ulterius determinare' (my translation—cf. William Heytesbury, *On 'Insoluble' Sentences*, 49–50).

24 F. Bottin, 'The Mertonians' Metalinguistic Sciences and the *insolubilia*,' in P.O. Lewry (ed.), *The Rise of British Logic*, Toronto, ON, 1985 (Papers in Mediaeval Studies, 7), 235–248.

25 See Thomas Bradwardine, *Insolubilia* § 7.2.5.

26 See also S. Read, 'Paradoxes of Signification,' *Vivarium* 54 (2016), 335–355, esp. 338.

27 See Pozzi, *Il Mentitore*, 238: '... si fiat casus de insolubili et cum hoc supponatur quod illud insolubile significet sicut termini illius praetendunt non tamen praecise ...' ('... if we take an insoluble scenario, and assume that the insoluble signifies in the usual way but not only like that ...'). For an alternative translation, see William Heytesbury, *On 'Insoluble' Sentences*, 49.

Given this distinction, Heytesbury presents rules for dealing with insolubles for which he became famous. First, he defines an insoluble scenario:

> An insoluble scenario is one in which mention is made of some proposition such that, if in that scenario it signifies only what its words usually mean, it follows that it is true and that it is false,[28]

and secondly, what it is to be an insoluble proposition:

> An insoluble proposition is one of which mention is made in some insoluble scenario such that, if in that scenario it signifies only what its words usually mean, it follows that it is true and that it is false.[29]

Then, he says in his second rule, if anyone presents an insoluble scenario, and specifies that the insoluble proposition signifies only what its words usually mean, the obligation should in no way be admitted. Only if the Opponent leaves open that the proposition might have an additional signification (that it signifies what the words usually mean, but perhaps not only in that way—see n. 27) should the obligation be admitted, and then the insoluble should be granted, and that it is true should be denied. Thus Heytesbury's position gives rise to the same Moorean paradox as Bradwardine's.[30] For example, in the case of Socrates' sole utterance of 'Socrates says what is false', we first grant that what Socrates says is false and immediately deny that what we granted is true. This situation is in fact not so very unusual in obligations where what is granted is often in fact false and something we certainly don't believe. Spade, in his study

28 Pozzi, *Il Mentitore*, 236: 'Casus de insolubili est in quo fit mentio de aliqua propositione quae, si cum eodem casu significaret praecise sicut verba eius communiter praetendunt, sequitur eam esse veram et eam esse falsam.' Some manuscripts and incunables give that last clause as 'from its being true it follows that it is false and vice versa' ('ad eam esse veram sequitur eam esse falsam, et econverso'): see William Heytesbury, *On 'Insoluble' Sentences*, 46 and William Heytesbury, *Regule solvendi sophismata*, Venice 1494, 6rb.

29 Pozzi, *Il Mentitore*, 236: 'Propositio insolubilis est de qua fit mentio in casu de insolubili quae, si cum eodem casu significaret praecise sicut verba illius communiter praetendunt, sequitur eam esse veram et eam esse falsam.' Again, some manuscripts give the alternative reading of the consequent mentioned in n. 28.

30 H. Field, 'Maudlin's Truth and Paradox,' *Philosophy and Phenomenological Research* 73 (2006), 713–720, at 715, notes that Tim Maudlin's solution to the Liar paradox commits him to believing simultaneously that the Liar is not true 'while believing that this belief of his is not true.' This is like Moore's paradox (see, e.g., G.E. Moore, 'Moore's paradox,' in T. Baldwin [ed.], *G.E. Moore: Selected Writings*, London 1993, 207–212), the apparent absurdity of making a first-person assertion of the form '*p* but I don't believe that *p*'.

of Heytesbury's theory, confronts Heytesbury with an insoluble that he claims the theory cannot deal with.[31] First, he defines a neologism, 'firm', to capture the notion Heytesbury expressed by the phrase *ita est sicut communiter pretendunt*, which I am translating as 'being as it usually means'.[32] Take the proposition *S*, namely, 'It is not as *S* usually means', for short, '*S* is infirm'. Suppose *S* is firm, that is, it is as *S* usually means. Since *S*'s usual meaning is that it is infirm, it follows that *S* is infirm. Hence, by *reductio*, *S* is not firm. But *S* signifies that it is infirm. So it is as *S* usually means, that is, *S* is firm. Contradiction.

Heytesbury appears to have no response to this, though it is very similar, if not identical, to an insoluble he levels at the theories of his contemporaries whom he criticizes in the first part of his treatise, albeit not by name. Gaetano da Thiene identifies these theories as those of the Calculators Roger Swyneshed, John Dumbleton, and Richard Kilvington respectively.[33] Swyneshed, for one, not only has a response to Heytesbury's purported counter-example, and Spade's variant, but in fact confronts it directly in the final part of his treatise, having laid the foundation for its solution in the very first paragraph of the work.

5 Roger Swyneshed

Roger Swyneshed is perhaps one of the lesser-known Calculators, not to be confused, as over the centuries he sometimes has been, with his better-known namesake, Richard Swineshead. Roger, who may have studied at Oxford under

31 William Heytesbury, *On 'Insoluble' Sentences*, 92.

32 William Heytesbury, *On 'Insoluble' Sentences*, 81.

33 See Gaetano da Thiene, *Declaratio* [= *Recollectae*] *super tractatu Hentisberi regularum*. In William Heytesbury, *Regule solvendi sophismata*, Venice 1494 (scattered within Heytesbury's text), 7rb–12rb, at 7va. J.A. Weisheipl, 'Ockham and some Mertonians,' *Mediaeval Studies* 30 (1968), 163–213, repr. in J.I. Catto and T.A.R. Evans (eds.), *The History of the University of Oxford, 2: Late Medieval Oxford*, Oxford 1992, 607–658, at 202–203, and P.V. Spade, *William Heytesbury, On 'Insoluble' Sentences*, 73 n. 28, cast doubt on the identification of Dumbleton as the author of the second opinion, since Dumbleton was some years Heytesbury's junior, his treatise on insolubles was arguably composed after Heytesbury's, and in it he discusses and rejects Heytesbury's solution. However, Heytesbury and Dumbleton seem both to have been at Merton in the mid-1330s, and when Dumbleton was named in the foundation statutes of The Queen's College in 1341, he is cited as *magister*, so it is reasonable to assume that he had been in Oxford for at least a decade and that he and Heytesbury were mutually aware of each other's ideas as they developed prior to each composing his own treatise. See also M. Hanke, 'Cajetan of Thiene on the Logic of Paradox,' *Studia Neoaristotelica* 13 (2016), 71–95, at 74 n. 10.

Bradwardine and Kilvington, wrote treatises on *Insolubles* and *Obligations* between 1330 and 1335 (and also a treatise on *Consequences* now apparently lost). He is also the author of *Descriptiones motuum* (or *De motibus naturalibus*), a treatise on natural changes, including locomotion. He subsequently became Master of Theology (though his lectures on the *Sentences* are also lost) and was another member of Richard de Bury's circle. He was a Benedictine monk at Glastonbury and died about 1365.[34]

Swyneshed is determined not to make any great play with the notion of signification, and although he often uses the phrases 'signify principally' or 'principal signification', by them he seems just to mean what it signifies, or in Heytesbury's phrase, 'what it usually means'. Instead, he varies the truth condition, introducing the idea of a proposition's falsifying itself. Consequently, a proposition might signify as it is and still be false, if it falsifies itself:

> A proposition falsifying itself directly is a proposition signifying principally as it is or other than it is, relevant to inferring itself to be false. And it is of two kinds. One kind is relevant sufficiently, another relevant insufficiently. Relevant sufficiently is a proposition signifying principally as it is or other than it is from which, signifying in this way, it directly follows or is apt to follow that it is false. An example: let the proposition 'This is false' signify principally that this is false, referring to itself. Then it directly follows 'This is false, therefore, this is false'. And in this way it is relevant sufficiently to inferring itself to be false.[35]

A proposition can also directly falsify itself insufficiently, if it requires added true premises to do so; and it can falsify itself indirectly, if it falsifies another

34 See J.A. Weisheipl, 'Roger Swyneshed, OSB, Logician, Natural Philosopher, and Theologian,' in *Oxford Studies Presented to Daniel Callus*, Oxford 1964 (Oxford Historical Society, New Series, 16), 231–252.

35 P.V. Spade, 'Roger Swyneshed's *Insolubilia*: Edition and Comments,' *Archives d'histoire doctrinale et littéraire du Moyen Âge* 46 (1979), 177–220, at 182–183, reprinted in P.V. Spade, *Lies, Language and Logic in the Late Middle Ages*, London 1988 (Collected Studies Series, 272): 'Propositio falsificans se immediate est propositio significans principaliter sicut est vel aliter quam est pertinens ad inferendum se ipsam fore falsam. Et illa est duplex. Quaedam est pertinens sufficiens, quaedam est pertinens insufficiens. Pertinens sufficiens est propositio significans principaliter sicut est vel aliter quam est ex qua sic significando immediate sequitur vel est natum sequi ipsam fore falsam. Exemplum: Significet illa propositio "Hoc est falsum" principaliter quod hoc est falsum, ipsamet demonstrata. Tunc sequitur immediate "Hoc est falsum; igitur, hoc est falsum". Et sic illa est pertinens sufficiens ad inferendum se ipsam fore falsam.'

which in turn falsifies the first. With these notions in place, Swyneshed can define truth and falsehood:

> A true proposition is a proposition not falsifying itself signifying principally as it is either naturally or by an imposition or impositions by which it was last imposed to signify … A false proposition is an utterance falsifying itself or an utterance not falsifying itself signifying principally other than it is either naturally or by an imposition or impositions by which it was last imposed to signify.[36]

Finally, Swyneshed defines an insoluble:

> An insoluble as put forward is a proposition signifying principally as it is or other than it is which is relevant to inferring itself to be false or unknown or not believed, and so on.[37]

One might be tempted to infer from these definitions that every proposition, whether or not it is an insoluble, is true or false. But this is to assume that every proposition either signifies as it is or other than it is. For Swyneshed lays down that

> a proposition is a well-formed indicative utterance significative either naturally or by an imposition or impositions by which it was last imposed to signify complexly.[38]

However, it turns out that, in his view, a proposition can signify in some way (*aliqualiter*) but still neither signify as it is nor other than it is. He says as much right at the start of his treatise:

---

36 Spade, 'Roger Swyneshed's *Insolubilia*,' 185–186: 'Propositio vera est propositio non falsificans se principaliter sicut est significans naturaliter aut ex impositione vel impositionibus qua vel quibus ultimo fuit imposita ad significandum … Propositio falsa est oratio falsificans se vel oratio non falsificans se principaliter aliter quam est significans naturaliter, ex impositione vel impositionibus qua vel quibus ultimo fuit imposita ad significandum.'

37 Spade, 'Roger Swyneshed's *Insolubilia*,' 186: 'Insolubile ad propositum est propositio significans principaliter sicut est vel aliter quam est pertinens ad inferendum se ipsam fore falsam vel nescitam vel non creditam, et sic de singulis'; corrected by Pozzi, *Il Mentitore*, 182.

38 Spade, 'Roger Swyneshed's *Insolubilia*,' 185: 'Propositio est oratio indicativa congrua naturaliter, ex impositione vel impositionibus qua vel quibus ultimo fuit imposita complexe ad significandum significativa.'

> A proposition neither signifying principally as it is nor other than it is, that is, which is neither true nor false, is a proposition signifying in some way and that so signifying is relevant to inferring itself not to signify principally as it is, for example, the proposition 'This proposition does not signify as it is', referring to itself, which principally signifies that it itself does not signify as it is. And this similarly, 'Every proposition signifies other than it is', which principally signifies that every proposition signifies other than it is.[39]

This stance allows Swyneshed to deal with the paradox of signification which, we noted, was launched at Heytesbury by Spade. To be sure, it does not fit Swyneshed's definition of an insoluble, which must signify as it is or other than it is (as stated in his definition, cited above), but it is still a problematic sophism, as Swyneshed notes in the fifth and final section of his treatise, which aims 'to solve some sophisms which appear to be insolubles but are not, e.g., "*A* is known", "This proposition signifies other than it is", "That proposition does not signify other than it is", "This proposition does not signify as it is", and similar ones.'[40]

How does Swyneshed deal with this last example? The core idea of Swyneshed's theory was that the criterion of truth be strengthened to exclude propositions which falsify themselves, that is, are 'relevant to inferring their own falsity,' and the criterion of falsehood weakened to include such propositions. This is recorded in Swyneshed's first iconoclastic thesis:

> Some false proposition signifies principally as it is.[41]

39 Spade, 'Roger Swyneshed's *Insolubilia*,' 180–181: 'Propositio nec principaliter significans sicut est nec aliter quam est, id est quae nec est vera nec falsa, est propositio significans aliqualiter esse et illa sic significando est pertinens ad inferendum se ipsam non significare principaliter sicut est, sicut haec propositio "Haec propositio non significat sicut est", demonstrata illa eadem, quae principaliter significet quod ipsa non significat sicut est. Et haec similiter "Omnis propositio significat aliter quam est" quae principaliter significet quod omnis propositio significat aliter quam est.'

40 Spade, 'Roger Swyneshed's *Insolubilia*,' 215: 'His vero completis superest solvere quaedam sophismata quae apparent insolubilia et non sunt, sicut sint "a est scitum", "Ista propositio significat aliter quam est", "Illa propositio non significat aliter quam est", "Ista propositio non significat sicut est", et his similes.'

41 Spade, 'Roger Swyneshed's *Insolubilia*,' 188: 'Aliqua propositio falsa significat principaliter sicut est.'

Socrates' lone utterance of 'Socrates says something false' is an example. It signifies as it is (it is false) because it satisfies Swyneshed's definition of falsehood—it falsifies itself. But the paradox of signification does not, or at least not obviously, falsify itself. Nonetheless, Swyneshed prepared himself for this right at the start of his treatise:

> A proposition relevant to inferring itself not to signify principally as it is, is one from which, with its being wholly as it is (*cum totaliter sic esse sicut est*), it follows or is apt to follow that it does not signify principally as it is. An example: let the proposition 'This signifies other than it is' signify principally that it signifies other than it is, referring to itself. Then it follows: this proposition signifies other than it is, and it signifies principally like that, hence, it does not signify principally as it is. And thus from it, with its being wholly as it is, it follows that it does not signify as it is. And so it neither signifies other than it is nor as it is.[42]

The same move works for 'This proposition does not signify as it is'. For, as he says, 'this follows directly: "it does not signify as it is, therefore, it does not signify as it is".'[43] So the second leg of Spade's argument against Heytesbury should be denied. To be sure, *S* (that is, '*S* is infirm') is not firm, and it signifies that it is infirm. But it does not follow that it is as it signifies. As Swyneshed says, 'it is necessary to add to the premise that it is not relevant to inferring itself not to signify as it is. And if that is added, it should be denied.'[44] So *S* neither signifies as it is nor other than it is, and is neither true nor false.

---

42 Spade, 'Roger Swyneshed's *Insolubilia*,' 181: 'Propositio pertinens ad inferendum se ipsam non significare principaliter sicut est est talis ex qua cum totaliter sic esse sicut est sequitur vel natum est sequi illam non significare principaliter sicut est. Exemplum: Significet haec propositio "Haec significat aliter quam est" principaliter quod haec significat aliter quam est, ipsamet demonstrata. Tunc sequitur "Haec propositio significat aliter quam est; et principaliter sic significat; igitur, haec non significat principaliter sicut est". Et ita ex illa cum totaliter sic esse sicut est sequitur quod ipsa non significat sicut est. Et ideo nec significat aliter quam est nec sicut est.'

43 Spade, 'Roger Swyneshed's *Insolubilia*,' 219: 'Nam sequitur immediate "Illa non significat sicut est; ergo, illa non significat sicut est".'

44 Spade, 'Roger Swyneshed's *Insolubilia*,' 219: 'Sed oportet addere antecedenti quod illa non foret pertinens ad inferendum se ipsam non significare sicut est. Et si illud addatur, illud est negandum.'

6 John Dumbleton

As noted, Heytesbury discusses, and rejects, the solutions to insolubles proposed by Swyneshed, by Dumbleton, and by Kilvington. He then adopts, reluctantly and with doubts as to there being any thoroughly satisfactory solution, a variant of Bradwardine's. If we turn to Dumbleton's own treatise, we find him discussing and rejecting the approaches of William of Ockham, of Bradwardine, of Swyneshed, and of Heytesbury.

John Dumbleton was probably a native of Dumbleton in Gloucestershire. He became Master of Arts at Oxford and Bachelor of Theology at Paris. He was listed as Fellow of Merton College in 1338 and was also a founding Fellow of Queen's College in 1341. He studied theology at Paris and was Fellow of the Sorbonne in the mid-1340s. Returning to Oxford around 1347, he was listed at Merton in 1348. His *Summa logicae et philosophiae naturalis* was left unfinished at his death, probably from the Black Death in 1349.[45]

Dumbleton's treatise on insolubles constitutes a very small part of his long treatise, running to some 400,000 words, entitled *Summa logicae et philosophiae naturalis*. It is preserved in 21 manuscripts, though two omit the first part, on logic. Part 1: *Summa logicae* opens with a short treatise on signification (*De significatione termini*). This is essential for underpinning and defending Dumbleton's approach to the insolubles, for he argues for a version of *cassatio*: insolubles signify in such a way that they fail to express propositions. The reason for this failure, he says, is that an utterance, spoken or written, is a proposition only if it communicates something from the speaker to the hearer.

The treatise opens with an at first surprising statement:

> No term signifies anything naturally.[46]

The manuscripts all agree in this reading, but in one manuscript the scribe inserts '*in voce vel scripto*' interlinearly after '*terminus*', that is, 'spoken or written'. Dumbleton accepts, along with most medievals, Aristotle's statement in *De interpretatione* (16a3) that 'spoken sounds are symbols of affections in the soul and written marks of spoken sounds.' But he restricts the word 'term' to spoken

45 See E.D. Sylla, 'Dumbleton, John,' in C.C. Gillispie, F.L. Holmes, N. Koertge and T. Gale (eds.), *Complete Dictionary of Scientific Biography*, vol. 7, New York 2008, 116–117.

46 *De significatione termini*, ch. 1: 'Nullus terminus aliquam rem significat naturaliter.' All citations from Dumbleton's *Summa logicae* are preliminary readings from an edition in preparation by Barbara Bartocci and myself.

sounds and written marks.[47] They are symbols of intentions (*intentiones*) in the mind (*anima*) or understanding (*intellectio*).[48] Consequently, given that they don't signify naturally, he says, we need

> to explain first in what manner written and spoken terms signify; secondly, in what manner they are imposed to signify; thirdly, in what manner a term is significative of any thing whatsoever at ⟨our⟩ pleasure; fourthly, how we come to know terms; and lastly in what manner any thing can be signified by any term.[49]

Dumbleton's answer is that a term signifies only when, on account of the proper intention (*intentio propria*) which is linked to the term, the hearer or the reader actualizes the true intention of the thing (*intentio vera rei*) which he has stored in his memory. This presupposes that there has already been a first imposition, namely, that a relation called 'signification' has already been established between the proper intention of the term (*propria intentio termini*) and the true or proper intention of the thing (*vera/propria intentio rei*). In other words, a term only signifies when and insofar as our understanding performs an act

47 Weisheipl, 'Ockham and some Mertonians,' 204, misunderstands Dumbleton's claim to mean that even concepts signify *ad placitum*, using this to cast doubt on the authenticity of some additional chapters on sophisms and other matters inserted in five of the nineteen extant manuscripts of Dumbleton's *Summa logicae* (Weisheipl cites Oxford, Merton College Library, Ms. 306, 1[3]ra, the first folio of the additional material, the third in the manuscript), where it is said: 'intentions in the mind do not signify arbitrarily, but naturally' ('intentiones in anima non ad placitum, sed naturaliter significant'). But Dumbleton agrees; he just decides not to call intentions 'terms'.

48 D.N. Hasse, 'Influence of Arabic and Islamic Philosophy on the Latin West,' in E.N. Zalta (ed.), The Stanford Encyclopedia of Philosophy (Fall 2014 Edition), URL = https://plato.stanford.edu/archives/fall2014/entries/arabic-islamic-influence/, writes: 'In Arabic-Latin translation literature, *intentio* is very often used to render *ma'nâ*, with the consequence that the term *intentio* took on a similarly broad semantic range as its Arabic counterpart. In the writings of Avicenna, *ma'nâ* may mean "concept", but also "meaning" of a word, or something "intelligible" by the intellect, or "perceptible" by estimation but not by the external senses … In Averroes' epistemology, the term *ma'nâ* has a specific meaning as the object of memory and a broader meaning as the abstracted content of sensory, imaginative or intelligible forms.'

49 *De significatione termini*, ch. 2: 'Cum ergo termini nec in voce nec in scripto ex natura eorum rem complexe et incomplexe significant, ut probatum est, restat exprimere primo qualiter termini significant in scripto et in voce, secundo qualiter imponuntur ad significandum, tertio qualiter terminus cuiuscumque rei ad placitum sit significativus, quarto quomodo in cognitionem terminorum deveniemus, et ultimo qualiter quelibet res a quolibet termino potest significari.'

which consist in recalling to mind or remembering the true intention of the thing by means of the proper intention of the term. In brief, all signification is mediated by intentions, or simple acts of understanding:

> For a term to signify something in a simple way is to actualize, i.e., call to mind that thing's intention by means of the term's proper intention, ⟨and⟩ that thing is said to be signified by that term in normal usage … For a term or terms to signify in a ⟨propositionally⟩ complex way is in the first place to actualize a thing's intention or different things' intentions by a simple act of comprehension.[50]

Consequently, Dumbleton draws the conclusion that

> a term's signifying to someone is nothing other than that term's calling to mind the true intention of something else by means of the ⟨term's⟩ proper intention.[51]

Indeed, what is (propositionally) complex or simple (*complexum et incomplexum*) exists only in the mind, for truth and falsity are there alone. Dumbleton cites Averroes' commentary on Aristotle's *Metaphysics* (1027b25–a1), where Aristotle notes that truth and falsity are accidental beings in the mind, and consequently, Averroes comments, composition and division are similarly only in the mind.[52] The term 'is' (*est*) is not itself the copula, but it prompts the mind to conjoin or compose (*copulare*):

50 *De significatione termini*, ch. 2: 'Primi membri descriptio talis est: terminum significare aliquam rem incomplexe est ipsius termini propria intentione intentionem rei ad actum sive ad memoriam reducere que res per talem terminum iuxta usum dicitur significari … Sequitur descriptio secundi membri que talis est: terminum sive terminos significare complexe est intentionem rei vel intentiones aliarum rerum ad actum simplicem intelligendi reducere.'

51 *De significatione termini*, ch. 2: 'Ex istis patet quod apud aliquem terminum significare non est aliud nisi illum terminum per intentionem propriam intentionem alterius rei veram ad memoriam reducere.'

52 *De significatione termini*, ch. 4: 'Ideo concluditur quod complexum et incomplexum solum habent esse in anima; ista sententia predicta habetur ab Aristotele sexto *Metaphysice* textu commenti ultimi, ubi ponit quod verum et falsum sunt entia per accidens et sunt in mente et non in rebus, super quo dicit Commentator quod illud quod est in cogitatione est compositio et divisio absque eo quod extra animam et illa entia non essent nisi esset anima.' See Averroes, *In Metaphysicam*, Venice 1562, repr. Frankfurt am Main 1962 (Aristotelis opera cum Averrois commentariis, 8 151^{v}–152^{v}).

> But just as sense-perceptibles prompt us to form a thought, the term 'is' does likewise for ⟨English speakers⟩, and yet it is not the copula, but it prompts the soul to couple for what it perceives by means of ⟨the thing's⟩ proper intention.[53]

This analysis of the notion of signification leads Dumbleton to the first conclusion of his treatise on insolubles:

> The first of the conclusions is this: any term signifies only while and insofar as there is actual comprehension by it.[54] This conclusion is clear like this: no term signifies by its own nature, nor by ⟨any nature⟩ introduced ⟨in it⟩ by an impositor, but terms only make us comprehend other things; therefore, if a term signifies, it signifies only while and insofar as ⟨there is actual apprehension through it⟩. Moreover, an intention only signifies while there is actual apprehension through it; but an intention signifies to us naturally-before and more than a term, and the signification of a term ⟨derives⟩ from the signification of the intention; therefore all the more so does a term only signify in so far as there is actual apprehension through it.[55]

The upshot is that Dumbleton's solution to the insolubles is part of a broader analysis of linguistic understanding and communication. Not only do insolubles fail to constitute propositions, so too do many other expressions.

This is in marked contrast to Bradwardine's and Heytesbury's solutions, which apply specifically and only to insolubles. Kilvington's and Swyneshed's

---

53 *De significatione termini*, ch. 4: 'Sed licet talis terminus "est" movet nos ad componendum, non tamen est copula, sed sicut sensibilia movent nos ut componamus, sic ille terminus "est" respectu nostri, et tamen non est copula, sed movet ut anima copulet pro illo quod percipit propria intentione.'

54 We could perhaps read between the lines something like 'only while and insofar as, through the proper intention of the term (or, the term's own intention–*propria intentio termini*), the term recalls to the intellect the true intention of the thing (*intentio vera rei*) …'

55 *Insolubilia*, ch. 13: 'Conclusionum prima est hec: nullus terminus significat nisi dum et quatenus est actualis comprehensio per eundem. Hec sic patet: nullus terminus significat ex natura propria nec inducta ab imponente, sed solum termini faciunt nos alias res intelligere; ergo si terminus significat, non significat nisi dum et quatenus etc. Item: intentio non significat nisi dum est actualis comprehensio per eandem; sed prius naturaliter significat nobis intentio et magis quam terminus, et termini significatio est ex significatione intentionis; ergo a multo maiori terminus non significat nisi quatenus habetur actualis comprehensio per illum.'

solutions are also a consequence of their reflections on wider issues: e.g., Swyneshed notes that his argument that sophisms such as 'This proposition signifies other than it is' are neither true nor false likens them to future contingents such as Aristotle's example of the possible future sea-battle, similarly neither true nor false.[56] Moreover, Kilvington's diagnosis of insolubles is an offshoot of his revision of the rules for obligations, realising that not only must one grant propositions one knows to be false and deny some that one knows to be true, one must also express doubt about some that one actually knows to be true or to be false, evaluating irrelevant propositions as if the *positum* was true.[57] Dumbleton defines insolubles as propositions which would be both true and false if one were to admit some apparently possible scenario.[58] He diagnoses the problem as arising from the presence of terms like 'truth', 'falsehood', 'proposition' and so on, which signify complex thoughts. Obligational scenarios should only be admitted if there is no circle of priority involved. Take 'A truth exists' (*Verum est*), given the scenario that this is the only proposition. The term 'truth' must stand in some way for a thought, according to his third and fourth conclusions. Although 'truth' is not itself a proposition, what it stands for must be a thought, or mental complex:

> The third conclusion is ⟨this⟩: no proposition in the soul is apprehended through a simple intention but through a ⟨propositional⟩ complex … The fourth conclusion, following from the third, is this: the subject or predicate of any proposition signifying ⟨in some way⟩ about a ⟨propositional⟩ complex is ⟨itself⟩ a proposition, and a proposition corresponds to one or both.[59]

But the proposition to which 'truth' corresponds must be a second proposition, different from the proposition 'A truth exists' itself, on pain of circularity or regress. Thus the scenario posited, where 'A truth exists' is the only proposition, must be rejected as impossible. Recall from § 3 that the Respondent in an obligation should admit a scenario and *positum* only if it is possible. This is not

56 Spade, 'Roger Swyneshed's *Insolubilia*,' 191.

57 See S. Read, 'Richard Kilvington and the Theory of Obligations,' *Vivarium* 53 (2015), 391–404, at 401–404.

58 *Insolubilia*, ch. 18: 'Descriptio insolubilis est talis: insolubile est propositio que ex apparenti casu possibili admisso concluditur esse veram et falsam.'

59 *Insolubilia*, chs. 15–16: 'Tertia conclusio est: nulla propositio in anima per intentionem simplicem, sed per complexum comprehenditur … Quarta conclusio sequens ex tertia est hec: cuiuscumque propositionis significantis pro complexo subiectum vel predicatum est propositio, et alteri vel utrique propositio correspondet.'

to deny that 'A truth exists' signifies in some way. But its signifying in that way is not sufficient for it to count as a proposition. Thus the proposed scenario is inherently contradictory and so, being impossible, should not be admitted.

The anonymous author of an epitome (*compendium*) of Dumbleton's *Insolubles* 'approved according to usage at Oxford' (*accepta secundum usum Oxonie*),[60] divides insolubles, or rather insoluble scenarios, into six groups (*ordines*), each with a rule for how to respond to them. In four of these groups, the scenario should simply not be admitted. These are the first group, where it is assumed in the scenario proposed that there is only one proposition, whose extreme (subject or predicate) supposits for a complex thought (*complexum*), e.g., 'A truth exists', 'A falsehood exists'; the second, where a proposition is referred to (*demonstrari*) by its own subject or predicate, e.g., 'It is not as this proposition signifies'; the third, where there is a circle of self-reference, e.g., where *A* refers to *B* and *C*, saying one of them is true, while *C* refers to *A* and *B*, saying that not both of them are true; and the sixth group, where it is proposed that someone believes just one proposition whose extreme supposits for a complex thought, e.g., where Socrates believes 'Socrates is deceived'. Although not spelled out explicitly, the idea is clearly that, although there is an utterance involved which signifies in some way, the order of priority results in a regress whereby its signification is not well founded and the utterance fails to constitute a proposition. So the scenario must be rejected in proposing that the utterance does constitute a proposition.

This threat of infinite regress is exactly in line with Dumbleton's diagnosis. Considering the proposition 'This proposition is false', he writes:

> If one supposes that the proposition 'This proposition is false', or a similar one, is spoken, ⟨call it *A*,⟩ take what is exactly comprehended through its subject—about which the coupling is performed in order to produce the proposition—call that *B*. *B* is either proposition *A* or another proposition. Not another, because then there would not be a composition about the self-same proposition, which contradicts the hypothesis. If what is signified is *A*, then *A* is not the proposition 'This proposition is false', but this one '"This proposition is false" is false'. The inference holds like this:

60 See F. Bottin, 'Gli "insolubilia" nel 'curriculum' scolastico tardo-medievale (con l'edizione di un anonimo compendio degli 'Insolubilia' di Giovanni Dumbleton),' in C. Giacon (ed.), *Logica, epistemologia, storia della storigrafia*, Padua 1980, 13–32, at 25–32. A revised edition and English translation, with discussion, can be found in B. Bartocci and S. Read, 'John Dumbleton on insolubles: an edition of an Epitome of his solution to insolubles,' *Noctua: La tradizione filosofica dall'antico al moderno*, 2022 (forthcoming).

> because *A* is a proposition equivalent to one where the subject refers to the proposition *A* itself, which signifies ⟨in some way⟩ about *A*—e.g., take the proposition: 'This proposition exists', ⟨call it *C*,⟩ referring to 'God exists'. ⟨*C*⟩ is the same as the proposition '"God exists" exists'—so, in the scenario proposed, to the subject of the proposition '"This proposition is false" is false' there would then correspond a ⟨further⟩ proposition and so on ad infinitum. And because there is successive reference ad infinitum, it follows that it is not fully comprehended.[61]

The solution presented by Dumbleton here is highly reminiscent of that put forward by Gilbert Ryle, sometimes referred to as the 'namely-rider' solution:

> When we ordinarily say 'That statement is false', what we say promises a namely-rider, e.g., '... namely that today is Tuesday'. When we say 'The current statement is false' we are pretending either that no namely-rider is to be asked for or that the namely-rider is '... namely that the present statement is false'. If no namely-rider is to be asked for, then 'The current statement' does not refer to any statement. It is like saying 'He is asthmatic' while disallowing the question 'Who?' If, alternatively, it is pretended that there is indeed the namely-rider, '... namely, that the current statement is false', the promise is met by an echo of that promise. If unpacked, our pretended assertion would run 'The current statement (namely, that the current statement (namely that the current statement ...'. The brackets are never closed; no verb is ever reached; no statement of which we can even ask whether it is true or false is ever adduced.[62]

The fourth and fifth groups are different. In these cases, the scenario may be admitted, but the paradoxical reasoning should be rejected. The fourth group consists of contradictory sets of propositions like the third, but where the circle

---

61 *Insolubilia*, ch. 13 (or 18): 'Cum ponitur quod talis sit in voce, hec est falsa, vel consimilis, ⟨que sit a⟩ capiatur illud quod precise intelligitur per subiectum eius pro quo est copulatio ut fiat propositio, et sit illud b. Vel est b ista propositio a vel alia. Non alia, quia tunc non esset compositio pro ipsamet, quod est contra ypotesim. Si significatum sit a, ergo a non est talis, hec est falsa, sed talis, hec est falsa est falsa. Consequentia tenet sic, quia a est propositio talis ac si a poneretur in loco subiecti demonstrantis a propositionem, que significat pro a, ut hec, hec est, demonstrando deus est, eadem est cum hac, deus est est, ita in casu proposito, tunc subiecto illius: hec est falsa est falsa, correspondet propositio et sic in infinitum. Et quia est demonstratio reciproca in infinitum, ideo non ad plenum intelligitur.'

62 G. Ryle, 'Heterologicality,' *Analysis* 11 (1951), 61–69, at 67–68.

of self-reference is not vicious or regressive, e.g., where *C* is 'Every truth is one of these', referring by 'one of these' to two true propositions *A* and *B*, by hypothesis, the only other propositions that exist. If *C* is true, then it is false, for *C* is not one of *A* and *B*; while if *C* is false, it must be true, for by hypothesis, it is the only proposition besides *A* and *B*. The solution proposed appears to be a kind of *restrictio*: even if *C* is true, it cannot be included under the signification of its subject 'truth'—else there would be a regress and *C* would not be a proposition at all. *C* is true, for all it really says is that each of *A* and *B* is one of *A* and *B*.

The fifth group concerns cases where someone says, or hears or reads, a proposition whose extreme supposits for a complex thought, e.g., if Socrates says 'Socrates says a falsehood'. Once again, the scenario may be admitted, but the reasoning is faulted by appeal to *restrictio*:

> When it is argued: 'Socrates says a falsehood and *A* ["Socrates says a falsehood"] signifies only that Socrates says a falsehood, therefore *A* is a truth', I deny the inference and one may respond in this way to every insoluble of the fifth group by admitting the ⟨whole⟩ argument until we reach the argument just denied or one like it, which should be denied. The reason why this inference is not valid is this, that in proposition *A* the predicate is a term suppositing for a ⟨propositional⟩ complex and no such term can supposit for a proposition of which it is the subject or predicate, hence it is required that it supposit for some other proposition. If the predicate 'falsehood' in that proposition 'Socrates says a falsehood' supposits for a proposition other than 'Socrates says a falsehood', the proposition 'Socrates says a falsehood' is false, because it signifies that Socrates says a proposition which he does not say.[63]

Thus what Socrates says is false, but the further inference that it is true is blocked. If a part could supposit for a whole of which it is part, we would have

63 Bottin, 'Gli "insolubilia",' 30, corrected against Ms. Padua Biblioteca Antoniana 397 Scaff. XVIII, 119^{r-v}: 'Et tunc ad argumentum quando arguitur: "Sortes dicit falsum et a sic precise significat quod Sortes dicit falsum, igitur a est verum", nego consequentiam et ita respondetur ad omne insolubile quinti ordinis admittendo totum usque quo deveniat ad illud argumentum iam negatum vel consimile ei, ⟨quod⟩ debet negari. Causa quare talis consequentia non valet est ista, quia in a propositione predicatum est terminus supponens pro complexo et nullus talis terminus potest supponere pro illa cuius est subiectum vel predicatum, ideo oportet quod supponat pro alia quacumque propositione. ⟨Si pro⟩ alia ab illa "Sortes dicit falsum" supponit illud predicatum "falsum" in illa propositione, "Sortes dicit falsum", falsa est illa propositio, "Sortes dicit falsum", quia significat Sortem dicere propositionem quam non dicit.'

a vicious circularity. That would lead to a breakdown in communication. But communication, we have seen, is for Dumbleton essential to utterances constituting propositions. Thus Dumbleton's solution is a subtle combination of *cassatio* and *restrictio*, by which paradox is averted.

7 Conclusion

Thomas Bradwardine's work initiated not just a revolution in mathematical physics, but a riot of responses to the logical paradoxes, the so-called 'insolubles'.[64] Most of the big names amongst the Calculators cut their teeth on their subtlety and difficulty. As Bradwardine (and others) observed, they are called 'insoluble' not because they are impossible to solve, but because of their sheer difficulty.[65] Their attempts at solution led them to explore fresh ideas about truth, signification (or meaning), and logic, the latter in a broad sense covering reasoning and argumentation, as developed in obligational disputations.

Bradwardine's 'multiple-meanings' solution built on the idea that each insoluble proposition is somehow implicitly contradictory, and by appeal to his closure postulate, that every proposition signifies anything that follows from what it signifies, he spelled out clearly how insolubles contradict themselves and must all be false. This involves a natural rethinking of the definition of truth. For the medievals, following Augustine, truth is an achievement, a zenith, such that any falling short results in falsehood.[66] Bradwardine infers that everything

64 Around 1360, Ralph Strode wrote: 'The opinions mentioned were in former times those of the old [logicians], who understood little or nothing about insolubles. After them there arose the prince of modern philosophers of nature, namely Master Thomas Bradwardine. He was the first to discover something worthwhile about insolubles' ('Predicte namque opiniones fuerunt antiquitus antiquorum, qui parum vel nihil de insolubilibus recte sapuerunt. Post quos surexit princeps modernorum physicorum videlicet magister thomas braduardyn qui aliquid quod valuerit de insolubilibus primitus adinvenit,' Ms. Erfurt, Universitätsbibliothek, Dep. Erf., CA Q. 277, 3vb, cited in P.V. Spade and S. Read, 'Insolubles,' in E.N. Zalta [ed.], *The Stanford Encyclopedia of Philosophy* [Fall 2017 Edition] URL = https://plato.stanford.edu/archives/fall2017/entries/insolubles/ § 3.1).

65 Thomas Bradwardine, *Insolubilia*, § 2.1: '"insoluble" is here taken [not in the first way, 'for what is in no way soluble' but] in this way: an insoluble is a difficult paralogism *secundum quid et simpliciter* resulting from the reflection of some act on itself with a privative determination' ('Insolubile autem sic acceptum describitur hoc modo: insolubile est difficilis paralogismus secundum quid et simpliciter ex reflexione alicuius actus supra se cum determinatione privativa proveniens').

66 Thomas Aquinas (*Summa theologiae* I, q. 16, a. 1, ad 1) quotes Augustine: 'Truth is a supreme likeness to a principle without any unlikeness' ('veritas est summa similitudo principii,

signified by a proposition must obtain for it to warrant the epithet 'true'. Insolubles signify both that they are false and that they are true, and nothing can satisfy that demand. That is what blocks the second leg of the paradox argument in § 1, from concluding from the first leg that the liar proposition is false to inferring that it must therefore be true. Not so: that it is false is not enough to show that it is also true. So the insolubles are simply false.

Although broadly influential, Bradwardine's solution itself commanded fairly narrow support. Robert Eland, for example, suggests the best solutions available were Bradwardine's and Heytesbury's, and one must choose between them.[67] Heytesbury's is a variant of Bradwardine's, dropping the closure postulate and thereby the proof that every insoluble signifies its own truth as well as its own falsity, relying instead on the simple but surprising conclusion that every insoluble must have some secondary, unspecified, signification, on pain of contradiction. Nonetheless, the upshot is the same: insolubles are false, since in some way they fall short of the demands of truth.

Like Heytesbury's, Kilvington's approach was also introduced in the context of obligational disputations. Given a certain scenario, e.g., where Socrates' only utterance is 'Socrates says something false', should we grant or deny what he says? It is not that simple, he replies. Strictly speaking, it is neither true nor false without qualification. But it is true in a certain respect, because what he said was false, so we should grant it, and deny that it is (simply) false; and it is also false in some respect, for we have granted what he said, so we should deny that it is (simply) true.

---

que sine ulla dissimilitudine est'), and John Buridan (*Questiones Elencorum*, eds. R. van der Lecq and H.A.G. Braakhuis, Nijmegen 1994 [Artistarium, 9], 91) writes: 'If there were a proposition which however it signified, so it was and signified in no way other than it was, it would be true. And as soon as it signified in some way other than it was, it would cease to be true and would begin to be false, so it is that, just as for something to be said to be the greatest heat, it is necessary that it does not have any degree of coldness, similarly for a proposition to be true, it is necessary that it signifies in no way other than it is' ('si esset aliqua propositio que qualitercumque significaret ita esset et nullo modo significaret aliter quam esset, ista esset vera. Et quam cito significaret aliqualiter aliter quam esset, desineret esse vera et inciperet esse falsa, ita quod sicut ad hoc quod aliquid dicatur summe calidum, requiritur quod non habeat aliquem gradum frigiditatis, ita ad ⟨hoc⟩ quod aliqua propositio sit vera, requiritur quod nullo modo significet aliter quam est').

67 See S. Read and M. Thakkar, 'Robert Fland—or Elandus Dialecticus?,' *Mediaeval Studies* 78 (2016), 167–180, at 173: 'So ⟨these⟩ two responses [sc. Bradwardine's and Heytesbury's] are better than the others for solving insolubles. So the respondent should choose one of them for his solution to the insolubles' ('Ideo duae responsiones sunt meliores aliis ad insolubilia solvenda. Eligat ergo respondens unam istarum pro sua solutione ad insolubilia').

Although Swyneshed had his own radical approach to obligations, he takes a much more direct approach to the insolubles, one which makes no play with any disputational context. Rather, he says, we should recognise that some propositions falsify themselves, and accept that if so, they really are false. This means revising the account of truth and falsity, so that a proposition is said to be true only if it does not falsify itself (and signifies as it is). But the paradoxes of signification require us to go further, and accept that some propositions are neither true nor false, and indeed neither signify as it is nor other than it is.

Finally, Dumbleton takes up Bradwardine's challenge where Heytesbury left off, and looks more closely at the very notion of signification. Propositions do not have meaning in themselves or naturally, but only by being used in communication. It is by that practice of communication and comprehension that they come to be signs of mental complexes. But ultimately, they must lead back to simple intentions, on pain of regress. Problematic in this regard are terms which are themselves signs of propositionally complex thoughts. If they were signs of propositionally complex thoughts of which they themselves are part, there would be a regress and their signification would be nugatory, failing to constitute a proposition. This is so in particular for the insolubles, which are either false, if, to avoid circularity, such terms are restricted to supposit only for other propositions; or else, without such a restriction, they fail to express meaningful propositions.

The disparateness of the varied solutions proposed by the Calculators might suggest that no progress was made. But this is to misconstrue the nature of progress in philosophy, which often comes not from narrowing the options available but by broadening them, and through the conceptual revisions and deeper understanding that come from this. The Calculators made the second quarter of the fourteenth century a lively and disputatious time for the discussion of insolubles.

Acknowledgements

The present work was funded by Leverhulme Trust Research Project Grant RPG-2016-333: 'Theories of Paradox in Fourteenth-Century Logic: Edition and Translation of Key Texts.'

CHAPTER 6

The Influence of the Oxford *Calculatores* on the Understanding of Local Motion: The Example of the *Tractatus de sex inconvenientibus*

Sabine Rommevaux-Tani

1 Dating of the Treatise

The *Tractatus de sex inconvenientibus* is an anonymous treatise written by someone familiar with the theories put forward by the Oxford *Calculatores*, including Thomas Bradwardine and William Heytesbury, both of whom are quoted by name.[1] Thus, the writing of the *Tractatus de sex inconvenientibus* can be situated after 1335, the date of Heytesbury's *Regule*, as noted by Curtis Wilson.[2] Furthermore, as Marshall Clagett points out, the *Tractatus de sex inconvenientibus* is quoted by John Dumbleton, in the *Summa logicae et philosophiae naturalis*,[3] meaning that the treatise was written no later than the late 1340s. Recently, Gustavo Fernandez Walker[4] has highlighted very strong similarities between a passage in the second version of a theological *Quaestio* by Nicholas of Autrecourt[5] and some arguments found in the *Tractatus de sex*

1 For example, an explicit quote from Thomas Bradwardine's treatise can be found in the question on local motion. See S. Rommevaux-Tani, *De sex inconvenientibus, un traité anonyme de philosophie naturelle au XIVe siècle*, édition, introduction et étude doctrinale, Paris 2022, 297: 'Prime due positiones demonstrative a pluribus improbantur … a magistro Thoma de Bradwardyn in tractatu suo *De proportionibus*' (all further references to the *De sex inconvenientibus* are to this edition). A quote from Heytesbury's *Regule* is found in the question on increase (262): 'Et illam positionem sustinet unus sollempnis magister potissimus et famosus Hethysbyry in suo tractatu capitulo de augmentatione.'

2 C. Wilson, *William Heytesbury: Medieval Logic and the Rise of Mathematical Physics*, Madison, WI, 2nd ed., 1960 (University of Wisconsin Publications in Medieval Science, 3), 7.

3 M. Clagett (ed.), *Nicole Oresme and the Medieval Geometry of Qualities and Motions. A Treatise on the Uniformity and Difformity of Intensities Known as* Tractatus de configurationibus qualitatum et motuum, Madison, WI, 1968 (University of Wisconsin Publications in Medieval Science, 12), 619.

4 G. Fernandez Walker, 'A New Source of Nicholas of Autrecourt's *Quaestio*: The Anonymous *Tractatus de sex inconvenientibus*,' *Bulletin de philosophie médiévale* 55 (2013), 57–69.

5 For this *Quaestio*, see Z. Kaluza, *Nicolas d'Autrécourt. Ami de la vérité*, Paris 1995 (Histoire littéraire de la France, 42/1), 195–204, and C. Grellard, 'L'usage des nouveaux langages d'analyse

 | DOI:10.1163/9789004512054_008

inconvenientibus. This suggests that Nicholas copied these passages from the copy of the anonymous treatise found today in manuscript Paris, Bibliothèque Nationale de France, Ms. lat. 6559. Christophe Grellard positions the writing of the second version of Autrecourt's *Quaestio* to around 1335, the date of his lectures on the *Sentences* and, undoubtedly, before 1339 when he was summoned for trial in Avignon.[6] In conclusion, the anonymous treatise *De sex inconvenientibus* was probably written between 1335 and 1339.

2 Writing Location

It is difficult to establish where the *Tractatus de sex inconvenientibus* was written. Anneliese Maier suggests that the anonymous author could be one of Heytesbury's students and that the treatise was written at Oxford.[7] Curtis Wilson notes that the author is familiar with the works of Oxford masters who are referred to as 'scola oxoniensis'.[8] This reference to the Oxford School may suggest that the author himself was not part of it. On the other hand, in a Prague manuscript (Národní Knihovna České Republiky, Ms. VIII. G. 19)[9] which contains a copy of the *Tractatus de sex inconvenientibus*, the *explicit* mentions questions disputed in Paris.[10] Finally, we have seen that Nicholas of Autrecourt read the treatise shortly after it was written. The anonymous author's evident familiarity with the Oxford School might suggest that the treatise *De sex inconvenientibus* was written by a member of Oxford University, who could then have gone to Paris and disputed some of the questions there. This could have been

dans la *quaestio* de Nicolas d'Autrécourt. Contribution à la théorie autrécourienne de la connaissance,' in S. Caroti and J. Celeyrette (eds.), Quia inter doctores est magna dissensio. *Les débats de philosophie naturelle à Paris au XIVe siècle*, Florence 2004 (Biblioteca di Nuncius, 52), 69–95.

6 I thank Christophe Grellard for this information.

7 A. Maier, *Die Vorläufer Galileis im 14. Jahrhundert* (Studien zur Naturphilosophie der Spätscholastik, 1), 2nd ed., Rome 1966 (Storia e letteratura, 22), 96.

8 Wilson, *William Heytesbury*, 7. For references to the Oxford School, see the third position in question 1 (*De sex inconvenientibus*, 147): 'Si in generatione formarum sit certa ponenda velocitas, igitur talis velocitas attenderetur solum penes latitudinem forme acquirende, sicut ponit tertia positio et tenet tota scola oxoniensis'; and the third position in question 2 (213): 'Si in motu alterationis velocitas sit signanda etc., igitur talis velocitas attenditur penes proportionem latitudinum intensivarum, sicut ponit tota scola oxoniensis.'

9 Daniel A. Di Liscia and I have produced a description of this manuscript, including the *Tractatus de sex inconvenientibus*; it is appended to this article.

10 Praha, Národní Knihovna České Republiky, Ms. VIII. G. 19, 46v: 'Expliciunt questiones de motu parisius disputate.'

someone from Richard de Bury's entourage, including among others Thomas Bradwardine, Walter Burley, and Richard Kilvington.[11] This suggestion has yet to be confirmed.

3 The Structure of the Treatise

The structure of the *Tractatus de sex inconvenientibus* was well-thought-out by the author, so that if some questions had been disputed in Paris, the text we have is not the faithful *reportatio* of this disputation. Thus, in the first lines of the text, the author provides the outline he wishes to follow in the questions; he then speaks of a *Tractatus*:

> With regard to the question asked and the doubts to be debated about the ratios between the velocities for the motion of generation, alteration, increase, and local motion, I will follow this outline: first of all, when I can, I will discuss the above subjects, and then address these subjects according to the mode of the treatise.[12]

The treatise is organized around four main questions, to each of which are attached three articles. The main questions relate successively to the determination of the velocity of the generation of substantial forms, the velocity of alteration, the velocity of increase or diminution,[13] and the velocity of local motion.[14] The articles present studies relating to the main questions: links between generation, place and form, the generation of colors, the generation

11 See E. Jung-Palczewska, 'Works by Richard Kilvington,' *Archives d'histoire doctrinale et littéraire du Moyen Âge* 67 (2000), 181–223, especially 186; N. Kretzmann and B.E. Kretzmann (transl.), *The 'Sophismata' of Richard Kilvington. Introduction, Translation and Commentary*, New York 1991, xxviii–xxix; N.W. Gilbert, 'Richard de Bury and the *Quires of Yesterday's Sophisms*,' in E.P. Mahoney (ed.), *Philosophy and Humanism. Renaissance Essays in Honour of Paul Oskar Kristeller*, New York 1976, 229–257.

12 *De sex inconvenientibus*, 135: 'Circa propositam questionem et cetera dubia disputanda de proportionibus velocitatum in motibus generationis, alterationis, augmentationis ac motus localis, presentem servabo processum: in primis, ut potero, disputabo materias antedictas, deinde materias illas tradam per modum tractatus.'

13 Cf. S. Rommevaux-Tani, 'La détermination de la rapidité d'augmentation dans le *De sex inconvenientibus*: comparaison avec les développements sur le même sujet de William Heytesbury,' in C. Grellard (ed.), *Miroir de l'amitié. Mélanges offerts à Joël Biard*, Paris 2017 (Études de philosophie médiévale, 106), 153–162.

14 Cf. S. Rommevaux, 'Six inconvénients découlant de la règle du mouvement de Thomas Bradwardine dans un texte anonyme du XIVe siècle,' in M. Malpangotto, V. Jullien and

of the first qualities by the celestial bodies, the alteration of a piece of iron by a magnet,[15] the alteration of a medium by light, the existence of reaction,[16] rarefaction, the motion of a heavy body, the motion of a sphere, the middle degree theorem. For each of the main questions, the author presents three opinions, the third being the one he claims he supports, without dwelling on the reasons for his support. For each opinion presented he gives six difficulties or *inconvenientia* which would result from it, except for the fourth main question which concerns local motion, where he proposes three series of six difficulties to the third opinion, which is also Bradwardine's, and which he supports.

For each of the articles, the author again proposes six difficulties that would result from a negative answer to the question asked (when he is too restricted by the number of six, to which he seems to have limited himself, he adds additional '*items*'). This series is followed by references to authorities and, in varying lengths, by the *determinatio* of the question. The treatise is thus clearly structured around questions and series of six *inconvenientia*. Hence the title of the work: *Tractatus de sex inconvenientibus*.[17]

E. Nicolaïdis (eds.), *L'homme au risque de l'infini. Mélanges d'histoire et de philosophie des sciences offerts à Michel Blay*, Turnhout 2013 (De diversis artibus, 93), 35–47.

15 Cf. S. Rommevaux, 'Un auteur anonyme du XIVe siècle, à Oxford, lecteur de Pierre de Maricourt,' *Revue d'histoire des sciences* 67 (2014), 5–33.

16 Cf. S. Caroti, 'Da Walter Burley al *Tractatus sex inconvenientium*: La tradizione inglese della discussione medievale *De reactione*,' *Medioevo* 21 (1995), 257–374.

17 The number six is omnipresent in this treatise: six *inconvenientia*, six possible causes for the acceleration of the motion of a heavy body, six opinions about the motion of a sphere, six proofs for the middle degree theorem, as we shall see. As it was already known in the Antiquity, the number six is a perfect number, i.e., a number the 'aliquot parts' of which sum up to itself. For Boethius, who (following Nicomachus of Gerasa) mentions in his *Arithmetica* the four first perfect numbers, the perfect number is 'aemulator virtutis' (Cf. A.M.S. Boethius, *Institutio arithmetica*, ed. J.-Y. Guillaumin, Boèce, *Institution Arithmétique*, Paris 2002 (Collection des Universités de France), 19. See also on 196–197 the comments by Guillaumin). For the Middle Ages this fact had a significance beyond the pure mathematical knowledge on number theory. Alcuin of York, for instance—to mention only a famous personality—, in a letter to Charlemagne from 798, explains the virtues of the numbers seven and five, and describes at length the various meanings of the number six in the context of the *computus* and the Bible exegesis: 'The number six expresses the perfecting of our life, for the almighty creator did his perfect work in six days, and the number six is perfect in its parts, as one of your learning will know ... Also there are six days each between Septuagesima and Sexagesima, Sexagesima and Quinquagesima, and Quinquagesima and Quadragesima, and three sixes are in the ratio of one to two, which signifies the single death of the divine Christ compared with our double death ...

4 The Three Articles on Local Motion

We will focus here on the three articles that relate to the fourth main question,[18] which concerns local motion:

1. Whether the acceleration of the motion of a heavy body occurs from some definite cause;
2. Whether the velocity of the motion of a sphere depends on a point or some space;
3. Whether the velocity of all uniformly difform local motion beginning at no degree is equal to its middle degree.[19]

These articles provide good examples of the reception of physics developed at Oxford.

5 The Acceleration of the Motion of a Heavy Body

In article 1 of question 4, the author wonders whether the cause of the acceleration of the motion of a heavy body can be determined. He presents six possible causes: a decrease in the resistance of the medium, the continuation of the motion, the proximity of the body to its natural place (this is Aristotle's thesis in *De caelo* I.8, 277a27–277b8), the propulsion (*propulsus*) of the medium, the accidental heaviness acquired during the fall, and finally the inclination (*appetitus*) of the body to reach its natural place. In each case, the author proposes a series of difficulties which would arise from the corresponding opinion. We will give one example of these arguments for each opinion; the others are often of the same order.

According to the first opinion, if the decrease in resistance is the cause of acceleration, during the fall of a heavy body from the orb of the moon, the motion would be decelerated, since the media that the body goes through—

For this speculation you can show your friends how enjoyable and useful is the study of arithmetic' (cf. S. Allott, *Alcuin of York: His Life and Letters*, York 1974, 97. For the Latin text see, Alcuin of York, *Epistolae*, ed. E. Dümmler, *Monumenta Germaniae Historica, Epistolae* IV.2., Berlin 1895, 224–227, at 226–227). I am grateful to Daniel Di Liscia for his remarks on this topic.

18 For a partial study of question 4, see Rommevaux, 'Six inconvénients.'

19 *De sex inconvenientibus*, 315: 'Utrum velocitatio motus gravis sit ab aliqua causa certa'; 326: 'Utrum velocitas motus spere cuiuslibet penes punctum vel spatium aliquod attendatur'; 342: 'Utrum velocitas omnis motus localis uniformiter difformis incipiens a non gradu sit equalis suo medio gradui.'

fire, air, and water, successively—become denser.[20] The author replies to this argument by explaining that partial resistances are not to be considered, but the total resistance from the position of the heavy body to the Earth, which decreases as the heavy body moves downward.[21] Second, if the acceleration is caused by the continuity of the motion, the Earth, which moves continuously due to the heat of the sun, would have accelerated ever since its creation, so that today it would have reached a significant velocity that would cause the destruction of all buildings.[22] In the same way, the motion of celestial orbs and clocks would accelerate.[23] Third, if the acceleration of the motion is caused by the proximity of the moving body to its natural place, in a scale, the weight closer to the center of the world would move down faster than the other weight that moves away from it.[24] If one considers fourthly that for a falling body, the air that pushes the body is the cause of acceleration, a vacuum could be created in the place of the air that follows the body.[25] Furthermore, the air

20 *De sex inconvenientibus*, 324: 'Item: si minoratio resistentie sit causa velocitationis gravis, ut ponitur, igitur grave existens in concavo spere ignis velocius ibi moveretur quam in spera aeris, et in spera aeris quam in spera aque, et in spera aque quam in spera terre, quia maiorem resistentiam habet grave in spera aeris quam in spera ignis, quia aer est medium densius quam spera ignis, et in aqua habet maiorem resistentiam quam in aere a consimili ratione, igitur tale grave non impeditum ab aliquo alio extrinseco quam a mediis elementaribus continue tardaret motum suum et nunquam velocitaret.'

21 *De sex inconvenientibus*, 324: 'Ad secundum concedo quod grave existens in concavo spere ignis velocius movetur quam in spera aeris, et sic deinceps, ex quo non sequitur quod tale grave continue tardaret motum suum, quia licet in motu gravis sic descendentis continue crescat resistentia partialis, minoratur tamen continue resistentia totalis, que est a spera ignis ad centrum mundi. Et penes minorationem totalem habet velocitatio huius gravis attendi.' See A. Maier, *An der Grenze von Scholastik und Naturwissenschaft* (Studien zur Naturphilosophie der Spätscholastik, 3), 2nd ed., Rome 1952 (Storia e letteratura, 52), 190.

22 *De sex inconvenientibus*, 317: 'Item: si continuatio sit causa velocitationis motus gravis, cum Terra ab initio sui et Solis fuit, et est in continuo motu propter calorem Solis, igitur Terra ab initio velocitabat motum suum, igitur nunc velocissime et sensibiliter movetur Terra. Et per consequens, continue erit motus Terre sensibilis, et sic ruerentur edificia, magna domus et castra.'

23 *De sex inconvenientibus*, 317: 'Item: si sic, cum motus celi et orbium planetarum sit continuus, igitur celum cum ceteris orbibus velocitaret motum suum continue … Item: si sic, cum motus horologii sit continuus, igitur motus talis esset intensior et intensior. Et per consequens motus talis per tempus esset sensibilis motus valde.'

24 *De sex inconvenientibus*, 318: 'Item, si ex appropinquatione istius gravis ad locum suum sequitur ipsum velocius moveri continue, contra … Item: si sic, sequitur quod equis ponderibus in equilibra appensis, si alterum deprimatur ab equidistantia orizontis versus centrum, velocius altero moveretur.'

25 *De sex inconvenientibus*, 320: 'Quarto, si velocitatio motus gravis esset ab aliqua causa certa, igitur propulsus medii esset causa velocitationis huius … Item: si sic, sit grave prope

under the body is compressed, so that the motion is more hindered than supported.[26] Fifth, if the accidental gravity acquired by the body during its fall is the cause of its acceleration, since a body of maximum heaviness—i.e., a piece of earth—cannot acquire additional gravity, its fall is not accelerated.[27] Finally, sixth, if the penchant of a body to reach its natural place causes the acceleration, a heavy body would have the same motion in a medium and in a vacuum.[28]

To these different opinions, the author answers that there is not a single cause explaining the acceleration of the motion, but several: 'the acceleration of a falling body, in its descent, comes from several causes, though one is more significant than the others.'[29] He then quotes a certain Adam of Pipewell, who was a fellow at Balliol College in 1321, then a fellow at Merton College by 1325, and was still there in 1327:[30]

> Then I say, with master Adam of Pipewell, that the decrease in the resistance is the principal cause whereas the continuation of the motion, the proximity, the impulse of the medium, the accidental heaviness, and the natural inclination, which is also the penchant, are partial causes. Indeed,

---

speram ignis et descendat. Tunc hoc grave in suo descensu pellitur ab aere medio; igitur aer insequitur. Et sequitur ultra: aer recedit a loco intermedio insequens motum gravis; nec ignis subsequitur; igitur ibi relinquitur vacuum.'

26 *De sex inconvenientibus*, 320: 'Item: corrupto medio supra lapidem, in vacuum non moveretur tardius descendendo, sed velocius, quia aer et medium superius pellens lapidem deorsum condensat medium inferius et per consequens facit quod medium inferius plus resistat. Et si sic, igitur talis pulsus medii plus impedit motum gravis quam promovet.'

27 *De sex inconvenientibus*, 321: 'Quinto, si velocitatio motus gravis etc., igitur gravitas accidentalis quam acquirit grave in descendendo esset causa velocitationis motus talis gravis ... Item: sit *a* aliquod summe grave et moveatur ad locum suum naturalem. Tunc *a* aut intendit motum suum aut non. Si sic, et non ab aliqua gravitate maiori accidentali vel per se, cum ipsum sit summe grave, igitur intensio huius motus est, et non a gravitate accidentali.'

28 *De sex inconvenientibus*, 322: 'Si velocitatio motus gravis etc., igitur appetitus esset causa velocitationis motus. Sed hoc non ... Item: quodlibet grave mundi appetit ita velociter moveri per medium plenum sicut per medium vacuum, ut patet ex dictis; et intensio motus sequitur appetitum; igitur ita velociter intendet motum suum per medium plenum sicut per medium vacuum.'

29 *De sex inconvenientibus*, 323: '... dico quod non, nam velocitatio motus gravis versus deorsum in suo descensu est a pluribus causis, licet una sit principalior ceteris.'

30 G.C. Brodrick, *Memorials of Merton College with Biographical Notices of the Wardens and Fellows*, Oxford 1885, 195; A.B. Emden, *A Biographical Register of the University of Oxford to A.D. 1500*, III: P to Z, Oxford 1959, 1484 (I thank Edith Sylla for this reference).

> none of these partial and auxiliary causes is necessarily required for the acceleration of the motion of the heavy body, as the arguments prove sufficiently.[31]

The anonymous author adds that this does not apply to all motions of a heavy body, but only to the descent to the center of the world. Indeed, when the lower limit of the body reaches the center, the motion continues while decelerating, until the center of the body reaches the center of the world,[32] as the author shows in the main question (the part of the body which is then under the center resists the motion and this part gets bigger and bigger so that the resistance increases[33]). Anneliese Maier notes that a very similar explanation can be found in the *Questions on Aristotle's Physics* found in a Venice manuscript (Biblioteca Nazionale Marciana, Ms. lat. VI. 72 [2810]), which she attributes to Richard Swineshead,[34] but which are now attributed to Richard Kilvington.[35] Kilvington became a Master of Arts at Oxford University in 1324/1325 and a Doctor of Theology around 1335. His *Questiones super Physicam* probably were compiled 1325/1326.[36] The *Tractatus de sex inconvenientibus* is later than this date. Anneliese Maier supposes that one of the authors copied from the other. In this case, our anonymous author would have copied Kilvington. But a third author plays a role in this story, Adam of Pipewell, with whom the author of the *Tractatus de sex inconvenientibus* says he agrees. The precise connection between Kilvington and Pipewell is unknown, but Adam of Pipewell was a fellow at Oxford when Kilvington probably wrote his *Questiones*.

31 *De sex inconvenientibus*, 323: 'Unde dico cum magistro Adam de Pipewelle, quod minoritas resistentie est causa principalis et continuatio motus, propinquitas, pulsus medii, gravitas accidentalis, inclinatio naturalis, que est appetitus, sunt cause partiales. Est enim quecunque illarum causa partialis et coadiuvans, non tamen est causa necessario requisita ad velocitationem motus gravis, sicut satis probant argumenta.'

32 *De sex inconvenientibus*, 323–324: 'Non tamen volo dicere quod aliqua illarum est causa principalis vel secundaria in velocitatione gravis per totum eius descensus, sed usque contingerit locum suum naturalem et centrum mundi. Nam deinceps grave continue tardat motum suum, cum continue post illud crescat resistentia, sive moveatur in vacuo vel in pleno.'

33 See the proof of the third difficulty of the first set of *inconvenientia* in the main question 4 (301–302).

34 Maier, *An der Grenze*, 190–191.

35 Jung-Palczewska, 'Works,' especially 182–183.

36 E. Jung, 'Richard Kilvington,' in E.N. Zalta (ed.), *The Stanford Encyclopedia of Philosophy* (Winter 2016 Edition), URL = https://plato.stanford.edu/archives/win2016/entries/kilvington/. See also Elżbieta Jung's contribution to the present volume.

6 The Motion of a Sphere

In the second article of the fourth main question, the author asks whether the velocity of a sphere depends on a point or some space (i.e., the space described during the motion, that can have one, two, or three dimensions).[37] Again, the author gives six possible opinions. He begins by examining the position, which he describes as common, according to which the motion of a sphere depends on its lowest point,[38] that is to say, in this argument, the center of the sphere. The author refutes this argument by explaining that, if one considers that the lowest point of the orb of the fixed stars is the Earth (considered as a point, following Ptolemy), the orb of the fixed stars would move as fast as the Earth and no faster, and therefore the motion of the Earth would be fast and perceptible.[39] But if one considers that the lowest point is the center of the world, which is motionless, the orb of the fixed stars and consequently all the orbs would not move.[40] It should be noted that what interests the author here is not the motion of spheres, conceived as mathematical objects, but the motion of orbs, which are physical objects.

The second opinion, for which the author refers to a treatise by Richard de Versellys, that is Gerard of Brussels' *De proportionalitate motuum et magnitudinum*[41] (cited by Bradwardine in his *Tractatus de proportionibus*[42]), is that

37 *De sex inconvenientibus*, 326: 'Utrum velocitas motus spere cuiuslibet penes punctum vel spatium aliquod attendatur.'

38 *De sex inconvenientibus*, 326: 'Si velocitas motus spere cuiuslibet penes punctum aliquod attendatur etc., igitur talis velocitas attendetur penes punctum infimum, sicut est positio aliquorum communis.'

39 *De sex inconvenientibus*, 327: '... tota Terra respectu orbis stellarum fixarum est quasi punctus, secundum Tholomeum in primo *Almagesti*, et est punctus ultimus mobilis saltem, ut patet; igitur secundum istam positionem motus orbis supremi attenditur penes illud punctum. Et constat quod iste punctus movetur continue, ut probatum est supra; igitur quam velociter precise movetur orbis supremus, tam velociter precise movetur orbis Terre, et econtra. Ex quo sequitur inconveniens adductum. Et per consequens cum motus celi sit velocissimus et sensibilis, igitur et motus Terre esset velocissimus et sensibilis.'

40 *De sex inconvenientibus*, 327: 'Item: si sic, cum punctus maxime infimus centrum mundi sit immobilis, sequitur quod penes non gradum motus attenderetur velocitas motus spere, et sic penes non motum attenderetur, quod est absurdum.'

41 See G. Eneström, 'Sur l'auteur d'un traité 'de motu' auquel Bradwardin a fait allusion en 1328,' *Archivio di Storia della Scienza* 2 (1921–1922), 133–136; M. Clagett, 'The *Liber de motu* of Gerard of Brussels and the Origins of Kinematics in the West,' *Osiris* 12 (1956), 73–175; M. Clagett, *Archimedes in the Middle Ages*, 5: *Quasi-Archimedean Geometry in the Thirteenth Century. A Supplementary Volume Comprising the* Liber de motu *of Gerard of Brussels, the* Liber philotegni *of Jordanus de Nemore together with its Longer Version Known as the* Liber de triangulis Iordani, *and an Appended Text of John Dee's* Inventa circa illam coni recti atque rectanguli sectionem quae ab antiquis mathematicis parabola appellabatur, Philadelphia, PA, 1984 (Memoirs of the American Philosophical Society, 157).

the velocity of a sphere would depend on its midpoint between the lowest point and the highest point.[43] In the first proposition of Book I of this treatise, Gerard of Brussels writes: 'Any part as large as one wishes of a radius describing a circle, [which part is] not terminated at the center, is moved equally as its middle point. Hence the radius [is also moved equally] as its middle point. From this it is clear that the radii and the motions have the same ratio.'[44] The author proposes several difficulties which would result from this opinion, namely that the orb of the fixed stars would move as rapidly as the midpoint of its radius and consequently as its middle orb, which is either the orb of the Sun, or an orb under the orb of the Sun. Therefore, the orb of Saturn or Mars would move faster than the orb of fixed stars, which goes against what all astrologers say.[45]

According to the third opinion, if the velocity of the motion of the sphere depends on a point, and since it cannot be either the lowest point or the middle point, it must be the highest point. And the author specifies that this is Bradwardine's opinion.[46] Then he proposes three difficulties that would follow from this position, each of which is related to the concept of uniformly difform motion. In particular, the author remarks that if the motion of the sphere depends on the motion of a point on the circumference, this motion is uniform. Yet if one considers all the points of the sphere, the motion is uniformly difform along a radius, because the further the point is from the center, the faster is its motion.[47] Since the author finally accepts this opinion, he explains that

---

42 Thomas Bradwardine, *Tractatus de proportionibus velocitatum in motibus*, ed. H.L. Crosby, Jr., *Thomas of Bradwardine. His* Tractatus de proportionibus. *Its Significance for the Development of Mathematical Physics*, Madison, WI, 1955 (University of Wisconsin Publications in Medieval Science, 2), 128.

43 *De sex inconvenientibus*, 328: 'Si velocitas motus cuiuslibet spere attenderetur etc., igitur attenderetur penes punctum medium inter punctum infimum et supremum, sicut est opinio et positio magistri Richardi de Versellys in tractatu suo.'

44 Clagett, *Archimedes*, 64: 'Quantalibet pars semidiametri circulum describentis ad centrum non terminata equaliter movetur suo medio puncto. Unde et semidiameter suo medio. Ex quo manifestum quod semidiametrorum et motuum, una est proportio'; English transl., 111.

45 *De sex inconvenientibus*, 328: 'Item: tunc orbis supremus stellarum fixarum moveretur eque velociter precise cum suo medio puncto, et per consequens cum suo orbe medio, puta orbe solari vel aliquo alio citra ipsum. Et per consequens motus suus non esset velocissimus, sed orbis Saturni vel Martis esset eo velocior, contra omnes astrologos.'

46 *De sex inconvenientibus*, 329: '… si velocitas motus spere attenderetur penes aliquem eius punctum, et constat quod non attenditur penes infimum vel medium, igitur velocitas motus spere cuiuslibet attendetur penes punctum supremum. Et ista est positio magistri Thome de Bradwardyn in tractatu suo *De proportionibus*.'

47 *De sex inconvenientibus*, 329–330: '… latitudo motus a centro spere usque circumferentiam est latitudo motus uniformiter difformis, quia latitudo motus cuius quilibet gradus

the argument is valid and that no difficulty arises from it.[48] After these three difficulties the author adds: 'Many things can be demonstrated on this subject, but I have left them aside to avoid the book being too long. I only touch on the subject briefly and give others subjects to study more thoroughly and to argue further.'[49]

The author has therefore examined all the possible ways in which the motion of a sphere could depend on a point. The fourth opinion, common and general, posits that the motion of a sphere would depend on the space described during the motion.[50] For the anonymous author, this opinion is the most erroneous.[51] He rejects it then, with a first argument that is close to the one we have just seen. He remarks that in a sphere there are infinitely many spheres that move at different velocities, since the corporeal spaces traveled are different; so the motion of the total sphere would be uniformly difform.[52] But if the motion of the total sphere depends on the corporeal space described by this sphere, the motion is uniform.

---

est remississimus qui non est supra et intensissimus qui non est sub, sicut patet per motum punctorum in spera. Motus enim cuiuscumque puncti in spera in alia et alia circumferentia est remississimus qui non est supra et intensissimus qui non est sub. Igitur tota latitudo motus spere est uniformiter difformis. Igitur, si spera data circumvolvatur in hora, hec spera movetur latitudine motus uniformiter difformis. Et tamen per eandem horam solum movetur uniformiter, quia ponatur quod punctus supremus tantum pertranseat de circumferentia maxima in uno tempore sicut in alio sibi equali. Tunc sic: punctus supremus illius spere per totam horam movebitur et movetur uniformiter.'

48 *De sex inconvenientibus*, 338: 'Conceditur tanquam possibilis et in casu supposito vera est. Et dico quod possibile est quod aliquid moveatur motu uniformiter difformi, et tamen uniformiter, sicut demonstrative probatur ex casu. Et est verum de omni spera sic mota.'

49 *De sex inconvenientibus*, 331: 'Multa possent hic argui, sed dimitto propter prolixitatem operis devitandam. Solum tango et breviter dans aliis materiam perscrutandi profundius et diffusius arguendi.'

50 *De sex inconvenientibus*, 331: '… si velocitas motus spere cuiuslibet penes aliquid attendatur et penes nullum punctum vel motum cuiuscumque puncti infimi, medii vel supremi attenditur, sicut patet et argutum est, igitur talis velocitas attendetur penes aliquod spatium et descriptionem alicuius spatii in tanto tempore vel in tanto. Sed hoc arguo esse falsum, quia si sic, igitur velocitas talis spere attenderetur penes spatium corporale descriptum a tali mobili vel a tali, sicut est communis positio et vulgaris.'

51 *De sex inconvenientibus*, 332: 'Infinita contraria alia possent adduci, sed transeo, quia reputo falsissimam istam positionem.'

52 *De sex inconvenientibus*, 331–332: 'Probatio: circumvolvatur aliqua spera. Tunc sic: in ista spera sic mota infinite sunt spere concentrice alie, quarum alique infinite pertranseunt maius spatium in equali tempore quam una certa data, et etiam quarum alique infinite pertranseunt minus spatium corporale in equali tempore quam una certa data, igitur in spera totali sunt infinite que moventur velocius et alie infinite que moventur tardius. Igitur totalis spera movetur difformiter.'

The fifth opinion is that the motion of a sphere depends on the surface traveled.[53] The author rejects this opinion by showing that, more generally, the motion of a body cannot depend on the surface traveled, because then the motion of a point *f* on a line *e* and the motion of this line cannot be compared, since the point *f* describes a line when the line *e* describes a square, and the line and the square are not comparable.[54] The same argument can be found in Bradwardine's *Tractatus de proportionibus*.[55]

The sixth and final opinion is that the velocity of the motion of a sphere depends on the point that moves most rapidly. The author remarks that this is the same opinion as the third he presented above and he recalls that it is Bradwardine's.[56] He accepts it as 'necessary and true', because he has rejected the other views.[57] However, following his outline, he proposes a number of difficulties to which this opinion could lead. We cannot go into detail concerning these difficulties, the explanations of which are often long.[58] We only point out here that one of them is based on a mathematical argument, consisting in asking whether the radii of two spheres, one of which is double the other, are in a double ratio.[59] The author uses a proposition from book XII of Euclid's *Ele-*

53 *De sex inconvenientibus*, 332: '... talis velocitas augeretur penes spatium superficiale descriptum a mobili in tanto tempore vel in tanto, sicut ponunt alii, et est secta quinta.'

54 *De sex inconvenientibus*, 332: 'Sed contra: ex isto sequitur quintum inconveniens adductum contra articulum. Probatio quod sequatur: sit *abcd* unum quadratum et superponatur per totum *ab* aliqua linea, et moveatur ista linea ab *ab* in *cd* describendo totum quadratum, sic quod equaliter per totum tempus motus equaliter a *cd*. Tunc accipio totam istam lineam, que sit *e* et punctum describentem, qui sit *f*. Et sequitur quod deduxi, quoniam in principio temporis *e f* mobilia equaliter distant a terminis suis fixis et per totum tempus equaliter distabunt in motu a suis terminis, et eque cito devenient ad suos terminos, ut patet ex casu, et tamen *e* improportionaliter velocius movetur per totum tempus quam *f*, quia in eodem tempore improportionaliter maius spatium describet quam *f*, quoniam *e* describet totum quadratum et *f* solum costam in equali tempore, igitur etc.'

55 Thomas Bradwardine, *De proportionibus*, 128.

56 Thomas Bradwardine, *De proportionibus*, 130: 'Cuiuslibet motus localis, velocitas secundum maximum spatium lineale ab aliquo puncto sui moti descriptum accipitur. Quorumlibet duorum motuum localium, velocitates et maximae lineae a duobus punctis duorum mobilium eodem tempore descriptae, eodem ordine proportionales existunt.'

57 *De sex inconvenientibus*, 333: '... velocitas motus spere cuiuslibet maxime mote circa centrum suum attenderetur penes spatium lineale a puncto velocissime moto descriptum vel penes spatia linealia a punctis velocissime motis in eodem tempore vel equali descripta, sicut tenet una opinio, et est positio magistri Thome de Bradwardyn, quam positionem reputo necessariam et veracem. Concordat tamen ista positio cum tertia, quas, ceteris abiectis, arbitror esse finaliter sustinendas.'

58 For more details, see the introduction to my critical edition of the *De sex inconvenientibus*.

59 *De sex inconvenientibus*, 334–335: 'Secundo, si positio esset vera, sequeretur quod aliqua spera moveretur in duplo precise velocius alia, et tamen motus istius ad motum alterius

ments, according to which the ratio between these spheres is the triplicate ratio of their diameters. He deduces that the ratio between the radii is smaller than the double ratio. So, mathematics is present in the treatise, especially in the reasoning that involves proportional parts of the subject or time, but the author of the *Tractatus de sex inconvenientibus* is not always thorough in his mathematical reasoning.

7 The Middle Degree Theorem

The subject of article 3 of the fourth main question is the middle degree theorem. Here, the author asks whether the velocity of a uniformly difform local motion beginning at no degree is equal to its middle degree.[60] In the statement

---

esset multo minor quam duplus. Probatio: sit *a* aliqua spera, et in spera *a* signo aliam speram concentricam per *b*, cuius dyameter sit precise subduplus ad dyametrum spere *a*, deinde signo punctum velocissime motum in *a* per *c* et punctum velocissime motum in *b* per *d* in eodem semidyametro, et circumvolvatur *a b* in equali tempore vel eodem. Tunc sic: *a* et *b* circumvolvuntur circa idem centrum; et *a* describet duplum spatium lineale in equali tempore vel eodem precise; igitur *a* in duplo velocius movetur quam *b* precise. Consequentia patet et antecedens similiter, cum equale spatium maximum lineale pertransietur in equali tempore ab *a b* et *c d*. Sed *c* pertransiet duplum spatium lineale in eodem tempore vel equali ad *d*, igitur et *a* pertransiet duplum spatium lineale etc. ad *b*. Igitur *a* precise in duplo velocius movebitur quam *b* in eodem tempore vel equali. Et tamen motus istius *a* erit multo minor quam duplus ad *b*. Quod arguo sic: motus ipsius *c* erit multo minor quam duplus ad motum *d*; sed motus *a* et *b* sunt equales motibus *c* et *d*, iuxta istam positionem; igitur motus istius *a* est multo minor quam duplus ad motum istius *b*. Assumptum arguo sic, scilicet quod motus ipsius *c* etc., nam arguo sic: *a* spera est precise dupla ad *b* speram, ut probabo, igitur dyameter *a* ad dyametrum *b* est multo minor quam dupla. Consequentia patet, quia proportio sperarum est proportio dyametrorum triplicata, ut patet 12 Euclidis, et per consequens dyameter *a* ad dyametrum *b* est multo minor quam dupla, ut patet. Et per consequens maxima circumferentia *a* ad maximam circumferentiam *b* erit multo minor quam dupla. Sed talis erit proportio motus *a* ad motum *b*, qualis est proportio maxime circumferentie *a* ad maximam circumferentiam *b*, ex positione data. Sed illa proportio circumferentiarum est multo minor quam dupla, ut dictum est; igitur motus *a* ad motum *b* erit multo minor quam duplus. Nunc, quod *a* spera sit precise dupla ad *b* speram, arguo sic: *a* secundum omnem dimensionem, secundum longitudinem, latitudinem et profunditatem est precise duplum ad *b*, igitur *a* est precise duplum ad *b*. Item: maxima dimensio *a* secundum longitudinem, latitudinem et profunditatem est precise dupla ad maximam dimensionem *b* secundum longitudinem, latitudinem et profunditatem, igitur *a* est precise duplum *b*. Utraque consequentia est satis nota et antecedens similiter notum, quia maxime dimensiones istius *a* vel *b* assignantur per dyametros eorumdem, ut patet; sed ex casu dyameter *a* ad dyametrum *b* se habet in proportione dupla; igitur etc. Et sic sequitur inconveniens adductum.'

60 *De sex inconvenientibus*, 342: 'Utrum velocitas omnis motus localis uniformiter difformis incipiens a non gradu sit equalis suo medio gradui.'

of the question, the latitude of the motion is not mentioned, but the velocity of the motion is, probably because this notion is the main subject of the treatise. However, the word 'latitude' is introduced immediately, in the heart of the question, and the theorem is stated in terms of an equivalence between the latitude or velocity of motion and its middle degree.[61] The author then explains that comparing velocities and latitudes of motion amounts to comparing distances traveled in equal times. One can read in the proof of the fourth difficulty:

> Moreover, the space traversed by the latitude *ab* is three time that traversed by the latitude *ac* in the same time or in an equal time, so that the latitude of the motion *ab* is three times more intense.[62]

Thus the question is whether the space traversed by a body moving with a uniformly difform motion with middle degree *b* is equal to the space traversed by the same body moving with a uniform velocity *b* for an equal time. This is what the author explains further:

> In local motion, when such a motion begins to intensify uniformly from no degree, the whole latitude of this motion corresponds or is equal to its middle degree, so what is exactly traversed by that which moves for one hour with the middle degree of this latitude is as much as what is traversed during the same hour if it moves with this latitude.[63]

This statement is similar to that given by the author of the *Probationes conclusionum tractatus regularum solvendi sophismata*, thought to be William Heytesbury:

> Any latitude of motion acquired or lost uniformly will correspond to its middle degree, that is, the same moving body uniformly acquiring this latitude, or losing it in a given time, will traverse a magnitude equal to that

61 See E.D. Sylla, 'The Oxford Calculators' Middle Degree Theorem in Context,' *Early Science and Medicine* 15 (2010), 338–370, especially 352–353.

62 *De sex inconvenientibus*, 344: 'Item, triplum spatium pertransietur per *ab* latitudinem in equali ⟨tempore⟩ ad illud quod pertransitur per *ac* latitudinem in eodem tempore vel equali, igitur latitudo motus *ab* est in triplo intensior; igitur etc.'

63 *De sex inconvenientibus*, 349: '… in motu locali, ubi a non gradu incipit talis motus uniformiter intendi, tota latitudo motus illius correspondet vel est equalis suo gradui medio, sic quod tantum precise est pertansitum ab illo qui moveretur gradu medio illius latitudinis per horam, quantum foret pertansitum ab eodem, si per eandem horam moveretur ista latitudine.'

> which it would traverse if it moved continuously for an equal time with the middle degree.[64]

Richard Swineshead explains the same conclusion when he states the middle degree theorem in treatise XIV, *De motu locali*, of his *Liber calculationum*:

> Any latitude of motion acquired or lost uniformly will correspond to its middle degree … And by 'the latitude, supposing it intensifies, corresponds to its middle degree,' I mean this: as much will be traversed precisely by means of this latitude thus acquired as by means of the middle degree of this same latitude, if during all the time it moves with the middle degree.[65]

Note also that in his answer to the fourth difficulty, the author of the *Tractatus de sex inconvenientibus* specifies what he means by the middle degree of latitude. It is not a degree that would divide latitude into two equal latitudes, but it is the degree acquired at the middle instant of time. He then remarks that this degree is equidistant from the two ends of the latitude, that is to say that it is the arithmetical mean between these two extremes:

> So, if you ask what I call the 'middle degree' to which the latitude is equal, I answer that if a latitude of motion begins to intensify from no degree to some degree, and that for some time, then at the middle instant of time it will acquire some degree to which the whole latitude is equal, and I call this the 'middle degree', which is equally distant, according to time, from the extremities of time and the extremities of latitude. I say 'according to time', because in an equal time it will reach the double degree as it will reach the half degree.[66]

---

64 M. Clagett, *The Science of Mechanics in the Middle Ages*, Madison, WI, 1959 and 1961 (University of Wisconsin Publications in Medieval Science, 4), 287: 'Omnis latitudo motus uniformiter acquisita vel deperdita correspondebit gradui medio ipsius, i.e., quod mobile idem ipsam latitudinem uniformiter acquirens seu deperdens in aliquo tempore dato equalem omnino magnitudinem pertransibit ac si ipsum continue per equale tempus moveretur medio gradu.'

65 Clagett, *The Science of Mechanics*, 298; Richard Swineshead, *Liber calculationum*, Venice 1520, 45vb: 'Omnis latitudo motus uniformiter acquisita vel deperdita suo gradui medio correspondet … Tunc voco latitudinem ut acquiretur correspondere suo gradui medio sic: tantum pertransiretur precise mediante illa latitudine sic acquisita sicut mediante gradu medio eiusdem, si per totum tempus illo gradu medio moveretur.'

66 *De sex inconvenientibus*, 355: 'Et tunc, si tu queras quid ego voco gradum medium, cui tota

Edith Sylla has rightly noted that for Heytesbury as for Swineshead, there is no geometric representation of latitudes, as would be the case later with Nicole Oresme.[67] We can make the same remark for the author of the *Tractatus de sex inconvenientibus*.

After having given six difficulties which would follow from the acceptance of the middle degree theorem, the author proposes six proofs for this same theorem which he recognizes as valid. Then, he makes the following statement:

> In any uniformly difform local motion beginning at no degree, the total latitude of motion is equal to its middle degree and only to it.[68]

The first proof is based on a property that the anonymous author attributes to Averroes in his commentary on the *Nicomachean Ethics*, according to which in all continua in which there is something bigger and something smaller, there is an equal. The author therefore considers a uniformly difform latitude *a*, with middle degree *b*. Then, he notes that *b* divides latitude *a* into two latitudes, one of which is more intense than *b* and the other less intense than *b*. He deduces that there is in *a* a particular latitude that is equal to the degree *b* and that it can only be the whole *a*, because any part of *a* produces a motion that is either greater or less than that produced by *b*.[69]

---

latitudo est equalis, dico quod si aliqua latitudo motus incipiat intendi a non gradu ad certum gradum, et hoc per aliquod tempus, tunc in medio instanti istius temporis acquiretur certus gradus cui tota latitudo motus est equalis, et istum voco gradum medium, qui equaliter secundum tempus distat a terminis temporis et a terminis latitudinis. Dico "secundum tempus", quia in equali tempore deveniet ad gradum duplum, sicut ad gradum subduplum.'

67 Sylla, 'The Oxford Calculators' Middle Degree Theorem,' 340.

68 *De sex inconvenientibus*, 346: '… in omni motu locali uniformiter difformi incipiente a non gradu tota latitudo motus sit equalis suo medio gradui et solum illi.'

69 *De sex inconvenientibus*, 346–347: '… signetur talis latitudo cuius medius gradus sit *b* et illa tota latitudo *a*. Tunc arguo sic: in *a* latitudine est aliquis gradus medius, puta *b*, quo est alia latitudo intensior et quo est alia latitudo remissior, igitur in *a* latitudine est aliqua latitudo sibi equalis, et nulla nisi tota; igitur tota est sibi equalis. Consequentia ultima est necessaria, sicut patet. Et prima similiter est formalis, ut patet per Commentatorem 2° *Ethicorum* commento 10, ubi dicit quod in omni continuo et divisibili ubicumque est invenire maius et minus, ibi est invenire equale. Et assumptum primum patet, quia latitudo ab *b* ad extremum sui intensius est intensior *b* gradu et latitudo ab *b* usque ad non gradum est remissior *b* gradu, ut satis patet. Et sic sequitur quod in *a* sit aliqua sibi equalis, et nulla nisi tota, quia, accepta quacumque latitudine que est pars illius latitudinis totius, illa vel est intensior *b*, vel remissior *b*, et per consequens sola latitudo totalis est sibi equalis; igitur etc.'

In the second proof he compares the motions of Socrates and Plato, where the former moves with a uniformly difform motion from no degree to degree *b*, and the latter, inversely, moves with a uniformly difform motion from degree *b* to no degree.[70] The author remarks that the two latitudes are equal and he adds that the degrees to which these two latitudes can correspond are their middle degrees, which are equal. Indeed they are not equal at their extremes, 'as is clear', he says. He then concludes rather too quickly that, the extreme degrees being excluded, the two equal latitudes can only correspond to their equal middle degrees; indeed, he says, 'one does not see to which other degrees they could be equal.'

It should be noted that in this argument, and in the ones that follow, the author considers only two positions as possible: either the latitude of a uniformly difform motion is equal to its middle degree, or it is equal to its extreme degree. So, to prove that it is equal to its middle degree, he takes it to be enough to show that it is not equal to its extreme degree. Here, the author has in mind the model of the motion of a sphere of which he spoke in the preceding article. Two hypotheses among those he examined made the motion of a sphere dependent on the motion of one point: Gerard of Brussels' opinion, according to which the motion of a sphere is measured by the motion of the midpoint of its radius, and that of Bradwardine for whom the motion of the sphere is measured by the motion of the extreme point of its radius. The author applies these two hypotheses to uniformly difform local motion which intensifies from no degree, the motion which he considers here. But while for the uniformly difform local motion beginning at no degree the author of the *Tractatus de sex inconvenientibus* accepts the middle degree theorem, we have seen that, as far as the motion of a sphere is concerned, he follows Bradwardine's opinion rather than that of Gerard of Brussels. Since this may seem contradictory, the anonymous author makes the following comment:

> So, to the article that asks whether velocity etc., I answer positively and I allow that in a uniformly difform local motion beginning at no degree, the whole latitude of the motion is equal to its middle degree, speaking of motion that intensifies continuously. And for every uniformly difform

70 *De sex inconvenientibus*, 347: 'Secundo arguitur sic: sit Sortes, qui incipiat a non gradu uniformiter difformiter intendere motum suum usque ad *b* gradum, et Plato ab eodem gradu vel sibi equali incipiat remittere motum suum uniformiter difformiter ad non gradum. Tunc iste latitudines motuum Sortis et Platonis sunt equales, et aliquibus gradibus sunt equale; et non extremis gradibus suis, ut patet; igitur mediis, quia non videtur quibus aliis corresponderent aut forent equales, igitur etc.'

> motion beginning at no degree, which intensifies continuously, I mean all this: in a certain time as much space is traversed as is traversed at the same time or in an equal time with the middle degree, and vice versa. And I speak expressly of the motion which intensifies continuously, of which no acquired degree remains at the same time as another during the time. And I say this insofar as in the extended motion of a uniformly rotating sphere, for which any degree remains with another, in such an extended motion the whole latitude of the motion corresponds to the most intense and extreme degree, but in the continually intensified and not extended motion this is not necessary and not true.[71]

In a sphere, all the points on a radius have different motions and, starting from the center and going towards the extremity of the radius, the motion of the points is faster and faster. This is why the author speaks of a uniformly difform extended motion. In a local motion that intensifies uniformly, the velocity of the motion of a single point increases continuously. As the natures of these two motions—the motion of a sphere and the uniformly difform motion—are not the same, it is reasonable to consider that their latitudes need not correspond to the same point.

The third proof links the latitude of motion and the space traversed. The author considers a case in which Socrates moves uniformly difformly from no degree to degree 8 and in which Plato moves with a uniform motion of degree 4. Then, he has to show that the spaces traversed by Socrates and Plato are equal. For this, he compares what happens in the two halves of the time involved. He proves that in the first half hour Socrates travels half as much space as Plato, and in the same first half hour, that Socrates travels one third of the space that he will travel in the second half. The conclusion follows immediately: in the two halves of the motion Plato traverses lengths 2 and 2 and Socrates traverses

71 *De sex inconvenientibus*, 352–353: 'Ad articulum igitur, cum queritur utrum velocitas etc., dico quod sic et concedo quod in motu locali uniformiter difformi incipiente a non gradu, tota latitudo motus est equalis suo medio gradui, et hoc loquendo de motu qui continue est in intendi. Et hoc totum sic intelligo: in omni motu uniformiter difformi incipiente a non gradu, qui continue est in intendi, tantum pertransitur de spatio in aliquo tempore, quantum in eodem tempore vel equali pertransitur gradu suo medio, vel contrario. Et signanter loquor de motu qui continue est in intendi, cuius nullus gradus acquisitus manet simul cum alio et per tempus; quod dico pro tanto, quia in motu extenso spere uniformiter revolute, cuius quilibet gradus motus manet cum alio, in tali motu extenso tota latitudo motus correspondet gradui intensissimo et extremo, sed in motu continue intenso et non extenso, non oportet nec est verum.'

1 and 3.[72] It should be noted that the proofs the author proposes for the two arguments leading to the conclusion again consist in excluding the possibility that the motions may correspond to their extreme degrees, which, in his opin-

72 *De sex inconvenientibus*, 347–349: 'Item ⟨tertio⟩: sit aliqua latitudo motus localis uniformiter difformis incipiens a non gradu et signata per *a* et eius medius gradus per *b*. Et incipiat Sortes moveri *a* latitudine et incipiat Plato moveri *b* gradu medio eiusdem. Tunc sic: Plato pertransiens aliquod spatium *b* gradu in aliquo tempore, tantum spatium precise pertransiet, quantum pertransiet Sortes tota *a* latitudine in eodem tempore vel equali; igitur *a* latitudo est equalis *b* gradui. Consequentia est manifesta. Et probo antecedens, ad cuius probationem sumo quod gradus terminans *a* in extremo suo intensiori sit *c*, signatus per 8, gradus medius inter *b* et *c* sit *d*, signatus per 6, *b* vero gradus medius latitudinis totalis per 4, gradus medius inter *b* et non gradum sit *e*, signatus per 2, et sit *f* tempus in quo Sortes pertransiet aliquod spatium, quod sit *g*. Et arguo tunc sic : per totam primam medietatem *f* temporis Sortes et Plato movebuntur, Sortes continue tardius Platone, igitur Plato continue velocius Sortem. Sit igitur quod in duplo velocius, et quod Sortes, per latitudinem que est a non gradu usque ad gradum medium exclusive qua solum movebitur in prima medietate temporis, pertranseat unam quartam tantum. Et sequitur: tunc in prima medietate *f* temporis pertransiet Sortes unam quartam de spacio suo, et Plato per idem tempus movebitur in duplo velocius; igitur Plato in eodem tempore pertransiet duas quartas de spatio suo. Igitur in prima medietate *f* temporis Sortes pertransiet solum unam quartam et Plato solum duas. Et tunc ultra sic: in prima medietate *f* temporis Sortes pertransiet unam quartam et in secunda medietate *f* temporis movebitur in triplo velocius; igitur in secunda medietate *f* temporis pertransiet Sortes tres quartas; igitur in toto *f* tempore pertransiet quattuor quartas. Et tantum precise in eodem tempore erit pertransitum a Platone, quia in prima medietate *f* temporis pertransiet Plato duas quartas et per totum *f* tempus movebitur uniformiter; igitur in secunda medietate *f* temporis tantum pertransiet quantum in prima. Sed duas pertransivit in prima medietate; igitur et alias duas pertransiet in secunda; igitur quattuor in toto; igitur tantum quantum Sortes. Et quod Sortes in secunda medietate *f* temporis movebitur in triplo velocius quam in prima, probo: signetur latitudo motus qua movebitur Sortes in secunda medietate *f* temporis, que erit latitudo terminata ad *b* et *c* gradus, cuius etiam gradus medius est *d*, sicut patet. Tunc sic: Sortes movebitur velocius Platone per totam secundam medietatem *f* temporis. Vel igitur secundum proportionem graduum mediorum illarum latitudinum, que est a *b* ad *c* et a *b* ad non gradum; vel secundum proportionem graduum extremorum latitudinum earumdem, quia non videtur penes quos alios deberet attendi velocitas Sortis in secunda medietate *f* temporis supra velocitatem eiusdem in prima medietate *f* temporis. Si penes proportionem graduum mediorum, cum gradus medius acquisitus in secunda medietate sit triplus precise ad gradum medium acquisitum in prima medietate, quia *d* ad *c* est proportio tripla, ut patet, igitur Sor movebitur in triplo velocius in secunda medietate quam in prima, et habetur propositum. Si detur alia pars, quod in secunda medietate *f* temporis movebitur Sortes, non secundum proportionem graduum mediorum, sed secundum proportionem graduum extremorum, cum unus sit precise duplus ad alium, ut est *c* ad *b*, igitur solum pertransiet Sortes in secunda medietate temporis duas quartas, et sic in toto *f* tempore plus pertransietur a Platone quam a Sorte, quia in toto tempore pertranseuntur a Platone quattuor quarte et a Sorte non nisi tres, quod falsum est, nec concedet adversarius. Relinquitur igitur primum, quod velocitas motus Sortis in secunda medietate

ion, leads immediately to their corresponding to their middle degrees, as was shown for the second proof.

At the beginning of the article, the author considers this same case and deduces the following difficulties: the motions are said to be equal even though in a portion of time Socrates moves faster than Plato and in all the instants of time, except one, the velocities are different.[73] Responding to these diffi-

---

f temporis supra velocitatem eiusdem in prima medietate *f* temporis attenditur penes proportionem graduum mediorum latitudinum predictarum. Et quod in prima medietate *f* temporis Plato movebitur in duplo velocius ipso Sorte, probo: latitudo motus Sortis, qua Sortes movebitur in prima medietate *f* temporis, et illa velocitas motus Sortis, aut attenditur penes gradum medium eiusdem latitudinis, puta penes *e* gradum, aut penes gradum extremum puta *b*. Si penes gradum medium, cum omnis gradus medius per totum tempus istius intensionis usque ad *b* gradum sit precise subduplus ad gradum extremum, ut patet de *c* et *b*, igitur per totam primam medietatem *f* temporis movetur Sortes precise in subduplo velocius Platone. Et per consequens Plato per totam primam medietatem *f* temporis movebitur in duplo velocius Sorte. Et habetur propositum et intentum. Si dicatur quod velocitas motus Sortis etc. attenditur penes suum gradum extremum, puta penes *b*, contra: igitur in prima medietate temporis tantum foret pertransitum a Sorte, quantum a Platone, et per consequens per totam primam medietatem *f* temporis ita velociter moveretur Sortes sicut Plato. Consequens falsum. Et consequentia patet, quia Sortes per totam primam medietatem *f* temporis moveretur aliqua latitudine motus terminata ad aliquem gradum et ad gradum equalem gradui motus quo movetur Plato, penes quem gradum attenderetur precise velocitas motus Sortis, igitur etc.'

73 *De sex inconvenientibus*, 342: 'Utrum velocitas omnis motus localis uniformiter difformis incipiens a non gradu sit equalis suo medio gradui. Et arguitur primo quod non, quia, si sic, sequitur ... Quinto, quod Sor et Plato equaliter iam moventur et Plato post hoc movebitur per horam et solum per horam in qua solum movebitur ita velociter sicut nunc movetur et Sor per eandem horam continue movebitur velocius et velocius Platone et nunquam tardius eo per illam horam et tamen in fine hore non movebitur Sor velocius quam nunc movetur sed precise equaliter. Sexto et ultimo pro hac vice, quod *a* et *b* moventur equaliter in *c* hora et tamen per totam *c* horam inequaliter moventur'; 345: 'Probatio quinti inconvenientis: si articulus foret verus, sequeretur quod talis latitudo motus foret equalis suo gradui medio, sic quod si Sor moveretur tali latitudine uniformiter difformi, tantum spatium pertransiret in tempore quantum Plato, si per idem tempus moveretur uniformiter suo gradu medio ... Sed contra: ex isto sequitur quintum inconveniens. Quod arguo sic: ... Sortes in prima medietate pertransiet solum unam quartam et Plato duas, in secunda pertransiet Sortes tres et Plato non nisi duas. Tunc sic: per totam primam medietatem temporis movebitur Sortes tardius Platone ; et per totam secundam medietatem temporis movebitur Sortes velocius Platone; igitur in instanti medio copulante illas duas medietates temporis, Sortes et Plato movebuntur equaliter ... Sit igitur nunc medium temporis in quo sic Sortes et Plato moventur equaliter. Et sequitur tunc inconveniens adductum : nam Sortes et Plato equaliter iam moventur et Plato post hoc movebitur per horam sic ita et solum per horam in qua solum movebitur ita velociter sicut nunc movetur et Sortes per eandem horam continue velocius et velocius movebitur Platone et nunquam tardius eo, sicut patet ex casu ... Et tamen in fine hore non movebitur Sortes velocius

culties, the author explains that a distinction must be made between what can be observed at each instant of time and what happens overall. Thus, he remarks that the two expressions 'to move unequally during all the time' (*per totum tempus moventur inequaliter*) and 'to move equally in all the time' (*in toto tempore movebuntur equaliter*) must be understood differently. In the first expression, 'during' (*per*) implies that 'all' is considered syncategorematically or in the divided sense, that is, the motion is unequal in each instant of time or in any part of time. On the other hand, 'in all the time' (*in toto tempore*) means that all the instants of time are considered together—'all' is considered categorematically or collectively.[74] So it is possible, as is the case here, that two motions are unequal at every instant but are equal in the total time.[75]

The fourth, fifth and sixth proofs are based on *reductio ad absurdum* arguments: demonstrating the absurdities that would result from choosing to make the latitude of a uniformly difform motion correspond to the uniform motion having for degree the extreme point of this latitude. Thus, in the fourth argu-

Platone, sed precise equaliter. Quod arguo sic: motus Sortis et Platonis erunt equales, et solum in fine hore, quia solum in fine hore et non citius erunt equalia spatia pertransita ab utroque. Igitur in fine temporis [movebitur vel] erit motus Sortis equalis motui Platonis …'; 346: 'Sextum inconveniens sequitur in hoc casu et eodem: nam Sortes et Plato in eodem tempore motus sui pertransient spatia equalia in equali tempore, igitur in toto tempore movebuntur equaliter. Et tamen per totum tempus inequaliter movebuntur, quia per totum prime medietatis temporis et per totum secunde medietatis temporis; igitur Sortes et Plato per totum tempus inequaliter movebuntur.'

74 *De sex inconvenientibus*, 355: 'Ad quintum dico quod conclusio ista non sequitur. Et nego quod in instanti medio istius temporis erit hec vera "Sortes et Plato moventur equaliter". Et cum arguitur contra: "Sortes per totam primam medietatem temporis movebitur tardius Platone; et per totam secundam medietatem movebitur velocius Platone; igitur in instanti medio temporis Sortes et Plato moventur equaliter", nego consequentiam. Sed sequitur "igitur in toto tempore moventur equaliter". Et per illud patet responsio ad sextum, quod non est inconveniens sed possibile et sequens in casu supposito, quod "aliqui duo per totum tempus moventur inequaliter et tamen in toto tempore movebuntur equaliter", quia aliud dicit ly "per" et aliud ly "in" et aliter dat intelligere et stat illa prepositio "per" et aliter hec prepositio "in", quia hec prepositio "per" facit illum terminum "totum" stare sincategorematice et dat intelligere quod per omnem partem temporis moventur inequaliter, et hoc est verum, sed ly "in" facit istum terminum "toto" stare categorematice et dat intelligere quod in toto tempore, collective sumpto ly "toto", Sortes et Plato equaliter moventur, et hoc est verum.'

75 This argument can be compared to the sophism 'Socrates et Plato incipient aeque velociter moveri' which Richard Kilvington discussed in his *Sophismata* and in which he exploits the confusions that can be made between instantaneous velocity and overall velocity. See Richard Kilvington, *Sophismata*, eds. N. Kretzmann and B.E. Kretzmann, Oxford 1990 (Auctores Britannici Medii Aevi, 12), 76–81, for the Latin text, and Sylla, 'The Oxford Calculators' Middle Degree Theorem,' 347–349, for an analysis of the argument.

ment, it is supposed that Socrates moves with a uniformly difform motion from no degree to degree *c*. It is assumed that the degree acquired at the end of the first half-hour is *b*, and the degree acquired at the end of the first half of the second half-hour is *d*. In the third argument, it was shown that in this case, Socrates acquires a quarter of the space traveled in the first half-hour and three quarters in the second. Since the ratio *d* to *b* is the sesquialterum ratio from 3 to 2, and if it is assumed that the velocity depends on the extreme point of the latitude, the velocity in *d* is three halves of that acquired in *b*; thus Socrates acquires in *d* three halves the space acquired in *b*, that is 1/4+1/8. In the second half of the second half-hour, he still has to acquire 3/4−1/4−1/8, that is 1/4+1/8, as much as in the first half. So, the velocity in the second half-hour is constant: the movement is uniform, which is absurd.[76] In the sixth argument,[77]

---

76 *De sex inconvenientibus*, 350: 'Quarto, si velocitas omnis motus localis etc. non sit equalis suo gradui medio, sed gradui intensissimo, contra, retento casu proximi argumenti et tertii, arguo sic: Sortes in prima medietate *f* temporis pertransiet unam quartam; et per totam secundam medietatem *f* temporis movebitur uniformiter difformiter continue intendendo motum suum; igitur in medio instanti secunde medietatis *f* temporis erit Sortes sub medio gradu inter *b* et *c*, qui est *d*. Et tunc intensio istius motus attenditur penes istum gradum. Sed tunc *d* gradus se habebit ad *b* gradum in proportione sexquialtera; igitur Sortes in medio instanti secunde medietatis *f* temporis movebitur precise in sexquialtera proportione velocius quam ipsummet movetur in medio instanti totius temporis. Igitur Sortes in prima medietate ⟨secunde medietatis⟩ *f* temporis acquiret sexquialterum spatium ad illud quod acquisivit vel pertransivit in prima; sed in prima solum acquisivit unam quartam; igitur in prima medietate secunde medietatis *f* temporis solum pertransiet quartam et dimidiam. Et non maius spatium sed precise tantum acquiret in secunda medietate secunde medietatis *f* temporis, ut patet ex casu. Igitur Sortes ita velociter movebitur in prima medietate secunde medietatis *f* temporis sicut in secunda medietate secunde medietatis *f* temporis. Et ultra: igitur per totam secundam medietatem *f* temporis movetur Sortes uniformiter. Et ex isto sequitur hoc impossibile, quod Sortes movebitur per aliquod tempus continue intendendo motum suum in isto tempore, et tamen per idem tempus continue uniformiter movebitur. Et patet demonstrative quod sequitur.'

77 *De sex inconvenientibus*, 352: 'Sexto, si per oppositum articuli intensio talis motus localis etc. non sit equalis vel correspondeat suo gradui medio, sed gradui suo intensissimo, contra: accipio igitur Sortem, qui a non gradu incipiat moveri localiter continue intendendo motum suum et hoc latitudine motus localis uniformiter difformi. Tunc sic: Sortes continue intendet motum suum; et hoc latitudine motus uniformiter difformi terminata ad non gradum, igitur Sortes continue movetur aliqua latitudine cuius gradus intensissimus continue est precise duplus ad suum medium gradum. Et intensio huius motus attenditur penes proportionem intensissimi gradus sui ad suum gradum medium; igitur Sortes continue intendet motum suum a proportione dupla. Ex quo sequitur ultra: Sortes continue intendet motum suum a proportione equali, igitur continue intendet motum suum. Et sequitur etiam: Sortes continue intendet motum suum; et hoc a proportione equali; igitur motus Sortis non plus intendetur in uno tempore quam in altero sibi equali quocum-

it is deduced from the same case, that Socrates increases his movement with the double ratio, because Socrates moves continuously with some latitude, the most intense degree of which is continuously double its middle degree, and the author assumes that the increase of the motion depends on the ratio of the most intense degree of velocity to its middle degree. So Socrates continually increases his motion with an equal ratio. As a result, Socrates does not intensify his motion, which is absurd.

In the fifth argument, which is particularly chaotic, two powers are compared, which both move bodies of the same resistance 4. It is assumed that the first power increases uniformly from 4 to 6 in the first half-hour and increases uniformly from 6 to 12 in the second half-hour. If the latitude of the motion corresponds to its extreme point, this power is twice as fast in the second half-hour as in the first half, because 12 is double of 6. Now, the second power is considered which, by hypothesis, increases from 4 to 6 in the first half-hour and increases from 6 to 9 in the second half-hour. The author notes that the ratio of 9 to 6 is equal to the ratio of 6 to 4, so that the ratio of 9 to 4 is twice the ratio of 6 to 4. He concludes that the second power moves twice as fast in the second half-hour as in the first. Thus, they are two different powers that both move twice as fast in the second half-hour as in the first half.[78] The author does

que. Et ultra: igitur Sortes non continue intendet motum suum. Igitur si Sortes continue intendet motum suum, non continue intendet motum suum, quod est contradictio, et per consequens impossibile. Ista sunt impossibilia; et ista sequuntur ex ista positione; igitur positio impossibilis.'

78 *De sex inconvenientibus*, 351: 'Quinto, si ex opposito articuli velocitas motus localis non attenditur penes gradum medium sed penes gradum intensissimum sui motus, contra: sit igitur aliqua potentia motiva, que sit equalis potentie sue resistentie, et crescat potentia motiva continue movendo resistentiam suam gradu intensissimo sui signato per 6, et sit quod sua potentia resistiva resistat sub gradu intensissimo resistentie sue signato per 4, qui sit *d*. Tunc sic: ista potentia motiva movebit suam resistentiam intensissimo gradu sui signato per 6, qui est *c*; et velocitas vel intensio huius motus attenditur penes intensissimum gradum suum; igitur duplato *c* gradu ad duplum precise, illa potentia motiva movebit precise in duplo velocius quam nunc movet. Intendatur igitur ista potentia ad duplum, puta ad gradum signatum per 12. Et sequitur tunc quod illa potentia movebit suam resistentiam precise in duplo velocius. Unde sit ita: acccipio igitur aliam potentiam motivam, que modo sit equalis precise potentie resistitive prime ipsius *a*, et volo quod ista potentia motiva intendatur post hoc uniformiter difformiter continue quousque fuerit sub gradu signato per 9, qui sit *b*, et potentia ista resistitiva signata per 4 sit *d*, et gradus quo prima potentia motiva movebit *d* in prima medietate temporis sit *c*, signatus ut prius per 6, quia ponatur quod in principio potentia motiva *c* et sua resistentia fuissent equales, et quod *c* cresceret in potentia sua continue intendendo motum suum, sic quod in prima medietate temporis sui motus intendat motum ad gradum signatum per 6. Et arguo tunc sic: qualis est proportio *b* ad *c*, talis est proportio *c* ad *d*; igitur proportio *b* ad *d* est dupla ad proportionem *c* ad *d*. Motus intensionis *b* sequitur istam proportionem, ut constat; igitur

not comment on this argument, but it can be shown that two points of view are opposed here: in the first case, the velocity of the motion is determined by the velocity of the extreme point, without taking into account the resistance in the calculation; in the second case, the resistance 4 is considered and the velocity is determined by the ratio of the power to the resistance.

For comparison, it is worth recalling how the middle degree theorem is demonstrated in the *Probationes conclusionum tractatus regularum solvendi sophismata*, whether Heytesbury is the author or not.[79] In this text, the author considers three moving bodies, *a*, *b* and *c*, such that *a* traverses a space of one foot in one hour with degree 4, during the first half of this hour *b* uniformly increases its motion from degree 4 to degree 8, and during this same half of the hour *c* uniformly decreases its motion from degree 4 to no degree. The author remarks that *b* and *c* traverse as much space in the first half-hour as *a* in the whole hour, because what *c* traverses less in relation to *a*, *b* traverses more in relation to *a*. If one now considers a moving body *d* with a uniformly difform motion from no degree to degree 8, it can be noted that in the first half of the time it traverses as much space as *b* and in the second half of the time it traverses as much as it traversed in the first half. Thus, in all the time it traverses as much space as *b* and *c* together, so as much as *a* in all the time. It is therefore true that the whole latitude from non degree to degree 8 corresponds to its middle degree 4. The author then demonstrates, as a consequence of this theorem, that in a uniformly difform motion, a moving body traverses in the second half of the time three times more space than in the first half of the time. In fact, the middle degree of the latitude of motion in the second half of the time is three times the middle degree in the first half of the time.

---

b movebit *d* precise in duplo velocius quam *c*. Et duplato *c* ad gradum signatum per 12, potentia cuius *c* est gradus intensissimus non movebit *d* nisi precise in duplo velocius, ut deductum est. Igitur hic sunt due potentie motive, que movebunt eandem resistentiam vel equalem, et hoc equaliter, et tamen una a proportione maiori, que est a 12 ad 4, et alia a proportione minori, que est a 9 ad 4, quia una est proportio tripla et alia dupla sexquiquarta. Sequitur etiam in hoc casu quod due potentie inequales, puta quarum una se haberet ad aliam in proportione sexquialtera, equaliter moverent eandem resistentiam vel equalem. Consequens est impossibile; igitur etc.'

79 Cf. Wilson, *William Heytesbury*, 122–123, and Sylla, 'The Oxford Calculators' Middle Degree Theorem,' 352. For a reconstruction of this proof, see D.A. Di Liscia, 'La "latitud de las formas" y la geometrización de la ciencia del movimiento,' *Mediaevalia. Textos e estudos* 36 (2017), 75–114, at 86–89.

Conclusion

The study of these three articles on local motion confirms that the author of the *Tractatus de sex inconvenientibus* had a good knowledge of the physics developed at Oxford by the *Calculatores* and that his thinking was in line with this tradition. The vocabulary he uses is common to these authors, particularly that of latitude. Following their example, he presents thought experiments to invalidate different opinions on the subject under discussion. He also uses logical tools, in particular the vocabulary of the theory of consequences or the distinction between categorematic and syncategorematic uses of a term. In particular, he uses mathematics, notably the theory of proportions, employing the proportional parts of space or time in his arguments, even if this is sometimes in an awkward manner.

In these three articles, the author shows his delight in producing cases that invalidate opinions he does not hold himself. These imaginary cases allow him to clarify fundamental notions at the heart of the physics of the time, such as the different types of resistance or the notions of uniformity and difformity of motion. He often confronts different points of view on the same case. The motion of the sphere is thus examined either in terms of the velocity of one of its fastest moving points or in terms of the relationship between power and resistance. In other cases, the author determines the velocity from the ratio of power to resistance (*quo ad causam*) or from the effect produced (*quo ad effectum*),[80] and asks what conclusion to draw if these considerations lead to contradictory results.[81]

Concerning the middle degree theorem, it is to be noted that the anonymous author does not take up William Heytesbury's proof, which he knew. He produced a series of arguments, which today can be considered unconvincing,[82] but whose motivation seems to be to establish a parallel with the determination of the velocity of the motion of a sphere. Just as the velocity of the sphere can be determined by the velocity of the midpoint (Gerard of Brussels' opinion) or by the velocity of the fastest moving point (Thomas Bradwardine's view), the latitude of a uniformly difform motion could be determined either

80 For the distinction between these two perspectives in determining velocity see D.A. Di Liscia, 'Velocidad *quo ad effectus* y velocidad *quo ad causas*: La tradición de los calculadores y la metodología aristotélica,' in D.A. Di Liscia, E. Kessler and C. Methuen (eds.), *Method and Order in Renaissance Philosophy of Nature. The Aristotle Commentary Tradition*, Aldershot 1997, 143–176.

81 See the introduction to my critical edition of the *De sex inconvenientibus*. I shall further elaborate on this important point of the treatise in a later study.

82 See Clagett, *The Science of Mechanics*, 264.

by the middle degree or by the extreme degree. Agreeing with Bradwardine's position for the motion of the sphere and accepting the middle degree theorem, the author of the *Tractatus* is forced to explain why there is no contradiction between these two positions. He does so by pointing out that the motions studied are different: he thus calls the motion of the sphere a uniformly extended motion, i.e., he consideres the motion of all points together (noting that all degrees of motion are present at the same time in the sphere, since along a radius all degrees of velocity can be found, from the non-degree to the extreme degree), whereas in determining the velocity of a uniformly difform local motion, only one point is considered, whose velocity varies continuously. This way of comparing the motion of a point on the sphere with the motion of the whole sphere, or, in other arguments, the velocity in an instant with the velocity over the whole duration of the motion, is characteristic of fourteenth-century century physics.[83]

Appendix: A Description of Manuscript Praha, Národní Knihovna České Republiky, VIII. G. 19

Daniel A. Di Liscia and Sabine Rommevaux-Tani

Physical description: Paper, ff. 81, 23×15 cm. Some folia are lacking: 9^{r-v}; 19^{r-v}; 29^{r-v}.

Hands: H1: ff. 1^v–2^r. H2 (= Gregorius): rest of the manuscript.

Numbering: There is only a modern numbering in the manuscript, which is used on 10^r (as 10), 20^r (as 20), and 30^r (as 30). However, this numbering is not consistent with the actual numbers of the folia contained in the manuscript. Since the manuscript does not include another numbering of folia, a solution for this problem—which was assumed by the scanned version in *Manuscriptorium*—is the loss of some folia after the modern numbering was introduced. It is evident that some pages were cut out from the manuscript between 5^v and 6^r. However, it does not seem that pages were cut out between 10 and 20, nor between 20 and 30, so we have to assume that the numbering from 10 until 30 is simply wrong. After 30^r the numbering

83 See, e.g., on velocity as measured at one point or velocity as a holistic concept, P. Souffrin, '*Velocitas totalis*. Enquête sur une pseudodénomination médiévale,' in S. Caroti and P. Souffrin (eds.), *La nouvelle physique du XIVe siècle. Actes du Colloque International (Nice, 3–5 septembre 1993)*, Florence 1997 (Biblioteca di Nuncius, 24), 251–275, and Sylla, 'The Oxford Calculators' Middle Degree Theorem,' 352–353.

becomes right. Thus, we assume for this description the numbering used by *Manuscriptorium*. In addition, it is to be noticed that: (a) the first text which we have included was copied on the inside of the cover; (b) 1r is blank; (c) 41 is an interpolation.

General Content: Collection of university textbooks on mathematics, natural philosophy, optics, and logic: geometry, proportion theory, motion, light, physiology of vision, *sophismata, Calculatores*.

Previous description: The former description by J. Truhlář (*Catalogus codicum manu scriptorum latinorum qui in C.R. Bibliotheca publica atque universitatis Pragensis asservantur*, Pars prior, Prague 1905, 594) has many areas needing improvement, especially regarding the content. Many texts were not recognised. The questions mentioned by Truhlář as *Questiones disputate de motu* are in fact several texts which have been combined, mainly from Bradwardine's *Tractatus de proportionibus* and from the anonymous treatise *De sex inconvenientibus*. Some texts contained in this manuscript are mentioned in D.A. Di Liscia, *Zwischen Geometrie und Naturphilosophie: die Entwicklung der Formlatitudenlehre im deutschen Sprachraum*, unpublished Ph.D. dissertation, University of Munich, 2003, 425, and D.A. Di Liscia, 'La "latitud de las formas" y la geometrización de la ciencia del movimiento,' *Mediaevalia. Textos e estudos* 36 (2017), 75–114 (at 83, n. 8 and 103). A full scanned version is available online in *Manuscriptorium*.[84]

Provenance and previous possessor: According to the references kindly provided by Martin Dekarli, this manuscript belonged to the private library of magister Jan Šindel or 'Iohannes Andreae dictus Schindel'.[85] Šindel, who donated his library to the *Collegium Carolinum* before the year 1440, was a student of the university of Prague during the years 1395 and 1399, and became rector in 1410. Furthermore, he was a mathematics professor in Vienna, where he also studied medicine, and came back to Prague as a professor of astro-

84 See http://www.manuscriptorium.com/apps/index.php?direct=record&pid=AIPDIG-NKCR__VIII_G_19____016CoRF-cs. We would like to thank to Petra Hofbauerová from the National Library of the Czech Republic (Department of Manuscripts and Old Printed Books) for her friendly and competent help when we, in April 2018, were able to study this piece *in situ*.

85 Email to Di Liscia, 16.04.18. We are very grateful to Martin Dekarli for the literature about Jan Šindel and the history of this manuscript. Especial thanks is owed to Harald Berger who identified text 7 and gave us decisive hints for the last two texts. Additionally, we would like to thank Lukáš Lička for confirming for us the main information of our survey regarding the difficult f. 41 on optics and for various details about this manuscript and its possessor, Jan Šindel. Of course, none of these colleagues is responsible for the mistakes or omissions that remain.

nomy. The famous 'Prague Orloj' constructed by Mikuláš z Kadaně (1350–1419) in 1410 is based on designs and calculations made by Šindel. Šindel became personal physician and astrologer of King Wenzel IV (1361–1419) of Bohemia and of Sigismund (1368–1437), the Holy Roman Emperor. Also in Nuremberg he worked as a physician during 1423–1436. Apparently, Šindel lectured on Ptolemy's *Almagest* at the university of Prague during the years 1412–1418 (Cf. H. Zepeda, *The First Latin Treatise on Ptolemy's Astronomy: The Almagesti minor* [*c. 1200*], Turnhout 2018 [Ptolemaeus Arabus et Latinus, Texts Volume I], 100. We thank Lukáš Lička for this bibliographical information). Additionally, it should be noticed that Šindel was also involved in important Church affairs. In 1418 he became canon at the St. Vitus Chapter of Prague. Having been firstly a supporter of John Hus, he later changed to the Catholic side. During the Hussite Wars he left Prague and moved to Olomouc. In 1441 he became Dean of the Vyšehrad Chapter in Prague. His correspondence with Enea Silvio Piccolomi, later Pope Pius II, has been preserved. He died between 1455 and 1458 (P. Spunar, *Repertorium auctorum Bohemorum provectum idearum post universitatem Pragensem conditam illustrans*, Wrocław 1985 [Studia Copernicana 25], 1: 134–135; see also A. Šolcová, 'Jan Šindel—pravděpodobný autor matematického modelu pražského orloje,' *Pokroky matematiky, fyziky a astronomie* 54 [2009], 307–317).

Special Remarks: 1. On the book cover (a music text) are mentioned three texts: (1) 'Questiones de motu parysius disputate' (surely referring to *De sex inconvenientibus*; text # 8); (2) 'Perspectiva Rygyderi Bachon' (text # 9) and; (3) *Tractatulus de uniformitate et difformitate*, which is not unambiguous since *de uniformitate et difformitate* could be a title of Oresme's *De configurationibus*; *tractatulus*, however, matches better the *Tractatus de latitudinibus formarum* (which is also included in the manuscript as text # 11). Besides these titles, the majority of the texts are conveyed as anonymous and most were probably used at the faculty of Arts. 2. For the sake of clarity we divide the content of the manuscript according to 'texts'. However, we warn that this denomination is questionable regarding several parts of a manuscript. In it, we find, in fact, (a) some complete texts (e.g., # 11); (b) some parts of texts which might belong together (e.g., # 2 and linked with it, perhaps # 4 and # 6); (c) Some very chaotic fragments on natural philosophy and logic (e.g., # 10); and (d) many *sophismata* (e.g., # 12, # 14, # 17, and # 18). 3. With one exception (text # 7) the first part of the manuscript seems to contain physico-mathematical texts, whereas the second part mainly contains logical texts. 4. These features and, in addition, the fact that there is an interpolation (see # 9), suggest that the volume has been rebound in a differing order with consequent changes in the order of some parts of the texts, fragments, etc.

Content:

1. Front cover: Anonymous: Logic fragment, possibly from some *Sophismata.*
   1^r^: blank.
2. 1^v^–8^v^ (2^v^: blank): Thomas Bradwardine: *Tractatus de proportionibus velocitatum in motibus. Inc.*: 'Omnem motum successivum alteri proportionari in velocitate contingit. Quapropter …' *Expl.*: '… proportio velocitatum in motibus cum istius motoris auxilio a quo motus cuncti procedunt ad suum mobile nulla proportio requiritur. Explicit.' Ed. Thomas Bradwardine, *Tractatus de proportionibus velocitatum in motibus*, ed. H.L. Crosby, Jr., *Thomas of Bradwardine. His* Tractatus de proportionibus. *Its Significance for the Development of Mathematical Physics*, Madison, WI, 1955 (University of Wisconsin Publications in Medieval Science, 2), without this Ms. The folia 1^r^–2^r^, written by H_1 and including the prologue and the beginning of chapter 3, finish abruptly in the middle of the definition of the superparticular proportion: 'pars autem aliquota est illa que aliquotans sumpta reddit.' The continuation of 2^r^ is blank; ff. 3^r^–8^v^, written by H_2, begins in the middle of the second part of chapter 1 'et differunt iste medietates …'
3. 8^v^–12^v^: Anonymous: ⟨*De uniformi et difformi*⟩. *Inc.*: 'Uniformiter difformis est tale quod est difformis aliqua latitudine ita quod quilibet gradus medius latitudinis correspondet puncto medio qualitatis …' This text (or rather: fragments of a text), which belongs surely to the *Calculatores* tradition, begins with a formulation of the 'Merton Rule'.
4. 12^v^: Anonymous fragment: *Inc.*: 'Si velis scire quantum una proportio maioris inequalitatis excedat aliam maioris inequalitatis …' Incomplete fragment of a text on proportions, within which a 'magister de proportionibus' is mentioned, most probably Bradwardine. Hence, this text could be directly connected to text # 2. The fragment could be connected to the previous text # 3.
5. 13^r^–17^r^: Anonymous: ⟨*Quaestiones de Geometria*⟩. *Inc.*: 'Dubitatur circa diffinitionem Euclidis an sit bona quando dicit "punctus est etc." …' This text shows a certain similarity with Bradwardine's *Geometria speculativa.* Kilvington, Jordanus and Campanus are mentioned.
6. 17^v^–20^r^: Anonymous: ⟨*De proportionibus velocitatum*⟩. *Inc.*: '… circa proportiones velocitatum in motibus …' *Expl.*: '… in questione sequenti faciliter innotescit.' This text includes a demonstration of the incommensurability of the diagonal with the side of the square. On this problem, see S. Rommevaux, 'Un exemple de *Question* mathématique au Moyen Âge,' *Annals of science* 63/4 (2006), 425–445.

7. 20^r–24^v: Richard Kilvington: *Sophismata*. *Inc.*: 'Sophismatum perscrutandorum utrasque partes contradictionis intendo discutere …' *Expl.*: '… sicut per aliquam earum signantur. Aliud require alibi' (this last remark emphasised with red in the manuscript). Ed. Richard Kilvington, *Sophismata*, eds. N. Kretzmann and B.E. Kretzmann, Oxford 1990 (Auctores Britannici Medii Aevi, 12), the text begins on p. xxi, lin. 3 of this edition.
8. 25^r–46^v: Anonymous: *Tractatus de sex inconvenientibus*. *Inc.*: 'Utrum in omni generatione formarum sit ponenda velocitas. Circa propositam questionem et cetera dubia disputanda de proportionibus velocitatum in motibus …' *Expl.*: '… erit totum pertransitum ab a ante finem hore et sic non sequitur inconveniens adductum et per illud clare patet quid in eodem casu ad alia sit dicendum. Expliciunt questiones *de motu Parisius disputate*.' After this explicit, follows the remark (indicated by a *manicula*) 'ista particula usque ad finem pertinet ad primam questionem de motu alterationis requirit …' Ed. Anonymous, *Tractatus de sex inconvenientibus*, Venice 1505, sigs. e1^{ra}–h4^{ra}; complete critical edition from all known manuscripts: S. Rommevaux-Tani, *De sex inconvenientibus, un traité anonyme de philosophie naturelle au XIVe siècle*, édition, introduction et étude doctrinale, Paris 2022. F. 41 does not belong to the text. This is an interpolated page on optics taken from a copy of Pecham's *Perspectiva communis* (Cf. John Pecham, *Perspectiva communis*, ed. and transl. D.C. Lindberg, *John Pecham and the Science of Optics*, Madison, WI, 1970 [University of Wisconsin Publications in Medieval Science, 14], prop. 5^a, 66–72, diagram at 69).
9. 47^r–64^v: Roger Bacon: *Perspectiva*. *Inc.*: 'Incipit Perspectiva magistri Rodygeri Bachon optimi perpectivi habens 3 partes principales. Tractatus perspective habens tres partes. Prima est …' (the incipit was probably added later). *Expl.*: '… non posset sustinere. Explicit perspectiva Rodigeri Bakkon' (same hand as the title above mentioned). See D.C. Lindberg, *A Catalogue of Medieval and Renaissance Optical Manuscripts*, Toronto, ON, 1965 (Subsidia Mediaevalia, 4), 41 (n. 67, D). Ed. D.C. Lindberg, *Roger Bacon and the Origins of* Perspectiva *in the Middle Ages*: a Critical Edition and English Translation of Bacon's *Perspectiva* with Introduction and Notes, Oxford 1996. On our manuscript, see 'Introduction', civ, Ms. n. 21 (siglum: G).
10. 64^v–65^v: Anonymous Fragments on natural philosophy. We find here several passages from different texts: 64^v: *Inc.*: 'Nota Ysydorius XI Ethymologicarum …'; 65^r: *Inc.*: 'Notandum secundum Commentatorem commento 13° quinti Physicorum quod in transmutationibus essentialiter ordinatis non est possibilis processus in infinitum …'; 65^v (a fragment that seems

to be related to the *sophismata* tradition): *Inc.* 'Lapis (?) est ... a quo non differt aliquid. Sor est homo a quo non differt aliquis homo ...'

11. 66^r–68^r: Jacobus de Sancto Martino: *Tractatus de latitudinibus formarum. Inc.*: 'Quia formarum latitudines multipliciter variantur que multiplicitas difficile discernitur nisi ad figuras geometricas consyderatas referatur. Ideo premissis ...' *Expl.*: '... Plura alia correlaria circa presentem materiam elici possent ex predictis, que considerantibus faciliter possent occurrere. Ideo transeo. Explicit tractatulus de uniformite et difformitate.' The text is written in two columns and the *explicit* in red. Ed. T.M. Smith, *A Critical Text and Commentary upon* De latitudinibus formarum, Ph.D. Dissertation, University of Wisconsin, Madison, WI, 1954, without this copy. Mentioned along with many other not known manuscripts in Di Liscia, *Zwischen Geometrie und Naturphilosophie*, 425 (for the sake of clarity we assume the usual attribution of this text to Jacobus de Santo Martino, which is by no means absolutely sure).
12. 68^r–72^v: Anonymous: ⟨*Sophismata*⟩. *Inc.* 'Mediate id est per mensuram temporalem ...' *Expl.*: '... et non sicut plures homines per casum. Ergo nulli sunt omnes homines similes.' The text was copied in two columns except for f. 69^r and 72^v. It is difficult to decide whether it is a unified text of a special collection or only different exercises which have been gathered over time for the sake of teaching.
13. 72^v–73^v: Anonymous: ⟨*Aliqua collecta de latitudinibus*⟩. *Inc.*: 'Nota primo quod aliqua latitudo est terminata exclusive ad aliquem gradum ...' *Expl.*: '... Expliciunt aliqua de latitudinibus que collegi ex quodam tractatu Erfordie tradito in quo alique conclusiones continebantur. Studeas: non enim omnia posui tanquam vera sed tanquam probabilia.' Above this explicit is stated 'Gregorius' (*explicit* in red).
14. 74^{r-v}: Anonymous: ⟨*Collectio sophismatum*⟩: 'Sophismata collecta a Gregorio ex quodam libro incorrectissimo.' *Inc.*: 'Restat aliqua sophismata et aliquas conclusiones de verbis istis incipit et desinit manifestare ...' *Expl.*: '... ad sophisma semper erit quod non semper erit ... Expliciunt quedam sophismata et conclusiones sophistice [above the line: 'a Gregorio'] collecte ex quodam libro incorrectissimo.' At the top of 74^r, by the same hand (Gregorius): 'nota de his verbis incipit et desinit conclusiones satis absurdas.'
15. 75^{r-v}: William Heytesbury: *De sensu composito et diviso. Inc.*: 'Nota de sensu compositionis et divisionis. Arguendo de sensu diviso ad sensum compositum vel econtra frequenter fallit argumentum ...' *Expl.*: '... Explicit quoddam collectum satis pulchrum de sensu compositionis et divisionis propositionum multiplicium, quod collegi ex quodam tractatu

novo, aliquantulum incorrecto. Videas si placet tibi' (title underlined with red). Cf. William Heytesbury, *Regule solvendi sophismata*, Venice 1494, 2ra. For an English translation (from the 1494 printed edition and some additional manuscripts) see: William Heytesbury, 'The Compounded and Divided Senses', in N. Kretzmann and E. Stump (eds. and transl.), *The Cambridge Translations of Medieval Philosophical Texts*, 1: *Logic and Philosophy of Language*, Cambridge 1988, 413–434.

16. 76^{r-v}: Anonymous: Fragment on natural philosophy. *Inc.*: 'Capiantur duo pondera *a* et *b* scilicet *a* sicut 12 et *b* sicut 6 ...' *Expl.*: '... probatur loco tertii a dicta, etc.'

17. 77^{r}–79^{v}: Anonymous: ⟨*Collectio sophismatum ex textu fallaciarum magistri Adolfi*⟩. *Inc.*: 'Nota disputando proprie dicitur ...' *Expl.*: 'Ista igitur scripta et recollegi breviter et faciliter ex quodam tractatu sophismatum unius magistri qui incipit ad laudem et honorem Dei et non moveri plus studendo ista. Explicit quedam colleccio sophismatum brevium et facilium circa textum fallaciarum Deo gratias que collegi ex textu fallaciarum magistri Adolfi. Studeas si placent tibi. Probabilia enim sunt et sophisticalia.' This seems to be a collection based on a text by master Adolphus on the fallacies. Assuming that this Adolphus is in fact Adolphus of Vienna, the pupil of master Ulrich of Vienna (see next text), this seems to be a hitherto unknown text. In any case, it was not mentioned in the well-documented entry by F.J. Worstbrock, 'Adolf von Wien,' in K. Langosch and K. Ruh (eds.), *Die deutsche Literatur des Mittelalters: Verfasserlexikon*, 1 (Aalen—Futerer), Berlin, 2nd ed., 1978, 68–71.

18. 80^{r}–81^{v}: Anonymous ⟨Ulricus de Wyenna⟩: ⟨*Fallacie*⟩. *Inc.*: 'Quanto aliquod est maius tanto minus videtur ...' *Expl.*: '... tunc est vera ubi nulla contradictoria istius logice non est scientia est falsa, ymo vera.' Magister Ulricus de Wyenna is mentioned at the beginning. On the superior margin (80^{r}): 'Nota pulchra sophisticalia que sunt antiqua et rara nec non profunda.' It seems that texts # 17 and # 18 are closely connected. P. Uiblein, *Die Universität Wien im Mittelalter. Beiträge und Forschungen*, Vienna 1999 (Schriftenreihe des Universitätsarchivs, Universität Wien, 11), 26, mentions that Ulrich of Vienna had a pupil named Adolphus. Whereas Ulricus and Adolphus were quite popular names, we assume (following a suggestion by Harald Berger) that Ulrich is not a contemporary but an earlier master who was active in the School of Saint Stephen at the turn of the thirteenth to the fourteenth century, that is long before the foundation of the University. That matches the characterisation of this *sophismata* collection as 'antiqua'. In his entry on Ulrich of Vienna for the *Verfasserlexikon*, Worstbrock mentions this text as being an excerpt of the *Falla-*

ciae by Ulricus (F.J. Worstbrock, 'Ulrich von Wien,' in K. Langosch and K. Ruh [eds.], *Die deutsche Literatur des Mittelalters: Verfasserlexikon*, 10 [Ulr--Z], Berlin, 2nd ed., 1999, 53–57, at 53–55). Several manuscripts in the Bayerische Staatsbibliothek in Munich (Clm 331, 4384, 4391, and 16520) convey texts by or connected to Ulrich of Vienna (see M. Grabmann, 'Der Anteil Deutschlands am Aristotelismus des Mittelalters,' in Id., *Mittelalterliches Geistesleben: Abhandlungen zur Geschichte der Scholastik und Mystik*, Munich 1956 [3 vols.], 3: 219–231, at 225).[86]

86 After a checking *de visu* we can provisionally add the following details: (1) Clm 4391 seems to convey the complete text of Ulrich's *Fallacie*: *Inc.*: 'Scribit Philosophus primo Elenchorum duo sunt opera sapientis …' (109ra). The explicit contains the attribution: '… expliciunt fallacie Ulrici de Wienna sub anno domini cccxliij [= 1343], vi kln. juliani' (122va). It is perhaps interesting to add that this copy is followed on ff. 124ra–133vb by a series of questions on logic copied by the same hand, specifically dealing with fallacies (ff. 122rb–123ra are blank and on f. 123^{va-vb} there is a German text written in two columns). (2) Clm 16520, 115^{r}–118^{r} seems to be an incomplete copy of the text contained in Clm 4391: *Inc.*: 'Secundum Philosophum circa librum primum Elenchorum duo sunt opera sapientis …' *Expl.*: '… ergo quando non unum includit etc. et quando duo includit nec excludit.' (3) Clm 4384 includes on ff. 104ra–122ra a *reportatio* of the *Fallacie* (described by Worstbrock, 'Ulrich von Wien,' 54, as a commentary). *Inc.*: 'Aristoteles scribit unam propositionem in primo Elenchorum opera sapientis sunt duo …' *Expl.*: '… Explicunt reportate super fallacias Ulrici … liber iste per me iudocum anno domini 1346 …' (4) Clm 331 contains a text on Ulrich's fallacies titled *Dicta magistri Ulrici super fallacias suas*: *Inc*: 'Secundum Philosophum in libro fallaciarum …' (40ra). *Expl.*: '… Expliciunt dicta magistri Ulrici super fallacias suas' (44va). A further text in the same manuscript and copied by the same hand which is in fact attributed to master Ulrich is *De clave intentionum* (46ra–47rb). All these details are, as we said before, provisional until further research on Ulrich of Vienna has been done.

CHAPTER 7

Wyclif, the Black Sheep of the Oxford Calculators

Mark Thakkar

Introduction

The familiar story of the Oxford Calculators has an arresting feature: the line of succession that begins with Thomas Bradwardine and Richard Kilvington in the late 1320s and continues via Roger Swineshead and William Heytesbury in the 1330s seems to stop abruptly with John Dumbleton and Richard Swineshead in the late 1340s. The wider story does not end there, of course; it is well known that the calculatory tradition had a long afterlife in continental Europe thanks to men like Giovanni da Casale in Bologna,[1] Nicole Oresme in Paris,[2] and John of Holland in Prague.[3] As far as Oxford itself is concerned,

1 Casale's disputed question *Utrum in moventibus ad qualitatem illud semper velocius moveatur quod in equali tempore acquirit maiorem latitudinem* (mid-1340s) survives in at least ten copies and was printed in Venice in 1505 under the generic title *De velocitate motus alterationis.* Daniel Di Liscia is working on a critical edition (see D.A. Di Liscia, 'Biagio Pelacani da Parma's Geometrisation of Latitudes and the Problems of the Mean Degree Theorem,' in J. Biard and A. Robert [eds.], *La philosophie de Blaise de Parme. Physique, psychologie, éthique*, Florence 2019 [Micrologus Library, 96], 129–152, at 131, n. 3). Meanwhile, see initially E.D. Sylla, 'Medieval Quantifications of Qualities: The "Merton School",' *Archive for History of Exact Sciences* 8 (1971–1972), 9–39, at 34–38, and D. Buzzetti, 'Intensione delle forme e strutture linguistiche negli scritti bolognesi di filosofia naturale del secolo XIV,' *Medioevo* 29 (2004), 133–176, at 135 and 151–161.

2 Oresme's *Physics* commentary (ca. 1346–1347) and the subsequent Parisian commentaries of John Buridan and Albert of Saxony are discussed in this connection in E.D. Sylla, 'Transmission of the New Physics of the Fourteenth Century from England to the Continent,' in S. Caroti and P. Souffrin (eds.), *La nouvelle physique du XIVe siècle. Actes du Colloque International (Nice, 3–5 septembre 1993)*, Florence 1997 (Biblioteca di Nuncius, 24), 65–110, esp. 92–110. Another early adopter in Paris was the eccentric theologian Peter Ceffons, whose *Sentences* commentary (1348–1349) is remarkably calculatory. See initially E. Mazet, 'Pierre Ceffons et Oresme: leur relation revisitée,' in S. Caroti and J. Celeyrette (eds.), Quia inter doctores est magna dissensio. *Les débats de philosophie naturelle à Paris au XIVe siècle*, Florence 2004 (Biblioteca di Nuncius, 52), 175–194.

3 John of Holland is known in this connection for his *De velocitate motuum* and, to a lesser extent, his *De primo et ultimo instanti* (1369). Cf. C. Wilson, *William Heytesbury: Medieval Logic and the Rise of Mathematical Physics*, Madison, WI, 1956, repr. 1960 (University of Wisconsin Publications in Medieval Science, 3), 26, 33–35, and the endnotes to chapter 4, and M. Clagett, *The Science of Mechanics in the Middle Ages*, Madison, WI, 1959 (University of Wisconsin Pub-

 | DOI:10.1163/9789004512054_009

though, the death of the tradition in around 1350 has achieved the status of a commonplace.[4]

My proposal in this paper is that we should extend the line of succession to one of Oxford's two most famous medieval philosophers, John Wyclif, whose role in the familiar story is currently confined to a solitary footnote.[5] Essentially, my contention will be that even though Wyclif is a conspicuous anomaly—a black sheep, as my title has it—there are good reasons to bring him into the fold. In anticipation of some resistance to this claim, I must begin with a historiographical question that has, I think, received insufficient attention: what entitles us to call someone an Oxford Calculator?

1 Quis calculabit ipsos Calculatores?

The concept of the Oxford Calculators is a modern one. This is nothing to be embarrassed about, but it is worth reminding ourselves how we came up with it. The spadework was done more than a century ago by Pierre Duhem, who had been scouring the Bibliothèque Nationale de France in search of 'the Parisian precursors of Galileo'.[6] The prime candidates proved to be John Bur-

lications in Medieval Science, 4), 247–250 and 268. We should add to the list his *Sophismata* (ca. 1370), whose sixth and final part, devoted to sophisms *de velocitate motuum*, incorporates a crash course in calculatory techniques 'for the sake of those without training in this area' (Leipzig, Universitätsbibliothek, Ms. 1360, 277rb: 'hiis ergo notabilibus cum suis correlariis propter in hac materia inexercitatos lucidissime declaratis'). The manuscripts of the *Sophismata* are listed in John of Holland, *Four Tracts on Logic* (*Suppositiones, Fallaciae, Obligationes, Insolubilia*), ed. E.P. Bos, Nijmegen 1985 (Artistarium, 5), *31*–*32*, albeit with an erroneous dismissal of Kraków, Biblioteka Jagiellońska, Ms. 704.

4 Thus e.g. R. Podkoński, 'Richard Swineshead's *Liber calculationum* in Italy: the Codex Bibl. Naz. San Marco, lat. VI.226 and its significance,' *Recherches de Théologie et Philosophie Médiévales* 84 (2017), 407–426, at 426, n. 72: 'there remains the compelling question why there was no more interest in the Calculators' method of argumentation in natural philosophy ... in England after 1350.' This question is usually asked without being answered. One exception is K. Snedegar, 'Merton Calculators (*act. c.* 1300–*c.* 1349),' in *Oxford Dictionary of National Biography* (2006) doi:10.1093/ref:odnb/95034: 'The black death and subsequent political and theological controversies ignited by John Wyclif ... may have led to the decline of Merton calculatory activity after 1350.'

5 To my knowledge, the extensive literature on the Calculators contains precisely two references to Wyclif's writings. The more recent one is in E.D. Sylla, 'The Oxford Calculators in Context,' *Science in Context* 1 (1987), 257–279, at 264, n. 6: 'For Calculatory material in John Wyclif, see Dziewicki 1893–1899.' For this vague reference, Sylla seems to have been relying on Marshall Clagett's first-hand account, published in 1959, which I discuss in § 3 below.

6 P. Duhem, 'La dialectique d'Oxford et la scolastique italienne,' *Bulletin italien* 12 (1912), 6–26,

idan and Nicole Oresme, but apparently Oresme had acknowledged a debt to some 'ancestors' (*veteres*) who were not to be found among the Parisians, prompting Duhem to dig up the Oxford material. He found the task distasteful,[7] and he might have been less thorough if he had known that the phrase '*ymaginationem veterum*' in his Oresme manuscript was a scribal corruption of '*ymaginationem meam*'.[8] Duhem's findings have since been corrected in almost every detail, but the material he unearthed is still recognizably the basis for the familiar story. It is therefore particularly striking in retrospect that Duhem himself did not identify the Oxford Calculators as a distinct group. His occasional mentions of 'calculateurs' were merely echoes of sixteenth-century references to writers in the tradition quite generally, including 'les *calculatores* parisiens',[9] and his phrase 'l'École d'Oxford' was merely a generic label for Oxford scholastics, despite his particular interest in those from 'le milieu du XIVe siècle'. The closest that Duhem came to the modern conception of the Oxford Calculators was in his identification of four treatises—Roger Swineshead's *De motibus*, the anonymous *Tractatus de sex inconvenientibus*, John Dumbleton's *Summa logice et philosophie naturalis*, and Richard Swineshead's *Liber calculationum*—as 'de la même famille' given their tables of contents.[10]

The Oxford Calculators as we know them are instead the brainchild of Anneliese Maier. Again, it is striking in retrospect that she did not identify them as such in her initial study, where she followed Duhem in writing of 'die Cal-

---

93–120, 203–223, 289–298, repr. in Id., *Études sur Léonard de Vinci*, 3: *Les précurseurs parisiens de Galilée*, Paris 1913, 405–481, esp. 405.

7 Duhem's disdain for Oxonian material is mercilessly documented in J.E. Murdoch, 'Pierre Duhem and the History of Late Medieval Science and Philosophy in the Latin West,' in R. Imbach and A. Maierù (eds.), *Gli studi di filosofia medievale fra Otto e Novecento. Contributo a un bilancio storiografico*, Rome 1991 (Storia e letteratura, 179), 253–302, at 262–272.

8 The error was noted by Anneliese Maier and confirmed in Marshall Clagett's critical edition: A. Maier, 'Die Mathematik der Formlatituden,' in Ead., *An der Grenze von Scholastik und Naturwissenschaft* (Veröffentlichungen des Kaiser Wilhelm-Instituts für Kulturwissenschaft im Palazzo Zuccari), Essen 1943, 255–380, at 275, n. 1; Nicole Oresme, *Tractatus de configurationibus qualitatum et motuum*, ed. M. Clagett, *Nicole Oresme and the Medieval Geometry of Qualities and Motions. A Treatise on the Uniformity and Difformity of Intensities Known as* Tractatus de configurationibus qualitatum et motuum, Madison, WI, 1968 (University of Wisconsin Publications in Medieval Science, 12), 158.

9 See esp. P. Duhem, 'La tradition de Buridan et la science italienne au XVIe siècle,' *Bulletin italien* 9 (1909), 338–360, repr. in Id., *Études sur Léonard de Vinci*, 3: *Les précurseurs parisiens de Galilée*, Paris 1913, 113–137, at 117, and Id., 'Dominique Soto et la scolastique parisienne (*suite*),' *Bulletin hispanique* 14 (1912), 275–299, repr. in Id., *Études sur Léonard de Vinci*, 3: *Les précurseurs parisiens de Galilée*, Paris 1913, 531–555, at 549 (see also 518, 528 f.).

10 P. Duhem, 'La dialectique d'Oxford', at 478 (see also 419, 452, 472). I have updated the names and titles to reflect subsequent scholarship.

culatoren des 14. Jahrhunderts' quite generally.[11] It was in a follow-up article published in 1946 that Maier first joined the dots between the seismic importance of Bradwardine's *De proportionibus* and the preponderance of Mertonians in the story. She now identified as a historiographical category 'those of Bradwardine's *Mertonenses* who were later known as *calculatores*' and who took his 'new teaching' ('neue Lehre') for granted in their work: Richard Swineshead, William Heytesbury, the author of the *De sex inconvenientibus*, and their many 'students, followers and imitators.' John Dumbleton was also to be included in this connection, albeit peripherally ('an sich etwas abseits steht'), because he devoted part of his *Summa* to solving problems of dynamics in a Bradwardinian fashion.[12]

Maier's paradigm-setting article is better known in the revised form that it took in 1949, where she spoke more boldly of 'the circle of the Bradwardine school' ('der Kreis der Bradwardine-Schule').[13] She imported into this context a reworked footnote that now identified Walter Burley as the only one of 'the Oxford circle of Bradwardine's followers' ('der Oxforder Kreis der Bradwardine-Nachfolger') with a surviving *Physics* commentary,[14] even though 'it/he ('der')

11 Maier, 'Die Mathematik der Formlatituden,' 276 (see also 271). Maier did once mention the '*calculationes der Oxforder Schule*' (286), but here she was merely paraphrasing Duhem before referring *in propria voce* to the 'Oxford circles' (*Oxforder Kreise*, plural) in which calculations were cultivated. For the *status quaestionis* before Maier, see C.B. Boyer, *The Concepts of the Calculus*, New York 1939, 72–88, and M. Clagett, *Giovanni Marliani and Late Medieval Physics*, New York 1941.

12 A. Maier, 'Der Funktionsbegriff in der Physik des 14. Jahrhunderts,' *Divus Thomas*, 3rd series, 24 (1946), 158–159. In support of her new picture, Maier cited the irrelevant facts that Bradwardine had dedicated the *De causa Dei* to 'his Mertonians' and that Thomas Buckingham had called Bradwardine '*doctor noster*' in his *Sentences* commentary. In her 1943 study, Maier had only mentioned Merton twice in passing, as being home to Heytesbury in 1330 and Richard Swineshead in 1348: 'Die Mathematik der Formlatituden,' 266, 356.

13 A. Maier, *Die Vorläufer Galileis im 14. Jahrhundert* (Studien zur Naturphilosophie der Spätscholastik, 1), 2nd ed., Rome 1966 (Storia e letteratura, 22), 1st ed., Rome 1949, 96–98. Clagett, *Giovanni Marliani*, 162, had already spoken of 'the Bradwardine school,' but this was merely an echo of Marliani's references to those who agreed with Bradwardine. Cf. e.g. *Questio de proportione motuum in velocitate*, Pavia 1482, sig. a 8va: 'contra aliquas conclusiones positas a Thoma Barduardini et quas tenet communis schola philosophorum.'

14 In the original footnote ('Der Funktionsbegriff,' 160, n. 6) Maier had also mentioned the lost *Physics* commentary of Richard FitzRalph. She later reinstated FitzRalph in this connection despite his unclear position 'between Ockhamism and the *via Bradwardini*' ('Bradwardine'scher Richtung'). See A. Maier, 'Verschollene Aristoteleskommentare des 14. Jahrhunderts', in *Autour d'Aristote. Recueil d'études de philosophie ancienne et médiévale offert à Monseigneur A. Mansion*, Leuven 1955, 515–541, at 529–532, repr. with addenda in Ead., *Ausgehendes Mittelalter. Gesammelte Aufsätze zur Geistesgeschichte des 14. Jahrhunderts*, 1, Rome 1964 (Storia e letteratura, 97), 237–264, 480–482, at 252–253.

almost belongs more in the intellectual atmosphere of Paris than Oxford.' As far as Maier knew, Burley had actually been in Paris since around 1320, but she included him in this 'Oxford circle' on the grounds that he was an ex-Mertonian and 'probably a personal friend of Bradwardine.'[15] She also noted that Richard Kilvington, 'who also belonged to Bradwardine's circle of friends,' had written a lost *Physics* commentary that apparently shared Bradwardine's approach to the *proportiones velocitatum*.[16]

This, then, was the picture that lay behind the coinage of the term 'Oxford Calculators' in 1951:[17] a school of thought based at Merton College and led by Thomas Bradwardine. Maier admitted that we knew 'little or nothing about the concrete teacher-pupil relationships.' As with the Buridan school that Duhem had discerned at Paris, however, it was 'the homogeneity (Einheitlichkeit) of

15 Maier later stated emphatically that Burley had 'no active part in the development of the science of *calculationes* that, taking its cue from Bradwardine's *Tractatus proportionum*, began in Oxford in the 1330s.' See A. Maier, 'Zu Walter Burleys Traktat *De intensione et remissione formarum*,' *Franciscan Studies* 25 (1965), 293–321, at 305, n. 28.

16 Maier's source was a citation in a *Sentences* commentary now attributed to Johannes de Burgo (see K.H. Tachau, 'French Theology in the Mid-Fourteenth Century: Vatican Latin 986 and Wrocław, Milich F.64,' *Archives d'histoire doctrinale et littéraire du Moyen Âge* 51 [1984], 41–80, at 47–59): 'proportio motuum non sequitur excessum potencie motoris super resistenciam rei mote … hoc probat Kilminton in questionibus suis super librum phisicorum'. Four such questions were identified in 1970 by E.D. Sylla, *The Oxford Calculators and the Mathematics of Motion, 1320–1350. Physics and Measurement by Latitudes*, Ph.D. dissertation, Harvard University, 1970 (repr. with preface and errata, New York 1991), 435–438. Elżbieta Jung argues that they predate Bradwardine's *De proportionibus*. See with caution (recourse to the cited literature and the manuscripts being advisable) E. Jung-Palczewska, 'Works by Richard Kilvington,' *Archives d'histoire doctrinale et littéraire du Moyen Âge* 67 (2000), 181–223, at 203–217. For cogent doubts, see E.D. Sylla, 'The Origin and Fate of Thomas Bradwardine's *De proportionibus velocitatum in motibus* in Relation to the History of Mathematics,' in W.R. Laird and S. Roux (eds.), *Mechanics and Natural Philosophy before the Scientific Revolution*, Dordrecht 2008 (Boston Studies in the Philosophy of Science, 254), 67–119, at 95–97. For further arguments, plus an update on the manuscript tradition, see E. Jung, 'The New Interpretation of Aristotle: Richard Kilvington, Thomas Bradwardine and the New Rule of Motion,' in the present volume.

17 E.A. Moody, 'Laws of Motion in Medieval Physics,' *Scientific Monthly* 72 (1951), 18–23. In his 'Galileo and Avempace: The Dynamics of the Leaning Tower Experiment,' *Journal of the History of Ideas* 12 (1951), 163–193, 375–422, esp. part v, 394–410, Moody spoke instead of the 'Mertonians', usually with scare quotes. Maier had appeared on the radar at Columbia and Wisconsin too late for her work to be taken into account in M. Clagett, 'Richard Swineshead and Late Medieval Physics,' *Osiris* 9 (1950), 131–140 (cf. 131, n. 1). The shock of the Americans' belated discovery is sympathetically documented in Maier's account of a letter from Moody in late 1949: A. Maierù and E.D. Sylla, 'Daughter of Her Time: Anneliese Maier (1905–1971) and the Study of Fourteenth-Century Philosophy,' in J. Chance (ed.), *Women Medievalists and the Academy*, Madison, WI, 2005, 625–645, at 642, n. 67.

the doctrinal tradition and of intellectual physiognomy' that afforded us a glimpse of the Bradwardine school at Oxford.[18] It was for this reason alone that Maier felt entitled to refer to Kilvington and Dumbleton as Bradwardine's 'pupils' ('Schüler').[19]

It was not long before cracks started appearing in Maier's picture. On the institutional front, James Weisheipl argued that Bradwardine had never taught Heytesbury, Dumbleton or Richard Swineshead, and that 'the adoption of Bradwardine's new method was by no means restricted to Fellows of Merton College.'[20] In particular, Kilvington was at Oriel by 1333, and there was no evidence to support the tradition of associating him with Merton.[21] Weisheipl also ended

18 Maier, *Die Vorläufer*, 3: 'auch wenn wir ... über die konkreten Lehrer-Schüler-Beziehungen in den einzelnen Fällen wenig oder nichts *wissen* [my emphasis]. Ausschlaggebend ist die Einheitlichkeit der Lehrtradition und der geistigen Physiognomie, die die beiden "Schulen" als solche erscheinen lässt.' Scholars have now rejected Duhem's 'Buridan school' in favour of a looser intellectual network: W.J. Courtenay, 'The University of Paris at the Time of Jean Buridan and Nicole Oresme,' *Vivarium* 42 (2004), 3–17, and esp. J.M.M.H. Thijssen, 'The Debate over the Nature of Motion: John Buridan, Nicole Oresme and Albert of Saxony. With an Edition of John Buridan's *Quaestiones super libros Physicorum, secundum ultimam lecturam*, Book III, q. 7,' *Early Science and Medicine* 14 (2009), 186–210. Thijssen's criticisms do not take into account Maier's later attempt to backpedal on the notion of a 'Schule': 'es soll damit ... kein konkretes Lehrer-Schüler-Verhältnis *gemeint sein* [my emphasis], sondern lediglich eine naturphilosophische *Richtung*, an deren Anfang Bradwardine (bezw. Buridan) steht' (A. Maier, 'Die *Quaestio de velocitate* des Johannes von Casale,' *Archivum Franciscanum Historicum*, 53 [1960], 276–306, repr. in Ead., *Ausgehendes Mittelalter. Gesammelte Aufsätze zur Geistesgeschichte des 14. Jahrhunderts*, 1, Rome 1964 [Storia e letteratura, 97], 381–411, at 400, n. 43).

19 A. Maier, 'Diskussionen über das aktuell Unendliche in der ersten Hälfte des 14. Jahrhunderts,' *Divus Thomas*, 3rd series, 25 (1947), 147–166, 317–337, at 328, repr. in Ead., *Ausgehendes Mittelalter. Gesammelte Aufsätze zur Geistesgeschichte des 14. Jahrhunderts*, 1, Rome 1964 (Storia e letteratura, 97), 41–85, at 75: '*Richard Killington*, dem Freund und Schüler Bradwardines.' Likewise *Die Vorläufer*, 174. See also A. Maier, Review of H.L. Crosby, Jr. (ed.), *Thomas of Bradwardine. His* Tractatus de proportionibus (1955), *Isis* 48 (1957), 84–87, at 85[b]: 'Dumbleton, der Schüler Bradwardines, soll ... die Dynamik seines Meisters ... interpretiert haben.'

20 J.A. Weisheipl, 'The Place of John Dumbleton in the Merton School,' *Isis* 50 (1959), 439–454, at 439–441. Weisheipl accordingly cautioned against his own habit of referring to the Oxford Calculators as 'the Merton School', though such an ambivalent warning was doomed to fail (witness e.g. W.R. Laird, 'The School of Merton and the Middle Sciences,' *Bulletin de philosophie médiévale*, 38 [1996], 41–51, at 41). The term was apparently coined by Robert Gunther in 1919 to refer to astronomers like Simon Bredon and William Rede, and had since been broadened to cover all scientific and logical activity at Merton ca. 1300–ca. 1360 (W.A. Pantin, *The English Church in the Fourteenth Century*, Cambridge 1955, 136–138).

21 Kilvington's alleged Merton connection probably arose from a misreading of Henry

centuries of confusion by disentangling three different Swinesheads: a lawyer named John, of no concern to us; 'the Calculator', Richard, who was at Merton in the 1340s and 50s; and the logician and natural philosopher (and later theologian) Roger, who was at Oxford but not Merton in the late 1320s and/or early 1330s, and who apparently died as a Benedictine monk at Glastonbury in 1365.[22] Roger Swineshead is counted among the Calculators on the strength of his *De motibus*. We may now add to his bibliography the *Sophismata logicalia*, a hitherto lost work that John Bale had seen in the library of Thomas Godsalve of Norwich after the dissolution of the monasteries;[23] one of the two

---

Savile's editorial preface to Bradwardine's *De causa Dei* (Thomas Bradwardine, *De causa Dei contra Pelagium et de virtute causarum ad suos Mertonenses libri tres*, ed. H. Savile, London 1618, sig. a 4^{r-v}). Before quoting William de Chambre's mid-fourteenth-century account of the scholars in Richard de Bury's household, Savile (himself a Mertonian) noted that the seven *clerici* listed besides Bradwardine had come '*ex sodalitio ferè Mertonensi*': *mostly* from the Mertonian fellowship. The actual figure seems to be three (Walter Burley, John Maudith, Walter Segrave) out of seven (the non-Mertonians being Kilvington, Richard FitzRalph, Robert Holcot and Richard Bentworth). On this 'circle', see W.J. Courtenay, *Schools and Scholars in Fourteenth-Century England*, Princeton, NJ, 1987, 133–137.

22 J.A. Weisheipl, 'Roger Swyneshed, OSB, Logician, Natural Philosopher, and Theologian,' in *Oxford Studies presented to Daniel Callus*, Oxford, 1964 (Oxford Historical Society, New Series, 16), 231–252. See now A.G. Molland, 'Swineshead, Roger (*d.* 1365?), Natural Philosopher and Benedictine Monk,' in *Oxford Dictionary of National Biography* 53, Oxford 2004, doi:10.1093/ref:odnb/52682, and E.D. Sylla, 'Swyneshed (Swineshead), Roger,' in N. Koertge (ed.), *New Dictionary of Scientific Biography*, 6, Detroit, MI, 2008, 563–564.

23 J. Bale, *Index Britanniae Scriptorum (ca. 1550)*, eds. R. Poole and M. Bateson, Oxford 1902, 403. Roger's *Sophismata logicalia* is the '*tractatus qui vocatur formishede*' found in Padova, Biblioteca Universitaria, Ms. 1123, 42ra–50ra, between an Oxford *proportiones* and (missing its first chapter) Heytesbury's *Regule*. An incomplete anonymous copy is found among logical texts in Wrocław, Biblioteka Uniwersytecka, Ms. IV. Q. 3, 241^{r}–249^{r}. These copies were recorded but not identified by Lambert de Rijk, who misread '*formishede*' (a corruption of '*swinishede*') as '*Formahede*' and later doubted its connection to the treatise: L.M. de Rijk, '*Logica Cantabrigiensis*: A Fifteenth Century Cambridge Manual of Logic,' *Revue Internationale de Philosophie*, 29 (1975), 297–315, at 308–310; Id. (ed.), *Some 14th Century Tracts on the* Probationes Terminorum (*Martin of Alnwick O.F.M., Richard Billingham, Edward Upton and Others*), Nijmegen 1982 (Artistarium, 3), *33*. Both copies are poor, but the beginning may be reconstructed as follows: 'Cum videmus logicalia sophismata acuere celeriter puerorum ingenia, ideo zelo veritatis pre ceteris me inclinante et iuvenum informatione hac parte me cogente, doctore summo primitus invocato, iuxta capacitatis mee †modicitatem indefectum [? *recte* modulum et decursum] stilo brevi quamvis rudi de sophismatibus logicalibus tractare temptabo cum effectu. In propositionibus universalibus de disiuncto subjecto vel predicato primo ponam agressum.' Roger's authorship may be confirmed by comparing the beginning of his *De motibus* (taken with modified punctuation from E.D. Sylla, 'Mathematical Physics and Imagination in the Work of the Oxford Calculators: Roger Swineshead's *On Natural Motions*,' in E. Grant and J.E. Murdoch [eds.], *Mathematics and its Applications to Science and Natural Philosophy in the Middle*

surviving copies proves to contain a digression on the minimal degree of a colour.[24]

Weisheipl's work paved the way for Edith Sylla's doctoral thesis, which examined the works of the Oxford Calculators in unprecedented detail and uncovered some unsuspected theoretical disagreements.[25] Not only were there differing conceptions of the fundamental concepts of degrees and latitudes,[26] for instance, but Roger Swineshead had even rejected the rule of motion for which Bradwardine's *De proportionibus* is most famous.[27] On closer inspection, then, the doctrinal uniformity that Maier had seen in the calculatory tradition proved to be an illusion.

To an extent, Maier's picture could be restored by speaking, as Sylla herself has often done, of the 'core' Oxford Calculators: Heytesbury, Richard Swineshead, and one or both of Bradwardine and Dumbleton.[28] These subgroupings

---

Ages. Essays in Honour of Marshall Clagett, Cambridge 1987, 69–101, at 100 n. 58): 'Motore primo primitus invocato, iuxta capacitatis mee modulum et decursum stilo brevi quamvis rudi tractare de motibus temptabo cum effectu. Plerisque siquidem pluries ad hoc excitantibus, zeloque veritatis pre ceteris inclinante, presentis negocii suppremis affectibus inducor ad agressum.'

24 In a chapter on sentences with a conjoint subject or predicate, after discussing inferences like '*Sortes est albus et coloratus, ergo Sortes est albus*', Roger argues that a body with a white half and a black half is coloured. In the Padua copy, this prompts the digression: 'Circa latitudinem coloris, moderni dubitant numquid sit dare remissimum gradum alicuius coloris qui potest esse in illa …' (43va–44ra). The main objection to a colour having a minimal degree (*quo nullus est remissior*) is that the lower bound of a latitude is the non-degree (*non gradus*). The Wrocław copy instead moves straight on to sentences like '*Nihil et chimera sunt fratres*' (243^{v}).

25 Sylla, *The Oxford Calculators*. Over half a century later, Sylla's thesis remains indispensable, not least for a large appendix (running to almost 250 pages) with outlines of works including Roger Swineshead's *De motibus*, Dumbleton's *Summa* and Richard Swineshead's *Calculationes*.

26 E.D. Sylla, 'Medieval Concepts of the Latitude of Forms: The Oxford Calculators,' *Archives d'histoire doctrinale et littéraire du Moyen Âge* 40 (1973), 223–283.

27 Sylla 'Mathematical Physics,' esp. 71. From his prefatory remarks about the reception of non-authoritative writings, Sylla portrays Roger as 'aware that a number of his positions are unusual or nontraditional' (83), but I think he is more likely to have been displaying his monkish humility as a modern author than to have been acknowledging the canonical status of a contemporary like Bradwardine.

28 Bradwardine and Dumbleton were both included in the 'core' in Sylla, 'Latitude of Forms,' 224. A new emphasis on sophismatic disputations saw Dumbleton relegated to the periphery in E.D. Sylla, 'The Oxford Calculators,' in N. Kretzmann, A. Kenny, J. Pinborg and E. Stump (eds.), *The Cambridge History of Later Medieval Philosophy. From the Rediscovery of Aristotle to the Disintegration of Scholasticism, 1100–1600*, Cambridge 1982, 540–563, at 540 (cf. 563), and Ead., 'The Oxford Calculators in Context,' 262. Bradwardine has since been relegated to the background and Dumbleton readmitted to the core: E.D. Sylla, 'The

dampen down the doctrinal variation and allow Merton to retain its institutional role; but they have a whiff of the *ad hoc* about them, and they leave it unexplained why Roger Swineshead should be accorded a peripheral status despite his nonconformist views. Given the problems with Maier's picture that had emerged by her death in 1971, what we really needed was a new characterization of the Oxford Calculators.

At this juncture, the process of correction, extension and refinement that had taken us from Duhem's field notes to Sylla's thesis was derailed by a side-on collision. In a string of influential papers on the sophistication of fourteenth-century philosophy, John Murdoch (Sylla's advisor) subsumed the innovations of the Oxford Calculators into an omnium-gatherum of 'new measure languages', deliberately stretching the word 'measure' to cover various non-quantitative techniques for solving limit-decision problems and problems involving infinity and continuity.[29] As it happens, this elastic interpretation of the word 'measure'—not, notice, of a medieval word meaning 'measure' that already covered such techniques—came with a conspicuously weak justification.[30]

Oxford Calculators' Middle Degree Theorem in Context,' *Early Science and Medicine* 15 (2010), 338–370, at 340–342.

29 The full manifesto for this project was launched in 1973: J.E. Murdoch, 'From Social into Intellectual Factors: An Aspect of the Unitary Character of Late Medieval Learning,' in J.E. Murdoch and E.D. Sylla (eds.), *The Cultural Context of Medieval Learning*, Dordrecht 1975 (Boston Studies in the Philosophy of Science, 26), 271–339 (plus discussion, 339–348), esp. 282–289. For the 1971 draft, see J.E. Murdoch, 'Philosophy and the Enterprise of Science in the Later Middle Ages,' in Y. Elkana (ed.), *The Interaction between Science and Philosophy*, Atlantic Highlands, NJ, 1974 (The Van Leer Jerusalem Foundation Series), 51–74 (plus discussion, 91–113), esp. 57–64. For a more cautious retrospective, see J.E. Murdoch, 'The Involvement of Logic in Late Medieval Natural Philosophy,' in S. Caroti (ed.), *Studies in Medieval Natural Philosophy*, Florence 1989 (Biblioteca di Nuncius, 1), 3–28. The term 'limit-decision problem', referring to the assignment of intrinsic and/or extrinsic limits, was first used for temporal intervals (*de incipit et desinit, de primo et ultimo instanti*) and subsequently extended to capacities (*de maximo et minimo*): J.E. Murdoch, 'Propositional Analysis in Fourteenth-Century Natural Philosophy: A Case Study,' *Synthese* 40 (1979), 117–146, and Id., 'Infinity and Continuity,' in N. Kretzmann, A. Kenny, J. Pinborg and E. Stump (eds.), *The Cambridge History of Later Medieval Philosophy. From the Rediscovery of Aristotle to the Disintegration of Scholasticism, 1100–1600*, Cambridge 1982, 564–591, at 585–588.

30 Murdoch, 'From Social into Intellectual Factors,' 282: '"measure" to medievals was most frequently considered a matter of *denominatio* and to elucidate the *denominatio* of a subject was to delineate accurately the predicate or predicates that belong to it under such and such circumstances.' Here (and at 288) he cited 'Philosophy and the Enterprise of Science,' 62: 'from the medieval point of view the problem of measure is better seen (as Anneliese Maier has cautioned, but few have heeded) as a logical problem of denomination. Thus, the measurement involved in saying that a uniformly difformly hot body is as hot as a body uniformly hot in its mean degree is exactly parallel to the logical problem of

More importantly for our purposes, though, it eroded the distinctiveness of the Calculators by lumping them together with pre-calculatory thinkers like Thomas Wylton and William of Ockham. This conflation soon led Murdoch to claim that John of Mirecourt's 'most extensive application of the calculatory tradition' in 1345 or so was to the problem of 'measuring' the acts of the will and their effects, even though it is clear from his quotations that these were limit-decision problems and that Mirecourt's solutions were not calculatory. It is no more than a play on words to say that he 'applied substantial portions of the *calculus* of first and last instants in determining the appropriate temporal limits.'[31]

One might wonder whether anything is really at stake here. After all, as Maier pointed out in defence of her use of 'Kreis' and 'Schule', a word means just what we choose it to mean.[32] Still, the question is whether we should make words mean so many different things. I myself think it is perverse to divorce our talk of the 'calculatory' tradition from the uses of '*calculare*' and its cognates that led to it in the first place. The obvious example is the use of '*Calculator*' as a title for Richard Swineshead, which goes back at least to 1400.[33] But this title too has an ancestry, and it has received surprisingly little attention. It is still not common knowledge that John of Holland had referred to *Calculatores*

denominating a body by (say) the predicate blue.' Maier herself said nothing about 'measure' in the cited passage, which was actually about the problem of attributing *motus* or *velocitas* to something that is not metaphysically informed by it (Maier, 'Die Mathematik der Formlatituden,' 278f.).

31 J.E. Murdoch, '*Subtilitates Anglicanae* in Fourteenth-Century Paris: John of Mirecourt and Peter Ceffons,' in M.P. Cosman and B. Chandler (eds.), *Machaut's World: Science and Art in the Fourteenth Century*, New York 1978 (Annals of the New York Academy of Sciences, 314), 51–86, at 56–59 (emphasis added). Murdoch's conflation is responsible for some otherwise inexplicable contortions in the historiography of the Calculators. Cf. e.g. E. Jung, 'Mathematics and the *secundum imaginationem* Procedure in Richard Kilvington,' *Przegląd Tomistyczny* 22 (2016), 109–120, at 112: 'The type of measuring presented in *Sophismata* is focused on establishing limits to the beginning and ending of continuous processes such as Socrates' whitening or increasing speed in motion. This type of measure, by first and last instance [*sic*], does not appear straightforwardly mathematical. Yet it deals with mathematical considerations, since it prescribes a measure for natural processes.'

32 Maier, 'Die *Quaestio de velocitate* des Johannes von Casale,' 400, n. 43: 'nomina significant ad placitum; es muss nur klar sein, wer oder was gemeint ist.' Cf. Sylla, *The Oxford Calculators*, 3: 'the works studied [in this thesis] do have sufficient mutual coherence to be classified as a group, however that group might be named.'

33 Angelo da Fossambruno, *De reactione* (ca. 1400), quoted most fully in G. Federici Vescovini, 'Il commento di Angelo di Fossombrone al *De tribus praedicamentis* di Guglielmo Heytesbury,' in A. Maierù (ed.), *English Logic in Italy in the 14th and 15th Centuries: Acts of the 5th European Symposium on Medieval Logic and Semantics, Rome, 10–14 November 1980*, Naples 1982 (History of Logic, 1), 359–374, at 365, n. 11.

(plural) in around 1370,[34] let alone that his older contemporary John Wyclif provides a much richer source of such vocabulary, speaking of '*materie calculatorie*' (calculatory topics), of '*casus calculatorii*' and '*imaginationes calculatorie*' (calculatory scenarios and thought experiments), and of the '*calculator*' or '*calculans*' as a generic figure.[35] As for the 'core' Oxford Calculators, I have not scoured all of their works, but Heytesbury used the word '*calculatio*' to refer to some relevant calculations in the sixth chapter of his *Regule*,[36] as did Richard Swineshead in the parallel discussion in his *De motu locali*.[37] In short, there is

---

34 Thus E.D. Sylla, 'Oxford Calculators,' in H. Lagerlund (ed.), *Encyclopedia of Medieval Philosophy*, Dordrecht 2011, 903–908, at 907b: 'Although the label "calculators" was not applied in the fourteenth century to Swineshead or the men surrounding him ... [it] began to be used in the sixteenth century.' But cf. John of Holland, *De velocitate motus localis*, in Clagett, *The Science of Mechanics*, 249: 'Motus localis uniformiter difformis describitur a calculatoribus sic.' Clagett translated '*calculatoribus*' as 'the [English] Calculators' (247), explained that John was referring to 'his Merton College predecessors,' and identified the text that followed '*sic*' as 'John's quotation of a common English definition of uniform acceleration' (248). I have not been able to trace the definition myself. John's use of the plural is confirmed by two other passages in his three-part *De velocitate motuum*: 'patet consequentia de se et iuxta omnes calculatores, et similiter antecedens' (*De velocitate motus augmentationis*, Città del Vaticano, Biblioteca Apostolica Vaticana, Ms. Vat. lat. 1108, 152^{v}); 'sumendo *alterari* stricte, scilicet †per [*recte* pro] intendi vel remitti, secundum quod calculatores ipso utuntur' (*De velocitate motus alterationis*, Città del Vaticano, Biblioteca Apostolica Vaticana, Ms. Vat. lat. 1108, 161^{v}).

35 I draw on Wyclif in § 2 and § 3 below, but here are two initial examples that hint at the penetration of calculatory material into theology: 'multe conclusiones impossibiles sequuntur ex potestate pape taxandi penas et peccata hominis suffragandi ad suum placitum et ex casibus calculatoriis quos ista scola habet admittere' (John Wyclif, *Tractatus de ecclesia* (1378/1379), ed. J. Loserth, London 1886, 574); 'Ceteras autem difficultates metaphysicas vel calculatorias de gracia debet fidelis respuere tamquam vanas' (John Wyclif, *Sermones* (ca. 1378–1382), vol. 3, ed. J. Loserth, London 1889, 33).

36 William Heytesbury, *Regule solvendi sophismata*, Venice 1494, 41^{rb-va}, on uniform intensification from/remission to a non-zero degree: 'nullus potest esse terminus universalis penes quem universaliter cognosci poterit quanto plus pertransiretur per primam medietatem huius intensionis vel remissionis quam per secundam ... Cognitis tamen gradibus extremis ... cognosci poterit per *calculationem* quantum pertransiretur in prima medietate et etiam quantum in secunda, quia ... haberi potest etiam gradus medius inter istos, et etiam gradus medius inter illum gradum medium et gradum intensiorem illam latitudinem terminantem. Sed huiusmodi *calculatio* †minorem [*recte* maiorem] sollicitudinem ageret quam profectum. Et ideo sufficit quod in omni huiusmodi casu respondeatur generaliter quod plus pertransiretur per medietatem talis latitudinis intensiorem quam remissiorem.'

37 R. Podkoński, *Richard Swineshead's Theory of Motion*, Łódź 2019 (Research on Science and Natural Philosophy, 2), 183–184: 'In certo casu, scito quantum transiretur per gradum terminantem extremum intensius, bene calcularetur per priorem regulam de uniformi inten-

surely scope for a characterization of the tradition that takes calculation as its starting-point.

Before I can propose such a characterization, though, I must deal with two attempts to interpret '*calculatio*' as meaning something other than calculation. The first of these was made by Weisheipl, who identified *calculatio* with the 'tedious' use of letters like 'a' and 'b' as generic terms, 'a method of argumentation common to all the disputations of the fourteenth century and even to commentaries on the *Sentences*.'[38] In support, he cited Maier's claim that the 1320s saw a significant expansion in the use of what she called '*argumenta in terminis*'.[39] Even if Maier's claim is true, though, it does not support Weisheipl's identification of *calculationes* with this kind of formalization, which is an accidental feature that calculatory arguments can (but need not) share with all sorts of arguments.[40] Indeed, the *calculationes* referred to by Heytesbury need not have involved this 'letter-calculus', as is clear from the worked example

tione a non gradu usque ad illum gradum quantum transiretur per illam partem datam. Pro ⟨regula⟩ generali in huiusmodi casu sufficit dicere quod plus transiretur … quam gradu subduplo ad gradum terminantem extremum intensius transiretur in equali tempore … non enim oportet quod dicatur †respectu [*recte* quantum] plus, quia requir⟨er⟩et unam [*v. l.* illam] specialem calculationem.' *Caveat lector*: the edition slavishly follows one of the two manuscripts on the grounds that the other copy is later and has some omissions (see 124–127, 133).

38 J.A. Weisheipl, 'Curriculum of the Faculty of Arts at Oxford in the Early Fourteenth Century,' *Mediaeval Studies* 26 (1964), 143–185, at 180 (= *Early Fourteenth Century Physics of the Merton 'School', with Special Reference to Dumbleton and Heytesbury*, unpublished Ph.D. dissertation, Oxford University, 1956, 40).

39 Maier, 'Die Mathematik der Formlatituden,' 257–261. Maier's account of arguments '*in terminis*' has been accepted in the literature even though she only provided one flimsy piece of evidence from Burley's *De intensione et remissione formarum* (ca. 1323). My own impression, taking into account the use of the phrase by thirteenth-century authors, is that an argument '*in terminis*' was a particular instantiation of a general argument, used for the purposes of exemplification. At any rate, the matter deserves investigation, and any account will need to be able to explain what Ockham says in his *expositio* of *Physics* VI, 239b14–18 (ca. 1323): 'Istam rationem facit Commentator in terminis, commento 78, dicens: "Sit velocissimum equus et tardissimum tortuca, et est non dubium quin habeant aliquam proportionem in motu; sit igitur una pars centum, et ponatur quod inter equum et tortucam sit una leuca …"' (William of Ockham, *Expositio in libros Physicorum Aristotelis. Libri IV–VIII*, eds. R. Wood, R. Green, G. Gál, J. Giermek, F. Kelley, G. Leibold and G. Etzkorn, St. Bonaventure, NY, 1985 [Opera philosophica, 5], 560).

40 Weisheipl's misreading of Maier is echoed without attribution but with helpfully illustrative detail in Sylla, 'The Oxford Calculators,' 562: 'In this broad sense … any of the familiar fourteenth-century arguments *in terminis*, in which the letters A, B, C, etc., were used to represent, for example, velocities, or degrees of quality, or propositions, or distances, or instants, may have been considered calculations.'

in the anonymous commentary that scholars have mistakenly attributed to Gaetano da Thiene.[41]

A more sophisticated attempt has been made by Sylla, who notes that 'even in the *Liber calculationum* most of the mathematics involved is not very numerical,' and who accordingly diagnoses a broad sense of *calculatio* as 'reasoning in which … variables and their relations are kept in mind and equivalences derived,' whether this reasoning is numerical or merely 'verbal'.[42] In support, she cites Heytesbury's mention of gratuitously elaborate scenarios that force the respondent to 'carefully calculate and run about from one proposition to another until it becomes clear which of them is insoluble.'[43] If this were the sole surviving occurrence of the verb '*calculare*', Sylla's interpretation would certainly be possible. In fact, though, '*calculare*' is an ordinary word in post-classical Latin, and since Heytesbury gives no examples that might rule out the standard arithmetical interpretation, it would be reckless lexicography to treat this quotation alone as proof of an otherwise unattested sense. As for the 'not very numerical' mathematics in Swineshead's *Liber calculationum*, the variables are quantitative and their relations are proportional, so I do not see why the operations should not count as calculations.

---

41 William Heytesbury, *Regule*, 41rb (ed. Mappelli). The table of contents in Mappelli's edition (1^{r}) gives no author for the '*Expositio litteralis supra tractatum de tribus*' interspersed in small type with the first two parts of chapter 6 of the *Regule* (37ra–45ra; 45rb–49va plus displaced paragraph on 52rb). But Duhem (*Les précurseurs parisiens de Galilée*, 494) conflated this anonymous commentary with Gaetano's *Recollecte* on chapters 1–5, printed at the end of each chapter in Mappelli's edition (4va–36vb), and the mistake has been repeated ever since, despite the laudable absence of the anonymous commentary in S. da Valsanzibio, *Vita e dottrina di Gaetano di Thiene, filosofo dello Studio di Padova (1387–1465)*, Verona 1948, repr. Padua 1949. P.V. Spade, 'The Manuscripts of William Heytesbury's *Regulae solvendi sophismata*: Conclusions, Notes and Descriptions,' *Medioevo* 15 (1989), 271–313, at 279, n. 20, has even claimed that although the commentary on chapter 6 is 'lacking', 'missing' or 'omitted' in the three copies of Gaetano's *Recollecte* that he describes (#1, #8, #15), it is nonetheless 'found in other manuscripts'. I have been unable to substantiate this claim. It would in any case be surprising given the concluding words of Gaetano's *Recollecte* on chapter 5 (quoted e.g. in Spade, 280, n. 21): 'Et hec … sufficiant pro introductione habenda ad materiam de maximo et minimo et aliorum tractatuum precedentium compilatorum per … Hentisberum per me Gaietanum de Tienis … hoc ordine compilata.'

42 Sylla, 'The Oxford Calculators,' 561–562.

43 William Heytesbury, *Regule*, 7rb: 'Multi etiam fiunt casus in hac parte prolixi nimis et inutiles in quibus oportet diligenter calculare et discurrere ab una propositione ad aliam quousque appareat que illarum sit insolubilis.' L. Pozzi's edition of this chapter from two Vatican manuscripts has '*discutere*' for '*discurrere*' (L. Pozzi, *Il Mentitore e il Medioevo. Il dibattito sui paradossi dell'autoriferimento. Scelta di testi, commento, traduzione*, Parma 1987 [Quaderni di filosofia, 10], 248), but this is just a transcription error.

With these obstacles out of the way, I will now try to set the historiography back on track. I propose that we should take the hallmark of the calculatory tradition to be *calculation*. Not just any calculation, mind, but specifically calculations involving the degree (*gradus*) of an attribute that may be intensified and remitted within a certain range (*latitudo*), and paradigmatically the application of the mathematics of proportion to the rate of various kinds of change (*velocitas motuum*). With this proposal in mind, I suggest that three distinctive features, plus an optional fourth, may be used as diagnostic tests of a given work's participation in the tradition:[44]

(1) The calculatory tradition revolves around the *theoretical question* of what some variable is 'attendant on' or determined by (*penes quid attenditur*).[45] What distinguishes the tradition is not the phrasing of the question, which goes back well into the thirteenth century,[46] but its centrality,

44 A trawl through the literature will reveal that the first and third features on my list have occasionally been noted, albeit without a focus on demonstrable distinctiveness; the second and fourth, to my knowledge, have not.

45 This curious phrase is only used in the passive, which discourages an interpretation in terms of a transitive sense of '*attendere*' like 'observe'. Since the passive can mean 'expand', F. Zanin, *L'analisi matematica del movimento e i limiti della fisica tardo-medievale. La ricezione della* perspectiva *e delle* calculationes *alla Facoltà delle Arti di Parigi* (1340–1350), Padua 2004 (Subsidia Mediaevalia Patavina, 6), 214, interprets '*attenditur penes*' as meaning 'increases in virtue of'. But this ignores non-quantitative cases, e.g. Thomas Aquinas, *Scriptum super II Sententiarum*, d. 42 q. 2, a. ii, ed. P. Mandonnet, Paris 1929, 1075: 'Ista divisio peccatorum ... attenditur penes ea ex quibus actus regulatur, scilicet praecepta legis.' I think the root sense of '*attenditur penes*' is actually that of being 'directed towards', and thus (via being 'attentive to') of subserviently accompanying or waiting upon, so that 'is attendant on' would be a literal translation. A close alternative is 'depends on', but this unfortunately suggests the possibility of a merely partial dependence. In this respect one might prefer 'is governed by', but this unfortunately fixes the causal direction. The loose translation 'is determined by' avoids both problems and is thus my suggested alternative; those who find 'determine' intolerably ambiguous might prefer 'is concomitant with'. For quantitative variables, the paraphrase 'is measured by' has some fourteenth-century precedent, e.g. in a gloss attributed to Albert of Saxony: 'velocitatem attendi penes aliquid non est aliud quam velocitatem mensurari et cognosci per illud recte sicut per ulnam mensuramus pannum et cognoscimus quantitatem eius' (Albert of Saxony, *Expositio et quaestiones in Aristotelis Physicam ad Albertum de Saxonia attributae*, ed. B. Patar, Louvain-la-Neuve, Leuven 1999 [3 vols.] [Philosophes médiévaux, 39–41], 3: 893). See also D.A. Di Liscia, 'Velocidad *quo ad effectus* y velocidad *quo ad causas*: La tradición de los calculadores y la metodología aristotélica,' in D.A. Di Liscia, E. Kessler and C. Methuen (eds.), *Method and Order in Renaissance Philosophy of Nature. The Aristotle Commentary Tradition*, Aldershot 1997, 143–176, at 161.

46 Here is just one example, from the anonymous accretions to Alexander of Hales' *Glossa in librum secundum Sententiarum*, d. 40, eds. PP. Collegii S. Bonaventurae, Quaracchi 1952 (Bibliotheca Franciscana scholastica medii aevi, 13), 390: 'Cum quantitas peccati

especially with regard to the rate of changes (local motion, qualitative alteration, quantitative augmentation/diminution, extensional rarefaction/condensation). Calculatory answers are also distinctive for employing the mathematics of proportion,[47] and for specifying whether the variables are said to be determined by their causes or their effects (*quoad causam/effectum*).[48]

(2) Two distinctive *terminological innovations* facilitate the calculations that give the tradition its name. One is the term '*non gradus*' (non-degree), which serves to designate the limit case of remission. Linguistically speaking, the construction is so idiosyncratic that its use guarantees a connection to the calculatory tradition.[49] The other is the '*gradus ut n*' con-

---

attendatur penes quantitatem contemptus, questio est penes quid attendatur quantitas contemptus, et propter quid quantitas peccati penes contemptum mensuretur.'

47 By 'the mathematics of proportion', I mean the Euclidean-Boethian framework that Thomas Bradwardine provided as background in the first chapter of his *Tractatus de proportionibus velocitatum in motibus*, ed. H.L. Crosby, Jr., *Thomas of Bradwardine. His* Tractatus de proportionibus. *Its Significance for the Development of Mathematical Physics*, Madison, WI, 1955 (University of Wisconsin Publications in Medieval Science, 2), 66–84. This involves not only the predictable array of specific proportions—'(*sub*)*duplus*', '(*sub*)*sesquialter*', and so on—but also the distinction between proportions *maioris inequalitatis* and *minoris inequalitatis*. For a guide to this territory, see Thomas Bradwardine, *Traité des rapports entre les rapidités dans les mouvements*, transl. S. Rommevaux, Paris 2010 (Sagesses médiévales).

48 The distinction is given in different terms at the start of the fourth chapter of Bradwardine's *De proportionibus*, 124: 'In comparatione ad potentias moventes et motas … in comparatione ad quantitates motus et pertransiti spatii.' Historians of science have often seen this as an exciting new demarcation between the fields of dynamics and kinematics, but Di Liscia, 'Velocidad *quo ad effectus* y velocidad *quo ad causas*,' has shown at some length that it was just the standard Aristotelian distinction between *scientia a priori*, gained via an explanatory *demonstratio propter quid*, and *scientia a posteriori*, gained via a less informative *demonstratio quia*.

49 The peculiarity of the '*non gradus*' construction is reflected in Nicole Oresme's *Livre du ciel et du monde* (1377): commenting on *De caelo* II, 6, 288a19–22, Oresme used '*degré*' as a French equivalent of '*gradus*', but his powers of translation were defeated by '*a non gradu*', which he used three times after first mentioning it. See Nicole Oresme, *Le livre du ciel et du monde*, eds. A.D. Menut and A.J. Denomy, Madison, WI, 1968 (University of Wisconsin Publications in Medieval Science, 11), 414. It is thus exceptional among the vestiges of Latin in this avowedly vernacular work, most of which are one-off mentions of names or technical terms (e.g. 'ceste duracion n'a pas nom apropriè fors que en latin puet estre dicte *evum*', 162; 'le soleil est pres de terre vers le point appellée *oppositum augis*', 352; 'une constellacion qui est vers aquilon, i.e. *maior ursa* que nous appellons le char', 534). Oresme did use '*ceteris paribus*' and '*a casu/fortuna*' three times (192, 438, 480; 406, 446) and '*rarum/dempsum*' once (464), but he had French equivalents for these too (*se les autres choses sont pareilles, a cas d'aventure/de fortune, rare/demps*). Although much has been

struction, as in '*gradus ut quatuor*' (degree 'as' 4).[50] This must be seen in contrast to the long-established practice of speaking of the *primus gradus* (first degree) and so on.[51] Perhaps the point of the new construction is to indicate that these degrees are not consecutive entries on a discrete scale but proportional calibrations on a linear scale.[52]

(3) A *conceptual innovation* that distinguishes the calculatory tradition is the demarcation of the uniformly difform (*uniformiter difformis*) as a theoretically tractable subcategory of difform change. The difformity will standardly be specified as either variation across time (*quoad tempus*), as with an accelerating or cooling body, or synchronic variation across space (*quoad subiectum/partes subiecti*), as with a rotating wheel whose rim moves faster than its inner parts.

(4) Many authors in the calculatory tradition employ an *unwitting catchphrase* that we may see as a kind of shibboleth: '(*ut*) *patet calculanti*' ('as/it is clear to anyone who does the maths'). Although it is obviously patterned on the more conventional phrase '*patet intuenti/inspicienti/consideranti*' ('it is clear to anyone who considers it'), careful research will confirm that

---

written on Oresme's translations, it seems only one scholar has discussed his occasional retention of Latin: S. Caroti, 'Nicole Oresme: dalla *questio* alla *glose*. La presenza del dibattito universitario nelle glosse di *Le livre du ciel et du monde*,' in N. Bray and L. Sturlese (eds.), *Filosofia in volgare nel medioevo. Atti del convegno della Società Italiana per lo studio del pensiero medievale*, Louvain-la-Neuve 2003, (Textes et études du Moyen Âge, 21), 155–190, at 178–179.

50 John of Ripa's *Sentences* lectures (Paris ca. 1355) use '*sicut n*' interchangeably with '*ut n*'; see e.g. John of Ripa, *Lectura super primum Sententiarum: prologi quaestiones I & II*, ed. A. Combes, Paris 1961 (Textes philosophiques du Moyen Âge, 8), 23–24. For background, see J. Coleman, 'Jean de Ripa, O.F.M. and the Oxford Calculators,' *Medieval Studies* 37 (1975), 130–189.

51 To take just one example, here is Duns Scotus discussing the augmentation of charity in his *Lectura* on Peter Lombard's *Sentences*, I, d. 17 (itself a *locus classicus* for the intensification and remission of forms): 'In forma accidentali naturali ut albedine, secundus gradus non educitur de priore sed de potentia subiecti, de qua etiam educitur primus gradus' (John Duns Scotus, *Lectura in librum primum Sententiarum. Distinctiones 8–45*, eds. C. Balić et al., Vatican City 1966 [Opera omnia, 17], 244).

52 If this suggestion is right, the '*gradus ut n*' construction may be seen as encoding the fourth implicit 'postulate' that C. Wilson discerned in Heytesbury's treatment of intensive variation: 'All the different possible intensities of a given quality … may be arranged according to a scale of degrees, of such character that it will be physically meaningful to speak of equal numerical differences between degrees as representing the same qualitative distance,' unlike a mere ordered series like 'the scale of hardness of minerals' (Wilson, *William Heytesbury*, 144–146).

the '*calculanti*' variant is fairly distinctive and that it remains so for the duration of the tradition.[53]

These features admittedly give a rather narrow characterization of the calculatory tradition. I should note, for instance, that Kilvington's *Sophismata* (early-to-mid-1320s) fails all four of my diagnostic tests, despite containing a string of sophisms involving the rate of change that one might have been tempted to label 'calculatory'.[54] In fact, Barbara Kretzmann has even suggested that this work 'may be taken as a paradigm of the philosophical problems and techniques that characterize the work of [the Calculators].'[55] As it happens, I think this is a mistake: as I see it, Kilvington's *Sophismata* represents the peak of an older Oxonian tradition of technical philosophy that gave rise to the calculatory tradition in the wake of Bradwardine's *De proportionibus* (which itself only passes my first test).[56] For present purposes, though, I needn't insist on this point, because my immediate aim in this paper is not to argue that any person or work should be *excluded* from the tradition, but that someone should be *included*.

---

53 The full phrase is found in Bradwardine's *De continuo* (no earlier than 1328, perhaps no later than 1335): 'sit ABCD quadratum cuius latus habet 3 puncta; tunc totum quadratum habet 9 puncta, quorum 8 sunt in lateribus, ut patet calculanti' (Thomas Bradwardine, *Tractatus de Continuo*, ed. J.E. Murdoch, *Geometry and the Continuum in the Fourteenth Century: A Philosophical Analysis of Thomas Bradwardine's* Tractatus de Continuo, unpublished Ph.D. dissertation, University of Wisconsin, 1957, 424). Later users in the tradition include Themo Judaeus, Blasius of Parma, Gaetano da Thiene, Giovanni Marliani, Oliviero da Siena, Jan Dullaert, Alvarus Thomas and Pietro Pomponazzi. John of Holland was a little out of tune with '*sicut patet computanti*' (*Sophismata*, Leipzig, Universitätsbibliothek, Ms. 1360, 274va) and Themo Judaeus was unusual in adding a variant with '*ymaginanti*'. The phrase is not improbable, so it is no surprise to find it occasionally outside the tradition, but I have only found two earlier versions, the first of which is not concisely formulaic: in the twefl th-century antidotary *Liber iste* ('ut clarissime patere potest cuilibet bene calculanti', s.v. *theriaca*) and in Henry of Harclay's *Questiones ordinarie* from the 1310s ('patet ergo calculanti quod hec fuit anno 657 post passionem', Henry of Harclay, *Ordinary Questions*, ed. M. Henninger, Oxford 2008 [2 vols.] [Auctores Britannici Medii Aevi, 17–18], 2: 934).

54 Richard Kilvington, *Sophismata*, eds. N. Kretzmann and B.E. Kretzmann, Oxford 1990 (Auctores Britannici Medii Aevi, 12), 64–94 (sophisms 29–35).

55 N. Kretzmann and B.E. Kretzmann, transl., *The Sophismata of Richard Kilvington*, Cambridge 1990, xix. Likewise, E.D. Sylla, 'Mathematics in the *Liber de Triplici Motu* (1509) of Alvarus Thomas of Lisbon,' in L. Saraiva and H. Leitão (eds.), *The Practice of Mathematics in Portugal*, Coimbra 2004 (Acta Universitatis Conimbrigensis), 109–161, at 116, has written that 'many features of the Oxford calculatory tradition are apparent' in Kilvington's treatment of '*Socrates est albior quam Plato incipit esse albus*' and '*Socrates est in infinitum albior quam Plato incipit esse albus*'.

56 Cf. Sylla, 'The Oxford Calculators,' 548: 'Kilvington's *Sophismata* exemplifies … how close

That said, I do think my tests might prove useful for the wider purpose of charting the diffusion of the calculatory tradition. For instance, since Gregory of Rimini is known as a transmitter of ideas from Oxford to Paris in the early 1340s, it is notable that his extensive *Sentences* commentary fails all four tests. This observation may be used to sharpen the claim that he 'was influenced at least by the early Calculators (specifically by Burley and Kilvington)'.[57] By contrast, an astronomical question that Themo Judaeus disputed at Erfurt in 1350 easily passes my three essential tests, and even runs to at least twelve occurrences of the optional catchphrase '*patet calculanti*'. By my standards, then, Themo emerges as a fully paid-up card-carrying calculator, albeit not (for the obvious reason) an Oxford Calculator.[58] As a final advertisement for the method, I should note that my tests are failed by the mid-fourteenth-century Merton logician Richard Billingham, whose occasional identification as a Calculator strikes me as providing a *reductio ad absurdum* of the looser characterizations that I wish to reject.[59]

---

disputations *de sophismatibus* might come to questions of mathematics and natural philosophy without ever leaving the confines of logic and semantics proper and without bringing in calculations to any extent.'

57 C.L. Löwe, 'Gregory of Rimini on the Intension and Remission of Corporeal Forms,' *Recherches de Théologie et Philosophie Médiévales* 81 (2014), 273–330, at 289. The copying of an extract on this topic from Gregory's *Sentences* commentary has been cited as evidence that 'calculatory doctrines' had spread to Padua by 1384. See F.P. Raimondi, 'Pomponazzi's Criticism of Swineshead and the Decline of the Calculatory Tradition in Italy,' *Physis*, New series, 37 (2000), 311–358, at 312–313 (a translation of a paper originally published in Italian in 1996 and quietly recycled yet again as 'Pomponazzi e la polemica anticalculatoria,' in F.P. Raimondi and J.M. García Valverde [eds.], *Tutti i trattati peripatetici*, Milan 2013, 9–86).

58 Themo's question *de difformitate motus lune supposita veritate Theoricarum* is edited in Themo Judaeus, *Question sur le mouvement de la lune* (1350), ed. H. Hugonnard-Roche, *L'oeuvre astronomique de Thémon Juif, Maître parisien du XIVe siècle*, Geneva 1973, 251–411. The calculatory material is mostly in article 4 (*de proportione velocitatis solis ad velocitatem lune*, 327–372) and article 5 (*numquid latitudo motus lune alicui gradui intrinseco correspondere possit*, 373–400). On Themo(n) more generally, see L. Petrescu, 'The Threefold Object of the Scientific Knowledge. Pseudo-Scotus and the Literature on the *Meteorologica* in Fourteenth-Century Paris,' *Franciscan Studies* 72 (2014), 465–502, at 469, n. 10, and 473–475. Note, however, that although Themo is said to have held *two* astronomical disputations at Erfurt, the other one was actually a determination of a sophism involving the final instant of rarefaction (cf. the opening lines transcribed in Themo Judaeus, ed. Hugonnard-Roche, 23).

59 N. Kretzmann once listed Billingham as an Oxford Calculator, apparently on the grounds that he had abbreviated Kilvington's *Sophismata*. See N. Kretzmann, 'Richard Kilvington and the Logic of Instantaneous Speed,' in A. Maierù and A. Paravicini Bagliani (eds.), *Studi sul XIV secolo in memoria di Anneliese Maier*, Rome 1981 (Storia e letteratura, 151), 143–178, at

2 Wyclif's Time of Reckoning

I turn now to John Wyclif—not the infamous heretic whose bones were dug up and burned forty-three years after his death in 1384, or the notorious theologian who courted controversy in the last decade of his life, but the neglected philosopher who became a doctor of theology in around 1372.[60] This Wyclif is attested in the spring of 1360 as the Master of Balliol, a position he apparently resigned a year later to take up a rectorship in Lincolnshire; he soon returned to Oxford to study theology, lodging at Queen's before being appointed Warden of Canterbury College, and graduating as a bachelor of theology in 1369. In his murkier formative years, he was recorded as a fellow of Merton in Billingham's bursarial account for the third term of 1355–1356. The college was for graduates, so he must have begun his undergraduate studies no later than 1351; 'circa 1350' seems a safe bet. That is, he was an arts student at Oxford in the immediate aftermath of the Black Death, when the university was still home to Heytesbury and Richard Swineshead. He may never have been a *socius* of either of these 'core' Calculators,[61] but he was certainly trained as a sophismatist before their heyday was quite finished.

---

145. Likewise, Sylla, 'The Oxford Calculators,' 540, listed Billingham's *Conclusiones* among calculatory works, even though it is a *sophismata* collection with no trace of calculatory content.

60 On this Wyclif, see J.A. Robson, *Wyclif and the Oxford Schools: The Relation of the* Summa de ente *to Scholastic Debates at Oxford in the Late Fourteenth Century*, Cambridge 1961 (Cambridge Studies in Medieval Life and Thought. New series, 8), xix–xxxviii; W.A. Pantin, *Canterbury College, Oxford*, 4, Oxford 1985 (Oxford Historical Society, New Series, 30), 16–35; J. Jones, *Balliol College: A History, 1263–1939*, Oxford 1988, 26; J.I. Catto, 'Wyclif and Wycliffism at Oxford 1356–1430,' in J.I. Catto and T.A.R. Evans (eds.), *The History of the University of Oxford, 2: Late Medieval Oxford*, Oxford 1992, 187–190; G.H. Martin and J.R.L. Highfield, *A History of Merton College*, Oxford 1997, 99–102; A.E. Larsen, 'John Wyclif, c. 1331–1384,' in I.C. Levy (ed.), *A Companion to John Wyclif. Late Medieval Theologian*, Leiden 2006 (Brill's Companions to the Christian Tradition, 4), 9–15. A few drops of the ink that flooded from his pen have been scrutinized by historians of philosophy; on these, see A. Conti, 'John Wyclif,' in E.N. Zalta (ed.), *The Stanford Encyclopedia of Philosophy* (Spring 2017 Edition), URL = https://plato.stanford.edu/archives/spr2017/entries/wyclif/.

61 Wyclif is said not to have completed his probationary year at Merton, in which case he cannot have joined the fellowship before the long vacation in the summer of 1355. Heytesbury is last attested at Merton in 1348, by which time he was a doctor of theology. He was probably chancellor of the university in 1352–1354, though, and is intermittently recorded at Oxford until his death in the winter of 1372–1373. Richard Swineshead left Balliol for Merton in 1344, and is last attested there in 1354–1355. See A.B. Emden, *A Biographical Register of the University of Oxford to A.D. 1500*, Oxford 1957–1959 (3 vols.), 2: 927–928 (Heytesbury); 3: 1836–1837 (Swineshead); Jones, *Balliol College*, 25.

Wyclif's philosophical training is reflected in his *Tres tractatus de probationibus propositionum*, a logic textbook probably written for advanced undergraduates in the late 1360s and early 1370s. The overarching concern of the *Probationes* (as I shall call it) is to teach students how to prove that a given sentence is true. The work is divided into three treatises that deal with (i) non-compound sentences, (ii) sentences that are somehow intermediate, and (iii) compound sentences.[62] Each treatise is internally structured according to a classification of sentences within each category, so that the *tractatus tertius*, for instance, deals in turn with compound sentences that are conjunctive, disjunctive, causal, comparative, conditional, locative and temporal. On the face of it, then, the work is just another contribution to the *probationes* genre associated with Richard Billingham.[63] Wyclif has, however, shoehorned an astonishing amount of tangentially related material into this framework; the chapter on conditionals, for instance, incorporates an extensive treatment of the topic of insolubles.

Our concern is with the chapter of the *tractatus tertius* that deals with locative compound sentences. This should have the modest goal of showing how to prove the truth of a sentence like '*Socrates currit ubi Plato currit*' (Socrates is running where Plato is running), but it turns out to be longer than the entire second treatise, occupying 132 pages in the printed edition.[64] Again, Wyclif has incorporated material that goes beyond his ostensible remit. After spending less than a page on the logic of locative compounds, he launches into an atomistic account of *locus* (1–11) that prompts a discussion of *motus localis* and occasionally other kinds of change (11–29). The rest of the chapter consists of two large subtreatises, the first on the atomistic composition of continua (30–85) and the second on how the rates of changes can be systematically determined by their effects (*penes quid ab effectu possunt regulariter attendi velocitates*

62 John Wyclif, *Tractatus de Logica*, ed. M. Dziewicki, London 1893–1899 (3 vols.): *tractatus primus*, 1: 75–120; *tractatus secundus*, 1: 121–234; *tractatus tertius*, 2–3. The *Probationes*, covering the 'higher part of logic' (*superior pars logice*, 1: 75), has often been conflated with the string of introductory *tractatuli* printed in the same edition (1: 1–74). See now M. Thakkar, 'Wyclif's *Logica* and the *Logica Oxoniensis*,' in L. Campi and S. Simonetta (eds.), *Before and After Wyclif. Sources and Textual Influences*, Basel 2020 (Textes et Études du Moyen Âge, 97), 1–31. Since Dziewicki's edition is notoriously unreliable, all of my quotations have been silently corrected against the manuscripts.

63 The fundamental study of this genre is A. Maierù, *Terminologia logica della tarda scolastica*, Rome 1972 (Lessico Intellettuale Europeo, 8), 393–498. Basic *probationes* texts do not cover compound sentences. For a sense of their scope, see the tables of contents in De Rijk (ed.), *Some 14th Century Tracts*, 5 and 46–47.

64 John Wyclif, *Probationes*, 3: 1–132. The remaining 95 pages are occupied by the chapter on temporal compound sentences.

motuum, 85–132). This second subtreatise falls into three parts, each dealing with a different kind of change: *de velocitate motus localis* (85–111), *de velocitate motus augmentationis* (111–118), and *de velocitate motus alterationis* (118–132).

Even from this overview, it is already clear that Wyclif shares the Calculators' theoretical concern with establishing rules for determining the rate of the various kinds of change. A closer look reveals that his treatment is idiosyncratic—on two occasions he avoids the standard phrase '*quoad effectum*' in favour of the novel phrase '*quoad signum*',[65] and his atomism must have deeper repercussions that will require careful investigation—but it is nonetheless recognizable as a counterpart of the sixth chapter (*De tribus predicamentis*) of Heytesbury's *Regule*. In particular, bearing in mind the other diagnostic tests that I proposed in section 1, we find Wyclif using the quintessentially calculatory '*non gradus*' and '*gradus ut n*'. He also distinguishes *motus uniformis* from *motus difformis* (both *quoad subiectum* and *quoad tempus*) and the *uniformiter difformis* from the *difformiter difformis* (27–29). Dedicated calculator-spotters are even treated to a '*patet calculanti*';[66] in fact, this unwitting catchphrase also occurs elsewhere in the third treatise, accompanying a piece of combinatorics[67] and a proportional calculation.[68]

---

65 John Wyclif, *Probationes*, 3: 85: 'de velocitate motuum quoad signum penes quid attendi debeat'; 3: 92: 'de mensura velocitatis motuum quoad signum.' The synonymy of '*quoad signum*' with '*quoad effectum*' is nonetheless clear from Wyclif's parallel use of '*ab effectu*' (3: 30); for his own explanation, see n. 71 below. Nor was the variant entirely unprecedented; recalling n. 48 above, cf. Francis of Marchia, *Quaestiones super Metaphysicam*, I, q. 1, a. iii, ed. N. Mariani, Grottaferrata 2012 (Spicilegium Bonaventurianum, 37), 75: 'nulla scientia probat suum subiectum a priori et propter quid et per causam; bene tamen potest probari a posteriori et per effectum sive per signum.'

66 John Wyclif, *Probationes*, 3: 85: 'Stat etiam columpnam rotundam 20 pedum in longitudine et 8 pedum in latitudine, sicut est hasta, descendendo uniformiter in hora describere situm secundum longitudinem duplum ad hastam; et aliam simillimam in figura volvi latitudinaliter precise eque velociter, describendo situaliter plus quam octogintuplum ad priorem, sicut patet calculanti.' If I understand the scenario correctly, at least one of the numbers must be wrong. A vertically-oriented cylinder of height H and width W, descending a distance d, will trace a volume of $d \times \pi(½W)^2$. The same cylinder, rolling the same distance sideways, will trace a volume of $d \times HW$. The latter is $4/\pi \times H/W$ times the former, so a volume ratio of over 80 would require H/W to be over 20π. Given the measuring units, the obvious solution is H/W = 64, which is achievable with H = 8 ft and W = ⅛ ft. These are plausible dimensions for a spear, so I suggest emending the text after '*rotundam*' to '*et 8* pedum in longitudine et *8^{ve} pedis* in latitudine'.

67 John Wyclif, *Probationes*, 2: 9, on possible combinations of the seven kinds of compound sentence: 'patet calculanti, si non fallor, quod sunt 120 combinationes distincte in septenario. Nam sicut iuxta conceptum Porphyrii [*Isagoge* 17.14–18.9] sunt 10 conjugationes quinque universalium, scilicet 4, 3, 2 et 1, sic in septem speciebus ypothetice sunt 21 combinationes binarii, comparando sexies copulativam ad sex species residuas, quin-

In short, Wyclif's chapter on locatives passes my tests with flying colours. For that matter, a diagnostic heatmap of the rest of the *Probationes* reveals other hotspots of potential interest to historians of the calculatory tradition.[69] In the absence of an argument to the contrary, therefore, the obvious conclusion is that we should extend the line of succession—indeed, that we should already have extended it—to include Wyclif as one of the Oxford Calculators.

This conclusion can be buttressed with new evidence from two sources. Firstly, Wyclif's unedited *Physics* commentary, perhaps his earliest known work, proves to contain a wealth of calculatory material.[70] For instance, in the

---

quies disjunctivam, et de quinque residuis habebuntur decem; combinando autem copulativam et disjunctivam ad quinque residuas, et post secundum alias combinationes trium, habebuntur 35 combinationes; combinando etiam quatuor adinvicem habebuntur 35; combinando autem quinque adinvicem habebuntur 21, et combinando sex adinvicem habebuntur 7; que conjuncta septennario constituunt 120.'

68 John Wyclif, *Probationes*, 2: 113: 'Ex isto videtur quod non sit possibile idem esse precise ita intensum ut remissum; quia tunc acquisita remissione dupla deperdetur precise intencio subdupla, et cum tantum deperdetur de intencione sicut acquiretur de remissione, patet calculanti quod sequitur intencionem fuisse duplam ad remissionem in principio. Non enim stat aliquod ens crescere ad suum duplum propter tantam latitudinem acquisitam quantam reliquum deperdendo decrescit ad suum subduplum, nisi primum fuerit subduplum ad secundum.' Earlier on the same page, Wyclif had used the phrase 'repugnat certe calculacioni'.

69 The first treatise contains no material that might be seen as calculatory. In the *tractatus secundus*, see esp. John Wyclif, *Probationes*, 1: 200–202 (on *incipit*). In the *tractatus tertius*, see esp. 2: 110–121 (on positive and privative latitudes); 2: 135–170 (on the limits of powers); and 3: 220–223 (on time and the motion of the world).

70 Wyclif's *Physics* commentary is preserved in Venezia, Biblioteca Nazionale Marciana, Ms. lat. VI. 173 (2675), 1[ra]–58[vb]. For basic details and a question list, see S.H. Thomson, 'Unnoticed MSS and Works of Wyclif II,' *Journal of Theological Studies* 38 (1937), 139–148, at 145–148, and (less usefully) W.R. Thomson, *The Latin Writings of John Wyclyf. An Annotated Catalog*, Toronto, ON, 1983 (Subsidia Mediaevalia, 14), 12–14. The only serious engagement with any part of the text is J.A. Weisheipl, *Early Fourteenth Century Physics of the Merton 'School', with Special Reference to Dumbleton and Heytesbury*, unpublished Ph.D. dissertation, Oxford University, 1956, 188–190 (see also 39, n. 87). See also the brief remarks in A. Robert, 'Atomism at Oxford after John Wyclif. The Cases of Robert Alyngton and Roger Whelpdale,' in L. Campi and S. Simonetta (eds.), *Before and After Wyclif. Sources and Textual Influences*, Basel 2020 (Textes et Études du Moyen Âge, 97), 57–94, at 68–69. Wyclif's authorship has been doubted, but I conjecture that an investigation of the remarks on density, coextension and mixture will confirm it (cf. n. 72 below). For instance, on f. 16rb the author argues that 'in liquacione cere cum viridi eris … ad quemcumque talem punctum [sc. cere] erit divisio vel penetracio et per consequens materierum coextensio.' Wyclif regrets in the *Probationes* (3: 68) that 'ad tantum desipui quod putavi ziniar, i.e. viride eris … cum cera liquata … coextendi.' (Owing to a regrettable mistake of mine, the word '*ziniar*' (from the Arabic '*zinjār*', verdigris) must be sought in the *DMLBS* under '*zingar*'.)

section on Book III, after establishing what *motus* is (26vb–31ra), he asks *penes quid attendi habeat velocitas in motu locali quoad causam* (31ra–36rb) and *quoad effectum* (36rb–38ra).[71] In the section on Book V, after discussing the intensity (44rb–48va) and the intensification (48va–49vb) of a quality, he asks *penes quid attenditur velocitas motus alterationis* (49vb–52vb). And we hit the autobiographical jackpot in the final question on Book I, where 'for the moment' (*ad presens*) he accepts it as probable that prime matter has infinitely many coextensive intensive parts, partly because this follows more readily from the view that elemental forms can be intensified and from 'other calculatory topics (*alias materias calculatorias*) that I am still persevering with.'[72]

Secondly, there is a tantalizing piece of evidence that Wyclif was viewed as a Calculator by the Italians who were keeping the flame alive a century later.[73] In the early 1460s, when Giovanni Marliani thought he had devised a killer argument against Bradwardine's rule of motion, he discovered that 'Dulmenton, Uuclif and Misinus'[74] had already answered it, albeit not to his satisfaction.[75]

---

71 Wyclif introduces these two questions by explaining the difference between them (*In Phys.*, 31ra): 'est ulterius tractandum penes quid attendi habeat velocitas in motu locali. Sed illud contingit dupliciter intelligere: vel querere causam quare mobile ita velociter movetur, vel querere ex quo signo iudicandum est mobile ita velociter moveri. Prima questio equivalet isti ac si queritur penes quid quo ad causam habet attendi velocitas motus localis; et ponitur communiter quod penes magnitudinem proportionis geometrice sic quod correspondenter in toto se habet ad invicem proportio motoris ad motum et velocitas motus causati ab eisdem. Secunda questio tantum valet ac si queritur penes quid attenditur quo ad effectum velocitas motus localis; et dicitur communiter quod penes lineam descriptam a puncto velocissime moto vel equivalenter.'

72 Wyclif, *In Phys.*, 20rb: 'Ad secundam reductionem de coextensionibus materierum, dico ad presens credens sensui et experimentis, etiam quia plus consequitur responsionem ponentem formas elementares intendi et alias materias calculatorias quibus adhuc insisto: teneo tamquam probabile quod materia prima habet infinitas partes intensivas coextensas; relinquo iudicio primi philosophi [sc. metaphysici] veritatem.' This whole question (ff. 14va–20va) will repay careful comparison with the account of density, coextension and mixture in John Wyclif, *Probationes*, 3: 63–85; for now, see n. 84 below.

73 More such evidence may be waiting to be discovered, as we are told with frustrating vagueness that Gaetano da Thiene cited '[Johannes] Euclif' in at least one work on natural philosophy: S. da Valsanzibio, 'La "via Scoti" in Gaetano di Thiene,' in C. Bérubé (ed.), *Regnum Hominis et Regnum Dei, 2: La tradizione scotista veneto-padovana*, Rome 1978, 199–205, at 199–200.

74 The enigmatic 'Misinus' has been confirmed, largely thanks to Graziella Vescovini, as Mesino de Codronchi (Mesinus de Coderoncho), who taught mostly at Bologna from 1384 to 1393; a useful starting-point is J. Celeyrette, 'Le mouvement selon la cause chez Messino da Codronchi et Angelo di Fossambruno,' in J. Biard and S. Rommevaux (eds.), *Mathématiques et théorie du mouvement, XIVe–XVIe siècles*, Villeneuve d'Ascq 2008 (Histoire des sciences), 15–29.

75 Giovanni Marliani, *Questio de proportione motuum in velocitate* (1464), Pavia 1482, sig. c 2ra:

He also felt obliged to engage with Wyclif's discussion of the quantity and quiddity of proportions, even though he found it a metaphysical irrelevance to mathematicians and natural philosophers.[76] I cannot find either of these passages in Wyclif's *Probationes*, and the only obvious candidate among his other known works is his *Physics* commentary. If the passages cannot be found there either, perhaps he wrote another calculatory work that has been lost.[77] At any

---

'Eadem conclusio potest ... roborari ... maxime per argumentum 14^{m}, quod maxime movit me ad oppositum communis opinionis ... quod argumentum postquam ego formaveram inveni etiam a dulmentone uuclif et misino esse formatum et ab eis solutum licet non bene.' For the argument, see sig. a 6^{rb-vb} (inc. 'Item, aliquis motus est alio precise in duplo velocior et tamen proportio a qua fit velocior non est in duplo maior proportione a qua fit remissior'). I owe this discovery to Clagett, *Giovanni Marliani*, 140–141, though both of his references to Wyclif are complicated by misinterpretations of Marliani's text that need not detain us. To Clagett's list of manuscripts (29–30) we should add Città del Vaticano, Biblioteca Apostolica Vaticana, Ms. Chigi E. VI. 197, 102^{v}–126^{r}, as described in A.F. Verde, *Lo Studio fiorentino 1473–1503. IV: La vita universitaria*, 1: *Anni scolastici 1473/74–1481/82*, Florence 1985, 174–179.

76 Marliani began his own response to the *questio* by acknowledging Wyclif's broader discussion of the metaphysics of proportions: 'Sciendum est primo quod quamquam Uuclif logice et metaphisice de quantitate et quicditate proportionis tractans ponat tres modos compositionis proportionis ... quia tamen hec tam subtilis consideratio de quicditate proportionis ... longa indigeret explanatione metas philosophi naturalis atque mathematici excedenti ... ideo non omnes hos compositionis proportionis modos methaphisice pertractabo' (*Questio*, sig. a 7vb). And in the fourth objection (sig. c 7^{ra-rb}) to his first conclusion (sig. c 1^{ra-rb}) he cited a view of Wyclif's on the quantity of proportions, dismissing it as irrelevant in his reply (sig. c 8vb–d 1ra): 'quamvis uuclif considerando proportionis quantitatem aliquo modo ex quantitate subiectorum, †mathematice [Città del Vaticano, Biblioteca Apostolica Vaticana, Ms. Vat. lat. 2225, 33vb: methephisice] procedens et logice, concedat unam proportionem equalitatis alia equalitatis proportione maiorem esse, secundum quem modum nequaquam mathematici considerant proportionis quantitatem, considerando tamen proportionis quantitatem ex parte denominationum, ut mathematici faciunt et philosophi magis, omnes proportiones equalitatis sunt equales, omnes etiam alie proportiones quarum denominationes sunt similes sunt equales, ut etiam uuclif dicit'.

77 Wyclif does touch on the composition of proportions in the section on local motion in his *Physics* commentary (*In Phys.*, 32ra): 'In illa igitur materia oportet diligenter advertere modum compositionis proportionis et in qua proportione una proportio est maior reliqua et permutatim se habebunt motus provenientes ex illis. Unde calculantes ymaginantur omnem proportionem maioris inequalitatis componi ex proportione maioris extremi ad medium vel equivalens ...'; and likewise in a discussion of aggregate dominion in John Wyclif, *De civili dominio* (ca. 1376), vol. 3, ed. J. Loserth, London 1903, 230: 'Sicut est dictum de composicione proporcionum quod omnis quarta [*recte* quadrupla] componitur primo ex duabus †nichil non communicantibus in extremis, licet talis omnis proporcio sit indivisibilis quoad molem.' The subtreatise on atomism in the *Probationes* suggests a third calculatory passage to hunt for: 'Raritatem autem primo modo dictam dixi quondam attendi penes magnitudinem situs quem rarum cum corpore alieno commixto occupat in comparacione ad paucitatem sue materie' (*Probationes*, 3: 64).

rate, the important point for my purposes is that Marliani thought to consult Wyclif as well as Dumbleton. To my mind, this is unsurprising, because Wyclif has more or less the same claim to be included among the Oxford Calculators.

3 The Continuity of the Tradition

Modern scholarship has produced just two assessments of Wyclif's relationship to the calculatory tradition, both of which have firmly segregated him from the likes of Dumbleton, but for wildly divergent reasons. In his landmark volume on medieval mechanics, Marshall Clagett grouped Wyclif together with Richard Feribrigge, John Chilmark and Edward Upton as 'continuators of the great tradition' who failed to make '*original* contributions'. It was 'not surprising' to find that he 'makes the usual Merton distinctions' and 'repeats [a] universally accepted opinion of Bradwardine'.[78] A quarter of a century later, however, Norman Kretzmann portrayed Wyclif as an *opponent* of the tradition who 'turn[ed] the Calculators' weapons against them'; he knew 'how to use the Calculators' techniques on examples of the sort they were particularly fond of … but he had turned against them before he wrote [the *Probationes*]'.[79] Evidently I need to defend my conclusion on two quite different fronts.

Clagett's assessment is easy to deal with. Even if we must treat the label 'Oxford Calculator' as an *appellation contrôlée* that guarantees originality—a stricture too far, in my opinion—it would take some audacity to dismiss Wyclif as unoriginal on the basis of the notoriously corrupt printed edition of the *Probationes*. Pending the availability of an intelligible text, it would be wiser to suspend judgement on this score.[80] As we shall see in the rest of this section,

78 Clagett, *The Science of Mechanics*, 629–634. In the light of what I say about Wyclif in the next paragraph, I should note that Rondo Keele has called for a properly informed assessment of Chilmark, who 'composed some seemingly original treatises as yet totally uninvestigated.' See R. Keele, 'The Logical and Scientific Treatises of John Chilmark (†1396),' *Bulletin de philosophie médiévale* 49 (2007), 119–137, at 135.

79 N. Kretzmann, 'Continua, Indivisibles, and Change in Wyclif's Logic of Scripture,' in A. Kenny (ed.), *Wyclif in His Times*, Oxford 1986, 31–65, esp. 51–63.

80 Wyclif's editor would have been the first to agree. Cf. e.g. M. Dziewicki (ed.), *Iohannis Wyclif Logica*, 3: 89: 'Of the whole of this paragraph and the following ones, I have been able to understand nothing relevant to the question discussed [sc. *penes quid attendi debeat velocitas motus localis*]. Readers should bear in mind that the MS. is unique, frightfully illegible, especially in this part, without any corrections from here to the end, and bearing upon a subject which is itself particularly hard to understand.' It may be useful for me to repeat that I have silently corrected my *Probationes* quotations against the manuscripts; I should add that I am preparing a new edition.

moreover, there is a good reason to expect Wyclif to have made some original contributions, so the smart money is not on Clagett.

The other assessment poses a stronger and more interesting challenge. According to Kretzmann, Wyclif formed his atomistic views 'in conscious, sometimes angry opposition to the Calculators' techniques of logical analysis and the continuism on which they are based.'[81] In other words, by adopting the idiosyncratic view that all continua were composed of a merely apparent infinity of indivisibles, Wyclif was *ipso facto* rejecting a fundamental tenet of the calculatory tradition. If Kretzmann is right about this, Wyclif's indivisibilism automatically disqualifies him from counting as a Calculator.

Alert readers may have raised an eyebrow at the mention of 'the Calculators' techniques of *logical* analysis'. There is indeed a complication here: Kretzmann was interested in the instant of change at a time when scholars had conflated limit-decision techniques with calculations,[82] so he took Kilvington's *Sophismata* as his reference point for the views of the Calculators, and his account must be read with this in mind. I must also correct a point of detail: where Kretzmann reported Wyclif as defiantly calling continuist views of time 'impossible', this was a misinterpretation, as the target of the cited passage was not continuism but presentism.[83]

Even so, Kretzmann was on to something: Wyclif *was* hostile to the continuism that he had to deal with in the calculatory tradition. The autobiographical remark that we saw in his *Physics* commentary, when he was 'still persevering' with certain calculatory topics, has a retrospective counterpart in the *Probationes* subtreatise on atomism:

> it is inconsistent with the composition of continua from non-quantities that something should, over a truly continuous period, become more or less dense throughout by continuously gaining or losing quantity. And the philosopher cannot give a basis for the calculatory scenarios belong-

81 Kretzmann, 'Continua,' 52.

82 See my remarks in the vicinity of nn. 29 and 55 above.

83 Kretzmann, 'Continua,' 36–37 (cf. John Wyclif, *Probationes*, 3: 184–185). Presentism is the view that only present things exist; Wyclif was an 'eternalist', holding that past, present and future things exist. This recurring theme in his philosophy has generally been neglected in the literature, but for a brief account see R. Pasnau, *Metaphysical Themes 1274–1671*, Oxford 2011, 388–390, and for a broader textual basis see (with caution) S. Penn, *Truth, Time and Sacred Text: Responses to Medieval Nominalism in John Wyclif's* Summa de Ente *and* De Veritate Sacrae Scripturae, unpublished Ph.D. dissertation, University of York, 1998, 91–96, 99–101.

ing to that topic (*fundare casus calculatorios illius materie*), which I once sweated over for no good reason.[84]

What's more, this lack of a basis in reality for scenarios involving infinite divisibility is something that Wyclif gratuitously brings up in his theological works for the rest of his life. 'The sayings of important figures in the church can be brought into harmony,' he says in a treatise on the Eucharist, 'which is better than inventing thoughts and words that are alien to the saints of antiquity, and from them, as from a false basis in a calculatory topic (*materia calculatoria*), deriving conclusions and opinions that will amaze the world.'[85] Likewise, defending the academic pastime of speculating about how things were before the Fall, he says 'this pursuit is not useless, unlike those calculatory thought experiments (*imaginationes calculatorie*) that we cannot know about because the matter is either impossible or inaccessible to our senses.'[86]

These quotations, and a few others,[87] might suggest that, far from being the black sheep of the Oxford Calculators, Wyclif was actually a wolf in sheep's

---

84 John Wyclif, *Probationes*, 3: 70: 'Ex quo patet quod repugnat composicioni continui ex non quantis quod aliquid per tempus vere continuum rarefiat vel condensetur per totum adquirendo vel deperdendo continue quantitatem. Nec scit philosophus fundare casus calculatorios illius materie in qua quondam multum inaniter insudavi.' For the counterpart in the *Physics* commentary, see n. 72 above. Wyclif laments his misspent youth for similar reasons in two passages of the *Trialogus* (1382/1383): 'Nec oportet nos sollicitari utrum componitur [sc. materia prima] ex suis partibus intensivis, quod alias in nostra stultitia diffuse tractavimus' (John Wyclif, *Trialogus*, ed. G. Lechler, Oxford 1869, 88); 'Partes autem intensivas non solum horreo in virtutibus sed in quibuslibet creaturis; nec solum horreo, sed quamvis in juventute mea circa istam materiam sollicitus fueram, modo tamen non intelligo' (ibid., 131). The notion of intensive parts (yet to be studied as such, but see e.g. Sylla, 'Latitude of Forms,' 273) also prompts a sideswipe at '*calculantes*' in the *De civili dominio*, 3: 232: 'Et patet quomodo dominium capit partes suas a dominabili secundum partes eius quantitativas vel aggregativas ex substanciis vel accidentibus absolutis vel respectivis; et per hoc tolluntur multe conclusiones quas calculantes inaniter elaborant.'

85 John Wyclif, *De eucharistia tractatus maior* (1380), eds. J. Loserth, London 1892, 76: 'Possunt concordari dicta gravium personarum ecclesie, quod est melius quam invenire sensum et verba ab antiquis sanctis extranea et ex illis ut in materia calculatoria ex falso fundamento deducere conclusiones et sentencias mundo mirabiles.'

86 John Wyclif, *Tractatus de statu innocencie* (1376), eds. J. Loserth and F. Matthew, London 1922, 493: 'Nec obest probabili racione convincere multa que fuissent in statu innocencie, licet in scriptura non fuerint expressata. Nec est hoc studium inutile sicut imaginaciones calculative [v.l. calculatorie] que propter impossibilitatem materie vel eius insensibilitatem a nobis necessario ignorantur.'

87 John Wyclif, *Trialogus*, 73: 'sicut in materia calculatoria multi labuntur perdendo seipsos

clothing. But I would like to offer a simple way round this suggestion: the hostility whose seeds were already visible in his *Physics* commentary was directed not at the calculatory tradition as a whole, but at a non-essential part of it. What he wanted to do in the *Probationes* chapter on locatives, it seems to me, was not only to defend atomism in its own right, which was the burden of the first subtreatise (30–85), but also to give an expurgated version of the usual calculatory material without the baseless excrescences permitted by continuism (85–132). This would explain the concern for preserving the rules *de velocitate motuum* that Wyclif later showed in his theological works;[88] it would also explain his affinity with 'mathematicians who are at the same time empirical philosophers (*sensibiles philosophi*),' who may find it necessary to redefine even a basic notion like 'half' to avoid some standard geometrical objections to atomism.[89] And if I am right about Wyclif's aims, then not only should historians of the calculatory tradition bring him into the fold, *pace* Kretzmann, but we should also recognize him as strikingly original, *pace* Clagett.

---

in pelagus intellectus'; 112: 'dicitur quod angeli possunt ex sua celeritate nunc esse in oriente et nunc in occidente, servando quantitatem sui limitati situs corporei, figura illius situs quantum expedit variata. Sed in ista materia oportet cavere de imaginationibus calculatoriis, ut non credatur superficiem vel corpus ad tantum dividi quod posset occupare simul quemlibet situm mundi.' See also nn. 35 and 84 above.

88 John Wyclif, *Tractatus de benedicta incarnacione* (ca. 1372), ed. E. Harris, London 1886, 207 f., against Christ's multiple assumption: 'posito quod [Christus] infinitas humanitates equales assumat ... conformis est consideracio de motu, ut infinitum intensus potest esse motus per motuum incommunicancium coextensionem, cum hoc quod mobile quantumcunque tarde ad velocissimum moveatur. Et ex istis potest calculator elicere quotlibet conclusiones nimis mirabiles, quia regularum de velocitate motuum destructivas.' Likewise *De eucharistia tractatus maior*, 234, against the multiplication of Christ's body: 'Et hic eliciunt quidam conclusiones mirabiles ... nam Petrus multiplicatus per mille loca ... videret in casu ad intra et extra se indifferenter secundum omnem differenciam posicionis, ut eliciunt illi qui particulariter sollicitantur in ista materia ad eliciendum conclusiones mirabiles, ut sophiste arguunt. Et hoc tollit regulas de velocitate motus, cum subito vel quantumcunque celeriter posset Petrus per viam multiplicationis quantumcunque tarde ⟨movendo⟩ acquirere quantumlibet longum situm, et sic de aliis inconvenienciis reductis.'

89 John Wyclif, *Trialogus*, 84: 'Sicut enim ternarius non est divisibilis in duas medietates equales simpliciter, sic sensibile continuum ex partibus indivisibilibus imparibus integratum. Et hinc mathematici qui cum hoc sunt sensibiles philosophi vocant tales partes compositi [? *recte* partes talis continui] eius medietates communiter intellectas quarum una insensibiliter vel indivisibiliter excedit aliam.' Cf. *Probationes*, 3: 52: 'quantamlibet lineam rectam dandam contingit dividere in duas medietates quarum neutra divisibiliter excedit reliquam. Et omnes tales vocat sensibilis philosophus, satis vere ad sensum suum, medietates equales.'

Knowing that he was facing an uphill struggle,[90] Wyclif concluded the locatives chapter by asking for his proposals not to be dismissed just because 'they disagree with many authoritative writers and destroy the arguments of the *Calculations* along with many opinions and many thought experiments of modern thinkers.'[91] Perhaps I should conclude with a similar plea. At the very least, historians of the calculatory tradition are now faced with a choice: either we can include Wyclif as an Oxford Calculator, as I have urged; or we can exclude him by providing an argument for the claim that continuism is essential to the tradition. Either way, we will have made some progress.

Acknowledgements

I am deeply indebted to the Leverhulme Trust for the Early Career Fellowship (2014–2018) that enabled the research for this paper. I must also thank three participants at the Munich conference in 2018: Daniel Di Liscia, who prompted me to look into Wyclif's *Physics* commentary; Magali Roques, who framed the choice that concludes the paper and encouraged me to stick to my guns; and Elżbieta Jung, who insisted on the second option. I am especially grateful to Rachel Farlie and Daniel Di Liscia for useful comments on my drafts.

90 For a clear account of both 'the historical improbability of Wyclif's indivisibilism' and 'its philosophical unattractiveness,' see Kretzmann, 'Continua,' 31–39. In an important preliminary study, however, Robert ('Atomism at Oxford') has now shown that Wyclif did succeed to an extent in reviving this unpopular view.

91 John Wyclif, *Probationes*, 3: 132: 'Sed de omnibus istis relinquens subtilibus logicis et naturalibus profundum scrutinium, rogo perlegentes istud capitulum non condempnare vel deridere hic dicta tanquam probabiliter opinata. Scio enim quod a multis auctoribus discrepant et ⟨multa?⟩ argumenta calculationum [v.l. calculantium] interimunt, sicut et multas opiniones et multas ymaginaciones modernorum.' If I am right to treat '*Calculations*' as a title, I believe this is the earliest known reference to Richard Swineshead's *magnum opus* as the *Calculationes*; if Dziewicki's manuscript was right to have '*calculantium*' instead, we might translate it as '(the) Calculators'.

CHAPTER 8

On the Reception of English Logic at Universities of Central Europe: Helmoldus de Zoltwedel (Prague, Leipzig) on the Liar-Paradox

Harald Berger

Introduction

The reception of English philosophy and logic, including that of the *Calculatores*, at the early University of Prague can be gathered from the fine manuscript catalogues of the main libraries in Prague as well as from old lists of books of those libraries.[1] In addition to these sources, I present the telling case of a neglected scholar, Helmold of Zoltwedel, who started his career at Prague in 1383 and composed his main philosophical work there around 1390. Helmold not only seems to be familiar with all the relevant literature of his time, but fortunately also frequently quotes authors by name, which makes him an outstanding source for modern research, including the special topics of English logic and its reception in Central Europe at the end of the fourteenth century.[2]

1 Helmold of Zoltwedel's Life and Career

Helmoldus Gledenstede de Zoltwedel (also spelled 'Solt-') was an outstanding member of the universities of Prague and Leipzig in the last years of the fourteenth and the first decades of the fifteenth centuries.[3] He held degrees of three

1 See mainly J. Truhlář, *Catalogus codicum manu scriptorum latinorum qui in C.R. Bibliotheca publica atque universitatis Pragensis asservantur*, Prague 1905–1906 (2 vols.); A. Patera and A. Podlaha, *Soupis rukopisů knihovny Metropolitní kapitoly Pražské*, Prague 1910–1922 (2 vols.); G.B. Korolec, *Repertorium commentariorum medii aevi in Aristotelem Latinorum quae in Bibliotheca olim Universitatis Pragensis nunc Státní Knihovna ČSR vocata asservantur*, Wrocław 1977; F. Šmahel, *Verzeichnis der Quellen zum Prager Universalienstreit 1348–1500*, Wrocław 1980 (Mediaevalia Philosophica Polonorum, 25); Z. Silagiová and F. Šmahel, *Catalogi librorum vetustissimi Universitatis Pragensis*, Turnhout 2015 (Corpus Christianorum. Continuatio Mediaevalis, 271).

2 See also S. Read's contribution to the present volume.

3 For details on life and works of Helmold, see H. Berger, 'Helmold of Zoltwedel (†1441): His

 | DOI:10.1163/9789004512054_010

faculties, namely Arts (master), Medicine (doctor), and Theology (bachelor). The first one he obtained in Prague, the other two in Leipzig, as it seems. As is indicated by the toponym, Helmold was born in Salzwedel, a German *Hansestadt*, now in Saxony-Anhalt. Therefore, Helmold belonged at both universities to the Saxon nation. The first known document regarding Helmold is on the occasion of his admission to the examination of the candidates for the bachelor degree at the Faculty of Arts of the University of Prague, dated to 1383. From this, it can be concluded that he was born before 1365, so maybe around 1360. The year of his death, 1441, seems to be safe, and so he has apparently reached a high age of about 80 years.

The Book of Deans of the Faculty of Arts of the University of Prague does not record the graduations (*determinatio, inceptio*) of Helmold, but only the admissions to the relevant examinations in 1383 and 1386, respectively. Fortunately, Helmold himself informs us about the identity of his master, namely Conrad of Soltau (d. 1407). Since Conrad left Prague for Heidelberg in 1386, we may conclude that Helmold obtained his degree of *magister in artibus* in that very year 1386.

At the University of Prague, Helmold was, among other things, dean of the faculty of arts (1394), rector of the university (1398–1399), and ambassador of the university at the council of Pisa (1409). As a teacher, he seems to have been quite popular, as is shown by the (at least) 37 graduations that took place under his supervision. Helmold was one of the many members of the non-Bohemian nations at the University of Prague who left this city because of the issuance of the so-called decree of Kuttenberg (Kutná hora) early in 1409, and he turned to Leipzig where a new university was about to be founded as a result of the events at Prague. Perhaps Helmold came to Leipzig directly from Pisa. At the University of Leipzig, Helmold was a professor of arts as well as a professor of medicine. Furthermore, he was rector of the university (1410 and 1416), vice chancellor (1413), dean of the faculty of medicine (1428), and member of the *Collegium Maius*. He died in Leipzig in 1441, as was already said.

As to his scientific and philosophical alignment, it is clear from his writings that Helmold was a staunch Aristotelian and an adherent to 'Buridanism' (as labelled by Mieczysław Markowski), a nominalist or conceptualist movement originating from the University of Paris, where John Buridan was active from the 1320s to around 1360, Albert of Saxony (d. 1390, active in Paris from 1351 to 1362) and Marsilius of Inghen (d. 1396, active in Paris from 1362 to 1377) being

---

Academic Career, Scientific Works, and Philosophical Alignment', in O. Pavlíček (ed.), *Studying the Arts in Late Medieval Bohemia*, Turnhout 2021 (Studia Artistarum, 48), 175–203.

the most famous other representatives of that movement. This means also, of course, that Helmold was an opponent of the new realist tendencies at Prague emerging from the study of John Wyclif's works there and culminating in Hussitism.[4] Even in his treatment of the Liar-paradox, which is the topic of the present chapter, there is a sideswipe at the realists.[5]

2 Helmold's Questions on the *parva logicalia*

Helmold's philosophical *opus magnum* is certainly the *Quaestiones parvorum logicalium* (hereafter: QPL) which he published in Prague, as is clear from the colophons of all three extant manuscripts (Kraków, Leipzig, Sankt Florian, see below) as well as from examples occurring in the text ('Pragae', 'rex Bohemiae'). The exact date of publication is not known, but it is certainly not before 1386 (the year in which Helmold obtained his master's degree) and not after 1393 (the date of the earliest manuscript copy of the work); this latter year 1393 may even coincide with the year of composition or publication.

QPL is a set of questions based on then well-known and esteemed treatises on the so-called *parva logicalia*. This set has nine parts each with a number of questions varying from four to twenty; all in all, it comprises 89 questions (at least in the Sankt Florian manuscript). The nine parts are *suppositiones, confusiones, ampliationes, restrictiones, appellationes, consequentiae, Billingham* (i.e., *probationes propositionum*), *obligationes*, and *insolubilia*. The corresponding authoritative treatises on which QPL is based are by Thomas Maulfeld (I, II, VI), Marsilius of Inghen (III, V), Richard Billingham (VII), William Buser and John of Holland (VIII), and John of Holland (IX). The basic treatise for part IV on restrictions could not be identified yet; it is definitely neither Marsilius' nor

4 Cf., e.g., F. Šmahel, *Die Prager Universität im Mittelalter: Gesammelte Aufsätze—Charles University in the Middle Ages: Selected Studies*, Leiden 2007 (Education and Society in the Middle Ages and Renaissance, 28), esp. part C, 465–598; A. Conti, 'Wyclif's Logic and Metaphysics,' in I.C. Levy (ed.), *A Companion to John Wyclif. Late Medieval Theologian*, Leiden 2006 (Brill's Companions to the Christian Tradition, 4), 67–125; V. Herold, 'The Spiritual Background of the Czech Reformation: Precursors of Jan Hus,' in F. Šmahel and O. Pavlíček (eds.), *A Companion to Jan Hus*, Leiden 2015 (Brill's Companions to the Christian Tradition, 54), 69–95, esp. 87–95.

5 See Leipzig, Universitätsbibliothek, Ms. 1348, 279^r: 'Et credo, quod isti de ista opinione sic respondentes ad insolubilia conformiter dicunt, sicut illi, qui dicunt, quod species reales universales sunt ponendae, quarum nulla est substantia nec accidens. Et quando quaeritur ab eis, ubi talia universalia sunt, respondent, quod non habent dare hoc, ubi sunt.'—To ask, where abstract entities are, is of course not a sound philosophical question, but it was quite common then.

Thomas', as one would expect. The complete set of QPL is extant in the following two manuscripts:

K Kraków, Biblioteka Jagiellońska, Ms. 696, 1ra–183vb (the first leaf is missing), Pragae 1396;

S Sankt Florian, Stiftsbibliothek, Ms. XI 627, 1ra–235ra, Pragae 1393.

The single question IX.5, the relevant text for our topic here, is copied also in a further manuscript:

L Leipzig, Universitätsbibliothek, Ms. 1348, 274^{r}–284^{v}, without date and place, but presumably from Leipzig around the middle of the fifteenth century.[6]

An extremely shortened version of Helmold's q. IX.5 is found in Berlin, Staatsbibliothek Preußischer Kulturbesitz, Ms. lat. fol. 206, 184rb–187vb, at 185va–186va, which figures in Spade's catalogue under Anonymous no. XXIII.[7] The Berlin codex was at least in parts written in 1433 at Rostock (as the University of Leipzig was a filiation of that of Prague, so the University of Rostock was a filiation of that of Leipzig, founded ten years after Leipzig, in 1419).

3 The Insolubles and the Liar-Paradox

Part IX of the QPL on insolubles[8] comprises the following six questions:[9]

1. *Utrum scientia insolubilium sit pars scientiae obligationum* (K 165vb–168rb, S 211rb–214va).
2. *Utrum omnis scientia de insolubilibus sit pars logicae scientiae* (K 168rb–170rb, S 214va–217rb).
3. *Utrum scientia insolubilium ex Insolubilibus Holandrini habita sit scientia apriori vel aposteriori* (K 170rb–172rb, S 217rb–219rb).
4. *Utrum definitio casus insolubilis et insolubilis posita ab Holandrino sit bona* (K 172rb–174rb, S 219rb–221vb).
5. See below.

6 At least according to M. Markowski, *Repertorium commentariorum medii aevi in Aristotelem Latinorum qui in Bibliotheca Universitatis Lipsiensis asservantur*, Kraków 2012, 215–217. According to the colophon on fol. 284^{v}, however, it could also date from Helmold's stay at Leipzig, i.e. 1409–1441.

7 P.V. Spade, *The Mediaeval Liar: A Catalogue of* Insolubilia-*Literature*, Toronto, ON, 1975 (Subsidia Mediaevalia, 5), 46–47, no. XXIII.

8 On insolubles in medieval philosophy in general see, e.g., P.V. Spade and S. Read, 'Insolubles,' in E.N. Zalta (ed.), *The Stanford Encyclopedia of Philosophy* (Fall 2017 Edition), URL = https://plato.stanford.edu/archives/fall2017/entries/insolubles/.

9 In the following text, quotations from QPL are based on a comparison of K and S (and, if

6. *Utrum intellectum per intentionem simplicem propositionem cuius est pars intelligere sit possibile* (K 180ra–183va, S 230ra–234vb).

Helmold quotes many contemporary authors throughout QPL, and in part IX in particular, namely: Thomas Maulfeld (here, however, only with reference to part I on suppositions), Richard Kilvington, John of Holland, Richard Billingham, William Heytesbury, Richard Brinkley, John Dumbleton, Thomas Bradwardine,[10] William Buser, Albert of Saxony, and Conrad of Soltau. So the most important English philosophers including Mertonians and Calculators are known to Helmold and received by him in his treatment of insolubles, as far as I can see with the only exception of Roger Swyneshed, whom he does not mention by name. In other parts of QPL, Walter Burley, William of Ockham, and John Wyclif, among others, are also quoted by Helmold.

The long title of q. IX.5—the 'Liar-question'—runs as follows:

> Utrum casus integer communis insolubilis sit admittendus, scilicet iste: Ponatur, quod Sor dicat istam propositionem 'Sor dicit falsum' et nullam aliam et quod solum unus homo vocetur hoc nomine 'Sor' et quod termini illius propositionis praecise sic significent sicut communiter praetendunt, seu quod ista propositio 'Sor dicit falsum' praecise sic significet quod Sor dicit falsum (K 174rb–180ra, L 274^{r}–284^{v}, S 221vb–230ra).

Already the title exhibits a dense concentration of technical terms: 'casus insolubilis' (qualified by 'integer' and 'communis')—'insolubile' should also be added here—, 'admittere', 'significare sicut termini communiter praetendunt', 'significare praecise' (wherefore the *casus* sometimes is called in short 'de praecise').

As has been said above, the basis for part IX of Helmold's QPL is John of Holland's *Insolubilia*. There we find at the very beginning definitions of 'casus insolubilis' and 'propositio insolubilis vel insolubile' to the effect that an insoluble case is a case (i.e., a setting or a situation, as Spade has it) in which a certain sentence is mentioned, so that it follows that this sentence is true and false if it exactly signifies as its terms 'pretend' to signify (i.e., if its terms have their usual

applicable, of L too), but they are presented only as 'the best text', not as a critical edition. The appendix, however, presents a critical edition of selected passages from QPL, part IX.

10 This assumes that the name 'Thomas Werbrik', which appears in the manuscripts at the start of q. 6, is a corruption for Thomas Bradwardine, not otherwise named. Bradwardine held the view on *insolubilia* attributed to the otherwise unknown Werbrik, viz. 'pars propositionis potest supponere pro propositione, cuius est pars.' Cf. Spade, *The Mediaeval Liar*, 105–110, no. LXIV, at 108. See Thomas Bradwardine, *Insolubilia*, ed. and transl. S. Read, Leuven, 2010 (Dallas Medieval Texts and Translations, 10), 96, § 6.3, *suppositio tertia*.

meaning). And an insoluble is just such a sentence as part of such a case.[11] The immediate source for these definitions is William Heytesbury, with some differences in the wording.[12] John also uses the phrase 'praecise significare sicut termini praetendunt', but without explanation. This phrase (actually it is two phrases), too, seems to have been introduced by Heytesbury.[13]

Question IX.5 contains two articles: the first one presents and discusses critically ten previous opinions, the second one is devoted to Helmold's own approach. The first article is historically interesting and valuable. It is true that most texts of the genre discuss previous opinions, but ten (his own being the eleventh) is rather unusual.[14] Thomas Bradwardine, e.g., has nine (including his own), but more common are lists of three or four opinions, as, e.g., in William Heytesbury,[15] who seems to be the favorite source of Helmold. The list of 16 opinions found in Paul of Venice (d. 1429) is especially well known due to Joseph M. Bocheński's widely read *Formale Logik*.[16]

In introducing the first article, Helmold states that there are 'valde multae viae' of dealing with this well-known insoluble, but that he will confine himself to discuss only the ten most famous ones (*famosiores*). He does not ascribe, however, these views to certain authors (apart from the tenth to his master Conrad of Soltau), because he apparently considers them as types. But in critically discussing the views, he quotes a lot of contemporary scholars, as has

---

11 John of Holland, *Insolubilia*, ed. E.P. Bos, *John of Holland, Four Tracts on Logic* (*Suppositiones, Fallaciae, Obligationes, Insolubilia*), Nijmegen 1985 (Artistarium, 5), 123–146, at 127.

12 Compare John of Holland, *Insolubilia* 127$^{8-16}$, with William Heytesbury, *On 'Insoluble' Sentences: Chapter One of his Rules for Solving Sophisms*, transl. P.V. Spade, Toronto, ON, 1979 (Mediaeval Sources in Translation, 21), 46, § 44 and 45. For Heytesbury's Latin text, see L. Pozzi, *Il Mentitore e il Medioevo. Il dibattito sui paradossi dell'autoriferimento. Scelta di testi, commento, traduzione*, Parma 1987 (Quaderni di filosofia, 10), 201–272, at 236, § 3.01 and 3.02.

13 Cf. William Heytesbury, *On 'Insoluble' Sentences*, 46, § 44 and 45, and Pozzi, *Il Mentitore*, 236, §§ 3.01 and 3.02. See also F. Pironet, 'William Heytesbury and the Treatment of *Insolubilia* in Fourteenth-Century England Followed by a Critical Edition of Three Anonymous Treatises *De insolubilibus* Inspired by Heytesbury,' in S. Rahman, T. Tulenheimo and E. Genot (eds.), *Unity, Truth and the Liar. The Modern Relevance of Medieval Solutions to the Liar Paradox*, Dordrecht 2008 (Logic, epistemology, and the unity of science, 8), 255–333.

14 See Spade's catalogue, *The Mediaeval Liar*, throughout.

15 Cf. Spade, *The Mediaeval Liar*, 116–118, no. LXIX, and William Heytesbury, *On 'Insoluble' Sentences*, 18–45.

16 J.M. Bocheński, *Formale Logik*, Freiburg 1956 (6th ed., Freiburg 2015), translated into English as *A History of Formal Logic*, transl. I. Thomas, Notre Dame, IN, 1961 (2nd ed., New York 1970). I thank Stephen Read for pointing out to me the correct number of views in Paul of Venice (16, instead of 15 mentioned by Bocheński, *Formale Logik*, 280–291). Cf. also Spade, *The Mediaeval Liar*, 82–84, no. L.

already been said, most often 'Hesbri', that is William Heytesbury (d. 1372/1373), whose famous work *Regule solvendi sophismata* with its first chapter on insolubles dates to 1335.[17] Other Mertonians or Calculators, respectively, referred to by Helmold include especially Richard Kilvington ('Climiton') and John Dumbleton ('Dulmiton' or 'Dolmiton'). In the following, I present an overview of the ten previous opinions in Helmold; a critical edition of the Latin texts (without Helmold's discussions, however) is given in the appendix.

The first five opinions agree insofar as they all admit the case, but differ in their respective answers. The first view states that two contradictories in an insoluble case may both be true, insofar as the one signifies precisely as is the case in reality, and the other verifies itself because of its self-reference. No advocate of this view is known to me, but it seems to be just a variant of the second view,[18] according to which two contradictories in an insoluble frame may both be *false*, which is definitely Roger Swyneshed's approach, listed as the first previous opinion in Heytesbury.[19] Both views are rejected by Helmold with reference to Aristotle: That two contradictories may be both true or both false can be uttered vocally, but not really thought or conceptually grasped—as Helmold states it: 'Ore dicunt, corde autem nequaquam affirmant, cum ratio humana hoc non capit.' Furthermore, he says, no rational argument can be given why the Rule of Contradictories should apply to sentences outside the frame of insolubles, but not within it.

The third opinion denies that the insoluble itself is a sentence and states accordingly that it is neither true nor false. This is the second previous opinion mentioned by Heytesbury. Gaetano da Thiene attributes this view to John Dumbleton, but Paul Vincent Spade has rejected this attribution.[20] Again, Helmold raises the problem of giving a rational criterion why it should suffice outside the frame of insolubles for the status of being a sentence to be an indicative expression signifying as is the case or as is not the case, but not so within such a context.

17 On Heytesbury see, e.g., M. Hanke and E. Jung, 'William Heytesbury,' in E.N. Zalta (ed.), *The Stanford Encyclopedia of Philosophy* (Spring 2018 Edition), URL = https://plato.stanford.edu/archives/spr2018/entries/heytesbury/.

18 Cf. Spade, 'Study,' in William Heytesbury, *On 'Insoluble' Sentences*, 76.

19 On Swyneshed's view see Spade, *The Mediaeval Liar*, 102–105, no. LXIII; Spade, 'Study,' in William Heytesbury, *On 'Insoluble' Sentences*, 72–79; P.V. Spade, 'Roger Swyneshed's *Insolubilia*: Edition and Comments,' *Archives d'histoire doctrinale et littéraire du Moyen Âge* 46 (1979), 177–220.

20 See, e.g., Spade, *The Mediaeval Liar*, 117, n. 219. On Gaetano, see Spade, *The Mediaeval Liar*, 53–54, no. XXVIII.

The fourth opinion (the third in Heytesbury) admits that an insoluble is a sentence, but denies that any insoluble is true and that any is false, whereas every insoluble is true or false. Gaetano ascribed this view to Richard Kilvington, which, however, does not seem to be completely justified.[21] Again, no good reason can be given why an insoluble should have the truth value 'true or false', whereas normal sentences have the truth values 'true' or 'false'. In arguing against this view, one has to avoid the use of the words 'true' and 'false', but should use instead the expressions 'is the case' and 'is not the case'.

The fifth view holds that an insoluble is neither absolutely true nor absolutely false, but only true or false in a certain respect, which is the Aristotelian-scholastic terminology of *simpliciter* vs. *secundum quid*. Helmold attributes this view to Aristotle and also to Kilvington. He argues against it, e.g., that it is not convincing to refer to Aristotle's example in *Nicomachean Ethics* III.1, 1110a4–12, of throwing goods overboard in a storm, which is neither absolutely voluntary nor absolutely involuntary, because in such actions two tendencies are concurrent, of which the one tends to the opposite of the other. But with respect to truth and falsity of sentences this is not the case, so that the simile is misleading (*similitudo hic claudicat*).

The discussion of the sixth opinion occupies by far the most space (in L, e.g., ff. 276^{v}–279^{r}). It is the first opinion which does not admit the case, because it is simply impossible, according to this view. The reason for this is that it is generally impossible for a part to stand for its whole (*quod pars supponeret pro suo toto*). According to Helmold, Dumbleton claims to have proven this a priori, insofar as every such insoluble case presupposes an impossibility, viz., that a part stands for its whole. Heytesbury, however, had shown this a posteriori, insofar as according to him no respondent could avoid contradicting himself once he has admitted the insoluble case. Therefore, such a case must in no way be admitted.

Albert of Saxony and his follower John of Holland also considered the insoluble case to be impossible and accordingly to be denied, but for a different reason, namely that every sentence signifies itself to be true. Since the Liar-insoluble also signifies itself to be false, it signifies that it is both true and false. Therefore, the insoluble 'Sor dicit falsum' is in effect equivalent to the conjunction 'Sor dicit falsum et Sor dicit verum' and its contradictory is the disjunction 'Sor non dicit falsum vel Sor non dicit verum'. Against this view, Helmold makes the interesting point that given this analysis the insoluble sentence 'Sor dicit falsum' would not even be intelligible and the corresponding thought would

21 Cf. Spade, *The Mediaeval Liar*, 117, n. 220, and for Kilvington 92–93, no. LV.

not be expressible; hence the sentence could not even be called 'false', because the understanding of a conjunction presupposes the understanding of its parts, which is apparently impossible here.

However, the main target in this passage is Heytesbury, who according to Helmold can only produce evasive words (*verba evasiva*) when it comes to the point of declaring in which way the signification of a sentence in an insoluble case is changed. Helmold takes the example 'Asinus videt falsum' which is written on a wall and which is first not seen by any donkey and then is inspected by one. Here is the central part of this long passage:

> [L 278r] Quemadmodum Hesbri[22] facit si quaeritur ab eo, si ista propositio 'Asinus videt falsum' ⟨sit⟩ scripta in pariete et nullus asinus inspiciat talem propositionem, utrum ista propositio scripta in pariete 'Asinus videt falsum' praecise et primarie significet quod asinus videt falsum—respondet quod sic, quod ista praecise sic significat quod asinus videt falsum ex significatione sui primaria.
>
> Sed quando Hesbri ponit in alio casu sic quod asinus inspiciat talem propositionem scriptam in pariete [L 278v] 'Asinus videt falsum', utrum tunc illa 'Asinus videt falsum' praecise ex primaria sua significatione significat quod asinus videt falsum vel non—respondet quod in dicto casu ista propositio 'Asinus videt falsum' quam asinus inspicit non significat praecise ex sui significatione primaria quod asinus videt falsum.
>
> Et quando quaeritur ab eo ulterius: 'et si sic est quod talis propositio ex sui significatione primaria non significat praecise quod asinus videt falsum, quomodo ergo aliter dicta propositio in dicto casu significat?'—respondet quod non habet dare quomodo aliter significat in dicto casu. Et per hoc dicit de se ipso bene respondisse ad insolubile. Quod ego non iudico aliter esse nisi respondere evasive, ut evadat manus opponentis ne contradictionem incurrat.
>
> Vel igitur Hesbri scivit quomodo ista propositio 'Asinus videt falsum' in dicto casu aliter significat plus vel minus ex significatione sui primaria quam ⟨quod⟩ asinus videt falsum vel non scivit quomodo talis propositio in dicto casu ex significatione sui primaria ⟨aliter⟩ significaret quam quod asinus videt falsum. Si Hesbri diceret quod ipse non sciret quomodo dicta propositio in dicto casu significaret significatione sui primaria aliter quam quod asinus videt falsum, quare tunc scienter diceret falsum, cum

22 Cf. Heytesbury's own discussion in William Heytesbury, *On 'Insoluble' Sentences*, 43–45, §§ 41–43.

> non esset obligatus per casum istum ad hoc nec hoc sequitur ex tali casu? Si vero Hesbri diceret quod ipse sciret quomodo ista propositio 'Asinus videt falsum' in dicto casu aliter significat ex sui significatione primaria quam quod asinus videt falsum, quare nobis hoc in scripto non relinquit, cum plura circa eandem materiam insolubilium in scripto pertractavit? Et utique credo, si Hesbri scivisset quomodo dicta propositio 'Asinus videt falsum' aliter ex significatione sui primaria significaret quam quod asinus videt falsum, utique dixisset. Ideo credo apud Hesbri talem modum respondendi non fuisse nisi gratia evasionis argumentorum insolubilium.
>
> Specialiter tamen ratione arguo contra talem evasionem Hesbri sic: Sit cum casu priori quod nihil in respondendo imponat de novo partem istius propositionis 'Asinus videt falsum' nec deponat[23] eius partem nec integram ipsam propositionem. Qui casus compossibilis est priori casui illi quod asinus inspiciat istam propositionem 'Asinus videt falsum', quod Hesbri et alii omnes admitterent. Isto casu stante sequitur quod ista propositio 'Asinus videt falsum' quam asinus inspicit non primarie praecise sic significat quod asinus videt falsum, et prius antequam asinus eam inspexit ipsa praecise primarie significavit quod asinus videt falsum. Ergo ista propositio 'Asinus videt falsum' novam significationem habet quando asinus eam inspicit vel ipsa in inspiciendo ab asino aliquid suae significationis amittit.

Helmold has 'Asinus videt falsum', whereas Heytesbury has 'Sortes videt falsum', but the example with *asinus* may be suggested by Heytesbury himself.[24] The important point is again that Heytesbury cannot explain the difference in the meanings of the sentence 'Asinus videt falsum' in the two cases assumed. This very point is also made by Paul Vincent Spade in his fine analysis of Heytesbury's theory and Heytesbury himself had admitted the difficulty.[25]

The seventh opinion admits self-reference, i.e. that a part may stand for its whole, and holds accordingly that the case is possible and is therefore to be admitted. Apart from this, this view is in line with that of Albert of Saxony and others, outlined under the sixth opinion, that the insoluble signifies both its

---

23 'Impositio' and 'depositio' are common kinds of the *obligationes* genre. Cf., e.g., Albert of Saxony, *Logik*, Lateinisch–Deutsch, ch. VI.6 and VI.11, ed. H. Berger, Hamburg 2010 (Philosophische Bibliothek, 611), 1216–1241 and 1308–1317.

24 Cf. William Heytesbury, *On 'Insoluble' Sentences*, 44, at the end of § 41: 'and indeed even an ass.' Pozzi, *Il Mentitore*, 234, § 2.18, has here 'ymmo unus alius (not: asinus) unico tali aspectu.'

25 See Spade's comments in William Heytesbury, *On 'Insoluble' Sentences*, 70–71, 80, 93, 94–95. Cf. Heytesbury's text in Spade's translation ibid., 40–45, §§ 37–43.

truth and its falsehood, so that a difference is only with respect to the admission or non-admission of the case.

The eighth opinion, however, denies this very assumption, viz. that the insoluble signifies both its truth and its falsehood, and does also not admit self-reference, wherefore the case is not admitted according to this view. Furthermore, this opinion holds that the insoluble signifies a sentence different from itself to be false. Helmold refers to Dumbleton here, but he postpones the discussion of this view to the next question (q. IX.6). I shall quote the interesting introduction to this question below.

The ninth view also rejects the case and states that no insoluble is true and none is false either. The reason offered for this view is that a sign naturally presupposes its significate; but the truth and falsehood of sentences are signs, so that they presuppose true or false significates. Now, in the case of the Liar there is, according to the proposed casus, nothing, which could count as the significate of the insoluble sentence. Since nothing can presuppose itself, it follows that such an insoluble is neither true nor false. Helmold does not ascribe this view to any author, but it seems to be Richard Brinkley's.[26]

Finally, the tenth and most recent (*magis moderna*) opinion is that of Helmold's master at the University of Prague, Conrad of Soltau. Conrad distinguishes between two meanings of the predicate 'false', which can either signify in the way of an object or in the way of a sentence (*per modum obiecti* vs. *per modum propositionis*). This terminology is left unexplained, but from the context it can be concluded that the predicate 'false' in insoluble, self-referential cases, such as the Liar, signifies by way of an object, whereas in normal contexts it signifies by way of a sentence. On the basis of this distinction, Conrad counts the insoluble 'Sor dicit falsum' as part of the insoluble case as false. Helmold tries to show that this view leads to accepting two contradictories as both true or as both false, which had been Conrad's own criticism with respect to the first two opinions. So if there were an extant anonymous manuscript copy on insolubles which makes use of this specific terminology 'significare per modum obiecti' vs. 'per modum propositionis', it could be ascribed to Conrad of Soltau on the basis of Helmold's report here. By the way, neither Conrad nor Helmold himself are listed in Spade's catalogue.

The second article of Helmold's question IX.5 is devoted to Helmold's own view as the eleventh one. First, he reports and criticizes some different modes

26 Cf. Spade, *The Mediaeval Liar*, 33–34, no. x; P.V. Spade, *An Anonymous Fourteenth-Century Treatise on* Insolubles: *Text and Study*, unpublished Ph.D. dissertation, University of Toronto, 1969, esp. 37–38, § 66; P.V. Spade, 'Richard Brinkley's *De insolubilibus*: A Preliminary Assessment,' *Rivista di storia della filosofia*, 46 (1991), 245–256.

of presenting the insoluble case. Then, he states three assumptions (*suppositiones*) and six theses (*conclusiones*). I confine myself to the latter here:

1. The common complete insoluble case, viz. assuming that Socrates said the sentence 'Some sentence is false' and no other one and that no other than this one existed at all and that this one precisely and primarily signified that some sentence is false—such a case is possible.—Note the shift from 'Socrates is saying what is false' (*Sor dicit falsum*) to 'Some sentence is false' (*Aliqua propositio est falsa*), which is founded in the *suppositiones*.
2. This case is to be admitted (because it is possible, according to thesis 1).
3. The insoluble case being stated this way and admitted, and further assuming that things happen to be just as they are now, viz. that there are many false sentences, then it must be conceded that the sentence 'Some sentence is false' is true.
4. The insoluble case as stated in the first *conclusio* still holding—if the insoluble sentence 'Some sentence is false' is presented then (*proponere*, a technical term of the *obligationes* genre), it has to be conceded as being true or false.
5. If then the further sentence 'This sentence "Some sentence is false" is true or this sentence "Some sentence is false" is false' is presented, this disjunction has to be doubted. (The reason being that each part of it is to be doubted.)
6. If then the insoluble 'Some sentence is false' is presented, it must not be conceded nor denied, but has to be doubted.—The reason for this thesis is, that such a sentence is irrelevant (*impertinens*) to the case and is neither known to be true nor known to be false, wherefore according to the rules of obligations it has to be doubted.

So much for a first overview of this very long question IX.5. The next and last question 6 of part IX, *Utrum intellectum per intentionem simplicem propositionem cuius est pars intelligere sit possibile*, which addresses the problem of self-reference, has an interesting prologue which may be quoted here:

> [K 180^{ra-rb}, S 230ra] Quamvis ista quaestio ex scientia insolubilium determinari non potest, ipsa tamen vel eius contradictorium a multis sumitur tamquam fundamentum in materia insolubilium ad solvendum argumenta insolubilium, quemadmodum faciunt Brincel (i.e. Richard Brinkley) et Dulminton (i.e. John Dumbleton), qui dictam quaestionem simpliciter dicerent esse falsam. Nam habent hoc tamquam fundamentum in materia insolubilium, quod pars propositionis pro propositione cuius est pars supponere non potest, in quo ipsi ambo fundant modum respondendi ad insolubilia et ad eorum argumenta.

> Illius in contrarium pro fundamento in materia insolubilium tenet Thomas Werbrik (i.e. Thomas Bradwardine?), Buserus et Albertus de Saxonia, quorum quilibet istorum diceret illam quaestionem simpliciter esse veram. Et habent hoc tamquam fundamentum in materia insolubilium, quod pars propositionis pro propositione cuius est pars supponere potest. Propter cuius controversiam circa materiam insolubilium dictam quaestionem proposui determinandum.

Concluding Remarks

Helmold offers us an informative overview of the discussions of the Liar-Paradox in particular and of logical and semantic matters in general from the end of the fourteenth century back over some decades. More specifically, he testifies that important new trends, as for instance English and Parisian approaches to logic and semantics, were available in Prague and Central Europe. In addition, we owe to him the special information that his teacher Conrad of Soltau introduced him to the field of insolubles and also composed a treatise on insolubles, which was unknown so far. On the basis of Helmold's QPL, one could write a history of fourteenth-century logic and semantics 'without any gaps' (to use Peter Adamson's words).

Appendix: The Ten Views Discussed by Helmold

This partial edition of Helmold's *Quaestiones parvorum logicalium*, pars IX, q. 5, a. 1, uses the following sigla and abreviations:

K Kraków, Biblioteka Jagiellońska, Ms. 696
L Leipzig, Universitätsbibliothek, Ms. 1348
S Sankt Florian, Stiftsbibliothek, Ms. XI 627
add. *addidit*
om. *omisit*

[K 174va, L 274^{v}, S 222rb] Quantum ad primum ⟨articulum⟩ primo nota quod, quamvis valde multae viae sunt et fuerunt de admissione casus insolubilis et de modo respondendi ad casum insolubilem et ad insolubile, tamen specialiter decem opiniones, quae in hac materia fuerunt famosiores, in praesenti articulo volo praesentare, quae sunt de admissione casus insolubilis et de responsione ad casum insolubilem et ad insolubile. Et postea aliam opinionem undecimam, quam ego credo esse veriorem, volo fundare in materia itaque insolubilium.[27]

[K 174va, L 274^{v}, S 222rb] Prima opinio scripsit quod integer casus insolubilis, in quo etiam insolubile ponitur praecise significare sicut eius termini communiter praetendunt, esset admittendus. Et ulterius dicit quod in materia insolubilium non est inconveniens duo contradictoria sibi invicem contradicentia in eodem instanti et in qualibet parte eiusdem instantis esse simul vera, ponens quod unum contradictoriorum in insolubilibus significat praecise sicut est, igitur unum illorum est verum; sed alterum contradictoriorum verificat se ipsum per reflexionem significationis suae super se ipsum; quare dicit ipsum etiam esse verum.

[K 174vb, L 275^{r}, S 222vb] Secunda igitur opinio conformis est opinioni primae quoad admissionem casus insolubilis. Admittit enim haec opinio casum

1 nota] est notandum S ‖ quod] *om.* S 2 valde] *om.* K ‖ multae] diversae KS 3 specialiter] ad *add.* K 5 praesentare] recitare L ‖ de responsione] ad responsionem S 6 et] *om.* K ‖ opinionem] *om.* L 7 itaque] *om.* K 9 etiam] *om.* K ‖ eius] *om.* K 11 duo contradictoria] *om.* K ‖ in] pro L 12 in qualibet parte eiusdem instantis] sic de aliis condicionibus K 13 contradictoriorum] contradictorium L ‖ in] *om.* L ‖ est] *om.* L 15 reflexionem] flectio (?) S 15–16 ipsum etiam esse verum] ipsum est verum L etiam ipsum se esse verum S 17 igitur] *om.* K 18 quoad] quo LS

27 *The phrase* 'in materia (itaque) insolubilium' *could also pertain to the following* 'prima opinio', *in my view even by far better so, but* KS *clearly put it to* 'fundare'; *L is neutral in this respect.*

insolubilem de praecise.[28] Sed in respondendo ad insolubile dicit quod quodlibet insolubile est falsum, et concedit quod suum contradictorium etiam est falsum. Rationem assignant istam dicentes, quia ipsum contradictorium insolubilis secundum significationem suam primariam et adaequatam significat sicut non est; igitur dicunt quod contradictorium insolubilis est falsum. Sed quod insolubile sit falsum assignant causam, quia ipsum, quamvis significat sicut est, tamen ipsum falsificat se ipsum; igitur concludit quodlibet insolubile esse falsum. Et consequenter ista opinio concedit quod duo contradictoria sibi invicem contradicentia in materia insolubilium sunt simul falsa et ambo talia duo contradictoria sunt neganda, quemadmodum prima ponit opinio quod in materia insolubilium duo contradictoria sunt simul vera et concedenda.

[K 174vb, L 275^{r}, S 222vb] Tertia opinio ponit conformiter sicut ambae opiniones praecedentes ponunt de admissione casus. Nam et haec opinio communem casum insolubilem admittit, in quo etiam ponitur quod insolubile praecise debet significare sicut termini eius communiter praetendunt. Sed consequenter dicit quod insolubile **[L 275^{v}, S 223ra]** est bene oratio indicativa de verbo praesentis temporis significans sicut est, aut significans sicut non est, sed non est propositio. Et sic dicunt quod insolubile nec est verum nec est falsum, eo quod dicunt quod insolubile non est propositio. Et ista opinio habet hoc pro uno suo fundamento quod ad hoc quod aliqua oratio in materia insolubilium sit propositio non sufficit quod sit oratio indicativa, quae secundum sui significationem adaequatam et primariam significat sicut est, aut significat sicut non est. Sed dicit quod hoc bene sufficeret ad orationem quae deberet esse propositio extra materiam insolubilium.

[K 175rb, L 276^{r}, S 223rb] Sed quarta opinio ponit quod casus **[S 223va]** integer insolubilis, in quo etiam ponitur quod insolubile debet praecise significare sicut termini eius communiter praetendunt, est admittendus. Sed ulterius ponit pro fundamento quod quodlibet insolubile est propositio, sed quod nullum tale insolubile est verum et quod nullum tale insolubile est falsum. Sed concedit quod omne insolubile est verum vel falsum.

[K 175va, L 276^{r}, S 223va] Quinta opinio ponit casum communem insolubilem esse admittendum, sicut et opiniones praecedentes ipsum admiserunt. Sed

---

1 in] *om.* S 4 et] *om.* L 5 insolubilis] insolubile L 7 tamen] sed quia LS 9 insolubilium] insolubili L 13 praecedentes] *om.* KS ‖ ponunt] dicunt KS quod *add.* L 15 sicut] ut KS 17 significans] *om.* K 18 insolubile] *om.* L ‖ est] *om.* KL 19 insolubile] *om.* K ‖ hoc] *om.* K 20 uno suo] *om.* K 22 et] *om.* LS ‖ significat] *om.* K 28 sed] et K 31–32 communem insolubilem] in quo insolubile est S 32 et] hae L

---

28 *The expression* 'de praecise' (*here as well as in opinions 8 and 9*) *is short for the condition that the insoluble sentence in the insoluble case signify precisely as its words commonly pretend.*

ultra hoc dicit quod nullum insolubile est simpliciter verum nec aliquod insolubile est simpliciter falsum, sed quod insolubile est secundum quid verum et secundum quid falsum.

[K 175vb, L 276^{v}, S 224ra] Sexta opinio ponit oppositum omnibus praecedentibus quinque opinionibus quoad admissionem casus. Dicit tamen haec opinio quod talis casus insolubilis simpliciter est impossibilis; et ideo nullo modo est admittendus. Quae inquam opinio duo habet principalia fundamenta ex quibus nititur hoc declarare. Quorum primum est quod nulla propositio per intentionem simplicem in anima potest apprehendi. Secundum fundamentum eius est quod cuiuslibet propositionis subiectum vel praedicatum stans pro complexo est propositio.

[K 177rb, L 279^{r}, S 225vb] Septima opinio ponit quod pars propositionis bene supponit pro propositione cuius est pars. Ex quo fundamento concludit quod casus communis integer insolubilis, in quo insolubile ponitur praecise significare sicut termini eius communiter praetendunt, est possibilis et admittendus. Sed ulterius capit pro fundamento quod quodlibet insolubile significat se esse verum et significat se esse falsum. Ex quo fundamento infert quod quodlibet insolubile est falsum.

[K 177rb, L 279^{r}, S 226ra] Octava opinio ponit quod numquam aliqua propositio significat se esse veram et significat se esse falsam; et addit quod numquam pars propositionis supponit pro propositione cuius est pars. Quod fundamentum etiam ponit Dolmiton; propter quod dicit Dolmiton casum communem insolubilem de praecise[29] esse negandum. Et addit haec opinio quod insolubile significat aliam propositionem a se esse falsam.

[K $177r^{b}$, L 279^{v}, S 226ra] Nona opinio ponit quod integer casus insolubilis de praecise[30] communis est admittendus, sed dicit quod nullum insolubile est verum et nullum insolubile est falsum. Quod sic declarat, quia signum prioritate naturae praesupponit suum significatum. Ex isto infert quod veritas propositionis, quae est signum, prioritate naturae praesupponit veritatem sui significati, a quo significato propositio vera est vera. Et ita conformiter dicit

---

1 aliquod] quod S 1–2 insolubile est] *om.* K 2 quod] aliquod *add.* L 4 opinio] ista *add.* S 5 quoad] quo S ‖ tamen] enim *praetulisset editor* 6 impossibilis] insolubilis S 7 admittendus] admittendum S ‖ inquam] *om.* L 8 nititur] innititur/innuitur (?) S ‖ declarare] declaratio S 13 pro] *om.* S ‖ fundamento] fundamentum S 14 insolubile] insolubili L 17 se] *om.* S 20 et] numquam *add.* S 21 propositionis] *om.* S ‖ quod] idem K et quod S 22 dolmiton] dulmiton LS ‖ dolmiton] dulmiton S 23 insolubilem] *om.* S 26 communis] non L 27 declarat] dicit L 28 praesupponit] supponit KL 29 prioritate] prioritatis S ‖ praesupponit] supponit KS 30 ita] *om.* S

---

29 *On this expression see above the relevant note to the second opinion.*

30 *On this expression see above the relevant note to the second opinion.*

quod falsitas propositionis, quae est signum, prioritate naturae supponit significatum falsum, a quo propositio falsa est falsa. Secundo ponit pro fundamento quod idem se ipsum prioritate naturae non praesupponit.

[K 177va, L 279^{v}, S 226rb] Decima opinio est magis moderna, quam tenet magister meus reverendus Conradus Soltow, qui dicit istum terminum 'falsum' duplici modo posse intelligi significare in insolubili. Uno modo, quod ille terminus 'falsum' significet falsum per modum obiecti. Alio modo, quod ille terminus 'falsum' significet falsum in propositione insolubili per modum propositionis. Ex quo fundamento ulterius dicit quod ista propositio 'Sor dicit falsum' est falsa in isto casu, quando ponitur quod Sor dicat istam propositionem 'Sor dicit falsum' et nullam aliam et quod unus Sor sit omnis talis qui vocetur hoc nomine 'Sor' et quod nullus vel nulla hominum alius vel alia a Sorte dicat talem propositionem et quod termini eius sic significent ut communiter praetendunt. Rationem assignat istam, quia haec propositio 'Sor dicit falsum' in isto casu significat quod Sor dicit falsum adaequate primarie aliqualiter qualiter Sor non dicit falsum in dicendo istam 'Sor dicit falsum'. Declarat hoc sic, quia ista propositio 'Sor dicit falsum' per modum obiecti significat in isto casu quod Sor dicit falsum et quod nulla propositio est falsa per modum obiecti, sed solum per modum propositionis, et quia iste terminus 'falsum' in hac propositione 'Sor dicit falsum' nullo modo significat per modum propositionis istam propositionem, quamvis ille terminus 'falsum' aliquo modo in certo casu significat ut propositio, quia alias pars propositionis esset propositio, quod secum credo esse falsum. Et quamvis magister meus [S 226va] reverendus me in materia insolubilium specialiter instruxit, ego tamen contra opinionem suam arguam reverentia sua semper salva, nam secundum Philosophum [L 280^{r}] primo Ethicorum 'Si veritas et amicitia dissentiant, magis est intendendum veritati quam amicitiae.'[31]

1 supponit] praesupponit *praetulisset editor* 3 praesupponit] supponit KS 5 soltow] sulto K zoltu S 8 in propositione insolubili] *om.* L 9–10 ista propositio ... dicat] illa propositio sor dicit falsum est falsa in dicto casu quando ponitur sor dicat S *om.* L 11 quod unus ... vocetur] solum unus sor sit quod vocatur S 12 talem] dictam S 13 sic] *om.* KS ‖ ut] termini *add.* L termini eius *add.* S 14 assignat] assignant L 15 adaequate] et *add.* K ‖ aliqualiter] aliquo modo K aliquo S quod sor dicit falsum *add.* KS 16 in] igitur (?) S 19 iste terminus] li K 21 ille terminus] li K 25 sua] sui S

31 Aristotle, *Nicomachean Ethics* I.4, 1096a11–15.

CHAPTER 9

Blasius of Parma on the Calculation of the Variation of Qualities and Aristotelian Physics

Joël Biard

Introduction

Blasius of Parma examines the problem of the intension and remission (i.e., weakening or reducing) of accidental forms several times. He discusses it in his *Questiones super tractatum de latitudinibus formarum* (Questions on the Treatise on the Latitude of Forms), of which there exist several redactions, among which is a manuscript dated to 1392;[1] in a long and rather confusing disputed question, the *Questio de intensione et remissione formarum* (Question on the Intension and Remission of Forms), which its editor dates between 1388 and 1396;[2] and in the second redaction of his *Physics* commentary, which is from 1397. In this last work, it is mainly question 10 on book V which is devoted to the problem, though some additional information can be found in book VII, question 6, which is about how to evaluate the speed of alteration.[3]

Though Blasius also uses the concept of *latitudo* with regard to substantial forms,[4] the *Questio de intensione et remissione formarum* is clearly about accidental forms. Moreover, the concepts of *intensio* and *remissio* had been used by Boethius with regard to certain qualities, both corporeal and mental;[5] and it is

1 On this text, see the articles by D.A. Di Liscia ('Biagio Pelacani da Parma's Geometrisation of Latitudes and the Problems of the Mean Degree Theorem') and G. Federici Vescovini ('Blaise de Parme et l'intelligibilité mathématique') in J. Biard and A. Robert (eds.), *La philosophie de Blaise de Parme. Physique, psychologie, éthique*, Florence 2019 (Micrologus Library, 96), 129–152 and 107–128. The 1392 copy is in Oxford, Bodleian Library, Ms. Canon. Misc. 177.

2 G. Federici Vescovini, 'La *Quaestio de intensione et remissione formarum* di Biagio Pelacani da Parma,' *Physis* 31 (1994), 433–535 (text 470–535).

3 An edition of the whole of book VII, by Sabine Rommevaux-Tani, can be found at the Blasius of Parma website, URL = https://bdeparme.hypotheses.org/questions-sur-la-physique.

4 See J. Biard, 'Nature et degrés de perfection chez Blaise de Parme,' in L. Bianchi, O. Grassi, and C. Panti (eds.), *Edizioni, traduzioni e tradizioni filosofiche (secoli XII–XVI): Studi per Pietro B. Rossi*, 2 vols., Cantenaro 2018, 1: 387–410.

5 See J.-L. Solère, 'Plus ou moins: le vocabulaire de la latitude des formes,' in J. Hamesse and C. Steel (eds.), *L'élaboration du vocabulaire philosophique au Moyen Âge. Actes du Colloque International de Louvain-la-Neuve et Leuven, 12–14 septembre 1998*, Turnhout 2000 (Ren-

 | DOI:10.1163/9789004512054_011

with regard to qualities in particular that the concept of *latitudo*—in the sense of a certain range within which variation is possible—appeared in the Latin world with the translations of medical works, in particular those of Galen and Avicenna. In the twelfth and thirteenth centuries, the issue was discussed at first with regard to charity (which has no contrary), and then very quickly with regard to natural qualities such as whiteness and heat.

The grounding of the concept of *latitudo* in physics is clearly underlined by Blasius. In the statement of the initial opposing arguments in the *Questio de intensione et remissione formarum*, he refers in a general way to Aristotle. One might assume that he is thinking of the *Categories*, in which some qualities seem be open to being predicated to a greater or lesser degree,[6] but here he is referring to the *Physics*, in particular to book v. In fact, Aristotle does not address this question explicitly, and the idea of quantifying qualitative forms presupposes a different perspective on change. Nevertheless, Aristotle presents the movement of alteration, with respect to quality, as 'change within the same kind from a lesser to a greater or from a greater to a lesser degree'; he also states that 'a thing's possessing a quality in a greater or in a lesser degree means the presence or absence in it of more or less of the opposite quality.'[7] In other passages, however, he argues that it is only the subject that varies, not the form

contres de philosophie médiévale, 8), 437–488, esp. 453; J.-L. Solère, 'D'un commentaire l'autre: l'interaction entre philosophie et théologie au Moyen Âge dans le problème de l'intensification des formes,' in M.O. Goulet-Cazé (ed.), *Le commentaire: entre tradition et innovation. Actes du Colloque International de l'Institut des traditions textuelles (Paris et Villejuif, 22–25 septembre 1999)*, Paris 2000 (Bibliothèque d'histoire de la philosophie), 411–424. The studies by Anneliese Maier remain fundamental: A. Maier, 'Das Problem der intensiven Größe,' in Ead., *Zwei Grundprobleme der scholastischen Naturphilosophie. Das Problem der intensiven Größe. Die Impetustheorie* (Studien zur Naturphilosophie der Spätscholastik, 2), 3rd ed., Rome 1968 (Storia e letteratura, 37), 3–109, and 'Die Mathematik der Formlatituden,' in Ead., *An der Grenze von Scholastik und Naturwissenschaft* (Studien zur Naturphilosophie der Spätscholastik, 3), 2nd ed., Rome 1952 (Storia e letteratura, 52), 255–384. For a general presentation of the issue, see also E. Jung, 'Intension and Remission of Forms,' in H. Lagerlund (ed.), *Encyclopedia of Medieval Philosophy*, Dordrecht 2011, 551–555. Among the many studies, see H. Shapiro, 'Walter Burley and the Intension and Remission of Forms,' *Speculum* 34 (1959), 413–427; E.D. Sylla, 'Medieval Concepts of the Latitude of Forms: The Oxford Calculators,' *Archives d'histoire doctrinale et littéraire du Moyen Âge* 40 (1973), 223–283; S. Kirschner, 'Oresme on Intension and Remission of Qualities in His Commentary on Aristotle's *Physics*,' *Vivarium* 38 (P), 255–274. Regarding the debate in Italy, see S. Caroti, 'La discussione sull'*intensio et remissio formarum* nelle università italiane (sec. XIV),' in L. Bianchi and C. Crisciani (eds.), *Forme e oggetti della conoscenza nel XIV secolo: studi in ricordo di Maria Elena Reina*, Florence 2014 (Micrologus Library, 61), 415–460.

6 Aristotle, *Categories* 8, 10b27–29.

7 Aristotle, *Physics* V.2, 226b1–2 and b8, transl. R.P. Hardie and R.K. Gaye, in *The Complete Works of Aristotle*, ed. J. Barnes, Princeton, NJ, 1984 (2 vols.) (Bollingen series, 71), 1: 383.

or affection itself.[8] It is precisely this latter alternative that became a central element in the medieval debate.

John Buridan tackles the problem in connection with book III of the *Physics*. After a discussion of whether or not it is necessary to posit a flux that is distinct from the mobile, in the case of alteration, there are long questions on whether it is possible for contrary qualities to coexist, on the ways in which qualities are acquired in alteration—whether by degrees or not—and on the problem of whether a quality is intensified by addition or by succession. Blasius of Parma on the other hand, like Oresme,[9] deals with *intensio* and *remissio* in book V.

1 Quality and Subject

Every time Blasius addresses the question, he begins by providing some clarifications and by stating hypotheses concerning the status of qualities. The major problem is whether qualities are ontologically independent of their subject. This is a standard question, but for Blasius it will take on a new importance.

When he addresses the question of whether it is possible for a quality to be intensified or weakened,[10] his discussions are structured at a deep level by the possibility of interpreting the question according to two ways of understanding the term 'quality'. A quality can be understood either as a form inherent in a subject or as the subject itself which is qualified in some way. In the resolution of the question in the *Physics* commentary, these two alternatives appear at the head of a series of distinctions presented as preliminary *evidentie*.[11] Understanding a quality as the qualified subject might seem counterintuitive. The existence of something called 'quality' is not in question here. However, the possibility is raised from the start whether intensification and weakening per-

8 See Aristotle, *Physics* V.1, 224b11–14. One can also think of the fact of being more or less just, since justice does not vary.

9 See Nicole Oresme, *Questiones super Physicam* (*Books I–VII*), V, q. 9 (*Utrum unum contrarium intendatur altero non remisso*), eds. S. Caroti, J. Celeyrette, S. Kirschner and E. Mazet, Leiden 2013 (Studien und Texte zur Geistesgeschichte des Mittelalters, 112), 622–631.

10 The translation of *remissio* as 'remission' is supported by certain usages of the abstract noun (as in an illness that is 'in remission'), but the verb lacks the corresponding sense; rather, one says that a quality 'is attenuated' or 'is weakened.'

11 Blasius of Parma, *Questiones super octo libros Physicorum*, V, q. 10, a. 2, tertium processus, Città del Vaticano, Biblioteca Apostolica Vaticana, Ms. Vat. lat. 2159, 157ra: 'Pro dicendis pono aliquas evidentias. Prima: cum queritur utrum qualitas sit intensibilis, aliquando capitur "qualitas" pro tali forma inherente subiecto, et aliquando capitur pro subiecto sic qualificato.'

tain not to an inherent form but to the subject itself. In the *Questions on the Logical Treatises*, Blasius had already conceded that from the point of view of physics it is the subject that becomes more and more intense with respect to a certain quality (e.g., it becomes more and more intensely white), and not the quality itself that is susceptible of the more and the less.[12]

The second distinction made in the *Physics* commentary is that the subject can be intensified in two ways. It can be the subject of *more* of the form in an extensional sense (a certain part of it was white, for example, and now a larger part is white), or it can be intensified because of the form itself (the subject is whiter than it was before).

The third distinction is that the intensification of a form can be conceived in two ways: either because it is less mixed with the contrary quality, or because it is further removed from the zero degree.

The two other distinctions introduce criteria for measuring the intensity, to which we will return below. A form can be intensified by a reduced mixture with contrary qualities in two ways: it can vary by an increased preponderance of one over the other (2 degrees more, 3 degrees more, etc.), where the ratio between the two intensities varies in an arithmetic progression (2, 4, 6, …); or it can vary in a geometric progression (the ratio between heat and cold is half, one quarter, one eighth, …).

Similarly, it can move away from the zero degree by the addition of degrees, or by the arrival of a new form (i.e., by succession). Here we find the standard division between the theory of addition (which was dominant among the Franciscans and then among the Oxford Calculators) and the theory of succession.

On the basis of these distinctions, Blasius addresses the initial question, first by supposing that 'it is only the subject that is intensified, not the quality'; then he adopts the opposite position, namely, that it is the quality itself that is intensified or weakened. But before laying out this twofold discussion, it should be noted that the same pair of alternatives is found in the *Questio de intensione et remissione formarum*. In article 2 ('De quesito'), there are also some preliminary *evidentie*. The first of these is that 'sometimes a quality is taken as an accidental form inhering in a subject, and sometimes it is taken as the qualified subject';[13] the second asserts that 'it is not the same to ask whether

12 Blasius of Parma, *Questiones super tractatus logice magistri Petri Hispani*, eds. J. Biard and G. Federici Vescovini, Paris 2001 (Textes philosophiques du Moyen Âge, 20), III, q. 16 (*Utrum termini de predicamento qualitatis sunt de se invicem predicabiles vel de suis inferioribus cum hiis adverbis 'magis' et 'minus'*), 322–325, at 325.

13 Blasius of Parma, *De intensione et remissione formarum*, 489: 'Aliquando capitur qualitas pro tali forma accidentali inherente subiecto et aliquando capitur pro corpore qualificato.'

a quality inhering in a subject is intensifiable and whether the subject of that form can be intensified.'[14] After this there is a short series of conclusions 'taking the quality as the qualified subject,'[15] and then some long discussions about 'the intensification of a quality as a form inhering in the subject.'[16]

By thus presenting the relationship to the subject as a determining element in any theorizing about the intension and remission of forms, Blasius to some extent simply continues a long tradition. Indeed, since Thomas Aquinas, one of the theses discussed had been that the amplitude or latitude (*latitudo*) of a form has to do not so much with the form itself (whiteness, charity) considered in its essence, but more with the subject that receives it and participates in it to some degree that is closer to or farther from the highest degree.[17] Of course, in fourteenth-century debates, the theory of participation no longer had much currency. Moreover, Blasius challenges the distinction between the degrees of a quality and the quality's being or essence.[18] But among the Oxford Calculators, the idea that qualities are individuated by their subject persisted,[19] even if the dominant tendency was to gravitate towards abstract considerations. However, in taking up the question about the relationship between qualities and their subject, Blasius sets it within his own conception of matter. Underneath the twofold approach that structures this examination, there is a more fundamental investigation into the mode of being of qualities, and into their very existence.

Thus, in the *Questio de intensione et remissione formarum*, Blasius begins by arguing about a presupposition: 'It is not evident that a quality exists or will exist.'[20] This preliminary argument becomes significant with the 'first difficulty'

---

14 Blasius of Parma, *De intensione et remissione formarum*, 490: 'Non est idem querere an qualitas aliqua inherens subiecto sit intensibilis, an subiectum illius forme possit intendi.'

15 Blasius of Parma, *De intensione et remissione formarum*, 491–492. Note that this discussion is quite brief: in the *Questio*, it is the opposite position that Blasius is interested in.

16 Blasius of Parma, *De intensione et remissione formarum*, 493–529.

17 See Thomas Aquinas, *In octo libros Physicorum Aristotelis expositio*, VI, lect. 7, n. 8, ed. P.M. Maggiòlo, Turin 1954, 825: 'Quod enim una et eadem pars sit magis vel minus alba, non est ex ipsa ratione albedinis (quia si esset separata, non diceretur secundum magis et minus; sicut neque substantia suscipit magis et minus): sed est ex diverso modo participandi albedinem ex parte subiecti divisibilis.' See also *Scriptum super libros Sententiarum*, I, d. 17, q. 2, a. 1, ed. P. Mandonnet, Paris 1929, 416.

18 In the course of the discussion, Blasius rejects one opinion because it would imply such a distinction: 'Hec opinio est falsa secundum partem, quia ponit gradus qualitatis distingui ab essentia ipsius qualitatis' (Blasius of Parma, *De intensione et remissione formarum*, 495).

19 This is the case with Roger Swineshead: see Sylla, 'Medieval Concepts,' 232 and 241.

20 Blasius of Parma, *De intensione et remissione formarum*, 470: 'Non est evidens aliquam qualitatem esse vel futuram esse.'

presented in the first article: 'Can one conclude from an evident cognition that some quality exists?' The conclusions of this discussion begin with the denial that there are qualities that inform an indivisible subject,[21] and then the denial that there are elementary qualities.[22] A doubt connected with this difficulty, however, leads to an additional clarification: 'Supposing by imagination that there were one or more qualities, what would their subject be?' The last conclusion pertaining to this doubt is then that 'if there were a quality, it would be subjectively in prime matter';[23] this refers to an entire debate carried out in other questions in the *Physics* commentary.[24] The underlying ontological problem is therefore perhaps not whether there are qualities at all, but whether there are qualities informing a subject that is a natural body (water, a tree, a man, etc.), or simply qualitative dispositions of prime matter.

Another passage in this question allows us to evaluate the plausibility of this thesis concerning the existence or non-existence of qualities. On the one hand, Blasius admits at the end of the question that one can concede the arguments that call into doubt the being of qualities;[25] on the other hand, however, one cannot leave it there, if only because the determination presupposes qualities:

> With respect to the second article, so as not to seem to be avoiding difficulties in view of the determination of the question, I now concede that there are many qualities, both spiritual and material, that are educed from the potency of matter. I therefore do not intend to remain in the difficulty which was the principle object of investigation in the previous article, in which difficulty it was concluded that no natural argument shows that any accidents must be posited.[26]

21 See Blasius of Parma, *De intensione et remissione formarum*, 481.

22 See Blasius of Parma, *De intensione et remissione formarum*, 484: 'Non videtur aliqua qualitas elementaris esse ponenda.' In this connection, Blasius writes: 'Plures rationes et magis efficaces sunt pro conclusione quam contra conclusionem.'

23 Blasius of Parma, *De intensione et remissione formarum*, 485.

24 See J. Biard, 'Matière, forme, qualités: Blaise de Parme et le statut de la matière,' in T. Suarez-Nani and A. Paravicini Bagliani (eds.), *Materia: nouvelles perspectives de recherche dans la pensée et la culture médiévales (XIIe–XVIe siècles)*, Florence 2017 (Micrologus Library, 83), 191–215.

25 Blasius of Parma, *De intensione et remissione formarum*, 530: 'Dico quod ad istas rationes que sunt de supposito non esset respondendum, quia probant conclusiones probabiliter positas in prima difficultate primi articuli.' The first of these conclusions was the following (481): 'Non est dare aliquam qualitatem informantem aliquod subiectum indivisible'; this conclusion was then extended to visual images and 'elementary qualities.'

26 Blasius of Parma, *De intensione et remissione formarum*, 489: 'Quantum ad secundum articulum, ne videar fugere difficultates gratia determinationis materie presentis, concedo

Blasius thus decides to disregard the doubts about the precise status of qualities for a very simple reason: the objective of the question is to treat of the intensification and weakening of accidental forms, whatever their mode of being—an objective set not only by Aristotle's text but also by experience.[27]

The approach Blasius adopts in his *Physics* commentary once more bears significant similarities to that in the *Questio de intensione et remissione formarum*, since the reasoning is also laid out against the background of a recurrent debate about the being of qualities. This time, however, it is clearly presupposed that 'contrary qualities exist.' Admittedly, this presupposition will not always be firmly maintained in the arguments, especially since certain qualities, such as light, do not have a contrary. However, this presupposition alone makes sense of the question of whether a quality can be intensified or weakened.

2 How Is the Intensification of a Quality to Be Evaluated?

As I have already indicated, Blasius' theory is laid out against the background of two possible alternatives.

2.1 *Quality as the Qualified Subject*

Suppose that it is the subject itself that is qualified more or less intensely in some respect: this is the first approach that Blasius adopts in his *Physics* commentary.[28] He considers the question from this perspective according to four different grounds of argument (*fundamenta*).

First, a thing can be said to have a given quality if more than half of the thing is of that quality and there is no contrary quality in the remaining part of the thing. For example, suppose that more than half of some thing is hot to degree 4, while the rest of it is not cold to the maximum degree (*summe frigidum*). It is thus the material parts of the subject body that should be taken into account. On this basis, one can state a first set of paradoxical conclusions—suggesting that one should not stop there. For example, a body can become hotter without

---

pro nunc multas qualitates esse, tam spirituales quam materiales, eductas de potentia materie. Non intendo igitur stare in difficultate que principaliter in remisso articulo querebatur, in qua difficultate conclusum fuit quod nulla ratio physica concludit aliqua accidentia esse ponenda.'

27 See Blasius of Parma, *Questiones Physicorum*, Ms. Vat. lat. 2159, 157^{ra-rb}, where Blasius appeals to experience with the example of hot iron, which becomes cold when plunged in cold water.

28 This is the third moment (*tertius processus*), 'De quesito,' in article 2.

acquiring heat, because of variations in the other part, transfer from one part to the other, or condensation.

Blasius then adopts a different perspective with a second ground of argument (*aliud fundamentum*). This time, something is said to be hotter than it was before because of a closer ratio between the heat and the cold. He calls this conclusion *multum probabile*, but then there follow once more some paradoxical conclusions, according to which something becomes hotter—or even infinitely hot—without acquiring heat, but by losing cold.

On a third basis of evaluation, which takes mixed bodies into consideration, it is not the ratio as such that is taken as a criterion, but by how much a quality exceeds its opposite. The consequence (the second conclusion) is that intensification can be thought of in terms of an arithmetic proportion or progression (and not a geometric progression).[29] For example, a body *b* will be said to be twice as hot as *a* if the heat in *a* exceeds its cold by 1 while the heat in *b* exceeds its cold by 2.[30] This way of measuring gives a result different from the result given by comparing the ratios: if *a* has 4 degrees of heat and 1 degree of cold, and *b* has 10 degrees of heat and 5 degrees of cold, *a* is not hotter than *b* because the excess in *b* is 5 whereas in *a* it is 3; and yet the ratio of heat to cold is higher in *a* than in *b*.

In a fourth way of evaluating, Blasius considers the relation of the quality to its maximum degree and to the zero degree. On the one hand, every quality has a certain latitude, which terminates at the zero degree since it can be weakened to zero; thus, it is not measured by its distance from the zero degree. On the other hand, it is not measured simply by its distance from its maximum degree, since whatever quality it is, it can make the qualified subject be qualified to the maximum degree. What does this mean? As the conclusion suggests, it depends on the amount of matter:

> Aliquid fieri intensius et intensius est illud magis approprinquari summo per qualitatem in tanta materia.[31]

---

29 See Blasius of Parma, *Questiones Physicorum*, Ms. Vat. lat. 2159, 158rb: 'Secunda conclusio: quod intensio mixti sit attendenda penes proportionem arismetricam et non geometricam est merito probabile.' Here Blasius is using concepts that are standard in the theory of proportions: four quantities, A, B, C, and D, are in arithmetic proportion if $A - B = C - D$ (with $A > B$); they are in geometric proportion if the ratios between A and B and between C and D are equal (e.g. $[A : B] :: [C : D]$).

30 See Blasius of Parma, *Questiones Physicorum*, Ms. Vat. lat. 2159, 158rb: 'Dum *a* est calidius in duplo quam *b*, hoc ideo est quia latitudo caliditatis in *a* supra frigiditatem est dupla ad latitudinem caliditatis in *b* supra frigiditatem suam.'

31 Blasius of Parma, *Questiones Physicorum*, Ms. Vat. lat. 2159, 158va.

The variation is measured by the greater or lesser proximity of the quality to its maximum degree in a certain amount of matter.

The *Questio de intensione et remissione formarum* also begins by considering —though rather briefly as compared with the other hypothesis—quality understood as the qualified subject. However, there are differences in the approach here. After conceding, as he does in his *Physics* commentary, the possibility of qualities thus understood being intensified or weakened, Blasius first states some conclusions that follow if one accepts that the ratio of intensities follows the *proportio geometrica* of a given quality to its opposite, 'as many would have it' (*ut multi volunt*).[32] I will restrict myself here to the first conclusion, which is as follows: 'Something can become hotter and hotter while continuously losing heat and not acquiring any.'[33] For this to occur it is sufficient that the thing's heat diminish less quickly than its cold: what matters is that the ratio of heat to cold is continuously increasing.[34] The same would hold for spiritual qualities, if it were accepted that they can have contraries (e.g., knowledge and ignorance). Blasius distances himself from this argument, even if he does not really come to a settled view:

> You should know that not every opinion of the philosophers says that intension in a mixed body follows the geometric ratio of a quality to its opposite. And whatever is the case here, you can hold whichever opinion you wish.[35]

To conclude this passage, Blasius notes that, in any case, this is not the main object of investigation, which is rather quality considered as an inherent form.

---

32 This set of conclusions also holds for opposite qualities that are present in the same subject in all its parts. It is different if one compares, as is often the case in the *Physics* commentary, the same qualities according to whether they are in different parts of a subject.

33 Blasius of Parma, *De intensione et remissione formarum*, 491: 'Stat aliquod per horam fieri calidius et calidius continue caliditatem deperdendo et nullam acquirendo.'

34 Blasius of Parma, *De intensione et remissione formarum*, 491: 'Continue per horam maioratur proportio caliditatis supra frigiditatem.' The text needs to be corrected, however: in the course of the explanation given, *frigiditas* and *caliditas* should be reversed, since in order to obtain the desired result it is heat that should be divided by 2, and cold that should be divided by 4.

35 Blasius of Parma, *De intensione et remissione formarum*, 492: 'Verumtamen sciatis quod non omnis opinio philosophorum dicit intensionem in mixto sequi proportionem geometricam unius qualitatis supra aliam oppositam. Et quicquid sit de hoc, teneatis opinionem quam vultis.'

2.2 *Quality as Inherent Form*

Secondly then, it is asked how the intensification of a quality understood as an accidental form is to be evaluated. Question 10 on Book v of the *Physics* formulates on this issue a series of theses (*conclusiones*) that can be divided into several groups. The first two theses posit that every quality that can be intensified can in principle also be weakened, and that every quality can be intensified. One can give natural examples, both of qualities in extended bodies and of dispositions of the soul. The third conclusion posits that a quality cannot be increased by another pre-existent quality being added to it. This thesis is understood first in relation to certain procedures previously mentioned, namely, the transfer of one qualified part of a subject to another part. Of course, now the result is considered to be produced in the quality itself and not in the subject; however, 'no quality can be separated naturally from its subject.'[36] This was also posited at the beginning of article 2 as a preliminary proposition: 'A quality cannot by nature be, or be going to be, or have been, without a subject.'[37] From an Aristotelian perspective, this assertion might seem banal, but the consequences drawn from it are significant: for even when one considers the quality itself as an accidental form, it is not possible not to take its subject into account.

The next conclusions are more specifically concerned with how to evaluate variations in qualities. They begin with two negative theses. First, the intensification of a quality does not result from a proportion or a 'geometric ratio of the quality to its opposite.'[38] Here one still has to understand that a certain intensity, expressed by a ratio between opposite qualities such as heat and cold, does not increase by a geometric proportionality, that is, by multiplication, such that the cold would be divided by going from half to a quarter, from a quarter to an eighth, and so on. Indeed, if the ratio of hot to cold were multiplied by itself, there would be no end to the progression: one would never obtain the maximal

36 Blasius of Parma, *Questiones Physicorum*, v, q. 10, Ms. Vat. lat. 2159, 158vb: 'Tertia conclusio: intensio qualitatis non est per naturam possibilis infundendo qualitatem ad aliam preexistentem. Patet conclusio, quia nulla qualitas potest per naturam a subiecto separari, quare etc. ... Non enim contingit per naturam quod aliqua qualitas informans medietatem pedalem ponatur in quarta parte illius pedalis nisi poneretur simul cum materia, sicut tu facis addendo aquam calidam cum alia aqua calida.'

37 Blasius of Parma, *Questiones Physicorum*, v, q. 10, Ms. Vat. lat. 2159, 156rb: 'Non contingit aliquam qualitem per naturam esse absque subiecto ⟨vel⟩ fore vel fuisse.' This assertion holds only *per naturam*, for a quality could exist separately by divine omnipotence, as in the case of the Eucharist.

38 Blasius of Parma, *Questiones Physicorum*, v, q. 10, Ms. Vat. lat. 2159, 158vb.

heat,[39] given that the division of the cold could carry on *ad infinitum*. Likewise, if one removed from water something that makes it warm, the water should then cool down endlessly; but this is contrary to experience.

Secondly, it is also not the case that the intensification of a quality results from a proportion or 'arithmetic ratio to the opposite quality that is coextensive with it.'[40] It can thus be seen that Blasius' comment in no way consists in privileging arithmetic proportionality over geometric proportionality. In other words, heat does not increase according to an arithmetic progression, such as 3 : 8, 4 : 8, 5 : 8, etc. Why not? There are counterexamples, such as the increase of light at sunrise, since in this case there is no contrary quality.

The theory of proportions thus does not seem at this stage to be an adequate tool for evaluating the intensification of a quality. Its inadequacy also explains the next thesis, which rejects the theory of mixture: 'That intension is produced by a lesser admixture of the contrary is not to be asserted without qualification.'[41] The reason given is that this would be the result of either an arithmetic progression or a geometric progression of the ratio between opposite qualities, which has just been disproven. This rejection is consistent with what was said in the second *processus* of article 2, in which Blasius asked: 'Supposing that heat and cold exist, I ask whether they differ from each other or not.'[42] The general thrust of the argument is to 'persuade' that heat does not differ specifically from cold, since cold is simply heat that is weakened.[43]

According to the next thesis, intensification seems not to be produced by moving away from (*per recessum*) the zero degree, nor by moving towards (*accessum*) the maximum. Paradoxically, the argument ultimately leads to a

---

39 This maximal heat (*summa gradus* or *summa caliditas*) is obviously not an *infinite* heat, but the greatest heat possible for a given body. This usage of the term *summus gradus* is standard; it became settled in the thirteenth century with a reference (taken out of context and with its meaning changed) to the *Liber sex principiorum*, which Blasius himself mentions in the argument in the question in his *Physics* commentary about the simplicity of form (V, q. 10, Ms. Vat. lat. 2159, 155vb).

40 Blasius of Parma, *Questiones Physicorum*, V, q. 10, Ms. Vat. lat. 2159, 159ra: 'Quinta conclusio: quod intensio qualitatis insequatur proportionem arismetricam ipsius ad qualitatem oppositam secum coextensam, ratio non convincit.'

41 Blasius of Parma, *Questiones Physicorum*, V, q. 10, Ms. Vat. lat. 2159, 159ra: 'Quod intensio fiat per minorem admixtionem contrarii non est simpliciter asserendum.'

42 Blasius of Parma, *Questiones Physicorum*, V, q. 10, Ms. Vat. lat. 2159, 156vb: 'Quero, supposito quod caliditas et frigiditas sint, utrum invicem differant an non.'

43 Blasius of Parma, *Questiones Physicorum*, V, q. 10, Ms. Vat. lat. 2159, 156vb: 'Ex hiis sequitur quod nulla est frigiditas, aut si aliqua sit, illa non est nisi remissa caliditas.' Recall, however, that in the case of the latitude of substantial forms, Blasius does seem, under the influence of medical texts, to envisage on the basis of *complexio* a certain ratio of the basic qualities of hot and cold, wet and dry, and heavy and light.

conclusion similar to what had previously been asserted from the perspective of treating quality as the qualified subject:

> Eighth and final conclusion: the intensification of quality is to be evaluated according to progression towards the maximum in some matter or other of a certain density.[44]

Indeed, the same heat that makes a one-foot body as hot as possible would not heat all the earth in the same way. Thus, considering quality as an accidental form and not as the qualified subject does not preclude (quite the contrary!) taking into account in the procedures of 'calculation' an indissoluble link to the subject and to matter.

This is why the same combination will be found in question 6 on book VII of the *Physics*, which is devoted to the evaluation of the speed of alteration. If this is considered from the perspective of effects rather than of causes, the evaluation will be centred not on the ratios between qualities but on the latitude. Now, one of the theses is that the velocity should be evaluated according to the form that has a certain intension in a certain amount of matter.[45] On the basis of his conclusions, Blasius emphasizes that we have to consider the quality that is in some matter, not to evaluate it purely and simply 'according to the latitude of the form in relation to extension'—for something that does not have extension (such as an intelligible) can also be altered—nor purely and simply according to latitude without extension, since extension *does* have a role in the case of natural qualities.[46] But how should we understand the evaluation of the intensity of an accidental form in a certain mass of matter without also considering extension?[47] Little help is provided by the examples given of non-extended subjects (the intellective soul and mathematical points). However,

44 Blasius of Parma, *Questiones Physicorum*, V, q. 10, Ms. Vat. lat. 2159, 159[rb]: 'Octava conclusio et ultima: intensio qualitatis attendenda est penes accesum ad summum in hac vel illa materia tante dempsitatis vel tante.'

45 Blasius of Parma, *Questiones Physicorum*, VII, q. 6, concl. 4, ed. Rommevaux-Tani (see note 3 above), Ms. Vat. lat. 2159, 196[rb]: 'Velocitas in alteratione est attendenda penes formam tante intensionis in tanta materia, aliis paribus ut requirat retentis.' Time in itself is not a criterion either.

46 Blasius of Parma, *Questiones Physicorum*, VII, q. 6, concl. 1, Ms. Vat. lat. 2159, 196[ra]: 'Velocitas in alteratione non est simpliciter attendenda penes latitudinem forme in ordine ad extensionem subiecti'; concl. 2, 196ra: 'Velocitas in alteratione non est simpliciter attendenda penes latitudinem forme sine extensione subiecti.'

47 See S. Rommevaux, 'L'ajout de conditions dans les règles du mouvement formulées par Blaise de Parme dans la seconde rédaction de ses *Questions sur la Physique*,' in J. Biard and

there are other examples which provide some pointers. Blasius mentions a one-foot piece of earth that has heat to the maximum degree—which, it should be recalled, is finite—and this piece of earth is condensed to an arbitrary extent, even to the point of being *non-quantum*. In such a situation, there is still the same amount of matter, and the same qualitative form can subsist to the same degree without being more intense or less intense, however much the body is condensed (or rarefied), since it is stipulated that the amount of matter remains the same. Therefore, in evaluating the velocity of an alteration, it is not the matter's extension that should be taken into account, but its quantity.

The *Questio de intensione et remissione formarum* differs in several respects from the text of the *Physics* commentary. First of all, Blasius presents two possible responses (one positive and one negative) to the question whether a quality, understood as an accidental form, is intensified or not. The argument in favour of the *pars negativa* considers the quality not as having a certain latitude that could vary, but calls 'quality' a determinate degree of whiteness or heat. Thus, if the quality is identified with a given degree, it cannot be intensified or diminished.[48]

However, Blasius has warned us that the opposite theory is 'more probable,' and so it is to the latter that he devotes more discussion.[49] First (conclusion 5), every quality can be intensified.[50] This is true both for material qualities such as heat and for spiritual qualities such as knowledge. The gradual character of qualities is also emphasized. This means that one can speak of parts of a quality, not only with regard to the extension of the subject (which is assumed, but is not pertinent here), but also with regard to the quality alone, which can have more or less intensity in one given point or part of the subject.[51]

This position has various consequences. First among these is that a form is intensified by the acquisition of a new degree, or weakened by the loss of a

---

A. Robert (eds.), *La philosophie de Blaise de Parme. Physique, psychologie, éthique*, Florence 2019 (Micrologus Library, 96), 153–174.

48 This tendency to identify a particular quality with a determinate degree was dominant in Oxford, and is also found in Oresme. Intensification is thus the appearance of a new quality, either by addition or by succession.

49 Blasius begins by laying out an opinion that distinguishes the degrees of a quality from its essence. I leave this discussion aside and reorder the conclusions that follow.

50 Blasius of Parma, *De intensione et remissione formarum*, 497.

51 Blasius of Parma, *De intensione et remissione formarum*, 496: 'Quelibet qualitas est divisibilis in partes distantes se habentes ad invicem, ut vellim dicere quod ultra partes extensas quas habet qualitas ratione subiecti, habet etiam partes cedentes totam caliditatem vel qualitatem tantam.' It is very important to be clear about when Blasius is speaking of 'parts distinct from one another' with regard to the intension itself and not with regard to the extension of the subject.

degree.[52] Second, this acquisition does not happen instantaneously, but successively.[53] Third, the motion by which a quality is acquired successively is not distinct from the quality itself.[54] Blasius draws a parallel with local motion, which according to him is not really distinct from the mobile. He thus rejects the idea of a flux that is distinct from the form itself.[55] Finally, Blasius seems here to explicitly reject the theory of succession of forms, of which Walter Burley was at the time considered representative, preferring the theory of addition, which was first argued for by Godfrey of Fontaines and John Duns Scotus, and became dominant in the fourteenth century.[56]

This set of theses can perhaps provide a basis for a quantitative measure of qualitative forms. The acquisition of degrees is also compared to the extension of a line, or to the extra space passed through in a motion.[57] In the determination itself, however, it is physical considerations that predominate. Procedures for quantifying appear instead in the discussion of new difficulties—which, Blasius points out, are no less useful than the question itself![58]

3 Difficulties in the Calculation of Qualitative Forms

On this issue I will focus on the *Questio de intensione et remissione formarum*.[59] There are five difficulties enumerated. The first concerns the correlative vari-

---

52 See Blasius of Parma, *De intensione et remissione formarum*, 498: 'Alia conclusio: non contingit aliquam formam intendi sine acquisitione novi gradus nec remitti sine deperditione preexistentis gradus.' This is the seventh conclusion.

53 See Blasius of Parma, *De intensione et remissione formarum*, 496 (third conclusion).

54 See Blasius of Parma, *De intensione et remissione formarum*, 496: 'Alia conclusio: motus et successio quo qualitas acquiritur successive non est aliud quam qualitas successive acquisibilis' (fourth conclusion).

55 In fact, it was standard to reject the idea of a flux that is distinct from the qualitative form in the case of alteration. John Buridan for example defended this position, but he posited a difference in the case of local motion. Blasius here defends a position more similar to that of Ockham and Albert of Saxony, and on this point treats local motion and alteration in the same way.

56 See Blasius of Parma, *De intensione et remissione formarum*, 498: 'Alia conclusio: non contingit aliquam qualitatem intendi per corruptionem presentium graduum et positionem sequentium' (eighth conclusion); and 499: 'Oportet intensionem forme fieri per additionem (*corr.* intensionem) graduum, gradibus prioribus remanentibus.'

57 Explanation of the seventh conclusion.

58 Blasius of Parma, *De intensione et remissione formarum*, 199, beginning of article 3: 'Quero difficultates sequentes determinationem questionis et non minoris utilitatis quam sit presens questio.'

59 In the question on the *Physics*, which is constructed differently, a good number of possible

ation of contrary qualities;[60] the second concerns the way of evaluating the speed of qualitative change, and raises the question of the relation between the amplitude of the quality itself and the extension of its subject; the third addresses once more the velocity of alteration and asks whether it follows a geometric progression; the fourth returns to the relation to the maximum quality and to the zero degree; and the fifth raises the question of the first instant in the acquisition of a gradual form. In what follows, the difficulties are not addressed in exactly this order, and some formulations are a little different.

It is not possible here to examine all these discussions, in which most of the problems raised by the tradition of the calculation of qualitative forms are mentioned. I will focus on the discussions which examine how to evaluate the speed of qualitative change, from the perspective of the effect and from the perspective of the cause.[61] Nevertheless, I should first point out a paradox that appears from the very beginning of the discussion of the first difficulty. Though it was posited that cold is nothing more than heat that is weakened, it is now presupposed that heat and cold are opposite qualities. Blasius is conscious of this slide. He explains:

> I presuppose that heat and cold are opposite qualities, and even though in the last of the previous difficulties I determined the contrary, I would like for now to have said that more in jest than as a determination.[62]

How is this to be explained? Perhaps it is because Blasius is now envisaging procedures of calculation, that is, conceptual instruments for operations and not for physical realities. My hypothesis is that from this point on, operational efficiency takes precedence over the ontological theses previously examined. These degrees will then be conceptual tools in the service of operational procedures.

Let us therefore turn to these procedures of calculation. The third difficulty that is discussed is as follows: 'Penes quid sit attendenda velocitas forme sicut penes effectum.' To my knowledge, the distinction between considering from

---

difficulties and solutions are presented in the responses to the initial arguments *quod non* (that is, against the possibility of the intensification and remission of forms); there are no fewer than thirteen arguments and responses.

60 Blasius of Parma, *De intensione et remissione formarum*, 499: 'Utrum intensio qualitatis fiat per remissionem qualitatis opposite et remissio per intensionem contrarie qualitatis.'

61 These discussions should be compared with what is presented in *Physics*, VII, q. 6.

62 Blasius of Parma, *De intensione et remissione formarum*, 500: 'Licet in ultima difficultate de precedentibus determinaverim oppositum, volo pro nunc istud magis solatiose dixisse quam per modum determinationis.'

the perspective of the effect and from the perspective of the cause was elaborated with regard to local motion. For local motion, analysis *quo ad effectum* consists in measuring the space passed through by two moving things in a given period of time. However, the analysis of displacement cannot be transferred wholesale to alteration. Concerning the question what would make it possible to evaluate or measure the speed of alteration, Blasius lists the following theses (I will quickly pass over the arguments for most of the negative theses):

1. 'Velocity of alteration should not be evaluated according to the time over which the alterable is altered, such that it would be said to be altered more rapidly than another thing that is altered over a greater time.'[63] It does not suffice that something is altered over a longer time for it to be altered more, since this could depend on other factors, such as the nature of that which receives the alteration.
2. 'Velocity in alteration should not be evaluated according to the subject, such that it would be said to be altered more rapidly than another thing that is altered with respect to a greater extension (*extensionem ipsius*).'[64] The intellective soul, which is not extended, can be altered; likewise, a small quantity of water can be heated more quickly than a large quantity.
3. 'Velocity should not be evaluated according to the subject in respect of time (*in ordine ad tempus*), such that it would be said to be altered more rapidly than another thing which is altered with respect to a greater extension over the same period of time.'[65]
4. 'Velocity should not be evaluated according to a form that is introduced or will be introduced in a determinate period of time (*in tanto tempore*), such that it would be altered more rapidly than something else which, in the same period of time or an equal period of time, acquires a more intense degree.'[66] This thesis is discussed at some length.

---

63 Blasius of Parma, *De intensione et remissione formarum*, 507: 'Velocitas alterationis non est attendenda penes tempus per quod alterabile alteratur, ut illud dicatur alio velocius alterari quod per maius tempus illo alteratur.'

64 Blasius of Parma, *De intensione et remissione formarum*, 507: 'Velocitas in alteratione non est attendenda penes subiectum, ut illud dicatur alio velocius alterari quod per maiorem ipsius extensionem alteratur.'

65 Blasius of Parma, *De intensione et remissione formarum*, 507–508: 'Velocitas etc. non est attendenda penes subiectum in ordine ad tempus, ut illud dicatur alio velocius alterari quod in eodem tempore per maiorem extensionem alteratur.'

66 Blasius of Parma, *De intensione et remissione formarum*, 508: 'Velocitas in alteratione non est attendenda penes formam introductam vel introducendam in tanto tempore, ut illud velocius alteretur alio quod in eodem tempore vel equali acquirat alio intensiorem gradum.'

5. 'Velocity etc. is not to be evaluated according to a form that will be introduced in a determinate time in respect of an extensible subject.'[67]
6. 'Velocity etc. is not evaluated according to the ratio of the acquired quality to the pre-existing quality.'[68] If one doubles two qualities that are different at the start, the one will grow (for example) by 4 and the other by 5.
7. 'Velocity in movement etc. should not be evaluated according to the ratio between (1) the aggregate of the acquired quality and the quality to be acquired, and (2) the acquired quality.'[69]
8. 'Velocity in alteration is not to be evaluated according to the latitude acquired absolutely in a determinate time.'[70] The argument against this appeals to the different parts of that which is altered. From this results the final conclusion.
9. *The affirmative conclusion, which is the answer to the question*: 'Velocity in alteration must be evaluated according to the latitude acquired in a determinate time, in relation to the parts of the subject or subjects, which are proportionate to their wholes.'[71] Thus, the amplitude (or latitude) of the quality acquired in a given time by two equal bodies, but taking into account what happens in the half, in the third, and in each of its parts.

The exposition of velocity *quo ad effectum* ends there. Two points can be emphasized. First, from this point on there is an appeal to the concept of *latitudo*, which was little mentioned before this point. Second, calculation should apply both to the alteration of the quality itself and to the alteration of the subject, even if it has been specified that alteration as such abstracts from extension: 'Whether alteration can occur without extension or cannot, the term "alteration" does not connote extension, and consequently you can understand alteration without extension.'[72]

67 Blasius of Parma, *De intensione et remissione formarum*, 508: 'Velocitas etc. non est attendenda penes formam in tanto tempore inducendam per subiectum extensibile.'

68 Blasius of Parma, *De intensione et remissione formarum*, 509: 'Velocitas etc. non attenditur penes proportionem qualitatis acquisite ad prehexistentem.'

69 Blasius of Parma, *De intensione et remissione formarum*, 509: 'Velocitas in motu etc. non est attendenda penes proportionem aggregati ex qualitate acquisita et acquirenda ad qualitatem acquisitam.'

70 Blasius of Parma, *De intensione et remissione formarum*, 509: 'Velocitas in alteratione non est attendenda penes latitudinem in tanto tempore absolute acquisitam.'

71 Blasius of Parma, *De intensione et remissione formarum*, 509: 'Velocitas in alteratione est attendenda penes latitudinem in tanto tempore acquisitam in ordine ad partes subiecti vel subiectorum proportionales suis totis.'

72 Blasius of Parma, *De intensione et remissione formarum*, 509: 'Sive alteratio possit fieri sine extensione sive non, tamen iste terminus "alteratio" non connotat extensionem, et per consequens tu potes intelligere alterationem sine extensione.'

It is with regard to the following difficulty that Blasius makes some clarifications about the term *latitudo*. This is the fourth of the difficulties discussed, though it is the third of those listed: 'Does velocity in alteration follow the geometric proportion as its cause'?[73] It seems thus to concern the evaluation of the velocity of alteration *quo ad causam*, which is why there is the appeal to the theory of proportions, which had been standard since Bradwardine. But the investigation gets displaced to the question of how velocity can be measured both according to latitude (taking into account the qualifications that have been introduced) and according to proportion.

In fact, Blasius is satisfied here with recalling a few definitions. The main text on *latitudo* is his *Questiones super tractatum de latitudinibus formarum*, which is probably commenting on the treatise by Jacobus de Sancto Martino.[74] In this work, Blasius begins by recalling the metaphorical character of the term 'latitude' as it is used here, where it does not mean width, in the sense of one of the dimensions of a body.[75] It is not the goal of this article to present Blasius' rather complex theory of *latitudo*, but let us recall two points. First, Blasius continues to use *latitudo* in the older sense, which treats it as the interval of variation between two degrees of intensity, and not as the measure of a particular quality; in this he differs from John Dumbleton, from most of the Oxford Calculators, and from Nicole Oresme.[76] The conception of latitude as an interval comes from the medical tradition via translations of Galen and Avicenna, and it is found in the treatise that Blasius is commenting on.[77] Secondly, and despite

---

73 Blasius of Parma, *De intensione et remissione formarum*, 499: 'Utrum velocitas in alteratione insequatur proportionem geometricam ut suam causam.'

74 See Di Liscia, 'Biagio Pelacani da Parma's Geometrisation' and Federici Vescovini's Introduction to her edition of the *Quaestio de intensione et remissione formarum.*

75 Blasius of Parma, *Questiones super tractatum de latitudinibus formarum*, Padua 1486, 12^v: 'Noto quod reperiuntur isti termini "longitudo", "latitudo", et "profunditas" quibus si proprie utamur sunt mensura quibus quodlibet corpus mensurari potest secundum omnem eius diametrum. Improprie enim hi termini [*corr.* tibi *ed.*] applicantur qualitatibus gradualibus, cum dicimus "latitudo summe caliditatis est ut *d*."' Here the term *diametrum* clearly means *dimensio*. In the medical tradition, and in the philosophical tradition before Oresme, the term *latitudo* meant simply the amplitude of the variation of a quality (or of a ratio of qualities). It is Oresme who reintroduced the geometrical connotation of the term by joining *longitudo* and *latitudo*. See Nicole Oresme, *Tractatus de configurationibus qualitatum et motuum*, ed. M. Clagett, *Nicole Oresme and the Medieval Geometry of Qualities and Motions. A Treatise on the Uniformity and Difformity of Intensities Known as* Tractatus de configurationibus qualitatum et motuum, Madison, WI, 1968 (University of Wisconsin Publications in Medieval Science, 12), 168–173.

76 See Sylla, 'Medieval Concepts.'

77 On Blasius' place in an entire tradition that existed in Italy, see G. Federici Vescovini,

this, one can detect echoes of certain works of Oresme (probably known indirectly via Giovanni da Casale or Simone del Castello), in Blasius' discussion of the rule of the mean degree and the adoption (though with some modifications) of geometrical figures.[78]

In the *Questiones super tractatum de latitudinibus formarum*, Blasius notes that, in an improper usage of the term for gradual qualities, one can trace two lines, one referring to intension and the other to extension. But from that starting point, he then makes an original usage of the term *latitudo*. On the one hand, while the mathematical metaphor might incline one to make latitude one of the lines, the other being called 'length' (*longitudo*), Blasius thinks of latitude as encompassing both dimensions: 'In latitude, there are principally two lines, one of which for us designates the intension, the other the extension in the subject.'[79] Blasius thus returns to the conception of certain masters (who are criticized by Oresme) who claim that latitude designates the whole figure or surface.[80] Thus, since 'the line of extension … represents for us the form that has parts external to one another insofar as it belongs to the subject,'[81] latitude is always in a certain way related to the subject, even when the quality that is intensified or weakened is taken for the form itself, and not for the subject.

The formulations in the *Questio de intensione et remissione formarum* are slightly different. Blasius emphasizes that as it is used here, the term *latitudo* in itself 'does not connote an extension.' This is shown by the fact that it is applied to certain qualities in the soul, which in itself does not have any extension:

'Simone di Castello e la medicina dei "moderni",' in Ead., *Arti e filosofia nel secolo XIV: Studi sulla tradizione aristotelica e i 'moderni'*, Florence 1983, 213–229.

78 See Di Liscia, 'Biagio Pelacani da Parma's Geometrisation.' See also Blasius of Parma, *Super tractatum de latitudinibus formarum*, q. 3, 17ʳ: 'Latitudines formarum per figuras geometricas representantur.'

79 Blasius of Parma, *Super tractatum de latitudinibus formarum*, 12ᵛ: 'Ita in latitudine eius sunt principaliter due linee quarum una designat nobis intensionem, reliqua eius extensionem in subiecto.'

80 See Nicole Oresme, *Tractatus de configurationibus qualitatum et motuum*, I, 3, 172: 'Quidam moderni non bene vocant latitudinem qualitatis ipsam totam, sicut abusio esset per latitudinem superficiei intelligere totam superficiem vel figuram.'

81 Blasius of Parma, *Super tractatum de latitudinibus formarum*, 12ᵛ–13ʳ (in the case of a rectangle of which *cd* is the base): 'Linea *cd* est linea extensionis que representat nobis formam quantum est ex parte subiecti habens partem extra partem, et quanto linea ista esset maior in longitudine, tanto representat formam esse extensive maiorem; linea vero *ac* representat intensionem [*corr.* intensiorem *ed.*] forme et quanto est in longitudine maior, tanto intensiorem formam representat.'

> The term 'latitude', insofar as it concerns the present issue, does not connote an extension, which is clear from the fact that someone can begin to take pleasure because of this or that lute from the zero degree up to a sign of a certain intension. This would not be the case if the term 'latitude' meant extension, since this latitude is said to be subjectively in the soul, which has no extension.[82]

However, it comes to connote an extension when this *latitudo* is classed and qualified insofar as it is uniform or uniformly difform (i.e., the quality changes but in a uniform way) in an extended natural body. So according to the definitions, 'a uniform latitude is that of which any part is of the same intensity as the whole'; and a latitude that is uniformly difform expresses 'an equal distance between the degrees of equal parts of the subject.'[83] Thus, both of them do indeed connote a certain extension of the subject; this is why '"latitudo uniformis" ultra *inextensionem gradualem, connotat* extensionem' (my emphasis).[84] Thus, latitude does not properly and directly designate extension, nor even (in this text) the two- or three-dimensional shape, but rather the intension (here referred to as *inextensionem gradualem*), whereas it connotes, and thus *secondarily signifies*, extension.

But if the main problem is that of the *variation* of a quality, one must also mention time, which does not appear in the first discussions. In the *Questiones super tractatum de latitudinibus formarum*, therefore, a distinction is made, concerning uniformity, between *quo ad subiectum* and *quo ad tempus*. Consequently, something can be said to be uniform or difform in three ways: *ratione subiecti et partium eius, ratione tempus in quo acquiritur*, or *ratione utriusque*.

---

82 Blasius of Parma, *De intensione et remissione formarum*, 510–511: 'Iste terminus "latitudo" quantum est in proposito non connotat aliquam extensionem, quod patet ex hoc quia quis potest incipere delectari propter hanc cytaram vel illam a non gradu ad certum signum intensionis. Et hoc non esset si iste terminus "latitudo" diceret extensionem, cum hec latitudo dicatur subiective esse in anima nullam habente extensionem.'

83 Blasius of Parma, *De intensione et remissione formarum*, 511: 'Latitudo uniformis est cuius quelibet pars est tante intensionis quante totum ... Dicitur enim latitudo uniformiter difformis cuius est equalis distantia graduum ad invicem equaiter distantium ... dicit equalem distantiam inter gradus equalium partium subiecti.' Cf. the description of uniform latitude, uniformly difform latitude, and difformly difform latitude in Blasius of Parma, *Super tractatum de latitudinibus formarum*, f. 13. In the third conclusion of question 1, however, Blasius specifies (13^{v}): 'Non cuiuslibet forme latitudo est uniformis vel difformis. Patet de forma substantiali humana que est indivisibilis.'

84 Blasius of Parma, *De intensione et remissione formarum*, 511: 'Sed hic terminus "latitudo uniformis" ultra inextensionem gradualem connotat extensionem, ut patet ex quid nominis huius termini "latitudo uniformis".'

On this basis, the subsequent questions will be able to explore the persistence or the introduction of a quality in a given subject by considering either the subject or the moments of this variation.

In the *Questio de intensione et remissione formarum*, these distinctions are made at a later stage of the analysis. In the consideration of a latitude that is uniformly difform, one can consider the intensity absolutely (a whiteness of *x* degrees), or one can consider the ratio of the intensity of one part to the intensity of another part; note that in this case what is considered is a quality that is uniformly difform *quo ad subiectum*, but variation according to time can be considered in the same way. One can then make a further distinction: one can consider either 'the ratio between the intensity of one part of the subject and an equal part of the subject that immediately follows it,' or 'the ratio of the preponderances, or differences, which are found in different parts of a given latitude.'[85] In the case of a two-dimensional representation, in the first case one would presumably compare the height of two parts that follow one another; in the second, one would compare the differences in height between different parts.

Some conclusions can then be drawn about proportionality in the variation in intensity of a quality. First, when alteration is considered by comparing the intensity of one part to the intensity of the part that follows it, this alteration follows a *geometric proportion*. The reason given is that otherwise—supposing that the place in which an agent produces a quality is uniform—it is not clear why it would act more intensely in the part that is closer than in the part that is farther away. (Let us take it that if a fire heats some body, the degree of heat that is induced will diminish proportionately to the distance.) Second, if one considers only the *differences* between degrees, one has an *arithmetic proportion*, that is, a proportion in which equal degrees are successively added (or removed, as the case may be).

Conclusion

From this brief examination, it becomes clear that the aspects of quantification employed to address the problem of the intensification and weakening

85 Blasius of Parma, *De intensione et remissione formarum*, 511: 'Sed nota quod in considerando proportionem intensionis, hoc est duplex: quia tu potes considerare proportionem intensionis unius partis subiecti ad intensionem equalis partis immediate sequentis. Tu etiam potes considerare proportionem excessus seu differentiarum, qui excessus reperiuntur in diversis partibus latitudinis date.'

of qualitative forms are not exactly the same as those used for the calculation of local motion. In the latter case, it is the theory of proportions that is at the heart of the analysis. In the former case, the theory of proportions is certainly present; this is shown by the repeated appeal to the difference between arithmetic proportion (or progression) and geometric proportion (or progression), since the issue is the ratio between different degrees and their variations in time, and their different intensities are related to the quantity of the subject (extension, density). This is why in Blasius' *Physics* commentary there is a positive reference to Bradwardine.[86] But the theory of proportions is not the dominant element.

In addition, only the *Questiones super tractatum de latitudinibus formarum* envisages the geometric representation of uniform or uniformly difform latitudes, which is not the case in either the question on book v of the *Physics* or in the *Questio de intensione et remissione formarum*. What is dominant in the latter texts are certain theses concerning the measure of gradual qualities, with two related problems:

1. whether the growth of a quality is measured in connection with the weakening of the opposite quality;
2. whether the variation is to be evaluated in relation to the zero degree or to the highest degree in a given subject.

Despite some variations, Blasius seems to accept the thesis that cold is not something other than either the absence of heat or a lesser heat. However, when the quality is considered in itself and not as equivalent to the qualified subject, one can state some rules putting opposite degrees (of hot and cold for example) into relation. It seems that there is a methodological decision, just as it is also by a methodological decision that, in order to measure them, one must consider qualities as accidents of a determinate substance which is their subject, whereas from an ontological perspective it is matter itself that will be the subject of qualitative dispositions. Here too, however, the problem is not that of the conditions of the emergence, persistence, and development of a form in matter (discussed elsewhere); rather, it is more specifically the uniform or non-uniform distribution of a quality in a subject and the change over time of this quality.

It is therefore not inaccurate to say that Blasius contributes to the use of (1) contrary qualities and (2) degrees of qualities as operative concepts. However,

86 Blasius of Parma, *Questiones Physicorum*, v, q. 10, 160vb–161ra: 'Ad hunc sensum loquebatur Magister Tomas Barduardinus a principio sui tractatus *De proportionibus*, dum dicebat "contingit alium motum alteri in velocitate et tarditate proportionari".'

this is only one aspect of these texts, which still take into account the physical and metaphysical aspects of the problem. Blasius does not really engage in working out methods of calculation, despite the fact that he examines the possibility of applying the rule of the mean degree to the analysis of qualitative variation,[87] comments on Bradwardine in his *Questions on the Treatise on Proportions*, and uses the rule in book VII of his *Questions on the Physics*.

Nonetheless, he is instructive as a witness to developments in Italy at the turn of the fifteenth century. In the problem of the intension and remission of qualitative forms, which was standardly discussed in the context of the *Physics*, besides the well-known influence of medical tradition,[88] the Italian masters have adopted many issues from the Oxford Calculators. Among these problems are:

1. the question of the approach to the maximum degree, and moving away from the zero degree;
2. the correlative variation of contrary qualities, which are discussed in an operative way;
3. the consideration of the speed of motion—in this case alteration—*penes effectum* and *penes causam*;
4. allusions, however brief, to the theory of proportions, with the alternative between expressing a qualitative variation by the calculation of ratios between contrary qualities, and expressing it by the calculation of the preponderance of a given quality over the contrary quality, centred on the concept of latitude.

In binding the quality and its subject closely together, Blasius does not go as far as the Oxford Calculators in the mathematization of intensive magnitudes. Nonetheless, he tries to integrate the new tools into his physics and his ontology, a key element of which is the immediate relationship between qualities and matter.

87 See Di Liscia, 'Biagio Pelacani da Parma's Geometrisation.'
88 See *supra*, n. 77.

CHAPTER 10

The Calculators' Tradition in Oresme's *De visione stellarum*

Aníbal Szapiro

Introduction

One of the most striking achievements of the Oxford Calculators was the discovery of the *mean-speed theorem* or Merton Rule, which establishes for kinematics that a uniformly accelerated body travels, in the same interval of time, the same distance as a body with uniform speed whose speed is half the final velocity of the accelerated body.[1] This theorem, first enunciated by Heytesbury in his *Regulae*, was geometrically demonstrated by Nicole Oresme in his *De Configurationibus qualitatum et motuum*.[2]

In a presumable earlier treatise that remained anonymous until 1960s—edited, translated and published by D. Burton in 2007 under the title *De visione stellarum*[3]—Nicole Oresme analysed 'whether the stars are seen where they

1 M. Clagett, *The Science of Mechanics in the Middle Ages*, Madison, WI, 1959 and 1961 (University of Wisconsin Publications in Medieval Science, 4) 255–337; E.D. Sylla, 'The Oxford Calculators' Middle Degree Theorem in Context,' *Early Science and Medicine* 15 (2010), 338–370.

2 Nicole Oresme, *Tractatus de configurationibus qualitatum et motuum*, ed. M. Clagett, *Nicole Oresme and the Medieval Geometry of Qualities and Motions. A Treatise on the Uniformity and Difformity of Intensities Known as* Tractatus de configurationibus qualitatum et motuum, Madison, WI, 1968 (University of Wisconsin Publications in Medieval Science, 12), 408–411. I will not discuss in this chapter whether Galileo later applied *this* theorem to falling bodies or not. To this respect, see W.A. Wallace, 'The Enigma of Domingo de Soto: *Uniformiter Difformis* and Falling Bodies in Late Medieval Physics,' *Isis* 59 (1968), 384–401; E.D. Sylla, 'Galileo and the Oxford Calculatores: Analytical Languages and the Mean-Speed Theorem for Accelerated Motion,' in W.A. Wallace (ed.) *Reinterpreting Galileo*, Washington, DC, 1986 (Studies in Philosophy and the History of Philosophy, 15), 53–108; D.A. Di Liscia, 'Die fallenden Körper und die Rätsel des Domingo de Soto,' in M. Segre and E. Knobloch (eds.), *Der ungebändigte Galilei: Beiträge zu einem Symposion*, Stuttgart 2001, 9–22; D.A. Di Liscia, 'Der Kommentar des Johannes Rucherat de Wesalia zur Aristotelischen Physik: seine Bedeutung und Überlieferung,' *Codices Manuscripti et Impressi* 99/100 (2015), 9–28 (at 16–19).

3 D. Burton, *Nicole Oresme's* De visione stellarum (*On Seeing the stars*)*: a Critical Edition of Oresme's Treatise on Optics and Atmospheric Refraction*, Leiden 2007 (Medieval and Early Modern Philosophy and Science, 7).

 | DOI:10.1163/9789004512054_012

are'.[4] To answer this question, Oresme considered astronomic and optic matters together with an original approach to some problems about atmospheric refraction, including a speculation on the behaviour of light in a *uniformly difform* medium.

The main goal of this chapter is to offer a contribution to an accurate characterization of the influences of the Calculators' tradition in the medieval studies of light. I will, thus, consider theoretical and empirical aspects involved in Oresme's attempt to *quantify atmospheric density*, discussing Burton's interpretation according to which Oresme applied the Merton Rule and drew a diagram that can be interpreted as those used for the configuration of qualities.

I first contextualise Oresme's approach to atmospheric refraction within the practice of astronomy according to Aristotelian cosmological premises. In section 2, I present Oresme's approach to atmospheric refraction with special attention to the role that experiences supporting it played. In section 3, I analyse Oresme's argument concerning the possibility of a *uniformly difform* atmosphere. In section 4, I discuss the thesis supporting that Oresme applied the Merton Rule in this treatise and the interpretation of his illustrations as some kind of configuration of qualities. Finally, in the concluding section, I offer a new interpretation of the reasons supporting Oresme's speculation without relying on an application of the Merton Rule.

1 Atmospheric Refraction in the Aristotelian Astronomical Tradition

Atmospheric refraction is a phenomenon that occurs because the variation of density of the medium affects the light coming from the celestial bodies when it passes through the atmosphere to reach the Earth. Due to this phenomenon, bodies that are observed at a certain *apparent altitude* above the horizon could be at a different (and lower) *true altitude*. The magnitude of the deviation due to atmospheric refraction varies depending on the altitude of the celestial body, the atmospheric temperature and pressure. It is certainly a not negligible phenomenon, since it can reach values around 35' under normal conditions of temperature and pressure.

Ancient and Medieval astronomers commonly accepted the difference between the atmosphere's and supralunary space's densities, as well as the difference of the densities of sublunary elements assumed in Aristotelian cosmology.

4 Burton, *Nicole Oresme's* De visione, 81.

Aristotle clearly explains these differences, for example, at the beginning of his *Meteorology*:

> We have already laid down that there is one physical element which makes up the system of the bodies that move in a circle, and besides this four bodies owing their existence to the four primary qualities, the motion of these latter bodies being of two kinds: either from the centre or to the centre. These four bodies are fire, air, water, earth. Fire occupies the highest place among them all, earth the lowest, and two elements correspond to these in their relation to one another, air being nearest to fire, water to earth. The whole world surrounding the earth, then, the affections of which are our subject, is made up of these bodies. This world necessarily has a certain continuity with the upper motions: consequently, all its power and order is derived from them.[5]

Astronomers were also aware of the optical implications of the variation of density in the behaviour of light, at least since Ptolemy, who, for example, studied the implications of the variation of density in the behaviour of light for sublunary phenomena in his *Optics*. To extrapolate this well-known sublunary phenomenon to the relation between sublunary and supralunary media was, for him, somehow obvious:

> It is possible for us to realize that, at the surface between air and ether, there is a refraction of the visual ray according to the difference in density between these two media ...
>
> We notice that [celestial] bodies that rise and set tend to incline toward the north when they are near the horizon and are measured by an instrument for measuring the stars ...
>
> This is a result of the refraction of the visual flux at the surface that separates the air and the ether, a surface that must be spherical, its center being the common center of all the elements, which is the center of the earth.[6]

---

5 Aristotle, *Meteorologica*. In *The Works of Aristotle*, ed. W.D. Ross, transl. E.W. Webster, 3, Oxford 1931, 339a (the pages of this volume are not numbered).

6 A.M. Smith *Ptolemy's Theory of Visual Perception: An English Translation of the* Optics *with Introduction and Commentary*, Philadelphia, PA, 1996 (Transactions of the American Philosophical Society, 86/2), 238–239. Likely, in the expression 'surface between air and ether', air is taken to include both air and fire; in the Aristotelian cosmos air does not reach up to the ether.

In his Almagest, however—and even if he seems to have been aware of this problem by the time he wrote it—Ptolemy considers astronomic data and calculations without raising this problem, as it appears to be the rule in Ancient times.[7] A reason for this disregard might be his conviction that it was impossible to determine all necessary parameters for making these calculations, as he states in his *Optics*:

> It would also be possible for us not only to examine the degree of such refractions, but also to analyze such refraction in the case of certain [celestial] bodies whose distance is given—e. g., the sun and the moon—and to determine the degrees [of refraction] toward the horizon as well as the amount by which the refraction of the visual ray shifts the apparent position upward if the distance of the interface between the two media [that is, air and ether] were known.
>
> But, although this distance lies nearer than the earth to the lunar sphere, where the ether stops, it is not known whether the [refractive] interface lies at the same distance as the aforesaid surface, or whether it lies nearer the earth, or whether it lies farther from [the aforesaid] surface. Therefore, it is impossible to provide a method for determining the size of the angles of deviation that occur in this sort of refraction.[8]

Ptolemy is suggesting in this passage that for bodies with a given distance (understood both as distance to them and as the angular distance to the zenith according to the model) it would be possible to determine the angles of deviation *at the interface* if the distance to it was known. As Figure 10.1 shows,[9] the true and apparent altitudes of the Sun (S_R and S_A, respectively) are different due to a refraction at point R. Angle α can be determined if angle β and the relative distances from O to R and to S_R are known. Nevertheless, as it is impossible to determine the distance to R, it is also impossible to determine α, and the problem was, thus, abandoned.

7 F. Bruin and M. Bruin, 'The Equator Ring, Equinoxes, and Atmospheric Refraction,' *Centaurus* 20 (1976), 89–111; G.J. Toomer, 'Hipparchus' Empirical Basis for His Lunar Mean Motions: A Historical Footnote to Olaf Pedersen, A Survey of the *Almagest*, 161–164,' *Centaurus* 24 (1980), 97–109; J.P. Britton, *Models and Precision: The Quality of Ptolemy's Observations and Parameters*, New York 1992 (Sources and Studies in the History and Philosophy of Classical Science, 1).

8 Smith, *Ptolemy's Theory*, 241–242.

9 All figures in this chapter are my own, unless stated otherwise.

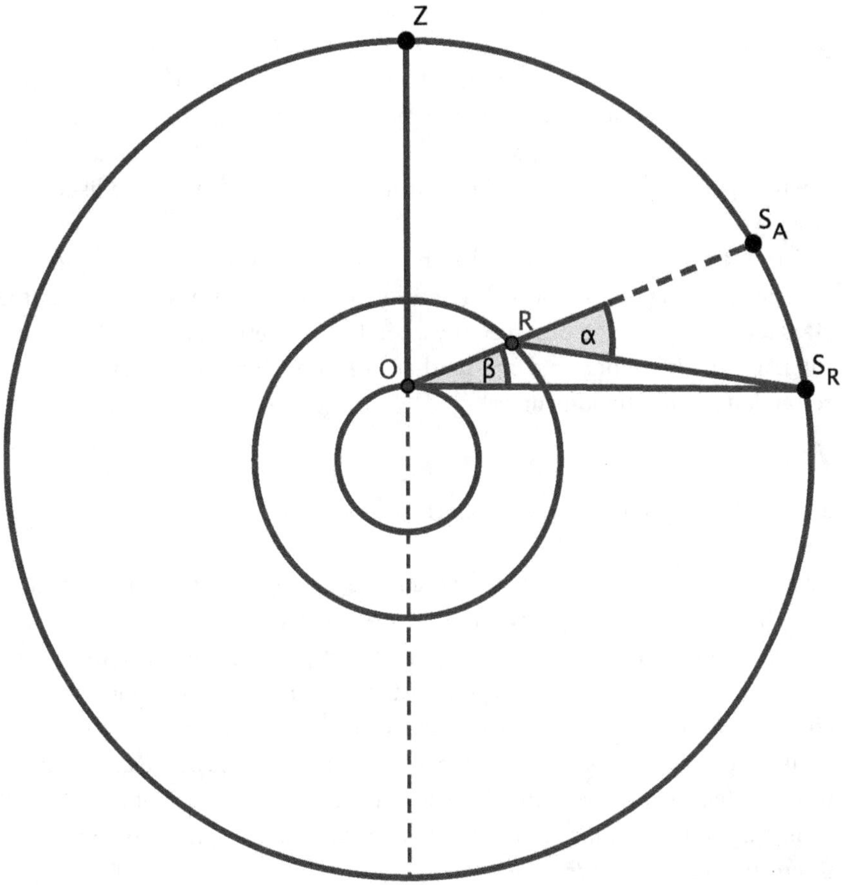

FIGURE 10.1 An observer O on the surface of the Earth sees the Sun in an apparent position (S_A) that is above its real position (S_R). The difference between S_A and S_R measured at O (β) depends on the amount of refraction α at R and of the ratio OR/OS_R.

Early Modern astronomers did not treat atmospheric refraction as a theoretical inquiry about the degree of refraction α *at the interface*, but as a problem about the amount of deviation β for an observer O on the Earth. That is the case, for example, of Tycho Brahe's first attempt to establish a refraction table for the Sun, an effort that Kepler would follow.[10]

10 T. Brahe, *Epistolae astronomicae*, Uraniborg 1596; J. Kepler, *Ad Vitellionem paralipomena, quibus astronomiae pars optica traditur*, Frankfurt am Main 1604.

A common feature in different Ancient, Medieval and Early Modern approaches to atmospheric refraction was their disregard of the fact that it was not necessary for light to go through two media only. The Aristotelian framework of the *Meteorology*, however, assumed a more complex situation, in which water, air, and fire were all present in the space under the lunar sphere. Thus, the problem was not to determine only a refraction at R, but more likely successive refractions at different points.

Oresme's treatise is probably the first to introduce the problem of *quantifying* the changing density of the path through which light passes between the stars and the eye of the observer. It was, in fact, something derived from Aristotelian principles, but analysed neither in Islamic inquiries on atmospheric refraction nor in Christian ones.[11]

2 Atmospheric Refraction in Oresme's Treatise

De visione stellarum is a treatise devoted to the analysis of a specific problem: 'whether the stars are seen where they are', or what in modern terms could be stated thus: whether the apparent and true positions of the stars are the same. Nicole Oresme answers negatively to this *quaestio*, considering two problems with different developments in the history of astronomy and optics: (1) in Book I, a purely geometrical problem of observation called *parallax* that arises from the fact that the astronomic observer is not at the centre of the celestial sphere but on the surface of the Earth; and (2) in Book II, the problem of our interest: *atmospheric refraction*.

Book II consists in the exposition and justification of seven *conclusiones*, arguments against the principal of them with their respective responses, and twenty-two corollaries. The first three *conclusiones* are optical, and simply reproduce the general knowledge of refraction represented in Figure 10.2: if an object O_R is observed from E_R through a surface separating two media with different densities through a path different than its normal, the ray is refracted towards the normal at point R in the denser one, and the image of the object

11 B.R. Goldstein, 'Refraction, Twilight, and the Height of the Atmosphere,' *Vistas in Astronomy* 20 (1976), 105–107; A.I. Sabra, 'The Authorship of the *Liber de Crepusculis*, an Eleventh-Century Work on Atmospheric Refraction,' *Isis* 58 (1967), 77–85; D.C. Lindberg, 'The Cause of Refraction in Medieval Optics,' *The British Journal for the History of Science* 4 (1968), 23–38; D.C. Lindberg, 'The Genesis of Kepler's Theory of Light: Light Metaphysics from Plotinus to Kepler,' *Osiris* 2 (1986), 4–42; W.H. Lehn and S. van der Werf, 'Atmospheric Refraction: A History,' *Applied Optics* 44/27 (2005), 5624–5636.

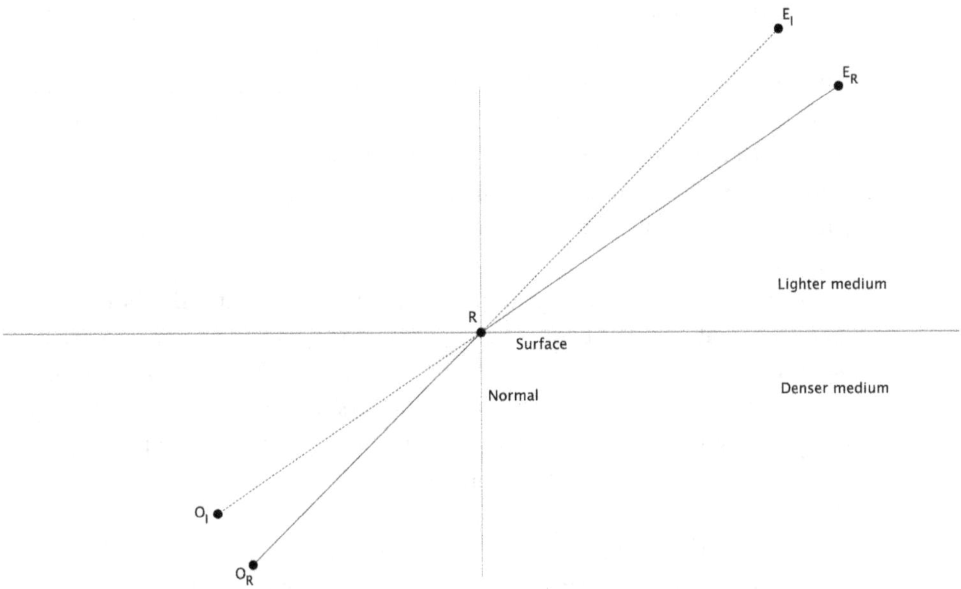

FIGURE 10.2 A ray of light passing through two media with different densities. An object at O_R would be seen at O_I when it is observed from E_R given that the refraction of the visual ray is towards the normal in the denser medium. An object at E_R would be seen at E_I from O_R given that the refraction of the visual ray is from the normal in the lighter one.

O_I will appear aligned to E_R and R in a straight line. The same will occur if an object at E_R is observed from O_R, being the line refracted from the normal in the lighter medium.

The other four *conclusiones* deal with astronomical problems. The fourth *conclusio* is strictly geometrical and depends on the Aristotelian notion of the spheres referred to in the previous section: 'every ray or line drawn from any star to our sight falls obliquely on the surface of the sphere of fire or air, unless the star is over the zenith'.[12] Since the Earth and all spheres are concentric, any ray falling on the surface of the Earth will fall obliquely on the interface separating media, except for any ray aligned with the line between the centre of the Earth and the observer—that is, the ray that passes through the observer's zenith. It is possible to appreciate this in Figure 10.1, where line OS_A is oblique to the tangent to the intermediate sphere at the point where it crosses the sphere (R), and the line OZ is not.

12 Burton, *Nicole Oresme's* De visione, 134: 'Quarta conclusio est quod omnis radius vel linea protensa de aliqua stella ad visum nostrum est oblique cadens super superficiem ignis spere aut aeris, nisi stella fuerit super zenith.'

The fifth *conclusio* is physical and also defined by the Aristotelian framework, and requires no further explanation: 'air is denser than a superior body, such as fire or the heavens'.[13] The sixth *conclusio* is that 'any star which is not over the zenith is seen through a line refracted from the perpendicular',[14] and is a strict consequence of the previous *conclusiones*. Finally, the seventh *conclusio* is derived from the sixth: 'any star which is not over the zenith is seen elsewhere than it is',[15] giving a negative answer to the *quaestio* on whether stars are seen where they are.

All *conclusiones* are not only supported by reason and authority, but also by experience, which is true both for the basic *conclusiones* and for those derived from others, as it is the case with the sixth *conclusio*. This is important because in the defence of the sixth *conclusio* appears the speculation about a *uniformly difform* atmosphere. Thus, I will focus on the three experiences supporting this *conclusio*. The first one establishes the following:

> Let one [of the circumpolar stars] … be observed on the meridian circle when it is near the zenith, and its distance from the pole of the world be taken by using an armillary sphere. Then at another time during the night, when the same star is at a point near the horizon, its distance from the pole should be observed through the instrument a second time. And one will discover its distance from the pole [to be] much smaller than it was the first [time], that is, when it was near the zenith.[16]

Figure 10.3 provides a representation of this experience. A circumpolar star S is observed twice when it passes through the meridian: first when it is near the zenith (S_1) and second when it is near the horizon (S_2). The first time subtends with the pole NP an angle α bigger than the one that subtends the second time, angle β, showing that S_2 is more affected by refraction than the Polar Star and than S_1.

13 Burton, *Nicole Oresme's* De visione, 136: 'Quinta conclusio quod aer est grossior quam corpora superiora, sicut ignis aut celum.'

14 Burton, *Nicole Oresme's* De visione, 140: 'Sexta conclusio quod quelibet stella que non est supra zenith videtur per lineam fractam a perpendiculari.'

15 Burton, *Nicole Oresme's* De visione, 146: 'Septima conclusio est quod omnis stella que non est supra zenith videtur alibi quam sit.'

16 Burton, *Nicole Oresme's* De visione, 142: 'Quoniam, si una illarum notetur in circulo meridiano quando est circa zenith, et per instrumentum armillarum capiatur eius distantia a polo mundi. Deinde alia vice, dum eadem stella fuerit in puncto medio noctis prope orizontem, iterum per instrumentum notetur eius distantia a polo. Et invenietur multo minor distantia eius a polo quam fuerit primo, scilicet, dum erat circa zenith.'

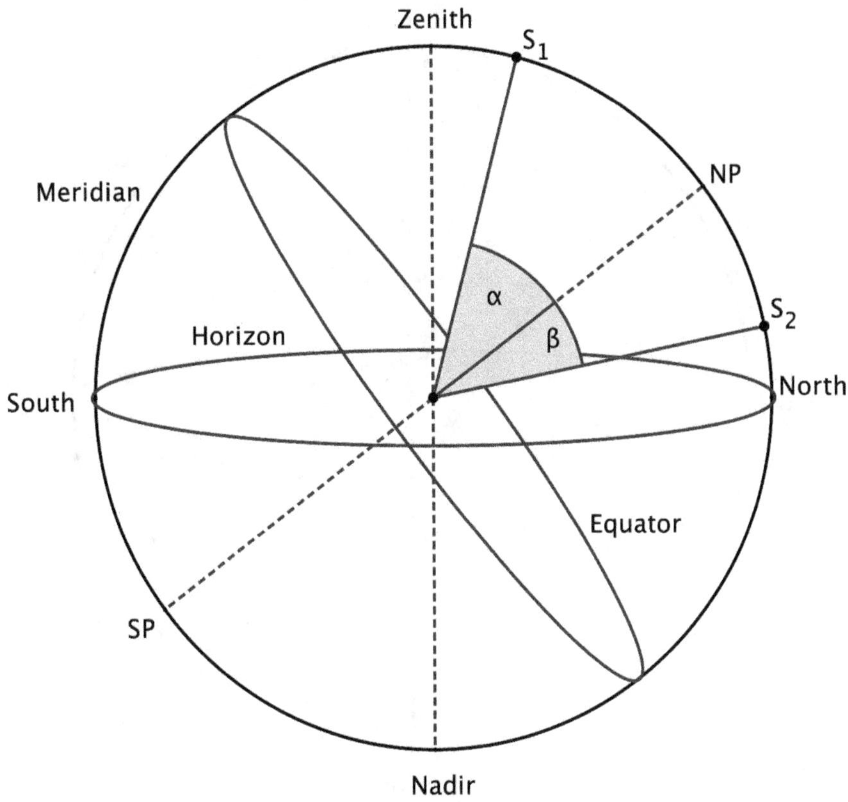

FIGURE 10.3 A circumpolar star S subtends different angles α (when it is at S_1 on the meridian near the zenith) and β (when it is at S_2 near the horizon), being $\alpha > \beta$. COPYRIGHT A. SZAPIRO

The second experience is the following:

> Let any of the stars which pass over or near the zenith be noted, and then, as before, let its distance from the pole be observed through an instrument when it is towards the east, and then again when it is near the zenith. Its distance from the pole will appear smaller when it is near [its] rising than when it is near the zenith.[17]

17 Burton, *Nicole Oresme's* De visione, 142: 'Si notetur aliqua de stellis que transeunt supra zenith, vel prope, et tunc, ut prius videatur per instrumentum distantia eius a polo, quando erit versus orientem, et deinde, quando erit circa zenith. Et apparebit per instrumentum minor distantia eius a polo, quando est versus ortum, quam dum est circa zenith.'

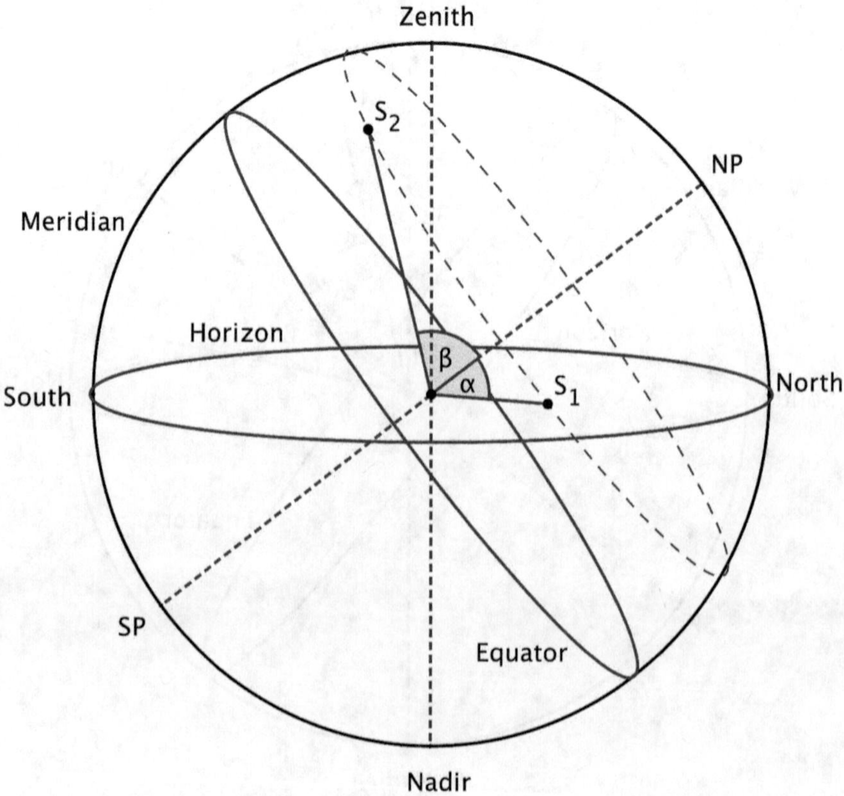

FIGURE 10.4 Any star S that passes near the zenith subtends with the North Pole NP different angles α (when it is at S_1 near the horizon) and β (when it is at S_2 near the zenith), being $\alpha < \beta$.

This idea is similar to the previous experience, but taking any star passing near the zenith and not a circumpolar one, as Figure 10.4 shows. Again, a star in two different positions (S_1 and S_2) and the pole will subtend different angles α and β, being smaller when the star is near the horizon (α), where it is more affected by refraction.

The third experience concerns the moon:

> Let the time of its rising be calculated using tables of its distance from the pole and its declination from the [celestial] equator, and then let [its position] be calculated again for when it is near the zenith. Then, using an instrument, the experimental observations will disagree with the data calculated from tables when the moon is rising, and [they will] not when [the moon] is near the zenith. From this it is clear that when [the moon] was rising it was not seen directly but through refraction.[18]

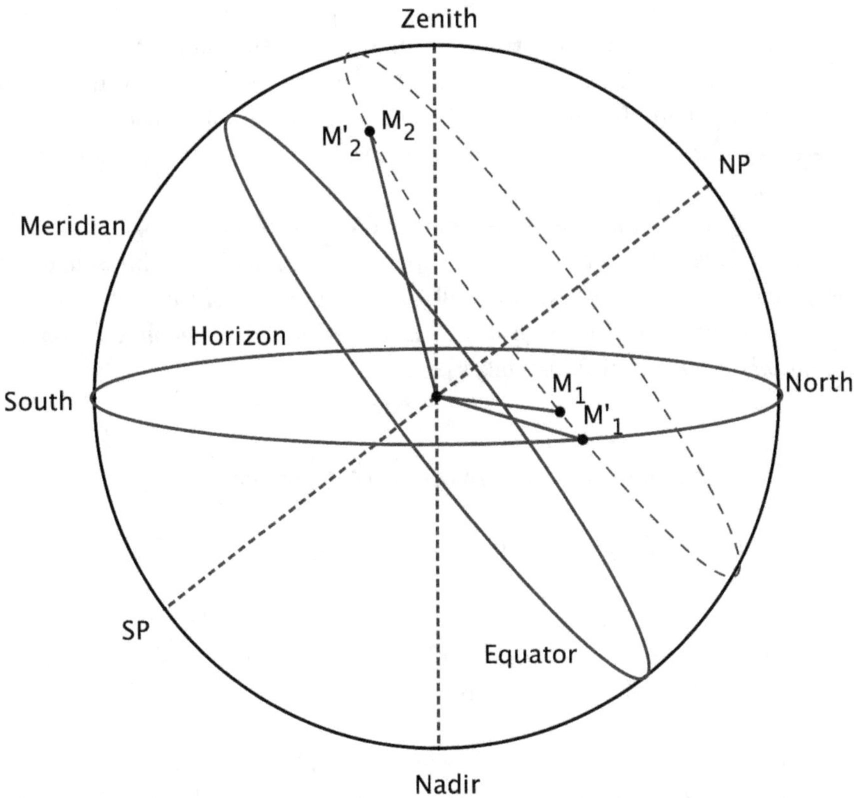

FIGURE 10.5 The position of the Moon M is calculated for two moments, one when it is near the horizon (M'_1), and the other when it is near the zenith (M'_2). The position of the Moon is measured with instruments in the same two moments (M_1 and M_2, respectively). There is a difference between M'_1 and M_1, while there is not between M'_2 and M_2.
COPYRIGHT A. SZAPIRO

This experience is not simply about measuring. Oresme is comparing observations and calculations made with tables. The calculated position of the Moon (M') for two times, when it is rising and when it is near the zenith (M'_1 and M'_2

18 Burton, *Nicole Oresme's* De visione, 143–145. The Latin original of this passage does not mention positions, but times. Thus, Burton's insertion between brackets '[its position]' seems inconsistent (142–144): 'Quia tempore ortus sui adequetur per tabulas distantia eius a polo et declinatio eius ab equinoc tiali, et deinde iterum adequetur, quando est prope zenith. Tunc experientia facta per instrumenta discordabit adequationi facte per tabulas, quando luna est in ortu, et non ita, quando est prope zenith. Ex quo, patet quod, dum oriebatur, non videbatur recte, sed fracte.'

respectively), will not disagree with observations of the Moon when it is near the zenith, but it will when it is near the horizon (M_2 and M_1, respectively). In other words, tables succeeded when they were used for observations near the zenith (where the amount of refraction is smaller), but failed when they were used for observations near the horizon (where amount of refraction is greater).

None of these experiences were new at the time of the writing of Oresme's treatise.[19] It is hard to say if Oresme experienced some of them himself, but it is important to say that they were all within the technical capabilities of the time, providing thus the empirical context in which the speculation about a *uniformly difform* atmosphere takes place.

3 The Problem of a Uniform Difform Atmosphere

Nicole Oresme considers the following possible objection:

> It could be reasonably maintained that air is rarified gradually as it ascends and [has] no limiting surface, [rather it undergoes] a continuous difformity all the way to [becoming] fire.[20]

This objection is rather important since common knowledge on refraction for sublunary phenomena considered in *conclusiones* 1–3 provides information on the behaviour of light passing a surface separating two media. Even when it was commonly accepted that a surface separated the spheres of air and fire, there was no experimental evidence for this. Thus, *conclusiones* 1–3 could be irrelevant to the problem of the position of the stars.

Oresme gives an assertive response to this objection:

> Such difformity destroys the straightness of rays. This is proven since just as some difformity [at a] limited [surface] causes refraction at an angle, so this [difformity] causes winding along a curve.[21]

19 Burton, *Nicole Oresme's* De visione, 143, notes 64 and 68.

20 Burton, *Nicole Oresme's* De visione, 154: 'Posset rationabiliter sustineri quod quamvis aer subtilietur ordinate ascendendo, et non sit superficies terminata, sed difformitas continua usque ad ignem.'

21 Burton, *Nicole Oresme's* De visione, 154: 'Talis difformitas tollit rectitudinem radiorum. Quod probatur, quia sicut alia difformitas terminata facit fractionem ad angulum, ita hic fit plicatio secundum curvitatem.'

To argue for this position, Oresme establishes an equivalence between the effect on light of a hypothetical *uniformly difform* atmosphere and one with the entire change of density occurring at one interface:

> If the whole [atmosphere], aggregated out of air and water, were made uniformly difform with such a density as it now has, then [it] would be equivalent to the original density. And it is seen that those rays which were previously perpendicular are still perpendicular, and [those] which [were] oblique [are still] oblique. And [a ray] previously seen by (a) refracted line ... will now be seen just where [it was] before [but] by (a) curved line ... and [it] will appear in the direction of the tangent line of such a curve.[22]

Allegedly against authorities, Oresme introduces an experiment to demonstrate that there should be refraction. This experiment, however and at least at the beginning, is not a speculation about the atmosphere, but about the experimental situation supporting *conclusiones* 1–3:

> Let there be water and air of the same quantity, between an eye *e* and an object seen *c*, and let the eye be in the water. And let the density of the air be uniform, for example, at an intensity [of] 3, and the water [at intensity] 7, and let *edc* be a refracted ray. Next, imagine that the density of the lower half of the water becomes, say, 8 in intensity, and the upper half [of the water becomes] 6, and the lower half of the air [becomes an intensity of] 4, and the upper [half of the air becomes] 2. In this way, through this conceptual image, one degree of density is removed from the upper half of the water and placed in the lower [half], and likewise concerning the air.[23]

---

22 Burton, *Nicole Oresme's* De visione, 156: 'Item, si totum aggregatum ex aere et aqua fieret uniformiter difforme tanta densitate quantam nunc habet, tunc equivaleret prime densitati. Et videtur quod illi radii qui prius erant perpendiculares adhuc sunt perpendiculares et qui obliqui. Et quod prius videbaturperlineam *edc* fractam, nunc videbitur ubi prius per lineam curvam *eghc*, et apparebit in directo linee contingentis huiusmodi curvitatem.'

23 Burton, *Nicole Oresme's* De visione, 156–158. Note that the original in Latin translated as 'through this conceptual image' is 'per ymaginationem'. This expression reinforces the connection with the Calculators' tradition pointed out in section 5. 'Et sint inter oculum *e* et rem visam *c*, aqua et aer eiusdem quantitatis. Et oculus sit in aqua sitque densitas aeris uniformis, sicut tria in intensione, et aque densitas, sicut 7, sitque fractio *edc*. Deinde, ymaginetur quod medietas aque inferior fiat sicut 8 in intensionem grossitiei, et medietas superior sicut 6. Et medietas aeris inferior sicut 4, et superior sicut 2. Ita quod, per yma-

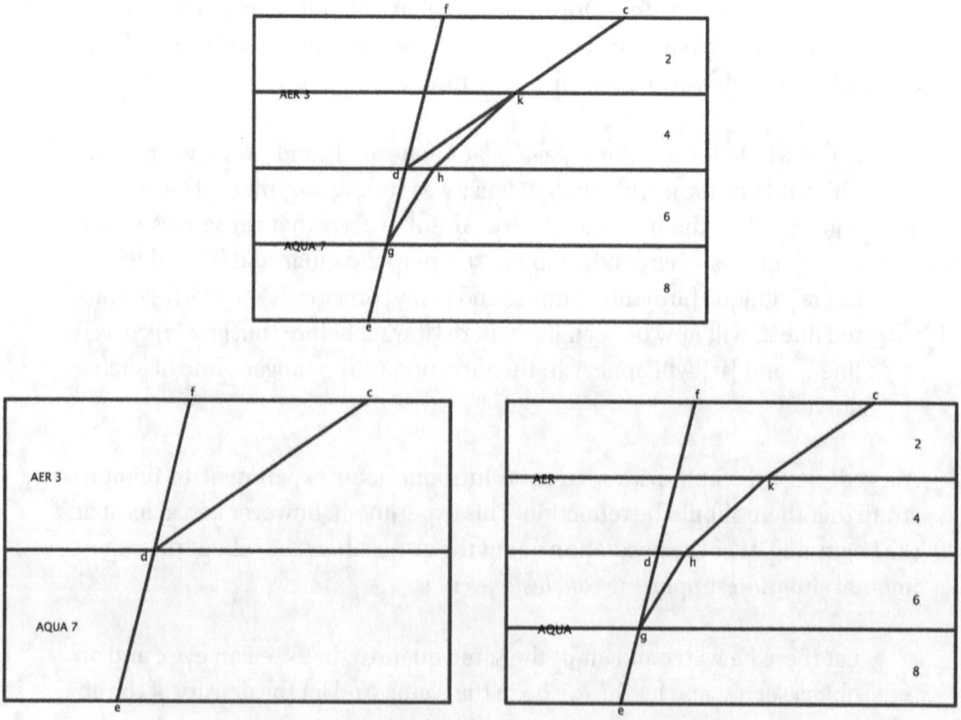

FIGURE 10.6 The top image shows the figure as appearing in manuscripts. The two bottom images show a decomposition of the first one: on the left, the first situation of two media separated by one surface, while on the right the situation of four media separated by three surfaces. In all of them, *e* is the eye, *c* the object, and *f* the image of the object.

This is a clear example of what is usually called a *thought experiment*.[24] The original situation of two media is changed in a way that was experimentally impossible: changing it to four media. This section presents an illustration that in extant manuscripts is rather similar, and which is, as Burton suggests, a superposition of two situations. Figure 10.6 reproduces Burton's images.[25]

The figure on the top can be understood as the superposition of the two figures at the bottom. On the left, air and water with arbitrary densities of 3 and

---

ginationem, unus gradus densitatis amoveatur a medietate superiori aque et ponitur in inferiori, et consimiliter, fit de aere.'

24 K. Ierodiakonou and S. Roux (eds.), *Thought Experiments in Methodological and Historical Contexts*, Leiden 2011 (Medieval and Early Modern Philosophy and Science, 15).

25 Burton, *Nicole Oresme's* De visione, 47–49.

7. On the right, both water and air are divided into two different media each. Where before were two media with uniform intensities of their densities of 3 and 7, respectively, now the figure presents four, with intensities of 2, 4, 6 and 8. Oresme, then, draws the consequence:

> Then there are three refractions which together are assumed equivalent to the single [refraction of the initial conditions], or they [are assumed to] make an even greater [total refraction]. This is clear considering the proportions of the media and the obliquities of the lines, since they change with respect to the sizes of the angles.[26]

This produces, thus, the result of having the same or even more refraction than before. Nevertheless, and since this proves nothing about the refraction in a *uniformly difform* medium, Oresme introduces a 'difficult and beautiful' speculation:

> It cannot be denied that, where previously the object was seen through one [refraction], the medium could be arranged (by the imagination) in such a way that the object will be seen through two refractions, or three, or four, or through any [number whatever of] smaller ones made in the same part, and together [they] are assumed equivalent to the first [refraction].[27]

To justify this situation, Oresme introduces a function between time and the amount of refractions:

> It could be imagined, therefore, that in the first half an hour it is seen through one refraction. And in the first half of the remaining half [an hour], the medium [is arranged] in such a way that the object is seen through two refractions, then, through four [refractions], later through eight [refractions], and thus to infinity, through the continuously proportional parts of an hour due to the alteration of the medium. It is not

26 Burton, *Nicole Oresme's* De visione, 158: 'Tunc, sunt 3 fractiones que simul sumpte equivalent prime vel faciunt adhuc maiorem deceptionem. Quod patet considerando proportiones mediorum et obliquitates linearum, quia secundum ista variantur quantitates angulorum.'

27 Burton, *Nicole Oresme's* De visione, 158: 'Non potest negari quin, per ymaginationem, medium possit taliter disponi quod res per duas fractiones videbitur, ubi prius videbatur per unam, vel per 3, vel per 4, aut per quotlibet ad eandem partem factas minores prima, et sibi equivalentes simul sumptas.'

> necessary for this to be infinite, but perhaps the whole is naturally possible. Then at the end of an hour, the medium will be difform without any uniformity. And the [resulting] line *ec* will be a curve without any straightness. This is clear, for example, if in the first proportional part of an hour, some one line may make an equilateral triangle, in the second [proportional part of an hour], out of the same [line], it might make a quadrilateral, in the third, a pentagonal. And thus, further, it is clear that in the end [the line] will have neither angularity nor rectilinearity, but it will be a circular line as may be easily demonstrated. And so, in the proposed, *ec* will be a curved line.[28]

This is a typical procedure in Oresme's analysis. A finite amount of time (one hour) is divided in halves to infinity. In the first 1/2 of the hour there is one surface and two media. In the following 1/4 of the hour there are two surfaces and three media. In the following 1/8 of the hour there are four surfaces and five media. Even when the positions e and c are conserved, this situation is not presented in the previous figure. In this situation the progression of surfaces is 1, 2, 4 … *n* instead of the former 1, 3. The new situation could be represented as in Figure 10.7 for the first three partitions of the hour.

Understanding the role time plays here is the key to understanding the argument. Oresme introduces a finite amount of time divided into infinite parts that allows him to introduce an infinite number of surfaces of refraction. The focus is on the last instant:

> It follows, therefore, that in [this] difform medium, *c* will either be seen through the curved line that is proposed, or that during the entire hour, it will be seen in the place *f* because of refraction, and in the end it will suddenly be seen where it is through a straight line and will sud-

28 Burton, *Nicole Oresme's* De visione, 158–160: 'Ymagineretur, ergo, quod in prima medietate hore videatur per unam fractionem. Et in prima medietate (mediante [?] *ed.*) medietatis residue medium taliter se habeat quod videatur res per duas fractiones, deinde per 4or, postea per 8, et sic in infinitum per partes hore continue proportionales propter alterationem medii. Quam non oportet propter hoc esse infinitam, sed forte totum est possibile naturaliter. Tunc in fine hore medium erit difforme, absque aliqua uniformitate et linea *ec* erit curva absque aliqua rectitudine. Quod patet exemplo, si in prima parte proportionali hore de una linea fiat triangulus equilaterus, in secunda fieret ex eadem quadratus, in 3a pentagonus. Et sic ultra, patet quod in fine non erit angulus nec etiam rectitudo, sed erit linea circularis ut posset faciliter demonstrari. Et ita in proposito erit *ec* linea curva.'

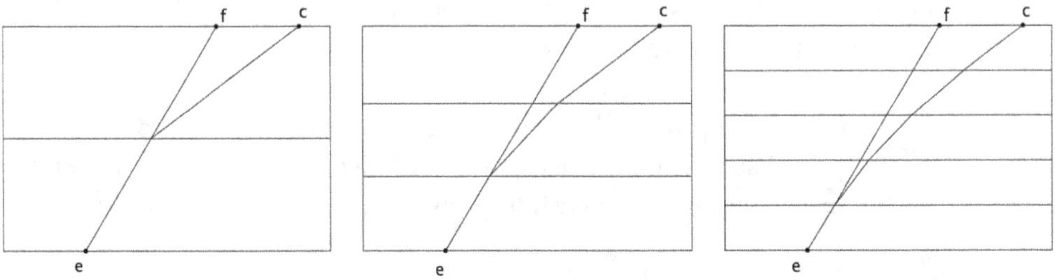

FIGURE 10.7 Refractions in the first three partitions of the hour. On the left, the first half of the hour when there are two media and one surface. In the middle, the third quarter of the hour when there are two surfaces and three media. On the right, the seventh eighth of the hour when there are four surfaces and five media

> denly appear to be changed [in its location]. And the same [result] follows if [the opposite] is posited, that *c* is an eye in air, and *e* is a thing seen.[29]

In other words, there are two alternatives at the last instant for the infinitely broken line: it either changes into a curved line, or it changes into a straight line. If the former, there would be refraction. If the latter, there would not. With regards to the second, Oresme claims:

> Indeed, it seems quite marvellous that the posited case [would be] naturally possible. The object would be seen in the place *f*, and later [*sic*] in the very distant place *c*, being changed neither by the thing itself, nor also by the [observer's] vision, but only because of the gradual alteration of the medium.[30]

What is impossible is that, at the end of the hour, the object that was seen in a position during the entire progression would be seen, instantaneously, in

---

29 Burton, *Nicole Oresme's* De visione, 160: 'Sequitur, ergo, aut quod *c* videbitur in medio difformi per lineam curvam quod est propositum, aut quod in tota hora videbitur in *f* loco propter fractiones, et in fine subito videbitur ubi est per lineam rectam et apparebit subito mutari. Et idem, sequitur, si ponatur primo quod *c* sit oculus in aere, et *e* sit res visa.'

30 Burton, *Nicole Oresme's* De visione, 160: 'Satis enim videtur mirabile quod casu, posito naturaliter possibili. Res videretur in loco *f* ; et postea subito in loco *c* distanti multum, nec ipsa re nec etiam visu aliqualiter transmutatis, sed solum propter alterationem medii successivam. Et quod radius vel umbra esset primo in loco *f*, postea subito in loco *c* valde distanti, et numquam in intermedio. Et hoc ipso luminoso non mutato, sed solum ut dictum est alterato medio successive.'

another one without its or the observer's position being changed. The core of the argument concludes as follows:

> Therefore, the [original] proposition holds, which is now proven [correct], since such a difform medium not only bends light [= *lumen*] directly, but [also bends it] obliquely into various parts.[31]

The influence of the Oxford Calculators' tradition in this speculation seems to be strong. The fact that Oresme discussed a special problem related to a *uniformly difform* medium and divides time into *proportional parts* suggests a direct contact with material proceeding from this tradition. Everything seems to indicate that Oresme had studied the corresponding passages of Richard Swineshead's *Liber calculationum* or parallel texts by other authors within the same tradition. As I mentioned in the Introduction, one of the most striking achievements of the Calculators was the discovery of the *mean-speed theorem*, or Merton Rule, which Oresme geometrically demonstrated in his *De configurationibus qualitatum et motuum*. In the next section, I will examine whether and how this demonstration could be related to both the Merton Rule and the geometrical representation that Oresme developed.

4 An Application of the Merton Rule?

Burton argues that the equivalence of one surface separating two media and infinite surfaces separating infinite media is a *reverse* application of the Merton Rule. To justify this statement, he says:

> Instead of beginning with a uniformly difform quality and then finding the mean uniform quality that would be the equivalent, Oresme does the opposite; he starts with a uniform quality and ends with an equivalent uniformly difform quality. And that quality is atmospheric density. Oresme concludes that the density of a uniform medium is equivalent to the mean density of a uniformly difform medium of the same substance.[32]

---

31 Burton, *Nicole Oresme's* De visione, 162: 'Restat, ergo, propositum, quod adhuc probatur, quia tale medium difforme reflectit lumen non solum recte sed oblique in varias partes.'

32 Burton, *Nicole Oresme's* De visione, 46.

As I explained in section 4, Oresme claimed that if the variation of the atmosphere's density were *uniformly difform* given an infinite number of surfaces separating media, this would produce the same effect (or *more*) than the variation between two media separated by one surface. This equivalence between two intensities of a quality, the density, is what leads Burton to interpret this situation as an application of the Merton Rule.

Nevertheless, the specificity of the Merton Rule (*reverse* or not) is that it establishes the *equivalence* between the *uniformly difform* quality and *one specific uniform degree* of it, the *mean*. Ignoring Oresme's consideration that refraction could be *more* in one of the situations, the *equivalence* in his argument is not with *one specific degree*. If we consider the steps of his speculation supporting his argument, we realize that his position is that *any* number of surfaces (one, two, four, and so on) would produce the same effect as a *uniformly difform* atmosphere with the same density. Thus, one aspect of the specificity of the Merton Rule is absent: the established equivalence is not with the *mean degree*, but with any degree.

Additionally, there is no reason to consider the starting point of two media separated by one surface as the *mean degree*. Oresme is not using this expression, and from his analysis one can conclude that this situation is not the *mean degree* but the *minimum*, since at least two different densities are necessary to produce refraction. This issue emphasizes an aspect of Oresme's discussion that is even more important than the previous one: difformity is a necessary condition for refraction. This is critical because it reveals that any *uniform density* (the *mean* or any other) would produce the same effect on light: there would be no refraction. Thus, Oresme is not establishing a relation between a *uniformly difform* and a *uniform degree*, but between a *uniformly difform* quality and a *difformly difform* one.

Thus, even if it is true that Oresme finds, as Burton claims, the *uniformly difform* quality instead of using it as a starting point, it does not seem to be true that he is using a *mean uniform* quality as his starting point. In fact, his starting point is neither the *mean* nor a *uniform* quality. Thus, if one also considers the peculiarity that the equivalence to the mean is only a consequence of the equivalence to any degree, it is evident that Oresme's argument is not an application of the Merton Rule.

The interpretation of the illustrations as a configuration of a quality is concomitant to the interpretation of the proof as an application of the Merton Rule. To this respect, Burton says:

> The illustration may be seen as the configuration of a quality, since the curve in the illustration not only describes the path of the light ray from

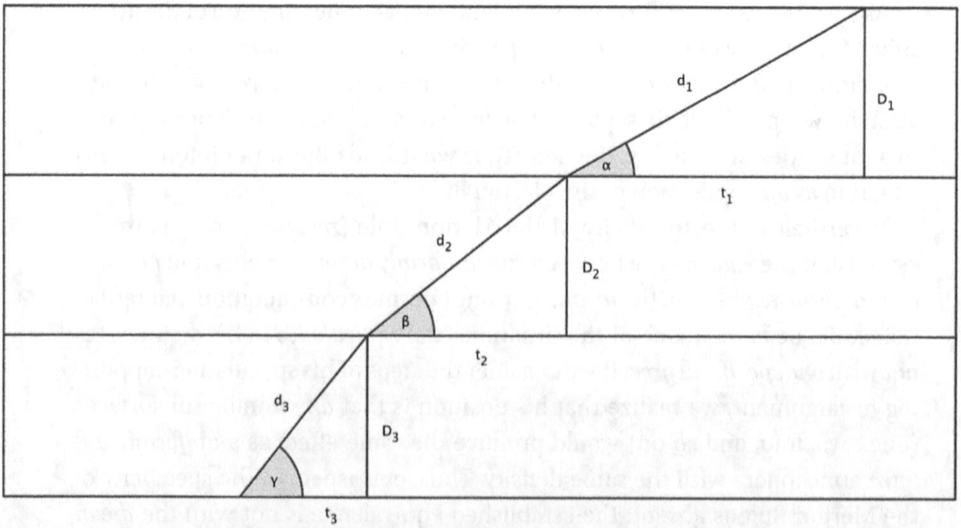

FIGURE 10.8 The visual ray has different obliquities in different media, being bigger in the denser media than in the lighter ones, $\gamma > \beta > \alpha$. The vertical distances between surfaces are equivalent ($D_1 = D_2 = D_3$). Thus, $d_1 > d_2 > d_3$ and, consequently, $d3/t3 > d2/t2 > d1/t1$
COPYRIGHT A. SZAPIRO

right to left over distance, but also over time. That is, this illustration may be seen as a crude graphing technique.[33]

The analysis I presented in section 4 clearly shows that the horizontal line should not be understood as representing time. Not only because it would imply assigning a finite velocity to light, as Burton acknowledges, but specially because of the implication in the relation between velocities and media. As Figure 10.8 shows, the example triangles formed in the third part of the hour imply assuming a greater velocity in a denser medium. The distances d are $d_1 > d_2 > d_3$ given that vertical distances D_1, D_2 and D_3 are equal. If time is uniformly represented in the horizontal axis, then $t_1 > t_2 > t_3$. Now, given that the velocities d/t are defined by the secant of the angles $\alpha < \beta < \gamma$, then $d3/t3 > d2/t2 > d1/t1$. In other words, the greater the angle, the greater the secant. Thus, the denser the media, the greater the velocity. Of course, this makes no sense neither for Oresme nor for any other philosopher.

Oresme's figure is a mere spatial representation of a ray of light. If time is implicit somehow in the figure, it is not increasing horizontally from right to

33 Burton, *Nicole Oresme's* De visione, 48.

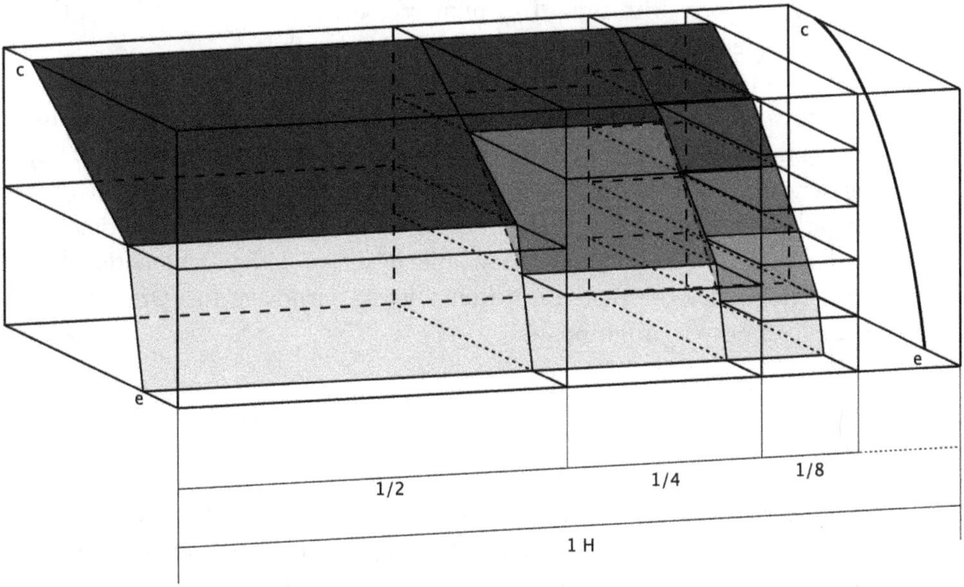

FIGURE 10.9 The division in proportional parts of the hour according to the relation 2:1 and their respective number of surfaces for each of them for the first three partitions (1, 2, and 4, respectively), and the last instant resulting of a *uniformly difform* medium
COPYRIGHT A. SZAPIRO

left, as Burton claims,[34] but in the *depth* of the figure, from back to front. If the figure is produced in perspective, something like Figure 10.9 could result, where I represent only the first 7/8 of the hour and the last instant (one needs to imagine the progression in between):

This kind of representation in perspective was not possible by means of technical knowledge in drawing in the fourteenth century. Nevertheless, it is the modern representation that fits better with the interpretation of the illustration that Burton offered, as one diagram being a superposition of two. My reconstruction makes it evident that time is not in the horizontal line of Oresme's illustration. It also reveals some aspects of the interpretation of the proof. It is easy to see that it is not true that 'as the medium becomes increasingly dense, the number of refractions increase to infinity over the remaining proportional parts of an hour,'[35] as Burton assumed after assigning time to the illustration's horizontal *axis*. Under the interpretation I provide here, the medium does not become increasingly denser with time; the medium becomes

34 Burton, *Nicole Oresme's* De visione, 48.
35 Burton, *Nicole Oresme's* De visione, 49.

uniformly difform with time. This reinforces the abovementioned idea that Oresme's theory is not an application of the Merton Rule.

If my interpretation of the passage and its illustration is correct, then there is no reason for expecting the mention of the Merton Rule, or to expect the display of the relations in terms of the configuration doctrine. The absence of any references to the configuration doctrine reinforces the idea that *De visione* is an earlier work only if the possibility and the advantages of representing this specific relation (and not the Merton Rule) by means of the configuration doctrine can be proved. Otherwise, it would not be 'surprising' that Oresme does not refer here to it, as Burton says.

Conclusions

Burton's valuable editorial and translation work of Oresme's treatise made available a significant piece for understanding the relation between optics and astronomy in the Late Middle Ages. The analysis I offered in this chapter was focused on a set of specific problems to contribute to a better understanding of the influences of the Calculators' tradition in medieval studies of light. I draw the following conclusions, some of which owe much to Burton's analysis as they are in opposition to his interpretation.

The first conclusion is that Oresme's approach to the problem of atmospheric refraction is—both in that it reproduces knowledge developed by previous thinkers and in that it offers an original approximation—a consequence of two connected, though independent, dimensions that were considered in section 2: (1) the Aristotelian premises about the structure of the cosmos, and (2) the empirical evidence of an astronomical phenomenon that astronomers observed since Antiquity.

The second conclusion is that Oresme's work shows, in its original inquiry about the possibility of a *uniformly difform* atmosphere, the necessity to give an answer to an existing problem in the Aristotelian tradition by means of theoretical elaborations that the Calculators developed, as I explained in section 2. The language that Oresme used, especially his notion of a *uniformly difform* medium, has an unquestionable Mertonian influence. The same goes for the *thought experiment* I analysed in section 4: its general structure and the strategies of quantification of the variation of the density of the atmosphere, space, and time, closely follow the Merton tradition.

The third conclusion is that, despite the unquestionable influence of the Oxford Calculators, the solution offered to the possible objection of a *uniformly difform* atmosphere was not an application of the *mean-speed theorem*, or Merton Rule, as I showed in section 5. Since Oresme's response was not constrained by that theoretical device, it seemed convenient to evaluate the influence of other factors in his speculation. One should not overlook the fact that Oresme was conducting a study of a physical problem associated with astronomical practice with a very specific empirical dimension, as I considered in section 3.

For all the above, Oresme's treatise is a valuable case for the study of the limits and challenges found in the application of theoretical speculative devices developed within the Calculators' tradition to the study of specific phenomena belonging to the physical world—specifically, and given the content of the treatise, to the studies of light. I hope the analysis I carried out in these pages makes a contribution to their understanding.

Acknowledgements

I am grateful to Daniel Di Liscia, Edith Sylla, Christián Carman, Diego Pelegrin and Gonzalo Recio, for discussing the ideas in this chapter with me. This research was possible thanks to the support of the *Deutscher Akademischer Austauschdienst* (DAAD, Jahresstipendium Nummer 57381410).

CHAPTER 11

Perfections and Latitudes: The Development of the Calculators' Tradition and the Geometrisation of Metaphysics and Theology

Daniel A. Di Liscia

Introduction

Due to its structural complexity, the wide range of problems treated, the sometimes unsolved and unfinished character of its questioning, and not least because of its profoundness, the science the principles of which Aristotle describes in his *Metaphysics* was destined to assume an important role in medieval thought from the twelfth century onward. This science, also called by the Philosopher 'first philosophy', is, besides physics and mathematics, the speculative discipline *par excellence*, which has to face the many difficulties of dealing with the *ens inquantum est ens* and which finally enters into the analysis of one special kind of being, i.e. 'substance', which justifies its further denomination as 'theology'.[1] No doubt, the problem of the exact determination of the *subjectum* of this science was noticed and assumed to be a difficult one.[2] However, despite these and other difficulties, it is obvious that here, at least in principle, an ideal starting point for further developments in medieval metaphysics and theology was available.

As the particular way of reading the Aristotelian texts proposed or tacitly encouraged by the Calculators is one of the major issues of the present volume, it should not be out of place to start by asking what the most obvious motivations are for commenting on a text belonging to the Aristotelian *corpus*. It seems that for those texts we call 'medieval–Renaissance commentaries' there

1 In William of Moerbeke's translation, the famous beginning of book IV runs as follows: 'Est scientia quedam que speculatur ens in quantum est ens et que huic insunt secundum se' (Aristotle, *Metaphysica* IV.1, 1003a21 transl. Guillelmus de Moerbeka, ed. G. Vuillemin-Diem, Leiden 1995 [Aristoteles latinus, 25/3.2], 67^{3-4}. For this volume I shall use the abbreviation AL XXV).

2 A. Zimmermann, *Ontologie oder Metaphysik? Die Diskussion über den Gegenstand der Metaphysik im 13. und 14. Jahrhundert: Texte und Untersuchungen*, Leuven 1998 (Recherches de Théologie et Philosophie médiévales. Bibliotheca, 1).

 | DOI:10.1163/9789004512054_013

are at least two principal reasons: either the 'commentator' thinks that his commentary will help the reader understand the text of Aristotle commented on, or he thinks that his commentary is a vehicle for exposing his own ideas. In any case, the reading of an Aristotelian text represents much more than an individual interest. The text often stands for the entire doctrine that roughly defines a scientific discipline, which is, consequently, the object of scholarly discussion. Of course, besides several personal or group-oriented preferences, between those two extreme approaches, the one more expounding and the other a more researching one, different degrees of commitment toward one or the other are possible; even a fortunate combination of both is not totally excluded.[3]

Assuming this organising criterion one can cluster a certain number of Aristotelian commentaries and distinguish them from others. Yet one will notice that this methodology has its limits at the very latest when encountering one of Agostino Nifo's commentaries on Aristotle's *Metaphysics*. Book VIII of Aristotle's *Metaphysics* is a key text for understanding the notions of matter and form, two of the most important metaphysical concepts with a direct link to physics. Looking at Nifo's discussion of, or 'comments' on, this book one finds (next to several other questions) that the final question runs: does the uniformly difform latitude correspond to its mean degree? It is clearly a question concerning the mean degree 'theorem' and already the questioning manner makes it clear that, for Nifo, the validity of this theorem is not without problems.[4]

We may not be surprised by the introduction of *calculationes*, *sophismata*, or 'analytical languages', into the commentary tradition on Aristotle's *Physics*, especially when going through the texts of the so-called Paris school of John Mair. Still, the case under discussion does not touch the *Physics* and the dis-

3 Thus, for instance, it is explicitly said by Albert the Great that he composed his commentary on the *Physics* for the sake of supporting his Dominican brothers in the understanding of Aristotle's philosophy. On the contrary, even if being in my opinion a great help to understand some aspects of the Aristotelian theories, Buridan's formidable commentary on Aristotle's *Physics* (*de ultima lectura*) is clearly not intended to this goal, at least not primarily. With his *Livre du ciel et du monde* Oresme was, to mention just a further especial case, acting on both stages at the same time: he translated into French and commented on Aristotle's *De caelo et mundo* for the King's court and produced, nevertheless, one of the most original commentaries ever written to this work. Giles of Rome's and Domingo de Soto's commentaries on the *Physics* are good examples of well succeeded mixed approaches.

4 A. Nifo, *Dilucidarium Augustini Niphi Suessani Philosophi solertissimi metaphysicarum disputationum in Aristotelis Decem et quattuor Libros Metaphysicorum* ..., Venice 1559, 260[b]: *Num latitudo uniformiter difformis correspondeat suo gradui medio* (see below, n. 78 for Nifo's work).

cipline associated with physics, natural philosophy, the subject of which is the concept of motion, but the *Metaphysics*, the foundational text of a speculative discipline which does not—according to the very nature of its objects—deal with motion and change. In fact, such a question Nifo poses has no apparent connection with the Aristotelian text, not even a minimal one. So one could argue that, since his text is a presentation *per viam disputationis*, this question should especially be relevant for presenting Nifo's own 'system of philosophy', since after all, we are dealing with one of the most significant professional philosophers in Renaissance Italy. Accordingly, the consequence that Nifo was a sort of 'Calculator' or a 'continuator of the Calculators' tradition' in the field of metaphysics, as for instance John Mair and some of his pupils were in the field of physics, seems to be unavoidable. That this is not the case is something I am going to address in the last section of this paper, where we will find some provocative declarations by Nifo himself concerning these matters. Thus, the question arises: what could have motivated Nifo to deal with the mean degree theorem and related matters in a *Metaphysics* commentary? And, above all: what is the historical background that makes Nifo's approach comprehensible?

Trying to take an attentive look at Nifo's utterances but without focusing on his own and original ideas—which might perhaps exist—I would like to describe some of the main lines of development that constitute the background of his commentary. This will be my focus. Nifo's attitude toward the 'calculatory way of philosophising' can be better appreciated by considering how the quantifying approach of the *Calculatores* had followed a path through the Aristotelian corpus until it was incorporated into the medieval metaphysical tradition—by the way, a fact Nifo reluctantly admits. At the end of this contribution, I hope to have shown convincingly that this is a main chapter in the history of medieval and Renaissance philosophy which deserves more attention than it has received until now.[5]

5 Even if some contributions belong to the best ever done in our field, like the famous paper by J.E. Murdoch, '*Mathesis in philosophiam scholasticam introducta*. The Rise and Developement of the Application of Mathematics in Fourteenth-Century Philosophy and Theology,' in *Arts libéraux et philosophie au Moyen Âge. Actes du quatrième Congrès international de philosophie médiévale, Montréal 1967*, Montreal 1969, 215–254. In the field of Renaissance philosophy, especially the *Metaphysics* commentary tradition deserves more attention. The overview by C.H. Lohr, 'Metaphysics,' in C.B. Schmitt, Q. Skinner, E. Kessler and J. Kraye (eds.), *The Cambridge History of Renaissance Philosophy*, Cambridge 1992, 537–638, and the comprehensive book by Galluzzo on Paul of Venice (see below, n. 75) offer a useful starting point. For a series of contributions regarding the medieval tradition, none of which, however, deals with our subject, see F. Amerini and G. Galluzzo (eds.), *A Companion to the Latin Medieval Commentaries on Aristotle's Metaphysics*, Leiden 2014 (Brill's companions to the Christian Tradition, 43). For the translations of the *Metaphysics* see in particular the contribution by M. Borgo,

Of course, a complete treatment of all the material involved will go beyond the scope of this chapter. For this reason, I present a selection of authors, texts, and issues considering one particular aspect of the mutual relationship between the *Calculatores* and the Aristotelian 'science of being'. My aim is to show that not only the 'physical' tradition—i.e., the theory of motion with all its subfields and variations—but also the metaphysical tradition was profoundly influenced by a corpus of ideas which can be linked unambiguously to the Calculators, though not only and not directly to Richard Swineshead and his Oxford associates. Hence, it is the development of the Calculators' tradition that I would like to discuss focusing on one branch and on a selection of aspects.

Focusing above all on the problem of the 'perfection of species', I will divide my subject matter along geographical lines, and by so doing I shall try to present some examples of how this tradition spread out nearly all over Europe, radically transforming the way of dealing with Aristotle's *Metaphysics*. After an introductory section on James of Naples, I shall refer to some texts and authors mainly from France, Germany, and Italy. In each case, I shall connect my discussion of the perfection of species (*de perfectione specierum*) with that of the latitudes of forms (*de latitudinibus formarum*), as the two notions and text traditions are intrinsically related. Characteristically, the posing of the questions itself and the favoured answers we will analyse often bear the stamp of a confidence in applying 'lines, angles, and figures' (to use Grosseteste's inspiring title) or in 'recurring to geometry' (*recurrendum ad geometriam*), to use Oresme's

'Latin Medieval Translations of Aristotle's Metaphysics,' 19–56. Especially relevant for the background of this paper are, in the same volume, the contribution by F.J. Kok, 'John Buridan's Commentary on the *Metaphysics*' (495–549), A. Conti, 'Paul of Venice's Commentary on the *Metaphysics*' (551–574), and P.J.J.M. Bakker, 'Fifteenth-Century Parisian Commentaries on Aristotle's *Metaphysics*' (575–630). See also J. Biard, 'God as First Principle and Metaphysics as a Science,' in R.L. Friedman and L.O. Nielsen (eds.), *The Medieval Heritage in Early Modern Metaphysics and Modal Theory, 1400–1700*, Dordrecht 2003 (The New Synthese Historical Library, 53), 75–97. In the present study I will quote only the before-mentioned 'recensio Guillielmi', edited by Gudrun Vuillemin-Diem (see above, n. 1). The first volume of this edition contains an impressive survey of the tradition of the text and the different translations of the Middle Ages. My paper aims to help completing the picture we have of philosophy between the fourteenth and the sixteenth centuries, as for instance presented in the valuable book by S. Swieżawski, *Histoire de la philosophie européenne au* XV*e siècle*, ed. M. Prokopowicz, transl. H. Rollet and M. Prokopowicz, Paris 1990, that contains on the one side a chapter on the Calculators (87–90) and, on the other side, a long part on metaphysics (205–298)—even including a chapter about the world according to a mathematical perspective (252–272)—but no references to the influence of the Calculators' tradition on late medieval metaphysical thinking. Some further references to this topic are to be found in two papers by E.P. Mahoney (see below nn. 24 and 74).

words.[6] Consequently, it is evident that this geometrical approach carries with it a good deal of critical tuning versus the more linguistic approaches of, for instance, Heytesbury and Kilvington. This geometrical approach seems to have been a branch of the *Calculatores* tradition especially fitting to metaphysical notions, not only because of including some nice illustrations but also because of its inner affinity with a major metaphysical-theological question: how are all the entities of the world, without exception of any kind, to be ordered?

1 James of Naples' *De perfectione specierum* and Its Background

This question is addressed by James of Naples in the only work we can attribute to him with certainty, the *Tractatus de perfectione specierum*. James of Naples (Jacobus de Napoli) was already discovered by Anneliese Maier, and yet, he is still hardly known.[7] We know almost nothing about him, except what he him-

6 In *De configurationibus* (henceforth: DC) Oresme is explicit about that: 'Pro mensuris igitur et proportionibus qualitatum seu velocitatum habendis remittendum est et recurrendum ad geometriam' (Nicole Oresme, *Tractatus de configurationibus qualitatum et motuum*, ed. M. Clagett, *Nicole Oresme and the Medieval Geometry of Qualities and Motions. A Treatise on the Uniformity and Difformity of Intensities Known as* Tractatus de configurationibus qualitatum et motuum, Madison, WI, 1968 [University of Wisconsin Publications in Medieval Science, 12], 404). In his previous questions on Euclid Oresme links his project *expressis verbis* to Grosseteste's geometrisation of a 'virtus activa' by a triangular surface (Nicole Oresme, *Questiones super geometriam Euclidis*, ed. H.L.L. Busard, Stuttgart 2010 [Boethius, 57], 138). A similar formula is included at the very beginning of *De latitudinibus formarum* (henceforth: LF): 'Quia formarum latitudines multipliciter variantur que multiplicitas difficulter discernitur, nisi ad figuras geometricas consideratio referatur' (Jacobus de Sancto Martino [?], *Tractatus de latitudinibus formarum*, ed. T.M. Smith, *A Critical Text and Commentary upon 'De latitudinibus formarum'*, Ph.D. Dissertation, University of Wisconsin, Madison, WI, 1954, 1).

7 A. Maier, *An der Grenze von Scholastik und Naturwissenschaft* (Studien zur Naturphilosophie der Spätscholastik, 3), 2nd ed., Rome 1952 (Storia e letteratura, 52), 371–375. James' name, however, had been mentioned before by Maximilian Curtze, who attributed the *De latitudinibus formarum* to Oresme and the commentary on this text contained in Wien, Österreichische Nationalbibliothek, Ms. 4953, to 'Jacobus de Neapoli' on the basis of the colophon, which confusedly states: 'Expliciunt latitudines formarum edite a venerabile viro fratre Iacobo de Florentia ordinis fratrum heremitarum Sancti Augustini' (17v). See M. Curtze, 'Eine Studienreise. Rechenschaftsbericht über seine mit Unterstützung der Königl. Preußischen Akademie der Wissenschaften zu Berlin in den Monaten August, September und Oktober 1896 unternommenen Forschungen zur Geschichte der Geometrie im Mittelalter,' *Centralblatt für Bibliothekswesen*, XVI/6–7, (1899), 257–306, 263. Both attributions are wrong. For an edition and commentary on this text see D.A. Di Liscia, *Eine Wiener* Expositio *zum* Tractatus de latitudinibus formarum. *Edition mit Einführung und Kommentar*, Vienna 2022 (forthcoming). Although only few manuscripts of *De perfectione specierum* have been reported (espe-

self tells us at the beginning of his treatise and, additionally, what some later historians of his order reported. From the first chapter of his treatise, written as Anneliese Maier says in a 'schwülstiger Stil', we learn that he was a member of the Order of the Hermits of Saint Augustine, which had been founded already in the mid thirteenth century and to which numerous personalities in philosophy and science from the Middle Ages and beyond belonged. James reports that he has been requested to explain to the students of the Naples monastery the subject matter of 'the series (or chain) of beings in the universe according to their perfection.'[8] There is neither in the manuscripts nor within the treatise any sure indication for the dating of the text. James treats the matter as if it were a well-established topic. I think that John Murdoch is right by saying that this text can be located intellectually in a context together with other texts which all discuss the same topic around 1340–1370.[9] The other texts Murdoch mentions are Hugolino of Orvieto's treatise *De perfectione specierum* (i.e., the corresponding parts from his commentary on Peter Lombard's *Sentences*), the commentary on the *Sentences* by Peter Ceffons and the anonymous commentary on the *Sentences* contained in Città del Vaticano, Biblioteca Apostolica Vaticana, Ms. Vat. lat. 986 (31^{v}–46^{v}, a manuscript that by the way on ff. 125^{r}–133^{v} contains a text *De perfectione specierum* probably by Oresme), and the

---

cially by A. Maier and J.E. Murdoch) this text was quite widespread, especially in Italy and Germany. It is conveyed in eleven manuscripts on the basis of which I am preparing a critical edition: (1) Bruxelles, Bibliothèque Royale de Belgique, Ms. 1631 (21.856), 68^{r}–75^{v}; (2) Erfurt, Universitätsbibliothek, Dep. Erf., CA Q. 387, 79^{r}–85^{v}; (3) København, Kongelige Bibliotek, Ms. Ny kgl. saml. 190 4°, 179^{r}–197^{r}; (4) Lüneburg, Ratsbücherei, Ms. Theol. 2° 99, 282ra–286vb; (5) München, Bayerische Staatsbibliothek, Clm 4377, 143^{r}–148^{v}; (6) Clm 26.838, 76^{v}–85^{v}, (7) Clm 26.838, 118^{r}–124^{v} (8); Clm 26.889, 206^{r}–$212v^{v}$; (9) Oxford, Bodleian Library, Ms. Canon. Misc. 177, 232^{r}–234^{v}; (10) Città del Vaticano, Biblioteca Apostolica Vaticana, Ms. Chigi F. IV. 66, 6^{v}–12^{v}; (11) Ms. Vat. lat. 4545, 32^{r}–48^{r}. I shall quote following only München, Bayerische Staatsbibliothek, Clm 26.889 206^{r}–$212v^{v}$ (henceforth: M), although I checked passages in all manuscripts (the few corrections I added between ⟨ ⟩ occur in at least one manuscript). The Copenhagen manuscript (= K_a), it is worth to be noted, conveys additionally a commentary on James' text and is surely to be dated later (see furthermore below, nn. 64 and 65).

8 James of Naples, *De perfectione specierum, M* 206^{r}: 'Cupientibus inquantum vobis venerabilibus studentibus Neapolitani conventus ordinis sanctissimi patris Augustini universi entium aliqualiter seriem perfectionis agnoscere, mihi fratri Iacobo de Neapoli professionis eiusdem vestro lectori indigno preces sepius porrexistis quatenus de dicta materia vobis aliqua enarrarem. Quibus salutis rogaminibus gratanter acceptis nec doctoris cuiuscunque hic librum aspiciens de hoc quod petitis diffuse tractarem vinculo caritatis astrictus ac ut vobiscum pariter veritatem investigarem aliorum doctrinam sequens quibusdam has cartas scripturis implevi de specierum narrantibus perfectione.'

9 Murdoch, '*Mathesis*,' 238.

Conclusiones by John of Ripa.[10] The discussion about the topic might, nevertheless, have begun even before and we know for sure that it continued for many years. For, as Marshall Clagett noted, also the Franciscan Peter of Candia (the later Pope Alexander V), who had first studied theology in Oxford and was lecturing on the *Sentences* at Paris around 1380, refers to a geometrical

10 For more information on John of Ripa see A. Combes, 'Présentation de Jean de Ripa,' *Archives d'histoire doctrinale et littéraire du Moyen Âge* 23 (1956), 145–242, and Z. Kaluza, 'La nature des écrits de Jean de Ripa,' *Traditio* 43 (1987) 257–298. Related to the topic of this chapter are A. Combes, 'La intensité des formes d'après Jean de Ripa,' *Archives d'histoire doctrinale et littéraire du Moyen Âge* 45 (1970), 17–147, and J. Coleman, 'Jean de Ripa, O.F.M. and the Oxford Calculators,' *Medieval Studies* 37 (1975), 130–189. The group of thinkers discussing this topic by then might be substantially larger and even taking a broader text basis for commenting than Peter Lombard's *Sentences*. To this group also belongs, for instance, Facinus de Ast, a member of the Hermits of Saint Augustine who also read the *Sentences* in Paris in 1362/1363. In his *Tractatus de maximo et minimo* the Calculators' tradition is clearly present (cf. I. Bodnár, 'Facinus de Ast. Opera Philosophica I. *Tractatus de maximo et minimo,' Archives d'histoire doctrinale et littéraire du Moyen Âge* 64 [1997], 405–437). In his commentary on the *Physics*, qq. 8–10, Facinus explicitly addresses the problem of the measurement of the species according to their perfection, studied by Murdoch in the above-mentioned texts (see I. Bodnár, 'Facinus de Ast. Opera Philosophica II. Questiones super libros Physicorum,' *Archives d'histoire doctrinale et littéraire du Moyen Âge* 65 [1998], 331–414, esp. 356–380). Moreover, q. 5 contains a brief rejecting reference to the system of representation using a sphere later proposed by Paul of Venice: 'Dico ergo ... quod non est sicud illi dicunt quod assimilatur tota latitudo perfeccionum figure spherice, ita quod omnes species incipiunt ab eodem et vadunt versus circumferencias, et specialiter si loquamur de latitudine perfectionis' (Bodnár, 'Facinus de Ast. Opera Philosophica II,' 364). On Paul of Venice, see below n. 77. After the mid fourteenth century, the perfection of species still seems to be one of the focusses of attention at Paris, as is evident in James of Eltville's commentary on the *Sentences*. See M. Brînzei, 'When Theologians Play Philosopher: a Lost Confrontation in the *Principia* of James of Eltville and His *Socii* on the Perfection of Species and Its Infinite Latitude,' in M. Brînzei and C. Schabel (eds.), *The Cistercian James of Eltville (†1393). Author in Paris and Authority in Vienna*, Turnhout 2018 (Studia Sententiarum, 3), 42–77. As for Oresme, it was also Anneliese Maier who already in her *Die Vorläufer Galileis im 14. Jahrhundert* (Studien zur Naturphilosophie der Spätscholastik, 1), 2nd ed., Rome 1966 (Storia e letteratura, 22), 190, n. 69, and later in *An der Grenze*, 305, referred to Oresme's authorship of the anonymous text contained in Ms. Vat. lat. 986. K.H. Tachau, who studied different aspects of this manuscript (including other texts) expressed her hesitation about this attribution. K.H. Tachau, 'French Theology in the Mid-Fourteenth Century: Vatican Latin 986 and Wrocław, Milich F.64,' *Archives d'histoire doctrinale et littéraire du Moyen Âge* 51 (1984), 41–80, at 72–73. During the refereeing process of this chapter, a recent publication was brought to my attention related to this manuscript: Nicole Oresme, *Écrits métaphysiques et théologiques. La résompte de 1362*, ed. A. Boureau, Paris 2021 (Bibliothèque scolastique, 17) (2 vols.). In general, for the application of mathematics to theology in the later Middle Ages, see M. Thakkar, 'Mathematics in Fourteenth-Century Theology,' in E. Robson and J. Stedall (eds.), *The Oxford Handbook of the History of Mathematics*, Oxford 2009, 619–638.

system of ordering the species according to their perfection.[11] In this context, Murdoch included the text on natural philosophy conserved in Paris, Bibliothèque nationale de France, Ms. lat. 6752—which was substantially later than the material discussed by Murdoch and, in general, requires a reinterpretation as I explain below. He discussed above all the question of the intension and remission of forms as applied to the special topic of species. From this point of view, 'the issue in question is that of the determination of the appropriate terminus relative to which any scale of measure for perfection must function.'[12] As an answer to this question one finds usually four possibilities, as listed by Murdoch: (1) *accessus vel approximatio ad summum (scilicet Deum)*; (2) *accessus ad speciem supremam et perfectissimam*; (3) *recessus a non gradu*, and (4) *recessus a materia prima*.

Although the focus on these questions makes it possible to carry out an analysis of the mathematical problems involved here, there is actually a more general question, which is more metaphysical than mathematical and which is determining for the whole discussion *de perfectione specierum*. This question is the following: if the universe, or 'all that there is' can be classified according to species and the individuals included in them, and if there has to be an order of some kind, according to which pattern are the species to be ordered in the best way? Or, simply formulated: what does the general order of a universe created by God and made up of species look like? The question is not easy to answer. For, first of all, the answer has to deal with the problem of the superior and inferior limits, God and 'non-being'. Second, it has to offer some rational way of symmetrical inclusion of all species and individuals belonging to the species ensuring a total completeness, so that nowhere gaps arise. Third—a requirement which seems to have been formulated sometime later—, the model to be used has to be a clear, i.e. an intuitive and well comprehensible pattern. As a consequence of the latter condition, a certain symmetry between the pattern or model and the modelled object is required.

Besides these questions, we have to pay attention to this general fact: the authors writing about these topics were not discussing a question of detail

11 Clagett, *Nicole Oresme*, 78. On Peter of Candia see F. Ehrle, *Der Sentenzenkommentar Peters von Candia des Pisaner Papstes Alexanders V. Ein Beitrag zur Scheidung der Schulen in der Scholastik des vierzehnten Jahrhunderts und zur Geschichte des Wegestreites*, Münster 1925 (Franziskanische Studien, Beiheft 9). A modern edition of Peter's commentary on the *Sentences* is available in Peter of Candia, *In Sententias*, eds. C. Schabel and S.F. Brown, URL = http://candia.ucy.ac.cy/index.htm. Peter's ideas on the geometrical representation of species are mentioned in the until now only known commentary on the treatise *De perfectione specierum* contained in the Copenhagen manuscript K_a (see below, n. 65).

12 Murdoch, '*Mathesis*,' 239.

about the one body being warmed more here than there, or about the possible motion of a body being accelerated to the double, the triple, the quadruple, and the likes. In their view, they were discussing the general form, the structure of the universe in its totality. That makes a considerable difference as compared to all other topics and to physics as a whole. By offering patterns of representation and discussing the chain of beings according to degrees of perfection, the comprehension of reality as a whole is at stake and not a subordinated discussion about a singular matter. Moreover, since the question concerns the universe as such, the first predicate of which is to be 'created', the notion of God, the principles behind his actions and decisions are an essential part of the discussion; and so, the religious perspective plays a crucial role. The Platonic and Neoplatonic background behind the question is evident: God has created the world exercising his own goodness and thus, whatever could exist, has to exist (no gaps are allowed), and whatever there is, has to be in some way good.[13] The point is that every existing being, and every class or group of them, exhibits a certain degree of perfection, so that we could very well imagine an ordering of the totality of the species according to the amount or quantity of their respective perfection.

Thus, concerning the content, the whole issue is clearly metaphysical-theological. Certainly, we cannot embark here on a complete survey of the background and the transmission of these ideas. But neither can we progress in this field without mentioning at least two sources which had a great impact on the late medieval debate. The first source is, of course, Aristotle, since he refers to the general order of beings not only in his biological investigations but also in the *Metaphysics*. Here, for instance, exploring the concept of perfection (*téleion*) in v.16, he distinguishes three chief meanings of the term *perfectum* (according to the translation of Moerbeke): (a) as something complete, lacking *nullam particullam*; (b) perfect *secundum virtutem* as the best exponent of a class of objects; and (c) according to the *finis*. The second meaning allows an interpretation in terms of intensity: inasmuch the 'perfect' one of a class is 'better' than the others, so it possesses the quality in question in a higher degree.[14]

13 Especially the Neoplatonic background is discussed by J. Dillon, *The Golden Chain. Studies in the Development of Platonism and Christianity*, Aldershot 1990 (Collected Studies Series, 333).

14 Using the terms *genus* and *species* in an undifferentiated way in this passage, Aristotle's examples are nonetheless clear: 'the best physicist' or 'the best flutist' are the more capable in each case. Thus, in this sense, reducing the meaning to the pure capacity of something or ability of someone, we could even speak of 'the best thief' or 'the best sycophant', i.e., the best of something which by its very nature is bad. It seems to me that Ross' translation of 'teleion' as 'complete' reduces its meaning to the first possibility considered by Aris-

With Platonic bias Aristotle asks the question about the order in the universe in the aporetic book XI, as follows (again according to Moerbeke's translation): 'Quomodo enim erit *ordo* non existente aliquo perpetuo et separabili et manente?'[15] Sometimes Aristotle's words can up to some extent be brought closer to Plato and to an (undynamic) Neoplatonic worldview, especially when speculating about the *summum bonum* in book XII. In this theological book *par excellence*, one can read this passage that may have had an inspiring influence on medieval metaphysicians:

> We must consider also in which of two ways the nature of the universe contains the good or the highest good, whether as something separate and by itself, or as the order of the parts. Probably in both ways, as an army does. For the good is found both in the order and in the leader, and more in the latter; for he does not depend on the order but it depends on him. And all things are ordered together somehow, but not all alike,—both fishes and fowls and plants; and the world is not such that one thing has nothing to do with another, but they are connected. For all are ordered together to one end. (But it is as in a house, where the freemen are least at liberty to act as they will, but all things or most things are already ordained for them, while the slaves and the beasts do little for the common good, and for the most part live at random; for this is the sort of principle that constitutes the nature of each.) I mean, for instance, that all must at least come to be dissolved into their elements, and there are other functions similarly in which all share for the good of the whole.[16]

totle (see Aristotle, Metaphysics, transl. W.D. Ross in *The Complete Works of Aristotle*, ed. J. Barnes, Princeton, NJ, 1984 [2 vols.] [Bollingen series, 71], 2: 1613). For Ross' commentary, see Aristotle, *Metaphysics*. A Revised Text with Introduction and Commentary, ed. W.D. Ross, Oxford, 1970 (2 vols.), 1: 331–332.

15 Aristotle, *Metaphysica* XI.2, 1060a25–26 (my italics), quoted according the *translatio Guillelmi*, AL XXV, 221$^{93-94}$. Ross' translation runs: '... for how is there to be order unless there is something eternal and independent and permanent?' (*Metaphysics*, 1: 853).

16 Aristotle, *Metaphysics* XII.10, 1075a11–24 (transl. Ross, 2: 1699). Moerbeke's translation runs as follows: 'Perscrutandum autem est qualiter habet totius natura bonum et optimum, utrum separatum quid et ipsum secundum se, aut ordinem. Aut utroque modo sicut exercitus? Et enim in ordine ipsum bene et dux exercitus, et magis iste; non enim iste propter ordinem sed ille propter hunc est. Omnia vero coordinata sunt aliqualiter, sed non similiter, et natatilia et volatilia et plante; et non sic se habent ut non sit alteri ad alterum nichil, sed est aliquid. Ad unum quidem enim omnia coordinata sunt. Sed quemadmodum in domo liberis non licet quod contingit facere, sed omnia aut plurima ordinata sunt, servis vero et bestiis parum quod ad commune, multum vero quod contingit; tale namque cuiusque principium ipsorum natura est. Dico autem puta ad discerni quidem necesse

The order (*táxis*) of the universe that Aristotle affirms in this passage is like the order of an army: each member of it is bound by itself to the leader and, at the same time, it is linked to the other members. The general 'design' includes random actions and demands also a sort of sacrifice or subordination of the parts for the good of the totality.

These are some of the pertinent passages we can find in the *Metaphysics*, one of the most important texts of the Aristotelian corpus which by the mid thirteenth century was integrated into medieval university teaching.[17] However, they are by no means the only ones in the Aristotelian *corpus*. Parallel passages are to be found in Aristotle's cosmological and biological works.[18]

If Aristotle's *Metaphysics* provided the immediate context for the discussion of the series of the entities in the world, the most general background, with it religious and ideological implications, especially for Christian medieval thinkers, was presented by Augustine, whose writings stood far on top of the list of authorities to be observed, particularly in theological matters.

Augustine was attracted to this topic from the very beginning of this intellectual activity, already in Cassiacum, where he composed his brief *De ordine*. He presents the subject matter of the book as a difficult intellectual task: to understand the order of things and, at the same time, to recognize the connec-

---

omnibus venire, et alia sic sunt quibus communicant omnia ad totum' (AL XXV, 266[463]–267[477]).

17 A.L. Gabriel, 'Metaphysics in the Curriculum of Studies of the Mediaeval Universities,' in P. Wilpert (ed.), *Die Metaphysik im Mittelalter. Ihr Ursprung und Ihre Bedeutung*, Berlin 1963 (Miscellanea Mediaevalia, 2), 92–102.

18 A.O. Lovejoy, *The Great Chain of Being. A Study of the History of an Idea*, Cambridge, MA, 1936, 340, n. 41, adds *De historia animalium* VIII.1, 588b, *De partibus animalium* IV.5, 681a, and *De generatione animalium*. 761a-15. In addition, at the beginning of *De caelo*, Aristotle discusses some of the fundamental features of material bodies, such as continuity and dimensionality, in terms of perfection. By so doing, he opens the door to a broad discussion about, as still Galileo said, the unity and perfection of the world (*De unitate mundi et perfectione*). See G. Galilei, *Opere*, 1, ed. A. Favaro, Florence 1890, 27–31. Thus, John Mair, for instance, deals with the problem of the perfection of the species only very briefly in his commentary on the *Metaphysics*, where he refers to the 'tota series universi nobilitate ac innobilitate' (*Octo libri physicorum cum naturali philosophia atque metaphysica Johannis Maioris hadigtonani theologi Parisiensis*, Paris 1526, 5ʳ, numbering from the beginning of the metaphysical questions). However, asking (in *De caelo*) *An mundus sit perfectus*, he defines *mundus* and *perfectus* and adds in the first conclusion: 'Perfectio rei non attenditur penes propinquationem ad summum [summam *ed.*] sed penes distantiam a non gradu entium. Prior pars patet sic: Deus non esset perfectus quia non propinquat gradui summo cum ipse sit gradus summus in tota latitudine entium. Secunda pars patet, nam quanto res est propinquior non gradui, etiam est imperfectior ...' (k5ʳ, numbering from the beginning of the questions on *De caelo*).

tion among the things, is something in which human beings rarely succeed.[19] Nor did he arrive at a definite solution or even a satisfactory position. However, Augustine formulated many of the major ideas which reappeared again and again under different aspects in the later medieval debate.

According to the doctrine of the *De ordine*, every creature pursues an order of perfection that is directed to the One; nothing occurs outside of this order, including evil. In fact, the problem of the existence of evil and of the theodicy is one of the main problems for Augustine, with which he tried to come to grips in this early text by means of his classical education. The fundamentals stated here—that *ordo* is complete, general, comprehensive, and in itself good—remain valid for later works, still after the turning point of his doctrine of grace.[20] Of course, nothing happens without a cause and there is nothing opposite to *ordo*, not even chance or *fortuna*, the very word Augustine later wished he had never written, as he declared in his *Retractationes*.[21] Other works by Augustine include further statements which were destined to have a broad reception in theology and philosophy. Especially important are the ideas of a hierarchical organization of all components of the chain which participate in a common propriety according to their degrees or realisation.[22] It is clear that the medieval doctrine of the transcendentals offers an excellent context for quantitative speculations on that topic: if *bonum* is to be somehow equalled to *esse*, then, the more reality can be attributed to a being, the more goodness can be attributed to it, and vice versa. For, again, all existing is in some way and according to a certain measure, good.[23]

---

19 Augustine, *De ordine*, ed. W.M. Green, Turnhout 1970 (Corpus Christianorum Series Latina, 29), 87–137, at 87: 'Ordinem rerum, Zenobi, cum sequi ac tenere cuique proprium tum vero universitatis, quo cohercetur hic mundus et regitur, vel videre vel pandere difficillimum hominibus atque rarissimum est.'

20 Concerning the notion of order in Augustine, see the study by A.-I. Bouton-Touboulic, *L'ordre caché. La notion d'ordre chez Saint Augustin*, Paris 2004 (Collection des études augustiniennes. Série Antiquité, 174). This comprehensive work contains a detailed lexicological survey of the notion of *ordo* and related concepts in Augustine and other authors of the same period. Though focused on *De ordine*, the earlier work by J. Rief, *Der Ordobegriff des jungen Augustinus*, Paderborn 1962 (Abhandlungen zur Moraltheologie, 2) is more directly connected with our subject.

21 Augustine, *Retractationes*, ed. A. Mutzenbecher, Turnhout 1984 (Corpus Christianorum Series Latina, 57), 12–13 (the expression used is 'fortunae vocabulum', 12^{9-10}).

22 See Rief, *Der Ordobegriff*, 123–126 and 138–141.

23 For the devastating ethical consequences of this conviction, one could take many works by Augustine himself about the doctrine of Grace or even Leibniz's *Theodicy*, despite the fact that Leibniz himself openly complained about Augustine's doctrine: 'On ne sera pas plus avancé pour dire avec saint Augustin, que les hommes étant tous compris sous la damna-

The Augustinian idea of order was developed in theology by Bonaventure and others; moreover, it gained ground as a legitimating tool in social and political theory.[24] That species could be understood as 'parts' of a general hierarchical system based on the notion of perfection was something only implied by Augustine but later explicitly addressed by John Duns Scotus, one of the figures from the turn of the century who knew best how to combine Aristotelian metaphysics and Augustinian theology.[25]

While the background of the discussion on the perfection of species goes back to ancient philosophy and theology, and seems to be almost all-embracing, the closer context for the topic of this chapter is more restricted. The

---

tion par le péché d'Adam …' (G.W. Leibniz, *Die Theodizee*, ed. H. Herring, Frankfurt am Main 1965 [Philosophische Schriften, 2.1], 32). Against the theodicy project of Leibniz and others, see the brief work by I. Kant, *Über das Mißlingen aller philosophischen Versuche in der Theodizee* (1791), ed. W. Weischedel, Frankfurt am Main 1974 (Werke, 11), 103–124. Without noticing any connection to the *Calculatores* and, in fact, without even mentioning any authors from the fourteenth century, A. Lovejoy wrote some valuable pages on Leibniz's ideas about the 'chain of beings' (*The Great Chain*, 144–150). Besides the fact that Leibniz was well aware of the *Liber calculationum* (see Edith Sylla's contribution in this volume), also Bradwardine's *De causa dei* (printed in 1618) could have been an inspiring source for his speculations on the *perfectiones rerum*. Leibniz mentions this text discussing the problem of the grace in his *Essais de Théodicée* § 95 (in the mentioned ed. on p. 350).

24 On Bonaventura's notion of order see J.A. Wayne Hellmann, *Divine and Created Order in Bonaventura's Theology*, St. Bonaventure, NY, 2001 (Theology Series). A. Lovejoy discusses some ideas of Thomas Aquinas in *The Great Chain*, 73–75. For a critical assessment of Lovejoy's book with some references to the Calculators, see E.P. Mahoney, 'Lovejoy and the Hierarchy of Being,' *Journal for the History of Ideas* 48 (1987), 221–230. The impact of some of these ideas on medieval political theory is discussed by J. Miethke, 'Kirchenstruktur und Staatstheorien im Zeitalter der Scholastik,' in B. Schneidmüller and S. Weinfurter (eds.), *Ordnungskonfigurationen im hohen Mittelalter*, Ostfildern 2006 (Vorträge und Forschungen, 64), 127–151. Some aspects of the international political dimension are discussed in connection with Beda's writings by D. Iogna-Prat, *Order and Exclusion: Cluny and Christendom Face Heresy, Judaism, and Islam (1000–1150)*, Ithaca, NY, 2002.

25 Scotus' *Tractatus de primo principio* is built upon the notion of order at logical, epistemological and theological levels. It is, apparently, a causal order, which unifies these three fields. See P. King, 'Scotus on Metaphysics,' in T. Williams (ed.), *The Cambridge Companion to Duns Scotus*, Cambridge 2003, 15–68, esp. 38–43. Before Scotus, the notion of order was discussed in Franciscan circles. I am grateful to Lydia Schumacher for calling my attention to the *Summa fratris Alexandri*, where the notions of 'ordo' and 'perfectio' are discussed in Book II, Inq. 1, Tract. 2, q. 4 (*De creatura secundum relationem seu de ordine universi*) and q. 5 (*De perfectione universi*) (Alexander of Hales, *Summa theologica*, Liber II, eds. PP. Collegii S. Bonaventurae [Quaracchi, 1928], 108–121). Additionally, she pointed out to me that early Franciscans also used the analogy of God as a center point and creatures as lines drawn out from him to explain the relationship of the finite to the infinite. See L. Schumacher, *Early Franciscan Theology: Between Authority and Innovation*, Cambridge 2019, 83.

question of the chain of beings according their hierarchical order and its corresponding mathematical treatment, was often discussed in commentaries on Peter Lombard's *Sentences*. This is evident from the studies by Murdoch and Clagett. It also holds true of the treatise *De perfectione specierum* by Hugolino of Orvieto, which is a repetition of the corresponding part of his *Sentences* commentary.[26] The treatise by James of Naples is, on the contrary, not a part of a commentary on Lombard's *Sentences* but a treatise on its own, as one can appreciate already in the first chapter. Besides, there is no mention in the text of a separate commentary on the *Sentences*. In fact, such a commentary by James is not transmitted.

The second chapter of James' *De perfectione specierum* brings a new approach to the matter, at least in comparison with Hugolino's treatise. It is true that in his text Hugolino introduced some vague ideas of a 'system' of representation using geometry. But his references are too incidental and unclear.[27] James, on the contrary, in the second chapter of his treatise, introduces geometry directly and purposefully. This chapter includes three groups of defin-

26 The text which was first edited by F. Corvino (Hugolino of Orvieto, *Tractatus de perfectione specierum*, ed. F. Corvino, 'Ugolini de Urbe Veteri Tractatus de perfectione specierum,' ACME *Annali della Facoltà die Filosofia e Lettere dell'Università Statale di Milano* 8 [1955], 119–204) contains the characterisation of 'tractatus' in all three manuscripts (which also contain the *De perfectione specierum* of James of Naples). It is evident, however, that the text is linked to the *Sentences* of Peter Lombard. Hugolino's commentary on the *Sentences* shows a verbatim textual congruence with the *De perfectione specierum* (see Hugolino of Orvieto, *Commentarius in quattuor libros Sententiarum*, 2, ed. W. Eckermann, Würzburg 1984 [Cassiciacum, Supplement, 9], 73–130). Interestingly, Hugolino uses the term *ordo* to refer to a system of gradual perfections which he seems to conceive in three dimensions (see the following passage from the *Tractatus*, 131–132: 'triplex est ordo gradualium perfectionum …'). It will be possible to build a kind of hierarchical structure from the lowest to the highest level of being and to represent it geometrically, but Hugolino's ideas are not self-explanatory and the emphasis of his discourse lies by no means on geometry but on theology. Moreover, Hugolino's expressions seem rather to fit to the kind of theologians Oresme approached critically in DC (see below, n. 29). It is noteworthy that the second part of Oresme's DC begins with a brief explanation of how a three-dimensional system of representation can be built (see DC, 270). In any case, all that (including Oresme's system) is far from being related to a 'sistema tridimensionale di coordinate cartesiane,' as it was formulated by F. Corvino, 'Il "De perfectione specierum" di Ugolino D'Orvieto,' ACME *Annali della Facoltà die Filosofia e Lettere dell'Università Statale di Milano* 8 (1954), 73–105, at 86.

27 For a recent reconstruction of Hugolino's system, see S. Roudaut, 'Hugolinus of Orvieto and the Controversies about the Perfection of Species,' *Augustiniana* 69 (2019), 299–331, and his *La mesure de l'être. Le problème de la quantification des formes au Moyen Âge (ca. 1250–1370)*, Leiden (forthcoming) (History of Metaphysics, Ancient, Medieval, Modern, 3), 321–373.

itions. (1) The first ten definitions are purely geometrical: point, line, straight line, surface, angle (including its special types). James ends with *superficies plana*, which allows us to guess that he is not going to deal with volumes. In this context, James makes the following philosophical remark, which is very similar to Oresme's statement in *De configurationibus*: these geometrical entities are mere fictional entities to which no existence *per se* can be attributed:

> It is to be noted for these and the following matters that here many things are posited to which no entity corresponds formally and regarding the reality (*formaliter a parte rei*), but which can be verified imaginatively, as 'point', 'line', 'surface' and many others. For I do not give a pure line nor a point, but only declare those through the imagination for the sake of being able to access to our purpose.[28]

(2) After the geometrical definitions, the definitions of the different sorts of *latitudines* follow. These are not defined ontologically but described geometrically: we do not know what they are, and we do not have to know it. We only have to know how they are to be represented. Here we find the usual distinction between *uniformis* (represented by a rectangular figure) and *dif-*

28 As we do not have any sure dating for the *De perfectione specierum*, the direction of the influence is uncertain. It is also possible that Oresme and James of Naples are referring to a common and popular view on mathematical entities. The fact, however, that in both works such a statement occurs in a treatise on the 'geometrisation of latitudes' is a strong indication of a link between the two texts:

Nicole Oresme: *De configurationibus*	**James of Naples: *De perfectione specierum***
Omnis res mensurabilis exceptis numeris ymaginatur ad modum quantitates continue. Ideo oportet pro eius mensuratione ymaginari puncta, lineas et superficies, aut istorum proprietates, in quibus ut vult Philosophus, mensura seu proportio per prius reperitur. In aliis autem cognoscitur in similitudine dum per intellectum referuntur ad ista. *Etsi nichil sunt puncta, indivisibilia aut linee, tamen oportet ea mathematice fingere pro rerum mensuris et earum proportionibus cognoscendi* (DC, 164, my italics).	*Et notandum pro istis et sequentibus quod multa hic ponuntur quibus formaliter a parte rei nulla entitas correspondet sed tamen ymaginative poterunt verificari, cuiusmodi sunt punctus, linea, superficies et talia plura. Non enim aliquam puram lineam do sive punctum sed tamen per ymaginationem ista declaro, ut ad nostrum propositum accedere valeamus* (M 206^r, my italics).

formis (with a non-parallel upper line). The latter can be either regular (*uniformiter difformis*) or irregular (*difformiter difformis*). In each case, the latitude, which is always represented by the whole figure, can be on the sides *ad gradum* or *ad non gradum*. James does not know the vital notion of *quantitas qualitatis* in Oresmian sense, nor his undeniably more suitable and better reasoned configurations system.[29] Moreover: by defining *gradus* in a somehow complicated way—as 'something indivisible existing *ymaginative* in the latitude'—James can occasionally give the impression of proposing that the *gradus* itself would be some kind of 'atom'. He uses a reformulation of Euclid's definition of 'point' which includes the notion of indivisibility. Where Euclid said, 'a point is that which has no part,' James says—referring to Euclid—'the point is something indivisible.' Given the symmetry between 'point' and 'degree', James understands 'degree' as something indivisible in the latitude, but

29 The notion of *quantitas qualitatis* is essential to Oresme's system as presented in his *De configurationibus* (esp. ch. 4: *De quantitate qualitatum*, ed. Clagett 172–177). In this work, Oresme paid special attention to a clear distinction between geometrical notions and the physico-ontological entities designated by them. Thus, according to his system of representation for only two dimensions (this is the standard case, but Oresme also mentions the possibility of representing in three dimensions; see DC, 270), the horizontal line called 'longitudo' represents the extension (*extensio*) of the quality either in the body (*secundum subiectum*) or in time (*secundum tempus*). On this line are erected perpendicular lines called *latitudines*, which represent the intensity (*intensio*) of the quality on one point of the body or in an instant of time. By the addition of a third line (*linea summitatis*) to this representation we obtain a geometrical figure. This figure, as it is meant to represent something, is called *figuratio* or, more usually *configuratio* (hence, one of the possible titles of this treatise. For other possibilities see Clagett's introduction, 135–139). The most important point to keep in mind according the very simplified description of Oresme's system I am giving here, is that the surface (*superficies*) delimited by this figure represents the total 'quantity of the quality' (*quantitas qualitatis*) distributed on a subject (for instance, the 'quantity of heat'). In the case of successive entities, like motion, this surface represents the 'total velocity' (*velocitas totalis*) of the process, which, still, in the special case of local motion can be expressed in terms of the transversed space. This notion of *quantitas qualitatis* occurs neither in *De perfectione specierum* by James of Naples nor in the treatise *De latitudinibus formarum*. In both treatises, the term 'latitude' is not used to refer to one intension in one point or in one instant, but to designate the total amount of the acquired quantity represented by the surface of the figure (that is, what Oresme called the *quantitas qualitatis*). For more details see, A. Maier, 'Das Problem der intensiven Größe,' in Ead., *Zwei Grundprobleme der scholastischen Naturphilosophie. Das Problem der intensiven Größe. Die Impetustheorie* (Studien zur Naturphilosophie der Spätscholastik, 2), 3rd ed., Rome 1968 (Storia e letteratura, 37), 3–109, at 89–109; Maier, *An der Grenze* 289–352; and the introduction by Smith to his edition of *De latitudinibus formarum*, XXXV. Incidentally, it is worth to be noted that in Oresme we find a criticism of the way *theologians* use the term 'latitudo' (DC, 170).

only according to its extension; according to the intensity a degree is of course divisible.[30]

(3) The third group of definitions is the theological-philosophical one: essential, accidental, and gradual perfections and the likes are defined (including the two terms of the chain, the *summum*, God, and the *non-gradus*). James explains that the essential perfection by which we compare the perfection of the species with each other is generally (*ab omnibus*) called *intensio*. On the contrary, the gradual perfection by which we compare the individuals within one species is generally (*ab omnibus*) called *extensio*.

Chapters three and four are the central chapters of the treatise, since they deal with the main opinions about how the chain of beings is to be constructed and represented. James warns about the fact that there are many opinions concerning the representation of the species according to their perfection. He only reports two views, which have 'magis colorem veritatis'. The first opinion, which we may call the 'arithmetical theory', states that 'the species or the specific latitude is to be imagined by means of numbers.'[31] The second one, which we shall call the 'geometrical theory', affirms that 'the specific latitude is to be imagined by means of a straight line which starts *a non gradu* and ends at the highest degree exclusively' (i.e., the *summum* degree does not belong to the line representing the perfections).[32] According to James, this latter theory comes very close to the truth. The geometrical theory can in turn be supported by 'measuring' the perfection of the species (a) according to the distance from *non gradus*, (b) according to the approximation to the *summum* degree,

---

30 M 206[r]: 'Punctus est quoddam indivisibile secundum quod ab Euclide haberi potest'; M, 206[v]: 'Gradus quoque dicitur in latitudine quoddam ymaginative existens, indivisibile quidem secundum extensionem, secundum vero intensionem divisibile, et hoc imaginandum ⟨est⟩ per lineam rectam perpendiculariter ascendentem totam latitudinem secantem quantum ad intensionem. Ex quo patet quod in qualibet latitudine gradus cui longior linea correspondet dicitur esse intensior et cui brevior dicitur esse remissior.' The question of the relation between *De perfectione specierum*, *De latitudinibus formarum*, and Oresme's *De configurationibus* is still unsettled. This will be the topic of a separate study.

31 M 207[r]: 'Opiniones quoque ad propositum facientes a pluribus pertractare sunt quamplures, quarum tamen duas pro nunc recitabo que magis colorem possident veritatis. Prima opinio est quod species sive latitudo specifica ymaginanda est per numeros et communiter conceditur quod species sunt immediate, quia numeri quilibet sunt immediati …'

32 M 207[r]: 'Secunda opinio in genere est quod latitudo specifica ymaginanda est per lineam rectam incipientem a non gradu et terminatam ad summum gradum exclusive, et cuilibet puncto illius linee correspondet una species … Hec autem opinio capit evidentiam ex sui intrinseca congruitate; et talis ymaginatio, ut michi apparet, satis ad veritatem appropinquare videtur.'

or (c) according to both criteria. This is not all, James explains. There is still an alternative group of two further opinions. The first opinion of this second set affirms that all individuals of the same species are equal in perfection. The second one sustains that the representation is to be made using a transversal line so that we somehow can appreciate the difference of perfection between different species on the same 'axis'. The decisive difference between the first set of opinions and the second set seems to be that the latter are both about the individuals within the species, whereas the former are about the species themselves. Thus, the second theory of the second set introduces an aspect, which is considerably different from the geometrical theory of the first set. According to these theories, the representation is always one-dimensional, no matter whether we take the lowest degree, the highest or both as reference for a quantitative consideration. The second opinion of the second set requires in addition the plane.[33] James' formulation is too concise and lacking clarity. For, it is confusing to present different theories concerning the 'perfection of species' using the word 'species' also for the theories themselves. We would like to know what effect this could have had on the students' minds. Some of the manuscripts include confusing representations, as if the scribes themselves were not completely sure about the matter. However, lack of clarity can sometimes act as a stimulating factor capable of setting in motion a series of further investigations.

2 Perfections and Latitudes in France

In the paper I have mentioned before, Murdoch continued the work started by Thorndike and included some useful indications on a text conserved anonymously in manuscript Paris, Bibliothèque Nationale de France, Ms. lat. 6752, 4^{r}–

33 M 207^{r-v}: 'Et dicta opinio secundum unam speciem triphariе utitur a diversis, secundum vero aliam speciem ab aliis | bifurcatur. Quantum autem ad illam speciem, scilicet quod una species alia perfectior est, ⟨a⟩ aliqui dicunt quod perfectio specierum attenditur penes distantiam a non gradu; ⟨b⟩ alii vero ⟨dicunt quod attenditur⟩ penes appropinquationem ad summum; ⟨c⟩ ceteri namque vero penes utrumque. Item secundum aliam speciem ⟨a⟩ quidam dicunt individua alicuius speciei esse in perfectione equalia. ⟨b⟩ Alii vero ⟨dicunt⟩ unum alio fore perfectius, et isti ultra lineam perpendiculariter ymaginatam ymaginantur alias lineas transversales, ⟨sic quod⟩ in quolibet puncto ⟨assignetur⟩ una linea, et illa linea perfectionem dicunt gradualem.' N.B. James' remark concerning the plane does not necessarily mean that *the surfaces* constructed are semantically relevant as in Oresme's system. For an explanation of Oresme's representation system, see above, n. 29.

236ʳ (= P).[34] Since this text deals thoroughly with *De perfectione specierum*, Murdoch put it into one and the same thematical context along as a set of other writings he dated around 1340–1370. However, later research enables us to provide a partial rectification of this decision. This text, as it had already been discovered some years before, is also extant in a second copy in Genova, Biblioteca Civica Berio, Ms. C.F. 53, 2ʳ–235ᵛ (= G). Further research was able to provide us with a detailed biographical and bibliographical account of its author. His name is Jacques Legrand (ca. 1360–1415/18), a famous writer, involved in the turbulent events of French politics at the beginning of the fifteenth century, besides being an important personality within his order, the same as that of James of Naples.[35] His text belongs therefore to the following generation, many years after 1370. It attests that the interest in the spectrum of ideas we find in the previously mentioned commentaries on the *Sentences* as well as in James of Naples' *De perfectione specierum* was still alive at the beginning of the fifteenth century.

Legrand was very well aware of the current philosophical trends. He knew his topics, no doubt, but he was not a career logician or a natural philosopher, like John Buridan or Blasius of Parma. He enjoyed a more than considerable fame as author of the *Livre des bonnes meurs* and of his *Sophilogium*, hence as theologian and moral philosopher. Most probably, his *Compendium utriusque philosophie* is a collateral work he produced to fulfil the requirements of the statutes of his order for the *Studium* of Paris.[36]

For, to offer an adequate tool to come to grips with 'Aristotle's loquaciousness and verbal difficulties,' so that students do not have to spend so much time on insignificant matters, was Legrand's general motivation to compose this text. Offering a kind of 'textbook', it is obvious that one of the central tasks he intended to accomplish was to summarise Aristotle's ideas. One cannot say

34 Murdoch, *'Mathesis,'* 239.

35 The Genova manuscript was reported by F. Roth, 'Jacques Legrand,' *Augustiniana* 7 (1957), 316–326, and incorporated later in the seminal paper by E. Beltrán, 'Jacques Legrand O.E.S.A. Sa vie et son oeuvre,' *Augustiniana* 24 (1974), 132–160 and 387–414 (here 395–396). For information about Legrand, I am basically following Beltrán's paper and the bibliography mentioned by him. For further bibliography and an edition of the first chapters of the text, see D.A. Di Liscia, 'The Subject Matter of Physics and Metaphysics in Jacques Legrand's *Compendium utriusque philosophie,' Revista Española de Filosofía Medieval* 24 (2017), 249–265, and D.A. Di Liscia, 'Transmutación y movimiento según el tiempo en Jacques Legrand (*Compendium utriusque philosophie* IV, 1–2),' in C.J. Fernández and M. Pérez Carrasco (eds.), Per philosophica documenta. *Estudios en honor de Francisco Bertelloni*, Buenos Aires 2021, 151–175.

36 Beltrán, 'Jacques Legrand O.E.S.A.,' 140.

that he did not deliver it but at the same time it is also true that Legrand did that according to his own criteria of selection, leaving enough room for the presentation of theories (*sententiae*) of other philosophers as well.[37] That means, he has something to say of his own, as is clear from the general way of discussion in the theoretical parts of the work.

His position regarding Aristotelian philosophy is not servile. Even if he tried to give at some passages a favourable description of Aristotle's opinion—which sometimes goes far beyond Aristotle's own intentions—he makes it clear enough that he was not really an Aristotelian 'to the core'. He defended, no doubt, many Aristotelian theories but he did not use the label 'Aristotelian' to approve or reject an opinion or to mean the affiliation to a philosophical group to which he felt especially committed. Other utterances dispersed at difference places of the work—certainly much more severe—, make it clear that he was not very fond of the Calculators, or at least of the particularly subtle treatment of some subjects we would not hesitate to ascribe to the Calculators. He repeatedly uses the kind of expressions we find in the Humanists or in some self-proclaimed Aristotelians tending to denounce the high degree of unnecessary sophistication philosophy was gaining by then, a sophistication which would not only be immoderate but also incompetent and crude. However, it is noteworthy that many of the analytical languages, especially that of *maxima et minima* and *intensio et remissio formarum*, constitute an essential component of Legrand's text. In addition, in contrast with the Calculators' logico-mathematical approach, he tried to integrate in his *Compendium* a significant amount of 'empirical science' (mineralogy, geology, alchemy, some botanic).

Thorndike has already described this text as a work 'on metaphysics and natural philosophy.'[38] For, the general orientation of the *Compendium* is indeed metaphysical, including an important theological component, as Legrand himself explains in the preface, where the metaphysical approach is explicitly announced. Legrand—in what looks like a conscious and deliberate deviation from Aristotle's classification—does not mention mathematics as a speculat-

37 P 1ʳ, G 12ʳ: 'Quia textus Aristotelis nimia prolixitate verborumque difficultate sepius inutili iuvenum proficientium studia retardant temporaque detinent nimium, ideo congruum apparet ipsius Aristotelis aliorumque philosophorum sententias summatim colligere, ut sub brevi compendio que prius extensa erant facilius comprehendantur.' For an edition of the *proemium* according to the Paris manuscript, see D. Lohrmann, '*Motus continuus* und *motus perpetuus* in der mittelalterlichen Technik und Physik,' in A. Speer and D. Wirmer (eds.), *Das Sein der Dauer*, Berlin 2008 (Miscellanea Mediaevalia, 34), 224–243.

38 L. Thorndike, *A History of Magic and Experimental Science*, New York 1923–1958 (8 vols.), 3: 568.

ive science. In fact, he does not mention it at all. Philosophy is either speculative or practical, but the speculative sciences are only two: natural philosophy, which considers its objects as linked to motion, and metaphysics or 'prima philosophia'. The *Compendium* will deal—he declares explicitly—with both disciplines at once, since both collaborate with each other. Especially when a deeper analysis of physical questions is concerned, metaphysics is required. Legrand does not remain at this general statement, but he gives us concrete examples. Thus, when discussing whether motion is a successive thing, Aristotle would have addressed—this question as such does not occur in Aristotle—a problem which belongs to natural philosophy. However, by including the question whether motion can be distinguished from the moved thing, Aristotle went over from physics to metaphysics. Hence there is a very elastic line between physics and metaphysics in the *Compendium*.[39] To the metaphysical-theological tendency contributes essentially the fact that almost in each part of the text Legrand writes on the perfection of species.[40] This topic opens the field to several attempts of quantification following the key notions of the Calculators' tradition.

While we were willing to accept—at best—a merely metaphorical meaning of the expression 'the quantification of Aristotle,' here we find, however, that quantifying is exactly what Legrand himself says he has been doing when discussing the question of the measurement of creatures, or, more exactly, the particular question of the criteria for the measurement of the perfection of forms. One of the discussed theses sustains that each being is to be quantified (*quantificari*) according to the imitation of the first Being.[41] The thesis,

---

39 P 4^{r-v}, G 12^{r}: 'Nam considerationes phisice et metaphisice, cum sint speculative possunt adunari, quinimmo difficulter possunt ab invicem separari quando precipue aliqua materia debet profundari. Nam cum philosophus naturalis considerat de motu locali utrum sit successivus, bene etiam congrueret investigare quid sit talis motus et utrum distinguatur a re mobili. Prima tamen consideratio pertinet philosophie naturali et secunda metaphisice.' Text in Lohrmann, '*Motus continuus*,' 239.

40 The *Compendium* is divided into six parts. For a useful list of the chapters according to the Paris manuscript, see Thorndike, *A History*, 3: 761–766 (Thorndike has already indicated many of the points I am including in this chapter). Unfortunately, Legrand does not mention his opponents by name. He frequently refers to the analytical languages of the Calculators using negative phrases such as 'modernorum indigesta subtilitas' (P 84^{r}) and 'inepta subtilitas' (P 17^{r} and 127^{r}). However, such negative expressions could also be directed at a more general target group.

41 This is discussed in Part II, ch. 14 (*Penes quid attenditur perfectio formarum*, P 35^{r}–36^{r}, G 44^{r}–45^{r}) and ch. 15 (*De aliis opinionibus circa eandem difficultatem*, P 37^{r}, G 45^{r}–46^{r}). See the following passage (P 36^{r}, G 45^{v}): 'Item Commentator 2° *Celi et mundi* dicit quod omnia entia per suas operationes nihil aliud intendunt nisi (quam P) ut assimilentur primo enti. *Sequitur igitur quod penes imitationem primi entis res quelibet debet quantificari*' (my italics).

however, is not to be immediately accepted, since, as the *oppositum* states, 'it seems ... that the perfection of the creatures cannot be quantified according to God, since he is of infinite perfection' and, as Aristotle warns in Book VIII of his *Physics*, *finiti ad infinitum nulla est proportio*.[42] Thus, beyond Legrand's own opinion on this question, it seems clear that 'to quantify' in this context means to bring the problem into the language of proportions, which requires a translation of the question into terms of finite magnitudes. More precisely, by 'a quantification' Legrand means a comparison in terms of more or less perfection, so that the notion of 'latitude' becomes a central one for the whole discussion.

What about the program of Oresme and James of Naples for geometrising qualities and latitudes? Did Legrand know it and was he interested in it? What can we expect from an author who does not even mention that mathematics is one of the three speculative sciences? It is evident that Legrand was by no means an accomplished mathematician, as Oresme was, and his interest in and knowledge of geometry was probably not better than that of James of Naples.[43] However, it is certain that he was familiar with the tradition of the treatise on the latitude of forms, which he picked up and expanded. An example of that is his analysis of different latitudes. Until the treatise *De latitudinibus formarum*, it was usual to classify motions, velocities, latitudes or even qualities—depending on the author—into *uniformis* and *difformis*, both cases according to the time and to the subject or 'magnitude of the body'. Within the latter category it could still be distinguished into *uniformiter difformis* or *difformiter*

42 P 36r, G 45v: 'Oppositum tamen argui potest, eo quod (eo quod] quia P) finiti ad infinitum nulla est proportio, ut dicitur 8° *Phisicorum*, 1° *Celi et mundi*. Videtur igitur quod penes deum, cum sit infinite perfectionis, creaturarum perfectio non debeat (possit P) quantificari. Verum est tamen quod istud argumentum contra sequaces Aristotelis non procederet (concluderet P), quia dicerent deum esse (*om.* P) finiti vigoris et consequenter finite perfectionis. In hoc tamen erraverunt, ut suo loco dicetur.'

43 Inquiring into the questio of 'quomodo est possibile, rem aliquam deo fieri conformiorem et cum hoc distare equaliter a deo', Legrand, introduces, however, the usual example from mathematics, the horn angle (P 36v, G 45v): 'Huius exemplum potest dari in mathematicis, quia angulus qui causatur in circumferentia per diametrum ab intra, qui alio nomine dicitur angulus portionis circuli, potest in infinitum augmentari per circulorum augmentationem.' By so doing, Legrand follows a well-established tradition which goes back at least to Campanus and was continued amongst others by Bradwardine and Peter Ceffons (see Murdoch, '*Mathesis*,' 242–246). For the Campanus passage see Euclid (Campanus), *Elementa*, ed. H.L.L. Busard, *Campanus of Novara and Euclid's Elements*, Wiesbaden 2005 (Boethius. Texte und Abhandlungen zur Geschichte der Mathematik und der Naturwissenschaft, 51/1), 120–122.

difformis. *De latitudinibus formarum* proposed a further division of the difform difformity into *uniformiter difformiter difformis* and *difformiter difformiter difformis*. Legrand, without referring explicitly to *De latitudinibus formarum*, takes up this distinction and proposes to keep on dividing *in infinitum* according to the same criterion, so that in the next step we would have a *uniformiter difformiter difformiter difformis* and a *difformiter difformiter difformiter difformis* latitude, and so on as long as we need.[44]

How should the complete reality be ordered? The discussion of this question could not be missing in the *Compendium*. Besides God as the maximum of perfection, Legrand proposes a universal order of eight main steps starting with prime matter and ending with angels and intelligences.[45] However, the question remains, how the different steps are to be ordered: is a geometrical model or a numerical one to be preferred? For Legrand that is not only a question of 'graphical representation', but also a question about the sort of ontological structure we can attribute to the order in the universe according to perfections. He considers the two opposite options whether the latitude of the creature is something continuous or discrete and, in fact, he seems to be substantially more favourable to the latter position following an ontological approach that already within the core of the Oxford Calculators has found at least one significant adherent, John Dumbleton.[46] Legrand rejects the continuity thesis on the

---

44 Quoted in Thorndike, *A History*, 575, n. 40 (from P 45^v); G 55^v.

45 Thorndike, *A History*, 575. 'Ideo melius dicendum apparet quod latitudo creabilium in octo generibus rationaliter (*om.* P) distinguitur (distinguitur] distingui potest G). Nec tamen sic est intelligendum quin illa genera ulterius in alia genera distingui (distingui] dividi G) possint, sed nihilominus sufficit quod latitudinem creabilium octo (octo] sex P) genera comprehendunt. Unde primum genus vocatur potentiale, in quo reperitur materia prima. Secundum genus est accidentium corporalium, in quo reponuntur omnia accidentia que corporibus (*ante* corporibus] rebus spiritualibus inhere possunt *scr. et del.* G) inherere apta nata sunt. Tertium genus est accidentium spiritualium, in quo ponuntur omnia accidentia que rebus spiritualibus (rebus spiritualibus] spiritibus] P) inherere possunt. Quartum est formarum non vitalium, in quo ponuntur omnes forme rerum inanimatarum. Quintum est formarum vegetabilium, in quo ponuntur omnes forme dantes vegetativam vitam sic quod non sensitivam. Sextum genus est formarum sensitivarum, in quo reponuntur forme omnes rerum sensum habentium, non tamen intelligentium (non tamen intelligentium] dummodo tamen intellectu careant G). Septimum genus (*om.* P) est rationale, in quo reponitur homo et (et] immo etiam G) quelibet res corporea habens intellectum, sive sit producta sive producibilis, si (si] sive G) intelligat cum fantasmate. Octavum genus (genus] *om.* P) est intelligentiale, in quo reponuntur omnes angeli universaliter et universaliter quecunque res intelligens sine fantasmate' (P 48^r, G 58^r; *P* repeats 'in quo reponuntur omnes angeli' after the first ocurrence of 'universaliter').

46 As pointed out by Edith Sylla, this opinion had been defended by Dumbleton, when he wrote: 'Every specific essence is simply indivisible in its nature and from indivisibles a

basis of three points. First, as he has shown discussing the principle of individuation in the first part of the work, this position leads to contradictions. Second, it involves an unacceptable interpretation of article 124 (147) of the Condemnation of 1277, which he considers in any case hardly acceptable.[47] Third, and more relevant to us, some authors believe that all individuals of the same species are equal in their essential properties (*omnia individua eiusdem speciei in essentialibus ⟨sint⟩ equalia*, P 46ᵛ, G 56ᵛ). Hence, each excess would vary the species. Supposing then the infinite divisibility of the continuum, we would always be able to find a new species between two other species and, thus, an infinity of species would follow. This quite popular position, Legrand admits, has something of the truth and something wrong. It is true that within a species can there be no 'essential excess' because in that case each excess would change the species. But in his view, it is fundamentally wrong that the latitude is a continuous magnitude, so that we could always insert an infinite number of species between two already existing ones.[48]

---

continuum can in no way be made. Therefore the perfections of things do not make a latitude nor are they properly comparable to a latitude …' (transl. E.D. Sylla, 'The Oxford Calculators and Mathematical Physics: John Dumbleton's *Summa logicae et philosophiae naturalis*, Parts II and III,' in S. Unguru [ed.], *Physics, Cosmology, and Astronomy, 1300–1700: Tension and Accomodation*, Dordrecht 1991 [Boston Studies in the Philosophy of Science, 126], 129–161, at 142; 156 for the Latin passage). Thus, Dumbleton, if I am interpreting this passage correctly, is here questioning or even inhibiting the very application of the notion of latitude to the problem of the great chain of being. For the authors I am discussing in this paper the question was rather *how* to apply this concept to the perfection of the beings in the universe. That the whole universe can be ordered in a great chain of latitudes was something previously accepted.

47 With article 124 (147) it should be ensured that the soul of Christ is always better than the soul of Judas: 'Quod inconveniens est ponere aliquos intellectus nobiliores aliis, quia, cum ista diversitas non possit esse a parte corporum, oportet quod sit a parte intelligentiarum; et sic animae nobiles et ignobiles essent necessario diversarum specierum, sicut intelligentiae. Error, quia sic anima Christi non esset nobilior anima Judae' (cf. R. Hissette, *Enquête sur les 219 articles condamnés à Paris le 7 mars 1277*, Leuven 1977 [Philosophes médiévaux, 22], 227). Legrand is critical of the article itself, as far as it permits a comparison, which is alone unacceptable. Referring to the previous authorities of Giles of Rome and Henry of Ghent, he preferred to avoid an in-depth discussion of the matter (P 46ᵛ, G 56ᵛ): 'Item aliter dici potest quod predictus articulus inconsulte seu minus diggeste fuit condempnatus (damnatus G). Hanc tamen responsionem dare non auderem nisi in doctoribus solemnibus etiam parisiensibus simile legissem, sicut in Egidio et Henrico de Gandavo. Credo tamen melius esse in predictis responsionibus sistere, quia presens responsio presumptione suspecta est.'

48 P 46ᵛ–47ʳ: 'Hec autem opinio | valde communis est. Aliquid tamen continet veritatis et aliquid falsitatis. Nam verum est quod in nulla specie reperitur excessus essentialis. Verum

On the opposite side of the discussion stands the point of view of discontinuity, firmly grounded on the statement most often quoted in this context: *forme sunt sicut numeri.* Legrand holds this Aristotelian principle without reticence. The species or forms are compartments of reality the perfections of which are among themselves incommensurable. Between one form and another, as between the numbers, there is nothing. His main argument for discontinuity is that it is enough to find only one part in the whole that is not continuous to be able to affirm the discontinuity of the whole. And this part is, Legrand affirms quoting Augustine and some passages from Aristotle, prime matter. For, this, as pure potency, stands at the lowest level in the universe. Its perfection has to be *simpliciter indivisibilis.* If this were not the case, it would be possible to find something of more imperfection, for instance by halving. Thus, 'the parts of the latitude of the created beings are punctual and indivisible and, consequently, the whole latitude is to be established arithmetically.'[49]

However, even if an arithmetical series would better reproduce the order of the species in the universe, Legrand does not renounce to a possible geometrical representation. To do that, he must return to the notion of latitude, which seems to be less applicable against an arithmetical background. Moreover, he devotes a complete chapter to explaining 'how the latitude of created beings is *uniformly difform.*'[50] With respect to his use of the notion of latitude, Legrand warns us that this term in a proper sense (*ex vi nominis*) is understood as involving a continuous quantity. However, in a broader sense, it can well be applied to discrete quantities and hence to the latitude of the created beings.[51]

---

est etiam quod quilibet excessus variat speciem. Sed falsum est quod talis latitudo sit continua. Et consequenter falsum est inter quascumque species infinitas posse produci ...'

49 P 47^r, G 57^r: '... materia prima est infima creatura posssibilis, ut patet per Augustinum 12° *Confessionum* dicentem eam esse prope nihil. Hoc etiam satis innuit Aristoteles 1° *Phisicorum* et 2° *Metaphysice* dicens quod materia prima est pura potentia. Videtur igitur quod sit natura imperfectissima non solum pro nunc existens imo etiam possibilis. Ex quo sequitur quod perfectio materie nullo modo est continua imo simpliter indivisibilis. Nam si esset divisibilis, tunc creatura inperfectior esset producibilis, ut puta illa que medietatem precise haberet sue perfectionis. Et sic materia prima non esset pura potentia nec esset ponenda prope nihil. Relinquitur igitur quod *partes latitudinis creabilium sunt punctuales et indivisibiles et consequenter tota latitudo ponenda est arismetica*' (my italics).

50 Ch. 23: *Quomodo latitudo creabilium est uniformiter difformis* (P 45^v–46^r, G 55^v–56^r). It should be pointed out that Legrand distinguishes three kinds of *uniformitas/difformitas*: *secundum excessum, secundum proportionem* and, especially important here, *secundum recessum a proportione.* I am preparing a critical edition of the complete *Compendium* with commentaries on each chapter.

51 P 43^v, G 53^v: 'Unde pro isto passu et pro (*om.* P) sequentibus notandum est (*om.* G) quod,

How can it be said that the totality of the beings in the universe is ordered according to their perfection in the way of a uniformly difform latitude? In explaining this notion, Legrand uses the model of light proceeding from a luminous body located in the centre of a circle.[52] Consequently, the geometrical model he proposes is as follows:

> Hence, it is to be imagined that as the light coming from a beaming body staying in the centre of a circle is diffused throughout the space of the circle, in the same way from God as the centre of a such perfection flow and are derived uniformly difformly all perfections, so that, in the same way as that part is more luminous which is closer to the beaming body, so it is also that this creature is more perfect which is more in conformity with the divine perfection. And with respect to distance the opposite has to be understood.[53]

Unfortunately, none of the extant manuscripts include diagrams showing us exactly how Legrand would draw a representation of this idea. With some interpretative permission, I think that Legrand's system could be reconstructed according to a standard model for the propagation of light (Fig. 11.1).

Since the latitude of all beings is not simply uniform but uniformly difform, one could represent it 'through the space of a circle.' However, in his example Legrand is probably thinking of a cone with God in its centre and the perfections deriving from Him, similar, he tells us, to light. Legrand goes back to a model which we know had been present a long time before in the Calculators'

licet iste terminus "latitudo" ex vi nominis dicat quantitatem continuam, tamen quandoque large accipitur et ita in quantitate discreta latitudines imaginantur sicut latitudo creabilium, licet sit discreta, tamen latitudo appellatur. De latitudine numerorum etiam pariter dicendum.'

52 P 44^{v}, G 54^{r}–54^{v}: 'Insuper alia est latitudo difformis ... cuiusmodi est latitudo luminosa procedens a corpore luminoso. Verbi gratia describatur circulus in cuius centro ponatur corpus luminosum. Hoc enim posito clarum est | quod lumen uniformiter difformiter dispanditur per spatium quod est a centro usque ad circumferentiam. Nam quanto spatium est corpori luminoso propinquius, tanto est luminosius, et quanto distantius est, tanto est minus luminosum.'

53 P 45^{v}, G 55^{v}: 'Unde imaginandum quod sicut a corpore luminoso existente in centro alicuius circuli lumen uniformiter difformiter in tali diffunditur per spatium circuli, sic etiam a deo tamquam a centro cuiuslibet perfectionis fluunt et derivant omnes perfectiones uniformiter difformiter, ita quod sicut illa pars est magis luminosa que corpori luminoso est propinquor, sic etiam illa creatura est perfectior que divine perfectioni est conformior. Et de distantia per oppositum est intelligendum.'

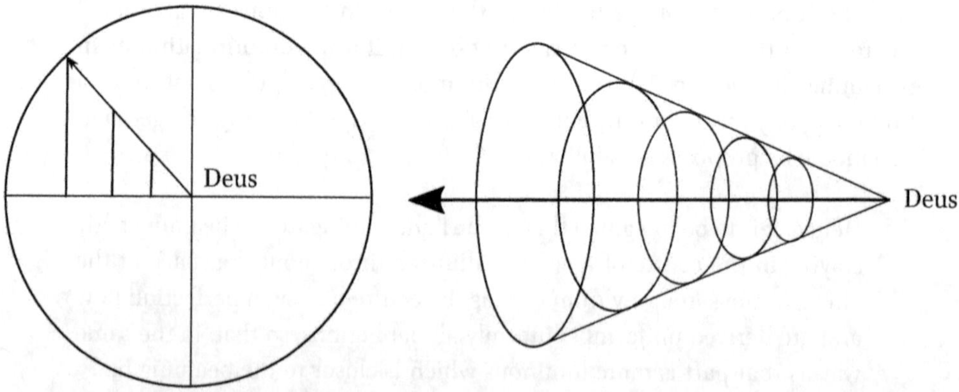

FIGURE 11.1 Interpretation of the representation system mentioned in Jacques Legrand's *Compendium utriusque philosophie*

texts.[54] Thus, the different perfections of the different beings were related to their origin, the Creator, as light to its source, according to their respective distances.

However, the concrete application which Legrand seems to think of is, already at a first glance, not unproblematic. If the lines are arising from the centre, where in fact God is located, the system seems to indicate that the lowest degrees of beings are closer to Him, the supreme being, and increase gradually as they move away from Him towards the circumference of the circle. That would be unacceptable in a system of continuous degrees with a supremum placed in the highest position. Yet, that is not Legrand's purpose. He does not affirm a continuous structure of the whole construction and, at the same time, he explicitly wishes to set God outside the system of degrees of perfection, so that an extreme difference of incomparable magnitude between Creator and creature emerges.

54 A fact that has been pointed out by Edith Sylla for the special case of Giovanni da Casale. See E.D. Sylla, 'Medieval Quantifications of Qualities: The "Merton School",' *Archive for History of Exact Sciences* 8 (1971–1972), 9–39, at 35. To avoid possible misunderstandings, it is necessary to say that Legrand has no idea of Oresme's notion of the quantity of a quality. See above, n. 29. Besides that, it is also clear that no trigonometry is involved in the whole speculation, even if we could be tempted to see the sinus of the angle coming out from the centre of the system, i.e., from God.

3 Perfections and Latitudes in Germany

Besides some copies of Oresme's *De configurationibus* and of his questions on Euclid, there are impressively many copies of the treatise *De latitudinibus formarum* produced in the German countries during the last decades of the fourteenth and the first half of the fifteenth centuries. Doubtlessly, this is not a coincidence since, based on *De latitudinibus formarum*, a new related middle science was taught at some of the Eastern European universities. Both facts are connected to each other and testify that the whole spectrum of problems involving *latitudines*, especially in their more geometrical approach, belonged to the standards of good scientific-philosophical training at that time.[55] Also the general problem of the perfection of the species received a great deal of attention in the same academic context. There is enough evidence that also here the reception of James of Naples' *De perfectione specierum* had a determining impact on this development. This statement can be very well illustrated by making reference to a small group of less famous authors among whom there is kind of connection—as yet not sufficiently investigated—through biographical and/or intellectual relationship.

The background and starting point here are not the Oxford *Calculatores*, at least not directly, but rather the later generation of the Parisian circle around Buridan who integrated the new quantitative approach into their Aristotle commentaries.[56] One of the most significant names in this case is Lawrence of Lindores, who followed Buridan commenting on the *Physics* and, at the same time, complemented his commentary with a more decided mathematical treatment of some problems related to infinity and continuity in local motion. We know for sure that Lawrence of Lindores was aware of the modern geometrising approach we are discussing in this chapter, since he mentions it as a middle science referring to both *configurationes* and *latitudines formarum*. Moreover, he intended to apply these notions to the analysis of the infinite like Oresme had done in his *De configurationibus*, even if mistakenly. Lindores made a simple mathematical error which shows that his understanding of the matter was lim-

55 See D.A. Di Liscia, 'The *"Latitudines breves"* and Late Medieval University Teaching,' *SCIAMVS* 17 (2016), 55–120; D.A. Di Liscia, 'La "latitud de las formas" y la geometrización de la ciencia del movimiento,' *Mediaevalia. Textos e estudos* 36 (2017), 75–114.

56 E.D. Sylla, 'Transmission of the New Physics of the Fourteenth Century from England to the Continent,' in S. Caroti and P. Souffrin (eds.), *La nouvelle physique du XIVe siècle. Actes du Colloque International (Nice, 3–5 septembre 1993)*, Florence 1997 (Biblioteca di Nuncius, 24), 65–110.

ited. However, his intent to apply Oresme's representations within a *Physics* commentary is noteworthy.[57]

In Germany, along with the general new tendencies coming from Paris, a group of 'Lindores followers' arose which was very diligent in spreading the texts of their master and in going in-depth into the mathematical approach presented in the analysis of some questions related to motion and qualities.[58] It is therefore not completely adequate to characterise these authors as 'Buridanians', since, even if they were followers of Buridan in many ways, they made more room for mathematics in their commentaries than Buridan himself.[59] In this group we may count at least three masters who were active at the universities of Basel, Cologne, Erfurt, Leipzig, and Rostock-Greifswald. The first is John Aurifaber, who has been considered a pupil of Lindores at Paris, though he was later active at the new University of Leipzig around the end of the fourteenth century. Aurifaber is the author of an extensive *questiones* commentary on the *Physics*, which is extant in only one manuscript.[60] One

57 T. Dewender, *Das Problem des Unendlichen im ausgehenden 14. Jahrhundert. Eine Studie mit Textedition zum Physikkommentar des Lorenz von Lindores*, Amsterdam 1999 (Bochumer Studien zur Philosophie, 36), 113, n. 119. The case is analysed in D.A. Di Liscia, *Zwischen Geometrie und Naturphilosophie: die Entwicklung der Formlatitudenlehre im deutschen Sprachraum*, unpublished Ph.D. dissertation, University of Munich, 2003, 86–95 (esp. 35 for the passage where Lindores mentions the configurations doctrine and the latitude of forms).

58 The question why the majority of copies of Lindores' *Questions on Physics* are German, whereas he himself studied in Paris and was active in Scotland (as Rector of the newly founded University of Saint Andrew) deserves more examination. See J.H.S. Baxter, 'The Philosopher Laurence of Lindores,' *Philosophical Quarterly* 5 (1955), 348–354. Nothing is known about a possible stay in Germany. The diffusion of his texts on natural philosophy could have taken place via some of his Paris pupils, like John Aurifaber (see below, n. 60).

59 Questions of 'group denomination' are difficult and, as far as such denominations tend to overvalue the significance of 'the group' over the individuals, always disputable. It is unquestionable that the enormous influence of Buridan in this context was, as Markowski has pointed out, a determining factor in academic life, so that to a certain extent it is adequate to speak of 'Buridanism'. Markowski's purpose seems to have been to offer an alternative to 'nominalism', the traditional category often used by Michalski and others (see for instance his 'Der Buridanismus an der Krakauer Universität im Mittelalter,' in: O. Pluta [ed.], *Die Philosophie im 14. und 15. Jahrhundert. In memoriam Konstanty Michalski [1879–1947]*, Amsterdam 1988 [Bochumer Studien zur Philosophie, 10], 245–260). For a discussion of this problem see J.M.M.H. Thijssen, 'The Buridan School Reassessed. John Buridan and Albert of Saxony,' *Vivarium* 42 (2004), 18–42. The relevant point here is that Buridan's approach is seldom a mathematical one whereas the approach of these authors we should according to Markowski call 'Buridanians' included more calculatory materials.

60 Leipzig, Universitätsbibliothek, Ms. 1423, 1^r–220^r. Some general information about it can be found in M. Markowski, 'Komentarz Jana Aurifabera z Erfurtu do *Fizyki* Arystotelesa

generation later, Nicholas of Amsterdam (Nicolaus Theoderici of Amsterdam) was active at the universities of Cologne, Erfurt, and Rostock-Greifswald. His wide commentary activity, including sometimes different versions of the same work, covers almost the complete Aristotelian corpus. Finally, John of Wesel (Johannes Rucherat de Wesalia) is to be mentioned, who for reasons which are not immediately related to our topic is perhaps a better-known personality.[61] John of Wesel commented on Aristotle's logic, on the *Physics* and, perhaps—as we will see—also on the *Metaphysics*. According to the current state of research, none of these *magistri* wrote mathematical works, neither on pure mathematics nor in the style of the *Calculatores*. However, they were all interested in problems of geometrisation as coming from the latitude of forms and the perfections of species tradition. They used these notions as innovative tools for commenting on Aristotle. By so doing, they contributed to a quantifying approach in the understanding of those traditional texts.

The first and perhaps more impressive witness we can mention of this approach, is the manuscript København, Kongelige Bibliotek, Ms. Ny kgl. saml. 190 4° (= K_a), containing a copy of *De latitudinibus formarum*. This manuscript, well-analysed and often mentioned in different studies on Nicholas of Amsterdam, includes also other related texts, among which a copy of James of Naples' *De perfectione specierum* with the only commentary to it known until now.[62] Since we know for sure that this manuscript belongs to the North-German area where Nicholas of Amsterdam and others were active, we can conclude that people were here not only copying the text of *De perfectione specierum* but also studying it in detail. But that is not all. Testifying to the same interest in this trend, also another relevant text is contained in the same manuscript. This is the text which Anneliese Maier had already referred to as *Tractatus bonus de uniformi et difformi* and which later was edited by Marshall Clagett as anonym-

w rękopisie 1423 Bibliotheki Univerwersyteckiej w Lipsku,' *Acta mediaevalia* 12 (1999), 335–343. Markowski affirms Aurifaber's connection with Lindores, and following him also C.H. Lohr, *Latin Aristotle Commentaries I.1. Medieval Authors A–L*, Florence 2013 (Corpus philosophorum medii aevi. Subsidia, 17), 239. Aurifaber's authorship of questions on *De caelo*, *De generatione et corruptione* and *De sensu et sensato* in the same manuscript is considered doubtful.

61 Already in the nineteenth century, John of Wesel was called a 'forerunner of Luther' by having questioned Papal authority and rejected indulgences. C. Ulman, *Johann Wessel, ein Vorgänger Luthers. Zur Characteristik der christlichen Kirche und Theologie in ihrem Uebergang aus dem Mittelalter in die Reformationszeit*, Hamburg 1834.

62 For a description and analysis of it in connection with Ms. Thott 583, see J. Pinborg, 'Die Aristoteles-Quaestionen des magister Nicolaus von Amsterdam,' *Classica et mediaevalia* 25 (1964), 244–261. I completed this description in Di Liscia, *Zwischen Geometrie und Naturphilosophie*, 422–424. See also above, n. 7.

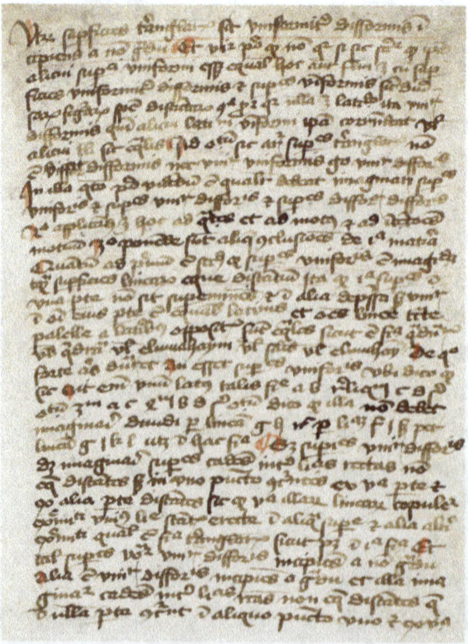

Utrum superficies triangularis sit uniformiter difformis incipiens a non gradu. Et videtur primo quod non, quia, si sic, sequeretur quod ipsa alicui superficiei unifomi esset equalis. Hoc autem falsum est, cum superficies uniformiter difformis et superficies uniformis sint diversarum figurarum specie distinctarum. Consequentia probatur, quia nulla est latitudo ita uniformiter difformis quin alicui latitudini uniformi ipsa correspondeat vel alicui tali sit equalis.
Ad oppositum sic arguitur: superficies triangularis non est difformiter difformis nec uniformiter uniformis; ergo uniformiter difformis.

In illa questio⟨ne⟩ primo videndum est ...
In illa questio⟨ne⟩ primo /MS Erfurt 325, 43r/ videndum est ... etc.

[Tractatus bonus de uniformi et difformi]

/ Notandum est diligenter qualiter debeamus ymaginare quid sit superficies uniformis et quid sit superficies /2/ uniformiter difformis et quid superficies difformiter difformis.

Unde sciendum quod superficies uniformis debet ymaginari /3/ tanquam superficies linearum equidistantium ita quod huiusmodi in una sui parte non sit supereminens et in alia parte depressa sed quod /4/ uniformiter in omni eius parte sit equalis latitudinis et extensionis ita quod omnes linee tracte parallele (/) a lateribus oppositis sint equales. /5/ Cuiusmodi est figura quadrangularis: vel quadratum [vel tetragonus longus] vel figura elmuain vel aliqua alia similis elmuain.

FIGURE 11.2 Beginning of the text until now wrongly called *Tractatus bonus*. Left: Reproduction from manuscript København, Kongelige Bibliotek, Ms. Ny kgl. saml. 190 4°, 167ʳ showing the hitherto unknown beginning of the question (*Utrum superficies triangularis* ...). Right (at the top): Transcription of the corresponding passage missing in Clagett's edition (From *Nicole Oresme and the Medieval Geometry of Qualities and Motions. A Treatise on the Uniformity and Difformity of Intensities known as* Tractatus de configurationibus qualitatum et motuum by Marshall Clagett. Reprinted by permission of the University of Wisconsin Press. © 1968 by the Board of Regents of the University of Wisconsin System. All rights reserved). Right (below): Beginning of the text—introducing brackets for the denomination—in Clagett's edition from the only until then known manuscript Erfurt, Universitätsbibliothek, Dep. Erf., CA Q. 325 (with 'notandum' for 'videndum' in Ms. Ny kgl. saml. 190 4°)
REPRODUCTION WITH KIND PERMISSION OF THE KONGELIGE BIBLIOTEK

ous from the one Erfurt manuscript as an appendix to his edition of Oresme's *De configurationibus*.[63] This manuscript clearly shows that the beginning part of the text was lacking and that it is not a treatise but a *Quaestio*, the beginning of which is clearly conveyed in this copy: *Utrum superficies triangularis sit uniformiter difformis incipiens a non gradu* (see Fig. 11.2).

Actually, the conveyed text is made up of two questions, the authorship of which is not immediately evident. According to the explicit, a possible can-

63 Maier, *An der Grenze*, 362–364; Clagett, *Nicole Oresme*, 576–621. The Erfurt copy is in Universitätsbibliothek, Dep. Erf., CA Q. 325 43ʳ–45ʳ.

didate is John Aurifaber, but even Lindores and—more likely in my opinion—Nicholas of Amsterdam could have written this text or a part of it.[64]

In addition, the manuscript includes among other texts a part of Nicholas' commentary on the *Physics* and an isolated question from Lindores' com-

64 In this respect it is important to underline that in Erfurt the latitude of forms was taught and there is a lot of material directly related to that issue in the *Amploniana*. The attribution to Aurifaber is by no means certain, so that the definitive determination of the authorship is still an open question. The question *Utrum superficies triangularis* ... (edited by Clagett as *Tractatus bonus*) is followed by another question *Utrum forma substantialis producatur in tempore vel in instanti* (174r–178v) which ends: 'Et hec sunt plura dicta Iohannis Aurifaber circa *primum Physicorum*' (Lohr, *Latin Aristotle Commentaries I.1.*, 240, reads 'probabiliter' instead of 'plura'). However, I was not able to identify this text in Aurifaber's commentary on the *Physics* (book I), mentioned above. A relevant indication in the text is that the second question includes a self-reference. In the second of both questions (*Utrum forma substantialis producatur in tempore vel in instanti*, K_a 174r), a series of conclusions are presented after the *oppositum*. The discussion context of the self-reference is given by the third conclusion: 'Est igitur tertia conclusio illa ⟨quod⟩ forma substantialis producitur successive in materia.' For the affirmation of one of the consequences, the argumentation moves to the context of the theory of the soul (177r): 'Et probatur consequentia eodem modo si anima successive produceretur. Probatur, quia forma accidentalis potest produci simul tota; igitur etiam forma substantialis. Consequentia probatur, quia plus videtur hoc de substantiali quam de accidentali. Et antecedens patet ex premissa conclusione. Item si in medio uniformiter subito irradiatur magnum luminosum, non videtur aliquod nisi quod tale medium totum subito illuminaret. Igitur lumen ibi subito producetur et tamen est qualitas extensa; igitur. *Et de hoc visum est in questionibus super secundum de anima*.' It is therefore clear that the author is referring to his own commentary on *De anima*. Lohr, *Latin Aristotle Commentaries I.1.*, 240, lists a text '*In libros De anima*' as a work by Aurifaber which apparently is unlocated or missing. That makes an attribution to Aurifaber difficult, except that one could take this sentence as a confirmation of the existence of such a commentary by Aurifaber which we, however, cannot verify. Lawrence of Lindores and Nicholas of Amsterdam are also possible candidates. Both commentaries are still unedited. Question 20 in Lindores' second book on *De anima* runs *Utrum species qualitatum per se et proprie sensibilium habeat generationem instantaneam vel successivam*. L. Moonan, 'Lawrence of Lindores on Life in the Living Being,' *Classica et mediaevalia* 27 (1966), 349–374, gives the *tabula quaestionum* according to manuscript København, Kongelige Bibliotek, Ms. Thott 583, 292r (this question on 371). Nicholas of Amsterdam has also a question related to that topic in both versions of his commentary on *De anima*: *Utrum lumen generatur in medio in instanti*. See O. Pluta, 'Nicholas of Amsterdam: Life and Works,' *Bochumer Philosophisches Jahrbuch für Antike und Mittelalter* 16 (2013), 185–265, 224 (in the first version this is q. II.13, in the second version q. II.14). On Nicholas' commentary, see additionally Z. Włodek, 'Le commentaire de Nicolas d'Amsterdam sur le *De anima* d'Aristote. Introduction—Textes inédits,' *Mediaevalia Philosophica Polonorum* 11 (1963), 23–42, and O. Pluta, 'Materialism in the Philosophy of Mind. Nicholas of Amsterdam's *Quaestiones De anima*,' in P.J.J.M. Bakker and J.M.M.H. Thijssen (eds.), *Mind, Cognition and Representation. The Tradition of Commentaries on Aristotle's* De anima, Aldershot 2007 (Ashgate Studies in Medieval Philosophy), 109–126.

mentary on the *Physics*, and not just any question, but exactly the question in which Lindores analyses (in the style of Oresme) the aforementioned case of intensification of a velocity *in infinitum*. The commentary on *De perfectione specierum* in the same manuscript could also be by Nicholas of Amsterdam himself. At least we can say that this stands completely in harmony with the place Nicholas devotes to the presentation of these problems in his commentary on the *Metaphysics*. At any rate, it is interesting to remark that the author of this commentary on *De perfectione specierum* knows the circular figure described by Legrand or a similar one which he attributes to Peter of Candia.[65]

As a matter of fact, Nicholas of Amsterdam has received the attention of outstanding scholarship in the last years.[66] His commentary on the *Metaphysics* shows us how dense the late medieval Aristotelian text tradition is, even to the point that in some cases it is difficult to distinguish works and authors from each other. Many of the authors were working according to the same or a similar pattern of learning and teaching, so that texts often contain striking similarities or even repetitions. Probably, the problem lies *in re ipsa*, not only in the fragmentary way in which these texts have survived after centuries of anti-scholastic philosophy.

According to the current state of research, Nicholas' commentary on the *Metaphysics* is found in at least two versions. From one of these versions, there is a copy with a clear attribution to John of Wesel in manuscript Basel, Universitätsbibliothek, F. VIII.7, 1^r–261^v (= B). However, it is probably a work by

65 K_a 192^r: '… Non gradus essendi simpliciter est circumferentia imaginarie infinita in qua est Deus a cuius centro egrediuntur genera per modum triangulorum infinitum protractum versus circumferentiam imaginarie infinitam in quibus multipliciter species se invicem excedent per modum latitudinis uniformiter difformis, de qua imaginatione magis videatur Petrus de Candia circa secundum *Sententiarum*.' On Peter of Candia, see above n. 11. Peter of Candia's *In sententias* could have been an important source for Jacques Legrand as well, as we have seen.

66 For Nicholas of Amsterdam's works see the seminal paper by Pluta, 'Nicholas of Amsterdam.' Specifically on Nicholas of Amsterdam's *Metaphysics* commentary see Bakker, 'Fifteenth-Century Parisian Commentaries'; P.J.J.M. Bakker, 'Nicholas of Amsterdam on Accidental Being: A Study and Edition of Two Questions from his Commentary on the *Metaphysics*,' *Bochumer Philosophisches Jahrbuch für Antike und Mittelalter* 15 (2012), 131–180; H.-U. Wöhler, 'Der operationale Konzeptualismus des Nikolaus von Amsterdam vor dem Hintergrund der Universaliendebatte,' *Bochumer Philosophisches Jahrbuch für Antike und Mittelalter* 15 (2012), 26–36, and H.-U. Wöhler, 'Nicolaus de Amsterdam, *Quaestiones circa libros Metaphysicae Aristotelis*, Liber VII, Quaestio 12: *Utrum ponenda sint aliqua universalia realia a singularibus secundum esse distincta*,' *Bochumer Philosophisches Jahrbuch für Antike und Mittelalter* 17 (2014), 221–250.

Nicholas of Amsterdam.[67] The link between these two *magistri*, for a certain time both at Erfurt, is not only external but also conceptual. John of Wesel was one of those teachers who considered the 'science of the latitude of the forms' to be a kind of middle science.[68]

In these later 'Buridanian' texts, we also find several references to the Oxford Calculators and to the different problems of quantifying Aristotle, references which are not present in Buridan's *Physics*. Commenting on Book VIII—we are coming closer to our mention of Nifo at the start—Nicholas asks different questions about the perfection of the species. The central one runs *Utrum entium universi perfectio specifica sit more numerorum ymaginanda*. The discussion of the topic starts with some previous 'notanda' about the question: Is the specific perfection in the universe to be imagined by numbers? Or perhaps better: 'which is the best form to imagine the specific perfection of the beings in the universe?'

The universe—as briefly described in the *notanda*—is an aggregate of all beings (including God or not), the parts of which are species, which are themselves not subjected to change. Supposing that this would be understood in the sense of 'all that there is,' then its form is the order according to which these beings are organised. This order is an order of 'priority' in dignity and perfection.[69] Now, we might ask, what does that have to do with mathematics or any kind of quantification? The answer is: species are 'parts', hence the universe is made up of parts and, as Aristotle himself had noticed in book V of the *Meta-*

67 Until now, two further manuscripts of this version have been reported: München, Bayerische Staatsbibliothek, Clm 28.395, 1ra–103ra (M^2), and St. Gallen, Stiftsbibliothek, Ms. 840, 1^{r}–205^{r} (S), both without any attribution at all. In a recent paper I proposed to call 'Nic2' the version I am going to use here, only to distinguish it from the other version contained in more manuscripts (among them, the Copenhagen manuscript 190 4° I have mentioned before), which I have called 'Nic1' (D.A. Di Liscia, 'The Questions on the *Metaphysics* attributed to Johannes Rucherat de Wesalia [A Textual Analysis]', *Filosofia. Revista da Faculdade de Letras da Universidade do Porto* 35–36 [2018–2019], 57–75). Nic2 is doubtlessly a work that stems from Nicholas of Amsterdam or the content of which essentially goes back to him (even if an attribution to John of Wesel would be possible on the basis of the Basel attribution). For this reason, I shall refer to this version as a work by Nicholas of Amsterdam.

68 See D.A. Di Liscia, 'Der Kommentar des Johannes Rucherat de Wesalia zur Aristotelischen Physik: seine Bedeutung und Überlieferung,' *Codices Manuscripti et Impressi* 99/100 (2015), 9–28.

69 B 187^{v}, S 144^{v}, M^2 70va: 'Sciendum quod universum capitur pro toto agregato ex omnibus entibus a Deo dependentibus excluso Deo, vel alio modo incluso Deo, cuius partes principales sunt species, quia tales perpetue et incorruptibiles manent ad modum expositionis prius date. Et forma autem universi est quidem ordo illarum partium universi secundum quem quedam sunt aliis priora, alia vero posteriora dignitate et perfectione.'

physics, parts belong to the category of quantity or *quantum*. This is, however, the object of mathematics, so that, consequently, this problem is usually to be discussed *per mathematicam*.[70]

That is what Nicholas tells us about the application of mathematics to metaphysics. In my view we here have a major change of attitude which is not obvious and which is very clearly connected to the Calculators' tradition. As already said, by this time the latitude of forms was conceived as a 'new middle science' which lies between physics and geometry. Here the mixing approach is touching upon the other speculative science, metaphysics. We have in this field—says our commentator—some questions which, as usual, are to be treated mathematically. The conceptual background is a kind of 'mereological' understanding of the species: because these are understood as parts of the universe, we enter the field of mathematics. The author of this text adds a further requirement by explaining that what we are looking for here is a well-knowable, perceptible or intellectual model or pattern expressing the mutual relationship between the perfections of the beings in the universe (including God).

With respect to the question concerning the best representation of the 'world system', three general positions are discussed: (a) numbers are the best (this is, of course, the arithmetical theory we have already mentioned); (b) geometrical entities, such as the line, are the best; and (c) the latitude *uniformiter difformis*, which we represent using a triangle. These are the same positions mentioned by James of Naples, albeit not so clearly. The third answer is considered the best:

> Third is the imagination through the way of a uniformly difform latitude starting from non-degree, which we describe using a plane triangular figure and, above all, a rectangular one.[71]

The mention of a rectangular figure could be misleading here: for this kind of latitude, a triangular figure is the usual representation. It is possible, that Nicholas of Amsterdam has the 'mean degree theorem' in mind, which also belongs to the discussed problems in this context and which he discusses in more detail

---

70 B 187ᵛ–188ʳ, S 144ᵛ, M² 70ᵛᵃ: 'Quia autem ex partibus constat universum, partes autem quantitati vel quanto conveniunt, ut patet per diffinitionem quanti 5° *huius* [= *Metaphysica* V.13, 1020a6–10], quantum autem vel quantitas obiectum est mathematice, ideo per mathematicam solet declarari quomodo se habeant species tamquam partes universi ad invicem.'

71 B 188ᵛ, S 145ʳ, M² 70ᵛᵇ: 'Tertia est ymaginatio per modum latitudinis uniformiter difformis a non gradu incipientis talis quam describimus per figuram planam triangularem et maxime rectangularem.'

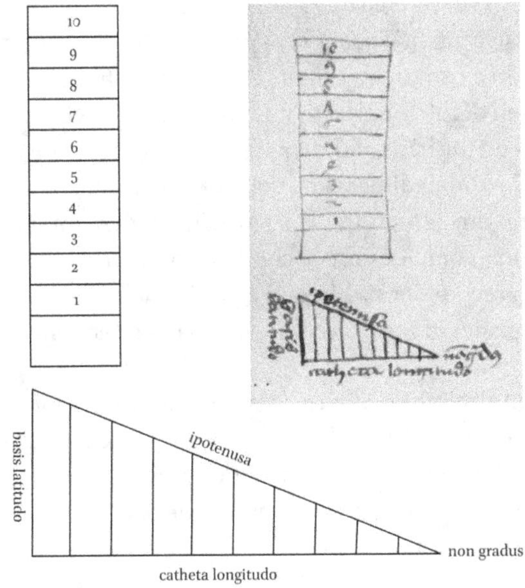

FIGURE 11.3
Figures in the *Metaphysics* commentary by Nicholas of Amsterdam according to Basel, Universitätsbibliothek, Ms. F. VIII. 7, 188v (right), and their corresponding reconstruction (left). Please note that in this manuscript the text is attributed to John of Wesel.

in his questions on the *Physics*.[72] However, as one can see in the figure included in one of the manuscripts with added explanations, the representation is built upon a triangle and is similar but not equivalent to those we know from Oresme and from the treatise *De latitudinibus formarum*.

Also the terminology is slightly different. By turning the figure 90°, we can, however, very well comprehend how the figure works. The lowest vertex represents the lowest degree of being, without perfection. As we ascend, we will find more and more perfect beings, the perfections of which are represented perpendicularly to their latitudes. Because this line is a continuous magnitude, it is possible to intercalate some other species in between, so that no gaps arise. That is of course an advantage over a numerical approach. The text includes a brief account of this representation system:

> Hence, the species of the universe are composed of degrees in latitude, to which correspond lines in the triangle. However, a latitude uniformly difform is one in which the excess of the degrees which lie at the same distances is precisely the same as ⟨the excess⟩ which the lines in the triangle have which are equidistant to the cathetus. However, a latitude

---

72 Question V.10 runs *Utrum omnis latitudo uniformiter difformis correspondeat alicui suo gradui intrinseco* (Pluta, 'Nicholas of Amsterdam,' 211).

> uniformly difformly intense is described by means of a triangle because of this equivalence: an intensive gradual quality lies on an extended subject and therefore intensively. In a certain way, to the extension of the subject corresponds one dimension of the triangle, that is, the longitude which is signified by the line drawn on the basis of the triangle. But the intension of the quality corresponds to another dimension, that is, the latitude, which we understand by the line erected on the basis which we denominate catheton. Thus, this opinion compares the species to degrees and lines. For, as there are no two degrees or lines which are equal in such a latitude, so there are neither two species in the universe ⟨equal⟩ in perfection.[73]

All points are always at different distances from each other and, what is more important, from the *supremum* of the scale, God, so that an 'ontological difference' is saved. Moreover, not only are there no gaps, but also this is a simple geometric pattern which regulates the proportions between the parts, i.e., species according to their respective perfections.

4 Perfections and Latitudes in Italy

Let me finally present some of the milestones in the reception and diffusion of the use of geometry and the concept of latitude in representing the degrees of perfection in Italy and by so doing arrive at Nifo, whose astonishing declarations we have seen at the very beginning of this chapter. Compared to the developments in France and Germany, the Italian context seems to be even more complex, for here the topic of the perfection of the species was not only discussed within the Aristotelian framework but also, so to speak, moving to its own natural place, it was rediscovered by Platonists. A good example of that is the incorporation of a long discussion about the perfection of the species in

---

73 B 188^{v}, S 145^{r}, M^{2} 70va: 'Unde species universi comparantur gradibus in latitudine, quibus correspondent linee in triangulo. Est autem latitudo uniformiter difformis in qua equalis est excessus graduum inter se eque distantium ut precise se habent linee in triangulo eque distantes catheto. Describitur autem latitudo uniformiter difformis intensa per triangulum propter talem convenientiam, quia qualitas intensiva gradualis in subiecto recipitur extenso et hoc modo intensive. Modo extensioni subiecti correspondet una dimensio trianguli, scilicet longitudo que per lineam tractam super basim trianguli significatur. Sed intensioni qualitatis correspondet alia dimensio, scilicet latitudo, quam per lineam erectam super basim quam catheton denominamus intelligimus. Comparat autem ista via hoc modo species, gradibus et lineis: sicut enim nulli duo gradus vel linee equales sunt in tali latitudine, sic nec due species in universi in perfectione.'

Ficino's *Theologia platonica*.[74] I will confine myself to two major figures who, in my opinion, represent very well the extent to which Italian natural philosophers were interested in latitudes and perfections within an Aristotelian framework. These two figures constitute, at the same time, a preparation for Nifo's disputations on the *Metaphysics*.

One almost ideal example of the impact of *De perfectione specierum* in Italy is Paul of Venice's inclusion of a long treatment of the complete subject matter into his handbook of natural philosophy, the *Summa naturalium*.[75] Like Hugolino of Orvieto, James of Naples, Facinus de Ast, and Jacques Legrand, he belonged to the Hermits of Saint Augustine, so that we really may conjecture a certain tendency within the order to prioritise this subject. The impact of his stay in Oxford, where he was sent by his order in the years 1390–1393, on his approach to many natural philosophical questions must have been considerable. Introducing in many passages logico-mathematical considerations, all analytical languages of the *Calculatores* tradition are well represented in his work. His quantifying approach is programmatic.

74 For this topic, with emphasis on the Italian Platonic tradition, see E.P. Mahoney, 'Metaphysical Foundations of the Hierarchy of Being According to Some Late-Medieval and Renaissance Philosophers,' in P. Morewedge (ed.), *Philosophies of Existence, Ancient and Medieval*, New York 1982, 165–257.

75 I am using the 1503 edition (*Summa philosophiae naturalis magistri Pauli Veneti noviter recognita … restituta*, Venice 1503, repr. Hildesheim 1974) which I have checked with Basel, Universitätsbibliothek, Ms. F. II. 9, 98^r–134^v, which according to the colophon was written by a copyist named Johannes de Basilea in 1412. This work was extremely widespread. Without considering some key European libraries from Germany and France, the list compiled by Charles Lohr arrives at 51 manuscripts and 10 printed editions (C.H. Lohr, *Latin Aristotle Commentaries I.2. Medieval Authors M–Z*, Florence 2010 [Corpus philosophorum medii aevi. Subsidia, 18], 54–55). Paul of Venice is a giant of late medieval scholastic thinking. His work covers theology, natural philosophy, and logic. He wrote several Aristotelian commentaries and independent works (mostly on logic). On the *Metaphysics* he composed also another, very comprehensive work, the so-called *Lectura super libros Metaphysicorum Aristotelis*, which is conserved in two Italian manuscripts. This commentary has recently been studied in detail and partially edited: Paul of Venice, *Lectura super libros Metaphysicorum Aristotelis*, ed. G. Galluzzo, *The Medieval Reception of Book Zeta of Aristotle's Metaphysics*, Leiden 2012 (2 vols.) (Studien und Texte zur Geistesgeschichte des Mittelalters, 110). For a well updated general presentation and bibliography of Paul of Venice, see A. Conti, 'Paul of Venice,' in E.N. Zalta (ed.), *The Stanford Encyclopedia of Philosophy* (Summer 2017 Edition), URL = https://plato.stanford.edu/archives/sum2017/entries/paul-venice/. In a specific contribution on Paul of Venice's *Metaphysics*, Conti underlines his connection to Wyclif and hence his realism: Conti, 'Paul of Venice's Commentary.' None of these valuable contributions, nor those mentioned by Conti in his entry for *The Stanford Encyclopaedia of Philosophy* includes any treatment of the problems I am discussing in this paper.

Paul of Venice's *Summa* covers all, or a great part of the Aristotelian books. In this sense, he has within his order a predecessor in Legrand's *Compendium*. But there is a vital difference regarding the organisation of the matters in both works. Legrand did not include in his work a part on 'Metaphysics'. Instead, he introduced all over the text metaphysical-theological speculations related to the *perfectio specierum*. Paul of Venice, on the contrary, having followed the way from the *Physics* to the *De caelo*, *De generatione et corruptione*, *Meteora*, and *De anima*, presents at the end of his work a commentary on the *Metaphysics*. Once there, he does not treat the *libri* of Aristotle's *Metaphysics* one by one but, instead, he organises the issues according to the Aristotelian categories. Focal points of his commentary are its six chapters on the category of quality, where he begins the discussion by establishing that the two 'principles' which regulate qualities are degrees and angles (two of the main topics Nicole Oresme had dealt with in his *Questions on Euclid's Geometry*). Moreover, the first *notandum* makes a discussion on the intension and remission of forms unavoidable, since—Paul points out—as quantity has quantitative parts, quality has *intensive parts*.[76] At the end of the *Metaphysics*, as the apex of the whole work, he discusses in three special questions the perfection of the species, something that originally neither belonged to the *Metaphysics* nor to the *Categories*. He offers in fact a summary of the previous tradition, focusing more on the criterion of measurement of perfection than on the system of representation. However, at the very end of his work, he finally proposes four possibilities for both measuring and representing the 'process up and down':

> And therefore it is first to be noted that such a process of genus and species, of substances and accidents was imagined by some ⟨a⟩ by means of an infinite straight line beginning from the matter and ending at the indivisibility of the first being. ⟨b⟩ Others were of the opinion that this process ⟨is to be represented⟩ by the way of a circle the centre of which is the non-degree of being absolutely and the circumference imagined as infinite is God himself, from whose centre the genera come out in the way of triangles in an infinite process toward the circumference imagined as infinite, in which genera the species are understood as being exceeding one to the others in the way of a uniformly difform latitude. ⟨c⟩ Others conceived the mentioned process in the way of an infinite latitude, ended

76 Paul of Venice, *Summa*, 105ra: 'Tertium est predicamentum qualitatis, cuius principia sunt gradus et angulus … Et sicut quantitas habet partes quantitativas, ita qualitas habet partes intensivas.'

> inclusively or exclusively. ⟨d⟩ Others, still, in the way only of a finite uniformly difform latitude, ended inclusively or exclusively.[77]

The four opinions discussed here, some of which we have already seen in the previous sections, are: (1) a simple infinite straight line going from matter to the indivisibility of the first being; (2) a circle, the centre of which represents absolute not-being and the circumference—which is thought as infinite—God, whereas from the centre of the circle toward the circumference the genera are to be imagined as triangles; within the genera (represented as triangles), the species exceed one another in the way of a uniformly difform latitude; (3) a uniform latitude; (4) a uniformly difform latitude (having in both latter cases the double possibility of being open or closed intervals). Again, the manuscripts and printed editions lack some diagrams showing us the exact way of representation. The third opinion, in particular, seems to be similar to Legrand's opinion I have referred to before. But this is only in appearance. For, instead of the model of the cone used by Legrand, God is placed not in the centre of the circle but at the circumference. A possible, interpretative, reconstruction could be as in Fig. 11.4 below.

This was the view of Paul of Venice, one of the most famous professors from Padua, where Agostino Nifo studied some years later. Between Paul of Venice and Nifo there took place in Italy an intensive discussion of the Calculators' ideas in the fields of natural philosophy and logic. It seems, however, that the discussion has inflated so that Nifo deemed it adequate to transpose the question about the sense and purpose of the 'calculations' to metaphysics.

After having exposed the *Metaphysics* of Aristotle and the commentary on it by Averroes, Nifo decided himself to give more room than usual to the treatment of questions like the remission and intensification of forms in the context of his metaphysical disputations. By so doing—it is important to underline—he is by no means trying to make further steps in quantifying Aristotle, but

---

77 Paul of Venice, *Summa*, 125^{rb-va}: 'Et ideo notandum primo quod huiusmodi processum generum et specierum, substantiarum et accidentium ⟨a⟩ quidam ymaginantur per modum linee recte infinite incipientis a materia et terminate ad indivisibilitatem primi entis. ⟨b⟩ Alii vero opinantur processum illum per modum circuli cuius centrum est non gradus essendi simpliciter et circumferentia imaginative infinita est ipse Deus a cuius centro egrediuntur genera per modum triangulorum in infinitum processum versus circumferentiam imaginative infinitam, in quibus generibus intelliguntur species se invicem excedere per modum latitudinis uniformiter difformis. ⟨c⟩ Alii concipiunt dictum processum per modum latitudinis infinite terminate inclusive vel exclusive. ⟨d⟩ Alii autem per modum latitudinis solum finite uniformiter difformis inclusive terminate vel exclusive' (Basel, Universitätsbibliothek, Ms. F. II. 9, 134ra).

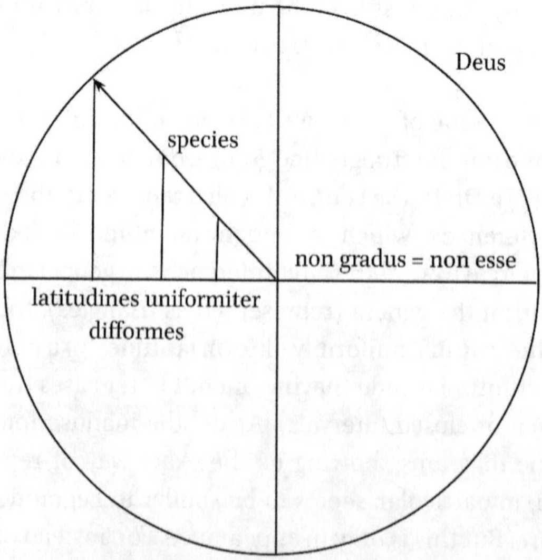

FIGURE 11.4
Interpretation of one of the representation systems mentioned by Paul of Venice in his *Summa philosophiae naturalis*
COPYRIGHT D.A. DI LISCIA

rather the opposite. What Nifo is really trying to do is to establish a clear stop signal against this development.[78]

When we pay attention to Nifo's metaphysical disputations, a further aspect emerges, which is significant to evaluate the level of the inquiry and the form the *Calculatores* tradition was assuming by this time in the Italian universities.[79] Nifo concentrates his efforts in a polemic discussion in which he is neither deep nor original at all. This fact, the root of which lies perhaps in Nifo's own psychology or in the 'professional ethics' of the time, is easily verifiable by attentively reading the question mentioned at the beginning of this chapter: the reduction of uniform difformity to uniformity, which is, or at least directly involves, a discussion of the mean degree theorem.

78 The decisive text is his *Dilucidarium* of metaphysical questions (see above, n. 4). The first edition mentioned by E. De Bellis in his Nifo bibliography is from Naples 1511, the last one from 1560. This text was printed seven times, which documents a considerable interest in it. See E. De Bellis, *Bibliografia di Agostino Nifo*, Lecce 2005, 87–93. The previous work, the exposition on Aristotle's *Metaphysics* and Averroes' commentary is less interesting for our scope. De Bellis (50–56) listed eight editions of this work, the first one being from 1505 and the last one from 1559. I do not know any study, paper, monograph or investigation in general on Nifo's metaphysical texts (for an exhaustive general bibliography until 2005, see De Bellis, 253–285). For the context, see E. De Bellis, *Nicoletto Vernia e Agostino Nifo. Aspetti storiografici e metodologici*, Lecce 2003, ch. IV: 'La formazione e l' opera speculativa di Agostino Nifo' (61–94).

79 For a recent discussion of Pietro Pomponazzi and the Calculators' tradition, see F.P. Raimondi, 'Pomponazzi e la polemica anticalculatoria,' in F.P. Raimondi and J.M. García Valverde (eds.), *Tutti i trattati peripatetici*, Milan 2013, 9–86.

Nifo divides the main question into two sub questions. First, one can ask whether there is a mean degree at all in such a latitude. As he points out, he does not want to discuss that question, since it belongs to '*calculatores* and arithmeticians.' Concerning the second question, however, which asks 'whether the whole latitude corresponds to this degree,' he wishes to give some summary *sine disputatione* and *per regulas*. Let us now briefly take a look at his second rule. This rule consists in a simple arithmetical consideration of the mean degree theorem: if the uniformly difform latitude goes from 0 to 8, the mean degree will be 4. But now, Nifo introduces a geometrical consideration quoting—without proof—a case of the intercept theorem as found in Euclid's *Elements* VI.4:

> The second rule referring to the uniform difform latitude which begins at no degree and ends at a degree of a certain number, for instance at eight: the mean degree is denominated by the half number of the number denominating the more intensive extreme, and since the number of the extreme is eight, the middle degree will be four. The geometers prove this rule through the fourth proposition of the sixth book of Euclid, which states that the line of any triangle which cuts the other two sides in equals is the half of the third, not cut, side. For instance, let ED be the line which cuts the sides AC and BC into equal ⟨parts⟩. ⟨This line ED⟩ is the half to the not cut side AB. Therefore, if the side AB is like 8, the line ED will be like 4.[80]

There is no diagram in the text, but the main idea can be easily reconstructed by drawing the following figure, a simple right-angled triangle representing a case of uniform difformity (Fig. 11.5).

Now, nothing of that is new. It is true, we will not find this version of the theorem in Oresme, neither in the *De configurationibus* nor in his *Questions on Euclid*. In addition, as we also know, the mean degree theorem does not occur in *De latitudinibus formarum*. Nevertheless, this latter work constitutes the starting point for Nifo's version of the mean degree theorem. How is that to be explained?

---

80 Nifo, *Dilucidarium*, 261^{a-b}: 'Secunda regula: in latitudine uniformiter difformi quae incipit a non gradu, et terminatur ad gradum certi numeri, v. g. ad octo, gradus medius denominatur a numero subduplo ad numerum intensius extremum denominantem, et quia extremi numerus est octo, gradus medius erit quatuor. Probant geometrae hanc regulam per quartam propositionem lib. 6 Euclidis, quae dicit quod cuiuslibet trianguli linea duo latera secans per aequalia est subdupla ad tertium latus non sectum. Verbi causa ED linea quae secat AC et BC latera per aequalia: ⟨illa linea ED⟩ est subdu | pla ad latus AB non sectum. Quare, si latus AB erit ut octo, linea ED erit ut quatuor.'

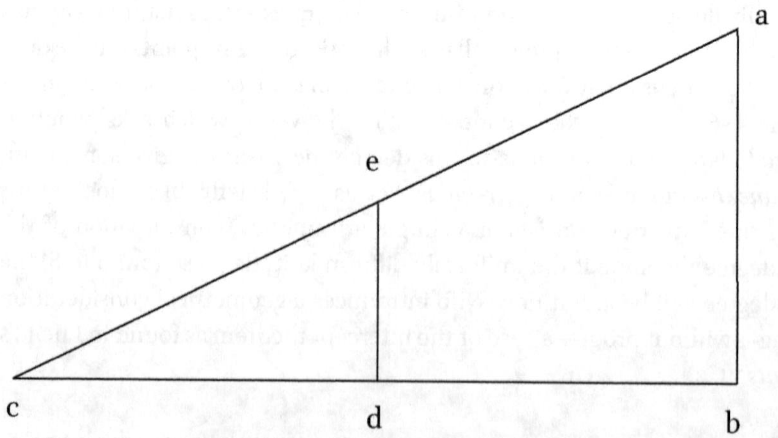

FIGURE 11.5 Reconstruction of the diagram corresponding with Nifo's passage. See also Fig. 11.6.

The doctrine of the latitude of forms was extremely successful in Italy, as many manuscripts and editions of the *De latitudinibus formarum* attest. On this text, the famous Blasius of Parma wrote a commentary in three questions, which is conserved in several manuscripts and was printed three times before Nifo.[81] This is the direct and immediate source of Nifo. In Blasius' questions we find the same remark nearly *verbatim*, with the same geometrical reference and including the same diagram (with exactly the same denomination for the sides of the triangle). And in fact, there is scarcely anything in Nifo's question *Num latitudo uniformiter difformis correspondeat suo gradui medio* which cannot be found in Blasius' previous questions, as we can see in a short text comparison:

Blasius of Parma, *Questiones super tractatu de latitudinibus formarum*, Padua 1486, 11^{v}–12^{r}	**Agostino Nifo, *Dilucidarium … metaphysicarum disputationum*, Venice 1559, 260^{b}–261^{a}**
	Disputatio duodecima De medio difformis latitudinis
Utrum quelibet latitudo uniformiter difformis correspondeat suo gradui medio, idest utrum talis latitudo sit suo gradui medio equalis intensive.	*Num latitudo uniformiter difformis correspondeat suo gradui medio.*

81 D.A. Di Liscia, 'Biagio Pelacani da Parma's Geometrisation of Latitudes and the Problems

(*cont.*)

Blasius of Parma, *Questiones super tractatu de latitudinibus formarum*, Padua 1486, 11^{v}–12^{r}	**Agostino Nifo, *Dilucidarium … metaphysicarum disputationum*, Venice 1559, 260^{b}–261^{a}**
	Caput dubitationum primum
Et arguitur primo quod non. Nam latitudo uniformiter difformis non habet gradum medium; ergo etc. Antecedens probatur, quia talis latitudo non habet extrema; ergo non habet medium. Consequentia tenet, quia ubi non convenit dare extrema, ibi nec medium …	Quod non arguitur: latitudo uniformiter difformis non habet extrema; igitur nec medium. Antecedens patet, quia saltem non habet remisissimum, quoniam omni tali dato, esset adhuc dabile minus …
Secundo principaliter arguitur sic: latitudo uniformiter difformis extensiva est unus triangulus, et non est equalis sue medie parti; ergo nec latitudo uniformiter difformis …	Secundo: latitudo uniformiter difformis est unus triangulus; sed triangulus non est aequalis sue mediae parti; igitur latitudo talis suo gradui medio non est aequalis …
Quarto ad principale sic: si questio esset vera, sequeretur hec conclusio impossibilis: *a* et *b* sunt mobilia que movebuntur precise per illam horam, et *a* movebitur in duplo velocius *b*; et tamen equalia spacia pertransibunt. Et quod conclusio sit impossibilis patet intuenti. Sed quod sequetur ex questione probatur. Et capio unam latitudinem motus uniformiter difformem a non gradu usque ad 8; et volo quod *a* moveatur continue illa latitudine circulariter, ita quod medius punctus ipsius *a* quiescat; et volo quod *b* continue per eandem horam medio gradu istius latitudinis	Quarto: data positione, tunc A et B erunt mobilia, quae praecisae (!) movebuntur per istam horam, et A movebitur in duplo velocius ipso B; et tamen aequalia spatia peragrabunt. Conclusio omnibus patet esse impossibilis. Et deducitur; et capio unam latitudinem motus uniformiter difformem a non gradu usque ad octo; et volo quod A movetur continuo illa latitudine circulariter, ita quod medius punctus ipsius A quiescat; et volo quod B continuo per eandem horam medio gradu istius latitudinis moveatur etiam circulariter. Illis positis, patet quod A in duplo velocius

of the Mean Degree Theorem,' in J. Biard and A. Robert (eds.), *La philosophie de Blaise de Parme. Physique, psychologie, éthique*, Florence 2019 (Micrologus Library, 96), 129–152.

(*cont.*)

Blasius of Parma, *Questiones super tractatu de latitudinibus formarum*, Padua 1486, 11ᵛ–12ʳ	**Agostino Nifo, *Dilucidarium … metaphysicarum disputationum*, Venice 1559, 260ᵇ–261ᵃ**
moveatur. Illo posito, patet quod *a* in duplo velocius continue movebitur quam *b*, quia omne mobile dicitur ita velociter moveri sicut aliqua pars eius et punctus velocissime motus ipsius movebitur gradu ut 8 et *b* continue gradu ut 4. Et quod tantum spacium pertransibit precise *a* ut *b* patet, quia omnis latitudo uniformiter difformis correspondet gradui suo medio per questionem …	continuo movebitur quam B, quia omne mobile dicitur ita velociter moveri, sicut aliquis punctus eius velocissimae (!) motus, at punctus velocissimae (!) motus movetur gradu ut octo, et B continuo gradu ut quatuor, et tantum spatii peragret (!) B quantum A; patet, quia omnis latitudo uniformiter difformis correspondet suo gradui medio per positionem. In contrarium sunt mathematici omnes. Et haec de dubitationibus …
Secunda suppositio: cuiuslibet trianguli linea duo latera secans per equalia est subdupla ad tertium latus. Patet per quartum sexti Euclidis, quoniam sit triangulus *abc* et linea secans per equalia duo latera sit *de*: dico quod linea *de* est subdupla ad lineam *ab*. Unde sicut *bc* ad *be*, ita se habet *ac* ad *ad* sic *ab* ad *de*; sed *bc* est duplum ad *be*; ergo *ab* est duplum ad *de* et per consequens linea *etd* (!) sive *de* est subdupla ad lineam *ab*.	Secunda regula: … Probant geometre hanc regulam per quartam propositionem lib. 6. Euclidis, quae dicit quod cuiuslibet trianguli linea duo latera secans per aequalia est subdupla ad tertium latus non sectum, verbi causa ED linea quae secat AC et BC latera per aequalia est subdupla ad latus AB non sectum: quare si latus AB erit ut octo, linea ED erit ut quatuor.

It is noteworthy that in Blasius' text (according to the 1486 edition) the following diagram is added that is not present in Nifo's corresponding text.

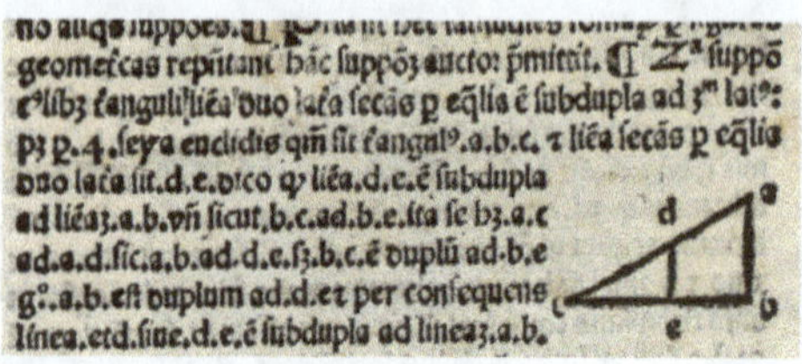
geometricas repūtam hāc supꝑōz auctor ꝑmittit. ¶ 2ª suppō
c⁹libz tanguli liēa duo lat̄a secās ꝑ eq̄lia ē subdupla ad 3ᵐ lat⁹:
pz ꝑ.4.sexti euclidis qm̄ sit tangul⁹.a.b.c. ⁊ liēa secās ꝑ eq̄lia
duo lat̄a sit.d.e.dico q̄ liēa.d.e.ē subdupla
ad liēaz.a.b.vn̄ sicut.b.c.ad.b.e.ita se hz.a.c
ad.a.d.sic.a.b.ad.d.e.sz.b.c.ē duplū ad.b.e
g°.a.b.est duplum ad.d.e⁊ per consequens
linea.etd.siue.d.e.ē subdupla ad lineaz.a.b.

FIGURE 11.6
Blasius of Parma, *Questiones super tractatu de latitudinibus formarum*, Padua 1486, 10ʳ

Finally, at the end of the question, Nifo introduces a general concluding remark excusing himself for having dealt with these matters within a metaphysical framework—after all, the work is about Aristotle's *Metaphysics*. This remark is extraordinarily unequivocal and important to grasp the reputation of the calculatory tradition in Renaissance Aristotelianism:

> This suffices for the 12th disputation, where I want to excuse myself for the questions I have disputed, which do not at all concern the metaphysician. I introduced them just because of an external question, which concerns all these matters. Thus, I wanted to dispute these questions here, so that in those matters which they are about, the truth comes out to light. Furthermore, through these questions we have illustrated in the natural ⟨books⟩ that many things are said in a confused way; that is why it is enough now. And if God prolongs our life, I shall compose an explanatory handbook (*Dilucidarium*) of calculatory disputations, in which we will object to all Calculators refuting their importunities. For, we know that these ⟨importunities⟩ are neither peripatetic, nor necessary for science, but only for the ostentations of humans. That is why we oppose them with principles which are contrary to their ⟨principles⟩ finding always the position of the Peripatetics in all matters.[82]

Nifo thus presents himself as 'the good Peripatetic' against the Calculators. As such, he is as good a fighter of (i.e., against) the *Calculatores* as for instance the humanist Juan Vives. But unlike Vives, he massively takes over the preliminary work of Blasius of Parma—without mentioning his source. By so doing, however, he might have contributed a good deal to the expansion of the calculators' tradition not only in physics, but also in metaphysics. To my knowledge, however, there is no such specific *Dilucidarium* of calculatory importunities by Nifo. He probably never wrote it. But he continued his anti-*Calculatores*

82 Nifo, *Dilucidarium*, 261ᵇ: 'Haec de disputatione duodecima; ubi volo me excusare, nam quaestiones disputavi quae nihil attinent ad metaphysicum; haec enim feci propter unam exterorum quaestionem, quae tetigit omnia haec. Ut igitur in his quae tetigerunt veritas dilucidetur, volui quaestiones hasce disputare. Praeterea per has quaestiones multa confuse dicta in naturalibus illustravimus, quare parcatur mihi. Et si Deus vitam producet, conficiam dilucidarium disputationum calculatoriarum, in quo omnibus calculatoribus obiiciemus eorum importunitates refellentes: scimus enim illas nec esse peripateticas, nec necessarias ad scientiam, sed ostentationes hominum. Propterea per principia contraria omnibus principiis eorum, illis obviabimus dilucidando semper peripateticorum in omnibus.'

programme in his *Dialectica ludicra*, an introduction to logic for beginners and veterans alike. Here, with similar strategy, he moves to the field of dispute with heavy munition against the *sorticolae* and *cavillatores* productors of ridiculous *sophisticulae* and deals at length not only with the more traditional logical topics but also with consequences, obligations, and even insoluble sentences.[83]

Concluding Remarks

The survey carried out in this chapter is mostly based on materials which were up to now almost or completely unknown. Perhaps that is not by accident. Already this very fact supports the general claim made at the very beginning, namely that this chapter in the history of philosophy was as such until now almost non-existent. Not only that the philosophy between the late medieval period and the Renaissance has been neglected, but also and above all this tradition was neglected, its richness and its complexity.[84]

For sure, it is undeniable that a great deal of outstanding research work has been done which is very helpful to understand the main ideas of the Calculators in their proper context. In a certain sense, much more is at stake than the old question of the precursors of Galileo. One of the main pictures we have learned to always keep in mind is that the interaction with the key notions of Aristotelian science and philosophy is an unavoidable background for the Calculators' tradition, however innovative the latter could appear to us. It will be a long process until we can finally get a more or less stable picture of the impact of the *Calculatores* tradition on late medieval, Renaissance, and early modern metaphysics. How general was the application of the quantitative approach to metaphysics promoted by the Calculators? Was it reduced only to the problem of the 'series of the entities in the world' or does it also find application within other questions? Which were the notions more strongly connected to scientific

83 *Augstini Niphi ... Dialectica ludicra tyrunculis atque veteranis utillima peripateticis consona: inuioribus sophisticantibus contraria*, Venice 1521. De Bellis, *Bibliografia*, n. 75, 117, mentions a previous edition from Florence 1520. A further investigation will have to establish to what extent Nifo's ideas on these issues are original. In his study on Nifo's logic, E. De Bellis provides some background information without touching upon the topics discussed in the present chapter: E. De Bellis, *Il pensiero logico di Agostino Nifo*, Lecce 1998.

84 A similar remark is made by Pluta, 'Nicholas of Amsterdam,' 185, regarding the philosophy of the fifteenth century.

fields and which ones were more clearly linked to theology? Which ideas, if any, were transmitted to the context of the scientific revolution of the seventeenth century? We cannot answer these or other similar questions, yet, but perhaps we can make some progress in this direction by proposing some general statements to be discussed.

1. It is evident that a not unimportant development took place within the *Calculatores* tradition which also influenced the metaphysical tradition. By that, I mean not only the tradition of metaphysical thinking in general, but the corresponding scientific discipline and also the way in which these ideas were expressed and conveyed by commenting on the main texts of Aristotle, including the *Metaphysics*. We know from many later commentaries for instance on the *Physics*, especially from those composed in the circle of John Mair, that a strong connection between the Calculators' quantifying approach and the more traditional physical doctrine was becoming usual, at least in certain centres of study such as in Paris. Here, John Mair, Luis Coronel, Juan de Celaya, John Dullaert of Ghent, Jacques Almain, and others transformed the appearance of the commentary tradition by introducing comprehensive discussions around quantification or quantifiable terms. They fundamentally transformed the physiognomy of Aristotle's *Physics*—and partially *De caelo*—and made it into a vehicle able to transport *maxima et minima, incipit et desinit, intensio et remissio formarum, proportiones velocitatum* and other issues of the analytical languages in the field of natural philosophy. The parallel development, of which this chapter describes some of the main features, was not based on the theory of motion and other physical concepts belonging to the Aristotelian *Physics* but built upon some of the metaphysical notions which were, perhaps because of their theological background, especially significant to late medieval thinkers.

2. The theological motivation is not only present but played a decisive role in these developments. The idea of a Christian God, designer of a world 'according to measure, number and weight' had been working already long before as an inducing idea bringing about from the very beginning a first level of transformation of Aristotle. Metaphysics is to be complemented by theology and the tradition of Aristotle's *Metaphysics* by the theological tradition of the Bible and *Sentences* commentaries.[85] As a consequence, the Neoplatonic view of species

85 Occasionally, however, especially to support the arithmetical theory, reference is made to the *Book of Wisdom*, which, as A. Maier said, is 'ein Freibrief für die weitestgehenden Calculationes'. See A. Maier, *Metaphysische Hintergründe der Spätscholastischen Naturphilosophie* (Studien zur Naturphilosophie der Spätscholastik, 4), Rome 1955 (Storia e letteratura, 52), 340. See, for instance, Legrand, *Compendium*, *P* 47^{v}, *G* 57^{v}: 'Praeterea *Sapientie*

emanating from God's essence, which are necessarily of less perfection than God, finds a further way into the Aristotelian tradition. In principle, there is nothing new in that, as we know how significant the role played by Platonism, especially in its Augustinian mould, was in the reception of Aristotelianism. Moreover, the idea of a certain order of species in a complete system of entities is not alien to Aristotle, as we have seen.

The Calculators' approach brought a new treatment of this subject and with it a new conceptual tension. The application of quantity to the perfection of species has to work somehow differently from the application of quantity to the velocity of motion in physics. In physics, it is about the mathematisation of changing entities. In metaphysics it is exactly the opposite. In this field, the new approach consisted in considering all these traditional notions and their implications—especially the idea of perfection—from the point of view of degrees of intensity. As the species do not change, they are quantified as (unchanging) parts of a whole which can be very well grasped using figures.

Consequently, as the new doctrine of the latitude of forms expanded successfully beyond all borders, a system which could be geometrically represented arose and found application in metaphysics. That is a simple but powerful consequence derived specifically from the new approach developed by the *Calculatores* and spread all over Europe at different levels of discussion. The original Oxford Calculators could surely not foresee such an expansion and it is doubtful that all those who later introduced such an approach into metaphysics were consciously and purposefully following a pattern of 'original thinking' once begun at an English university. Nevertheless, it was this approach that made it possible to reformulate an old Christian-Neoplatonic fundamental idea in quantitative terms.

3. As a new wave of development within metaphysics was taking place, we cannot, however, affirm that the intentions and contexts were the same everywhere. I have focussed only on a selection of authors, texts, and topics, which I considered adequate for the aim and scope of this chapter. And already here we know that there are enormous differences between them. Legrand, for instance, did probably not consider himself to apply mathematics, even though he was purposefully 'quantifying'. The texts written in the German cultural setting by Nicholas of Amsterdam or John of Wesel let us appreciate that their authors were well aware of applying mathematics to a metaphysical problem and, by so doing, quantifying it. They had fewer scruples at mixing sciences.

3° dicetur domine omnia fecisti in numero pondere et mensura. Si ergo in numero omnia facta sunt, ergo latitudo creabilium ad modum numerorum debet poni discreta. Unde Plato in numeris posuit rationem secundum quam mundum deus fabricavit ...'

Nifo, on the contrary, saw in all that an attack against the true doctrine, which he identified with the Aristotelian doctrine (as interpreted by himself). The best of his efforts was aimed at ending the tradition kept alive and developed before by Blasius of Parma and Paul of Venice. His utterances are historically important if we wish to understand in which direction this tradition, started in Oxford, was moving in the Renaissance. I do not think Nifo alone is representative of this direction, but he does constitute an essential and not isolated case within it. There is no single way we could consider without distortion his attitude as a step to any positive approach to the scientific revolution of the seventeenth century. It seems to be far more likely that a diversity of approaches was at stake. Perhaps it is more adequate to consider that it was not a relay race with a single line of descent, but a complicated interaction of traditions within which perfections and latitudes occupied a significant place.

Acknowledgements

I am grateful to the DGF for funding the project 'Die Geometrisierung der Metaphysik im Spätmittelalter: Jacobus de Neapoli und die Tradition *De perfectione specierum*' (Deutsche Forschungsgemeinschaft / German Research Foundation, project number 282682744). I would like to thank Edith Sylla for revising my English and commenting on my text, and Paul Bakker for his editing work and several useful remarks on the content of this paper.

CHAPTER 12

Decline of the Calculators in Paris c. 1500: Humanism and Print

Richard J. Oosterhoff

Introduction

One usual account of why the calculatory tradition barely survived past the early sixteenth century turns to humanism. When mentioned by grand humanists such as Bruni, Petrarch, Erasmus, Barbaro, and Vives, the Calculators are in for a good kicking.[1] The mood is set by Thomas More's letter to Martin Dorp of 1518, in which he decried the 'supersophistical trifles' of modern logic. He identified a better future in the textbooks of the eminent Parisian scholar Jacques Lefèvre d'Étaples (c. 1455–1536).[2] And indeed, Lefèvre often presented his textbooks as an antidote to the bad philosophizing of sophists. Moreover, the Calculators emblematized the worst kinds of pathological thought:

> What about this tramp who, like some sort of trickster or fraud, long ago was rightly exiled from Italy and found shelter among us—who calls herself 'calculation' while perverting all rational calculating? First off, remove the too rough and unlearned absurdity of speech, which confounds men and angels with asses and mingles gods above with realms below. The absurdity propagates every day in the poisonous manner of weeds; indeed they ensnare and ruin tender wits, spoiling the bread of all teaching.[3]

---

1 Studies of this reputation include C. Dionisotti, 'Ermolao Barbaro e la Fortuna di Suiseth,' *Medioevo e Rinascimento: Studi in onore di Bruno Nardi*, Florence 1955, 1: 219–253; C. Vasoli, *Studi sulla cultura del rinascimento*, Manduria 1968 (Biblioteca de studi moderni, 5), 139–177; D.A. Di Liscia, 'Kalkulierte Ethik: Vives und die "Zerstörer" der Moralphilosophie (Le Maistre, Cranston und Almain),' in S. Ebbersmeyer and E. Kessler (eds.), *Ethik: Wissenschaft oder Lebenskunst? Modelle der Normenbegründung von der Antike bis zur frühen Neuzeit*, Münster 2007, 75–105.

2 T. More, 'Letter to Martin Dorp,' ed. D. Kinney, New Haven, CT, 1986 (The Complete Works of St. Thomas More, 15), 22–23. Basic studies of Lefèvre include A. Renaudet, *Préréforme et humanisme à Paris pendant les premières guerres d'Italie, 1494–1517*, 2nd ed., Paris 1953; E.F. Rice, Jr. (ed.), *The Prefatory Epistles of Jacques Lefèvre d'Étaples and Related Texts*, New York 1972.

3 J. Lefèvre d'Étaples (ed.), *Egregii patris et clari theologi Ricardi* [*Sancti Victoris*] *De superdivina*

 | DOI:10.1163/9789004512054_014

By taking aim at the Calculators, Lefèvre echoed a standard critique of certain tendencies in medieval university thought. The critique targeted a genre that began in the undergraduate exercises of disputation known as *sophismata*, devoted to fast-paced unknotting of logical problems.[4] Students often deployed calculatory techniques in the course of disputing sophisms, as Edith Sylla once pointed out.[5] In the works of William Heytesbury, Richard Swineshead, and Thomas Bradwardine, many *sophismata* aimed to apply mathematical models to physical change, using what John Murdoch called 'languages of analysis' that drew on proportions, the intension and remission of forms, first and last instants, and maxima and minima.[6] Over the fourteenth century, at Oxford, and then at Paris and elsewhere, these techniques and exercises became popular.[7] Although the word *sophista* could simply describe students engaged in such exercises, without reprobation, the term's pejorative meanings were convenient for university critics.

These mathematical *sophismata* already seemed pernicious to late medieval critics within the university. In fact, despite the angle of humanist attack, it is misleading to see it as coming from outside the university, for it was university men who set the trajectory of the humanist critique—a first hint that the usual humanist story is oversimplified. The critique had been given par-

---

Trinitate theologicum opus hexade librorum distinctum. Commentarius artificio analytico, Paris 1510, sig. a2r (ed. Rice, *The Prefatory Epistles*, 226): 'Quid quod inculcatoria quae pridem tamquam circulatrix quaedam et subdola iure ab Italis exulat apud nostros invenit asylum et se calculatoriam nominat quae omnem rationis pervertit calculum? Tolle insuper rudem nimis et indoctam sermonis absurditatem quae homines et angelos cum asinis confundit et superos Acherontaque miscet, loliorum more sese in dies noxie propagantem et tenella implicantem immo perdentem ingenia, omnis doctrinae panem inficientem.'

4 E.g. the recent study of S. Ebbesen and F. Goubier, *A Catalogue of 13th-Century Sophismata*, I: *Introduction and Indices*; II: *Catalogue of* Sophismata, Paris 2010 (Sic et non).

5 E.D. Sylla, 'The Oxford Calculators,' in N. Kretzmann, A. Kenny, J. Pinborg and E. Stump (eds.), *The Cambridge History of Later Medieval Philosophy. From the Rediscovery of Aristotle to the Disintegration of Scholasticism, 1100–1600*, Cambridge 1982, 540–563, esp. 542–547; E.D. Sylla, 'The Oxford Calculators in Context,' *Science in Context* 1 (1987), 257–279.

6 J.E. Murdoch, 'From Social into Intellectual Factors: An Aspect of the Unitary Character of Late Medieval Learning,' in J.E. Murdoch and E.D. Sylla (eds.), *The Cultural Context of Medieval Learning*, Dordrecht 1975 (Boston Studies in the Philosophy of Science, 26), 271–348.

7 On this expanding popularity, see J.E. Murdoch, '*Mathesis in philosophiam scholasticam introducta*. The Rise and Developement of the Application of Mathematics in Fourteenth-Century Philosophy and Theology,' in *Arts libéraux et philosophie au Moyen Âge. Actes du quatrième Congrès international de philosophie médiévale, Montréal 1967*, Montreal 1969, 215–254, esp. 216–227; Sylla, 'The Oxford Calculators in Context'; D.A. Di Liscia, *Zwischen Geometrie und Naturphilosophie: die Entwicklung der Formlatitudenlehre im deutschen Sprachraum*, unpublished Ph.D. dissertation, University of Munich, 2003.

ticular force by Jean Gerson, chancellor of the University of Paris in the years around 1500. He worried that even theology had become, over the last generation, filled with unhealthy habits, so that theologians were jostling for position over logical puzzles rather than preparing to serve a public in need of pious teachers. Gerson used the word 'sophist', Zénon Kaluza has suggested, both to describe students engaged in certain disputations as well as those espousing the 'English' style of inquiry after Bradwardine, Heytesbury, etc.[8] In Gerson's view, the problem was partly that students gained bad habits, valuing obscurity over common sense, a kind of 'curiosity' that hungered after novelty more than truth. The problem was also partly one of boundary crossing. Rather than leaving these techniques in logic or mathematics, he thought students carried them into domains of theology where other forms of reasoning should prevail. At Paris, such criticisms can be traced through the works of university reformers from Gerson to Jean Standonck at the Collège de Montaigue. As this rhetoric also became a standard line among critics who took up stances outside the university, these charges have set 'scholasticism' against 'humanism', opposing an ideal of philosophy as driven by technical school-logic to one rooted in common-sense language.[9]

But did this rhetorical opposition cause the decline of the Calculators' project? Quite rightly, historians have noted that calculatory techniques received a burst of interest in Paris shortly after 1500, in the circle around John Mair at the Collège de Montaigu.[10] A key figure here is Alvarus Thomas, who published in 1509 the most sophisticated contribution to the calculatory tradition of the sixteenth century.[11] Others included Luis Coronel, Gaspar Lax, Pedro

8 Z. Kaluza, *Les querelles doctrinales à Paris: nominalistes et réalistes aux confins du XIVe et du XVe siècles*, Bergamo 1988 (Quodlibet, 2), 43–45.

9 This definition of humanism is given by L.W. Nauta, 'Latin as a Common Language: The Coherence of Lorenzo Valla's Humanist Program,' *Renaissance Quarterly* 71 (2018), 1–32; C.S. Celenza, *The Intellectual World of the Italian Renaissance: Language, Philosophy, and the Search for Meaning*, New York 2018.

10 Perhaps the most forceful view has been advanced by William Wallace, following Pierre Duhem, that this was a key link between the Calculators and Galileo: W.A. Wallace, 'The *Calculatores* in Early Sixteenth-Century Physics,' *The British Journal for the History of Science* 4 (1969), 221–232 (republished in Id., *Prelude to Galileo: Essays on Medieval and Sixteenth-Century Sources of Galileo's Thought*, Dordrecht 1981 [Boston Studies in the Philosophy of Science, 62], 78–90). On this generation, see H. Élie, 'Quelques maîtres de l'Université de Paris vers l'an 1500,' *Archives d'histoire doctrinale et littéraire du Moyen Âge* 18 (1950–1951), 193–243; A. Broadie, *The Circle of John Mair. Logic and Logicians in Pre-Reformation Scotland*, Oxford 1985; Di Liscia, 'Kalkulierte Ethik.'

11 A. Thomas, *Liber de triplici motu proportionibus annexis magistri Alvari Thome Ulixbonensis Suisseth calculationes ex parte declarans*, Paris 1509. Besides E.D. Sylla, 'Alvarus Thomas

Ciruelo, Juan Martínez Silíceo, and Juan de Celaya—who all wrote on mathematics, and would go on to teach and publish mathematical works in Spain, nourishing the mathematical culture that encouraged Domingo de Soto to put the mean-speed theorem to work on falling bodies, much as Galileo would a couple of generations later.

In this picture, one missing piece is the pedagogical scene. What relation did these analytical languages have to usual physics teaching on the *cursus*? How far, in other words, was Alvarus Thomas from the reach of most students? Was this burst of interest in Calculators a sudden flash among a few high-powered intellects or was it part of everyday intellectual experience at Paris? In this chapter, I argue that, to understand the demise of the calculatory tradition, we need to take account of how this discipline was framed in university textbooks, as well as the changing mathematical culture of the early sixteenth century.

1 Parisian Printed Textbooks c. 1500

Notwithstanding the strong words he would aim at the Calculators, Lefèvre himself wrote one of two new contributions to the genre to be printed in Paris in the 1490s: his *Dialogus difficilium physicalium introductorius* (*Introductory Dialogue on More Difficult Physics*).[12] (The other was a collection of extracts from various sources, by a young Scotist theologian in training, Pierre Tartaret.[13]) Lefèvre's contribution concluded his first publication, a textbook of Aristotelian natural philosophy intended to help undergraduate students through all the topics they needed to navigate natural philosophy at Paris. The book began with a paraphrase of Aristotle's *Physics*, continued through a sur-

---

and the Role of Logic and Calculations in Sixteenth-Century Natural Philosophy,' in S. Caroti (ed.), *Studies in Medieval Natural Philosophy*, Florence 1989 (Biblioteca di Nuncius, 1), 257–298, see now S.P. Trzeciok, *Alvarus Thomas und sein* Liber de triplici motu, 1: *Naturphilosophie an der Pariser Artistenfakultät*; 2: *Bearbeiteter Text und Faksimile*, Berlin 2016, URL = https://edition-open-sources.org/media/sources/7/Sources7.pdf. See also Edith Sylla's contribution to the present volume.

12 J. Lefèvre d'Étaples, *Dialogus difficilium physicalium introductorius*, Paris 1492. The present study grows out of an account of this work in R.J. Oosterhoff, *Making Mathematical Culture: University and Print in the Circle of Lefèvre d'Étaples*, Oxford 2018 (Oxford-Warburg Studies), 181–199. I have expanded on the concerns of genre around these works in R.J. Oosterhoff, 'The Dialogue of Ingenuous Students: Early Printed Textbooks at Paris,' in S. Berger and D. Garber (eds.), *Teaching Philosophy in the Seventeenth Century: Text and Image*, Dordrecht 2021, 11–30.

13 P. Tartaret, *Tractatus de intensione, rarefactione et condensatione formarum utilis ad totam phisicam intelligendam extractus a Gregorio de Herimino et aliis doctoribus*, Paris 1493.

vey of *On the Heavens*, *On Generation and Corruption*, the *Meteorology*, *On the Soul*, and a selection from the *Parva naturalia*. The volume closed with two dialogues: the first introduced the basic terms of the *Physics*, preparing the reader to approach the second, 'more difficult' dialogue, which introduced techniques for reasoning about latitudes of forms, maxima and minima, and so on.[14]

My task will be to see what the context of printed textbooks at Paris may suggest about the place of such studies in regular teaching. As Edith Sylla pointed out, attitudes about and interest in the texts of the Oxford Calculators varied considerably from place to place; Daniel Di Liscia has uncovered important evidence that the study of the *latitudines formarum* retained a place in the curriculum of the universities of Cologne and Vienna.[15] A fuller answer for Paris would have to take account of manuscripts; but in 1490s Paris, university teachers were beginning to experiment with new printed books, a fact which would swiftly reshape the university curriculum over the next decade.

In 1492, then, a printed handbook to natural philosophy was new. There had been medieval *summae naturales*, such as those by Robert Grosseteste and (ps.) Albert, which covered much of the same ground.[16] But even though rhetor-

---

14 J. Lefèvre d'Étaples, *Totius Aristotelis philosophiae naturalis paraphrases*, Paris 1492.

15 E.D. Sylla, 'The Fate of the Oxford Calculatory Tradition,' in C. Wénin (ed.), *L'homme et son univers au Moyen Âge. Actes du septième congrès international de philosophie médiévale (30 août–4 septembre 1982)*, Louvain-la-Neuve 1986 (Philosophes médiévaux, 26), 2: 692–698; Di Liscia, *Zwischen Geometrie und Naturphilosophie*, 173–212; D.A. Di Liscia, 'The "Latitudines breves" and Late Medieval University Teaching,' *SCIAMVS* 17 (2016), 55–120.

16 The classic study is M. Grabmann, *Methoden und Hilfsmittel des Aristotelesstudiums im Mittelalter*, München 1939 (Sitzungsberichte der Bayerischen Akademie der Wissenschaften, Philosophisch-historische Abteilung, 1939/5); see also A. Cunningham and S. Kusukawa, 'Translators' Introduction,' in A. Cunningham and S. Kusukawa (eds.), *Natural Philosophy Epitomised: Books 8–11 of Gregor Reisch's* Philosophical Pearl *(1503)*, Farnham 2010, ix–lxxiv; D.A. Lines, 'Teaching Physics in Louvain and Bologna: Franz Titelmans and Ulysse Aldrovandi,' in E. Campi, S. De Angelis, A.-S. Goeing and A. Grafton (eds.), *Scholarly Knowledge: Textbooks in Early Modern Europe*, Geneva 2008 (Travaux d'humanisme et Renaissance, 447), 183–203. For examples of late fifteenth-century Parisian handbooks of Aristotelian natural philosophy, see P.J.J.M. Bakker, 'Natural Philosophy and Metaphysics in Late Fifteenth-Century Paris. I: The Commentaries on Aristotle by Johannes Hennon,' *Bulletin de philosophie médiévale* 47 (2005), 125–155; P.J.J.M. Bakker, 'Natural Philosophy and Metaphysics in Late Fifteenth-Century Paris. II: The Commentaries on Aristotle by Johannes le Damoisiau,' *Bulletin de philosophie médiévale* 48 (2006), 209–228; P.J.J.M. Bakker, 'Natural Philosophy and Metaphysics in Late Fifteenth-Century Paris. III: The Commentaries on Aristotle by Johannes de Caulaincourt (alias Johannes de Magistris),' *Bulletin de philosophie médiévale* 49 (2007), 195–237. For Louvain, see B. Bartocci, S. Masolini and R.L. Friedman, 'Reading Aristotle at the University of Louvain in the Fifteenth Century: A First Survey of Petrus de Rivo's Commentaries on Aristotle (I),' *Bulletin de philosophie médiévale* 55 (2013), 133–175; B. Bartocci and S. Masolini, 'Reading Aristotle at the Uni-

ics and classics were some of the first books to be printed in Paris, followed quickly by theological works, it took decades for the usual university textbooks of logic and natural philosophy to appear from the new presses.[17] For whatever reason, it seems that those *libraires* to whom the university had given copying privileges did not have the first printing presses. This would explain why in the 1480s such introductions were first published *outside* of Paris, notably Thomas Bricot's abbreviations of Aristotle's natural philosophy, which he had interspersed with *quaestiones* from George of Brussels.[18] Gradually, the Parisian market opened up. In the late 1480s, two editions of Bricot's abbreviations of logic were printed in Paris.[19] In 1491 Wolfgang Hopyl published Bricot's edition of *quaestiones* on physics, and then in 1494 Hopyl and his new partner Johann Higmann printed another edition with Bricot's abbreviations too.[20] Early in 1495, a new printer, André Bocard, also published a similar *Expositio totius philosophiae nec non metaphysicae Aristotelis* by Pierre Tartaret, who would stand alongside Lefèvre and Bricot (with George of Brussels) as the main textbook authors printed for the University of Paris over the next two decades.[21] When Hopyl and Higmann printed Lefèvre's new introduction and dialogues in 1492, it was part of this first wave of printed handbooks to Aristotelian natural philosophy.

In the 1490s, however, Lefèvre's dialogues represent the high point of the calculatory techniques. The combined handbook by Thomas Bricot and George of Brussels did not mention such techniques at any length, or in any focused way; neither did Pierre Tartaret's list of *quaestiones* cover the same ground. Only Lefèvre's textbook dealt with these topics—and then, only within a distinct

---

versity of Louvain in the Fifteenth Century: A First Survey of Petrus de Rivo's Commentaries on Aristotle (II),' *Bulletin de philosophie médiévale* 56 (2014), 281–383.

17 On the omnipresent *summulae*, absent in print until the 1480s, see S. Corsten, 'Universities and Early Printing,' *Bibliography and the Study of 15th-Century Civilization* 5 (1987) 83–123.

18 Thomas Bricot and George of Brussels, *Textus abbreviatus in cursum totius physices et metaphysicorum Aristotelis*, Lyon 1486, ISTC ib01202000.

19 Thomas Bricot, *Textus abbreviatus in cursum totius logices Aristotelis*, Paris 1489, ISTC ib01199400; Paris 1489, ISTC ib01199500.

20 George of Brussels, *Cursus quaestionum super philosophiam Aristotelis*, ed. Thomas Bricot, Paris 1491, ISTC ig00146900.

21 P. Tartaret, *Expositio totius philosophiae nec non metaphysicae Aristotelis*, Paris 1495, ISTC it00044000. As for Bricot, the first edition of Tartaret was published outside Paris, in Poitiers in 1493—Bocard, who was not yet a *libraire-juré* with university privileges, seems to have published the first Paris edition for Jacques Bezanceau in Poitiers. An immediate measure of the place of Bricot, Tartaret, and Lefèvre can be taken from comparing their entries in the ISTC and USTC, which confirms the claims by Renaudet, *Préréforme et humanisme*, 95–99 and *passim*.

dialogue, rather than integrated into the flow of his introduction to physics. The one other Paris example, Pierre Tartaret's extracts from earlier philosophers, similarly treats the intension and remission of forms as a topic outside the usual introduction to physics.[22] On this evidence, it seems that the latitude of forms was taught at Paris, as Daniel Di Liscia has shown for Cologne and Vienna. But when the students of Paris did encounter such techniques as a distinct genre, they encountered it—unlike, it would seem, at Cologne and Vienna—apart from the rest of the natural philosophy course.

The format of these experimental printed textbooks shaped how a student encountered these calculatory techniques. At a glance, Lefèvre's version remained unusual in comparison to competitors.[23] It offered a single column of text printed on pages in small quarto format, reminiscent of the first classical texts printed at Paris and Venice. All other textbooks of physics printed at Paris were in larger quartos and folios. These typographical choices were deliberate: all of Lefèvre's works used the same format of single columns—even when students later added commentary to new editions. And when Lefèvre's printers, Hopyl and Higmann, printed the works of Bricot and George of Brussels, for example, they immediately switched format. Moreover, Lefèvre's dialogues on physics were—like his other books—full of visual aids that other books on these topics did not include, such as printed diagrams, lines labelled with letters. A student who picked up these books would immediately notice the difference in size, format, and visual layout.

These differences went beyond the surface, as a dense paratextual apparatus guided students into the subject.[24] Lefèvre prefaced each of his introductory works by condensing the longer paraphrase into a shorter list of theses; this was itself further condensed into a prefatory diagram of the text's main elements. Therefore the reader, from the outset, was coaxed into visualising the text at multiple levels of abstraction, first catching a birds-eye view of terms before focusing on general statements, and finally zooming down to the level of specific arguments. Indeed, as discussed below, Lefèvre's first dialogue on phys-

22 Tartaret, *Tractatus de intensione*. One might also compare Alvarus Thomas, *Liber de triplici motu*, which is pitched as a specialist study, well beyond a beginning student.

23 For an overview of the subgenres found within Lefèvre's textbooks, see D.A. Lines, 'Lefèvre and French Aristotelianism on the Eve of the Sixteenth Century,' in G. Frank and A. Speer (eds.), *Der Aristotelismus in der Frühen Neuzeit: Kontinuität oder Wiederangeignung?*, Wiesbaden 2007, 273–290.

24 This use of prefatory elements is analysed at length in Oosterhoff, *Making Mathematical Culture*, esp. ch. 4.

ics opened with an example of a student using such an introductory diagram to memorize the main lineaments of the discipline, before discussing specific claims with his teachers. Such visual pedagogy remained a constant presence in textbooks Lefèvre published in later years.

The reader also navigated layers of more discursive notes to the reader, poems, and dedicatory notes—these were like the prefatory figures, lists of theses, and diagrams, simply absent from Tartaret and Bricot's introductions to natural philosophy, and from Tartaret's brief introduction to the intension and remission of forms. Lefèvre and his circle were sharply aware of the novelty of their prefatory texts. They used these textual fragments to critique current practices in the university, and to offer an alternative vision of how intellectual discourse should unfold.[25]

The medium was intended to perform a message. In fact, the various prefatory notes and explicits of Lefèvre's books are strikingly chatty: they are by, to, or about a whole community of students and colleagues. This display of the friendship binding this community together underpinned the pedagogical claim driving these works. Lefèvre wrote the preface of the 1492 volume to a certain Stephanus, describing their longstanding friendship. This 'holy oath of friendship' held together reform-minded insiders against bad philosophizers. 'Outsiders may marvel,' Lefèvre exclaimed, 'at how great the goodwill is among those who cultivate the liberal arts in our Paris studium, where this experience is well known.'[26] Those outsiders, Lefèvre went on to clarify, are 'envious, malevolent men who cut each other up with their teeth,' and therefore are dogs, not lovers of wisdom who merit the title philosopher.[27] Friendship was not simply a conceit to frame the production of books (though it was that); in those books, intellectual friendship exemplified what good philosophy was all about. Aristotle had pointed out that real friendship was characterised

25 The next two pages are lightly revised from Oosterhoff, *Making Mathematical Culture*, 184–189.

26 Lefèvre d'Étaples, *Totius Aristotelis philosophiae naturalis paraphrases*, sig. J viii^r (ed. Rice, *The Prefatory Epistles*, 15): 'Mutua nos multos annos astrinximus benevolentia. Carissime Stephane, quanta sit animorum benevolentia inter liberalium artium cultores in hoc nostro Parisio studio (ubi res cognita esset) exteri mirarentur.' Presumably 'hoc nostrum studium' refers to their Collège du Cardinal Lemoine.

27 Lefèvre d'Étaples, *Totius Aristotelis philosophiae naturalis paraphrases*, sig. J viii^r (ed. Rice, *The Prefatory Epistles*, 16): 'Quo fit ut cum primum invidos, malevolos et sese mutuis dentibus lacerantes sentiunt, non amplius philosophos reputent, sed eos pro flagiti conditione in Pythagoreos canes versos putant.' The reference to 'Pythagorean dogs' is surely to transmigrated souls; philosophers of such ill will are canine souls in human bodies.

by shared goodwill (*mutua benevolentia*).[28] Therefore, what could good philosophers share more deeply than the love of wisdom?[29]

What rescued all this talk of friendship from being mere self-indulgent flattery was the way it informed the dialogues on physics. The choice of dialogue matched pedagogical goals. Lefèvre explained this choice. Guillaume Gontier, a student who accompanied him to Rome, had suggested dialogues because 'if you do, you will advise those who are learning how they should ask questions and how they should answer them; at the same time you will usefully counsel both student and teacher.'[30] Lefèvre later recalled how when visiting Rome he had marvelled at two youths pleasantly disputing in a mode of dialogue they evidently had learned from George of Trebizond's rhetoric.[31] The dialogues on physics therefore present a model of how discussion *ought* to happen. The sorites and sophisms of adversarial disputation were to be contrasted with a sociable search for the truth.

The two dialogues self-consciously exemplify this ethos of philosophical friendship. The interlocutors bear names that match their function. Thus the first dialogue is led by the teachers Hermeneus ('interpreter') and Oneropolus ('conjector'), who respectively interpret the subject and offer conjectures for reasoning about it. Enantius ('contrarian') presents arguments from the opposing viewpoint, while Homophron ('harmonizer') suggests ways to find common ground between all the perspectives on offer: the *personae* perform a kind of concordance of opposites.

The two dialogues take different approaches. In the first dialogue, the father Polypragmon places his son Epiponus in the care of several teachers. Lefèvre translates the father's name as *negociator*; we might see him as a 'busybody' or man of action, who regrets that his business ties leave him unable to pursue

---

28 E.g. Aristotle, *Nicomachean Ethics* VIII.2, 1155b30–1156a5; J. Lefèvre d'Étaples and J. Clichtove, *Artificialis introductio per modum Epitomatis in decem libros Ethicorum Aristotelis adiectis elucidata [Jodoci Clichtovei] commentariis*, Paris 1502, 31ᵛ.

29 I have also tried to come to terms with these themes in R.J. Oosterhoff, 'Lovers in Paratexts: Oronce Fine's Republic of Mathematics,' *Nuncius* 31/3 (2016), 549–583. See further U. Langer, *Perfect Friendship: Studies in Literature and Moral Philosophy from Boccaccio to Corneille*, Geneva 1994 (Histoire des idées et critique littéraire, 331).

30 J. Lefèvre d'Étaples, *Introductio in metaphysicorum libros Aristotelis*, ed. J. Clichtove, Paris 1494 (Johann Higmann), sig. b1ᵛ (ed. Rice, *The Prefatory Epistles*, 22): 'Si ita feceris, [Gontier] inquit, admonebis qui docturi erunt quo pacto interrogare debeant, interrogataque docere, et simul utiliter discipulo consules et docenti.'

31 J. Lefèvre d'Étaples (ed.), *Georgii Trapezontii dialectica*, Paris 1508, sig. a1ᵛ-a2ʳ (ed. Rice, *The Prefatory Epistles*, 190–191).

the greater contemplative wisdom he wishes for his 'intellectual' son.[32] As the teachers talk with Epiponus, they set him the task of reading a book of introductory physics much like Lefèvre's own: 'he should read it three or four times over again, and set it in his memory. Meanwhile, we will go for a walk as we wait' until he finishes.[33]

On his return, the teachers point to a diagram of circles—the same diagram introducing Lefèvre's own paraphrases on physics: 'Do you see this figure placed at the beginning of our introduction?'[34] And so the lesson begins, glossing the various parts of the diagram: *Natura, causa, motus, infinitum, locus, vacuum*, and *tempus*.

Sometimes the student prods his teachers for glimpses of future delights. He conjectures about how the union of form and matter can be compared to a line and a surface, which fuse together to make an object. Hermeneus swiftly intervenes:

> Hermeneus: 'Don't tease the boy with analogies!'
> Epiponus [student]: 'Do you mean then that all natural things arise whole and composite from lines and surfaces, since they are fused from matter and form?'
> Hermeneus: 'See whether or not I'm right to think you're teasing the boy.'
> Oneropolus: 'I speak, boy, but I do not speak as you suppose, for now you cannot understand. But keep on and you will understand eventually.'[35]

At such points in the dialogue, mathematics threatens to enter but is successfully rebuffed—until the end of this first dialogue, where the discussion turns

---

32 The term *polypragmon* has an uncomplimentary history: M. Leigh, *From Polypragmon to Curiosus: Ancient Concepts of Curious and Meddlesome*, Oxford 2013.

33 J. Lefèvre d'Étaples and J. Clichtove, *Totius philosophiae naturalis paraphrases, adiecto commentario* [*Clichtovei*], Paris 1502, 119^{v}: 'Noere, huic coaequali tuo Physicam introductionem procura. Quam ter quarterve repetitis vicibus legat, memoriaeque mandet. Nos obiter expectantes deambulabimus.'

34 Lefèvre d'Étaples and Clichtove, *Totius philosophiae naturalis paraphrases*, 119^{v}: 'Figura igitur huic nostrae introductioni praepositam vides?'

35 Lefèvre d'Étaples and Clichtove, *Totius philosophiae naturalis paraphrases*, 121^{v}: 'Her. Noli o Oneropole puerum analogiis ludere. Epi. Visne ergo rem naturalem omnem ex lineis et superficiebus consurgere perfectam, compositamque esse, quandoquidem ex materia et forma conflata sit? Her. Vide an ne non recte sentirem te puerum ludere. One. Dico fili, et non dico ut concipis, nunc autem intelligere non possis. Sed tu aliquando intellecturus reserva.'

to time. Aristotle had described time as an example of continuous magnitude, since it can be properly considered as a continuous series of instants, just as a line is a series of points.[36] Together, Oneropolus and Epiponus labour through a list of analogies. It is here that we find the intensification and remission of forms: just as extension characterises change in the quantity of objects, so qualities such as whiteness change by *intensio* or intensification; so also, the succession of time is *latio*. Here also, we see Lefèvre's editorial favouring of diagrams. Oneropolus teaches his student how to visualise the change of time on a line from A (past) to E (future).

The second dialogue—which now takes these calculatory techniques as its focus—is framed as a 'difficult' one, beginning with its title: *Dialogus difficilium physicalium introductorius*. The student is no longer the *puer* Epiponus, but now the *adolescens* Neanias ('youth'). The dialogue's difficulty is thematised in a pair of tantalising *notae*—which also explain the choice of dialogue as genre. First, Proteus is presented as an ancient sage who would only willingly foretell the truth if overcome and bound. Second, Milo of Croton is the Greek wrestler who 'killed a bull with a barehanded blow in an Olympic contest, carried it for a hundred yards, and then ate the whole thing that very day.'[37] Both stories would become powerful metaphors for the difficulty of pursuing natural knowledge. Over the sixteenth century, the metaphor of binding Proteus would become a telling image of nature itself, to be wrestled into revealing its secrets, or else of the investigator, chasing down barehanded experience.[38]

---

36 E.g. *Physica* IV.13, 222a28–b7. This misses the finely nuanced debates on the relation of indivisibles to *continua* worked out in the fourteenth century. See *inter alia* E.D. Sylla, 'Thomas Bradwardine's *De continuo* and the Structure of Fourteenth-Century Learning,' in E.D. Sylla and M.R. McVaugh (eds.), *Texts and Contexts in Ancient and Medieval Science. Studies on the Occasion of John E. Murdoch's Seventieth Birthday*, Leiden 1997 (Brill's Studies in Intellectual History, 78), 148–186.

37 Lefèvre d'Étaples and Clichtove, *Totius philosophiae naturalis paraphrases*, 127^r: 'Proteus filius Oceani et Tethyos, vates maximus, nonnisi coactus, victusque volens ora veridica soluere. Milo Crotonensis atheleta fortissimus, qui nude manus ictu in certamine olympico taurum interfecit, stadio uno spiritu retento portavit, quem totum die illo comedit.'

38 Proteus appeared in the *Odyssey* 4.382–569; Plato, *Euthydemus*, 288b; Virgil, *Georgics* 4.387–529; Ovid, *Metamorphoses* 8.731, 11.221–256; Diodorus Siculus, *Antiquities of Egypt* 1.62. The identification of Proteus as a *vates maximus* suggests that Lefèvre had Diodorus at least partly in mind, while a later reference to Aristaeus reveals Virgil as another source. For the early modern period, see *inter alia* E. Wind, *Pagan Mysteries in the Renaissance. An Exploration of Philosophical and Mystical Sources of Iconography in Renaissance Art*, London 1958, 158–175; W.E. Burns, '"A Proverb of Versatile Mutability": Proteus and Natural Knowledge in Early Modern Britain,' *The Sixteenth Century Journal* 32 (2001), 969–980; P. Pesic, 'Shapes of Proteus in Renaissance Art,' *Huntington Library Quarterly* 73 (2010), 57–82.

Here in Lefèvre's dialogue, however, Proteus is a friendly conversationalist. Near the end of the dialogue Oneropolus claims that he has finished the subject, but Enantius and Homophron together plead with him to continue. 'Just as Aristeus once bound up the varied and multiformed truth-telling Proteus with arms and chains, and so compelled him to tell the truth, so Homophron and I bind you in our arms, and we compel you to teach us what we asked.'[39] Oneropolus acquiesces: 'you're a friend, Enantius, and nobly you compel me—fearing the worthy arms of Milo—to assent to your requests.' The toil of knowing, the wrestling of the natural philosopher is not in the savage cut and parry of disputatious one-upmanship, but the encouraging conversation of friends who share the goal of knowledge.[40] Friendship does not eliminate struggle. Rather, friendship makes the struggle for truth *fruitful*. Throughout the dialogues, then, the motif of friendship counters the association of calculatory techniques with the putative pathologies of sophisms.

Wrapped in layers of paratext, the techniques of the Calculators could be taught safely—perhaps too safely. We might suppose that such ginger handling would put off some readers. A master or student looking for strategies to deploy in disputations would have found Lefèvre's dialogue unnecessary work, since it did not straightforwardly set out arguments *pro* and *contra* on given *quaestiones*. What other options were available to a young scholar interested in calculatory techniques?

The limited number of introductory printed options may help explain why such techniques became less prominent over the next decades. As already mentioned, Tartaret's excerpts on the intention and remission of forms (1493) was far from basic, and did not gain enough readers to be reprinted; similarly, Alvarus Thomas' *Liber de triplici motu* (1509) was sophisticated and difficult.[41] Jacques Almain's *Embammata phisicalia* (1506) seems more promising, at first glance. It was published under conditions much closer to those of Lefèvre's textbooks: one of Almain's students, Peter Heymeric, published the book in 1506, when Almain was an arts regent before he became a doctor in the fac-

39 Lefèvre d'Étaples and Clichtove, *Totius philosophiae naturalis paraphrases*, 148^{v}–149^{r}: 'Ut Aristaeus olim varium, multiformemque veridicum tamen Protea brachiis et vinculis implicuit, compulitque verum fateri, ita quoque ego et Homophron brachiis te implicabimus, cogemusque quod quaerimus te nos docere. One. Comis es o Enanti, et pulchre me cogis, emerita Milonis brachia timentem, vestris petitionibus acquiescere.'

40 Elsewhere, Lefèvre, Clichtove, and Alain de Varennes expand on the Aristotelian view that true friendship involves mutual goodwill around shared aims: Oosterhoff, 'Lovers in Paratexts,' 565–566.

41 Sylla, 'Alvarus Thomas.'

ulty of theology in 1512.[42] Though printed in the traditional double columns, it shares with Lefèvre's books a generous apparatus, including dedicatory poems, Almain's own preface thanking his student-cum-editor, and a list of arguments at the end. The first two pages make some introductory moves, quickly surveying the theory of proportions. But the contents of the book swiftly swing into a very different direction. Instead of gradually building up a familiarity with the contents of Aristotle's *Physics*, the bulk of the 228 quarto pages surveys a series of philosophical 'difficulties' thrown up by these analytical languages, picking up themes from Book I and Book III of the *Physics*, on the divisibility of points, the difference between mathematical and physical accounts of magnitude, and the maximum and the minimum. Each difficulty is argued from several angles, and ultimately resolved from a nominalist perspective.[43]

Just how much the book differs from Lefèvre's introduction can be measured from Almain's preface, where he explains the enigmatic title *Embammata* (literally, 'sauces'). The letter begins in a posture of humility often found in Lefèvre's letters. It is addressed to his student-editor, Peter Heymeric, and stresses Peter's role in bringing the work to print: Almain had abandoned these *indigesta* long before, but Peter had compelled him to offer them for public consumption. In fact, like Lefèvre, Almain describes philosophical love as a motivation for Heymeric's labours—but with a different emphasis. 'You were so filled with love of philosophy that, unless you had taken a spiked stick to drive out the shackles and chains from the various flowers of physics, you would have died with Diodorus, who they say ended his life over the shame of a philosophical

42 For an overview of Almain's life, see J.K. Farge, *Biographical Register of Paris Doctors of Theology, 1500–1536*, Toronto, ON, 1980 (Subsidia mediaevalia, 10), 15–17. An incomplete manuscript of the *Embammata* was copied by one Antonius of Toledo (Lugdunensis), now in Sevilla, Biblioteca Capitular y Colombina, Ms. 7-6-12. For this reference, my thanks to Daniel Di Liscia, who has rediscovered Almain's work; see the first fruits in D.A. Di Liscia, 'Velocidad *quo ad effectus* y velocidad *quo ad causas*: La tradición de los calculadores y la metodología aristotélica,' in D.A. Di Liscia, E. Kessler and C. Methuen (eds.), *Method and Order in Renaissance Philosophy of Nature. The Aristotle Commentary Tradition*, Aldershot 1997, 143–176, and D.A. Di Liscia, 'La *conclusio pulchra, mirabilis et bona*: una ingeniosa demostración atribuible a Nicole Oresme,' *Mediaevalia. Textos e estudos* 37 (2018), 139–168, at 163–167. We can anticipate an edition of work by Almain from Di Liscia and Sabine Rommevaux-Tani.

43 E.g. on the maximum and minimum, see J. Almain, *Embammata phisicalia*, Paris 1506, 23^{v}–42^{v}. Oddly enough, the table of theses appended to the end includes intriguing topics such as 'Bertha is the mother of two, who endure even when Bertha herself no longer exists as their mother' (sig. a iii^{v}: 'Bertha est mater duorum et ipsis manentibus cum Bertha ipsa illorum duorum non erit amplius mater').

question.'[44] Like Lefèvre's letters, this note is alert to the passions in philosophical debate; but it draws a different lesson about pride and shame. Almain does not present a solution to harmful passions, but rather gives a charmingly self-deprecating account of his 'sauces'. He observes that, although drawn from his 'pained breast, moved by cares—like a tender kid goat from its mother's teats,' nevertheless his work may not nourish like fresh milk. But, 'even if the taste does not please, perhaps it may find favour in sight or smell. For the *libanotis* (as they call the plant growing on the seaside), even though it does nothing for the tongue, is pleasing to see and smell when green and watered, and so is often gathered.'[45] These sauces are a sensual pleasure. They may not nourish, but like a bouquet of flowers they are still good to have around.

Even more than Lefèvre's dialogues—which, recall, are outside the main paraphrases of the textbook—Almain's *Embammata*, like Alvarus Thomas' *Liber de triplici motu*, is not so much a main course as a pleasure for particular appetites, an elective study that was self-consciously distinct from the regular cursus in natural philosophy. In the 1510s, two books did begin to frame calculatory techniques in pedagogically sensitive ways: the *Physice perscrutationes* of Luis Coronel (1511) and Juan de Celeya's *Expositio in octo libros Physicorum* (1517).[46] But the two previous decades seem to have set in place a printed textbook tradition that undercut the role of the calculatory techniques among the introductions to natural philosophy.

2 A Changing Mathematical Culture

The rather spotty presence of the Calculators in the printed textbook tradition that came to dominate instruction at Paris around 1500 surely goes some way towards explaining the decline of their analytical tools as an intellectual

44 Almain to Peter Heymeric, preface to *Embammata phisicalia*, sig. a2r: 'Presertim cum tanto philosophie amore infiltratus fueris, ut nisi ferrula variis physices flosculis dentata numellas torquesque ipsos elim[in]averis, cum Dyodoro (quem aiunt) philosophice questionis pudore vitam terminasse.' The story about Diodorus was passed on by Diogenes Laertius, *Lives of the Eminent Philosophers*, § 111.

45 Almain, *Embammata phisicalia*, sig. a2r: 'At ubi expetitas nostrorum quamvis immaturorum videris Embammatum taleas, se motis tristi a pectore curis (tanquam caper tenellus ad matris ubera) covagies. Quas quidem (ut spero) tantis tamquam frequentibus ictulis saturaturus mulgebis, ut si non lacteos emanarint liquores; attamen nescio quid stillantis succi avidus exhauries. Aut si non gustulum voluptaverint forsitan visum vel nares ipsas confovebunt. Libanotis enim (quem rorem marinum appellitant) quamquam linguam non decorat, olfactum tum visum placent viredine irrigando, componere solet.'

46 L. Coronel, *Physice perscrutationes*, Paris 1511; J. de Celaya, *Expositio in octo libros*

tradition. But it may be even more significant that these techniques were not a priority—barely even present—in the later generations of mathematically minded Parisian practitioners who built on Lefèvre's mathematical teaching. These included the theologians Juan Martínez Silíceo and Pedro Ciruelo, as well as the arts masters Oronce Fine and Jean Fernel. Considering why these figures did not make calculatory techniques central, despite their fascination with mathematics and its potential for modeling nature, may let us hazard two alternative answers.

The first answer is that the mathematical interests of these scholars were in fact somewhat disconnected from calculatory techniques. They framed their mathematical projects more in relation to the pedagogy of liberal arts than in relation to natural philosophy. Wallace cited Silíceo and Ciruelo among the group of mathematically sophisticated members of Mair's circle, pointing out that they would go on to play a crucial role in setting up the mathematical culture of sixteenth-century Spain, and so 'are entitled to be called the first Spanish "Calculatores".'[47] Both Silíceo and Ciruelo did publish relevant works by Swineshead and Bradwardine.[48] But both more thoroughly connected their mathematical expertise to the tradition begun at Cardinal Lemoine. Lefèvre had published, with students such as Josse Clichtove, David Laux, Charles de Bovelles and others, introductions to mathematics as liberal arts, the Boethian quadrivium. His main works, therefore, included epitomes of Boethius' arithmetic, a reworking of Jordanus' *Arithmetica*, the astronomy of Sacrobosco, a brief theorica, and eventually a new edition of Euclid.[49]

This vision captivated Silíceo and Ciruelo, and formed the basis of their *translatio* of Parisian mathematics to the Spanish universities in the 1510s. In 1513, Silíceo published his *Arithmetica practica*, with the evocative subtitle 'very useful for astrologers, physicists, and calculators' (*Liber arithmetice practice astrologis, physicis, et calculatoribus admodum utilis*). But the contents of the book itself make no reference to the calculatory tradition in physics. Instead, Silíceo dedicates the book to the bishop of his hometown in Portugal with a long list of commonplaces about the value of arithmetic for civic rulers. The context for this exercise becomes clearer in the preface, which gives a

---

Physicorum Aristotelis, cum questionibus eiusdem secundum triplicem viam beati Thomae, realium, et nominalium, Paris 1517.

47 Wallace, 'The *Calculatores*,' 225. See now also V. Navarro-Brotons, *Disciplinas, saberes y prácticas: Filiosofía natural, matemáticas y astronomía en la sociedad española de la época moderna*, Valencia 2014, 26–28, 58–72.

48 Thomas Bradwardine, *Arithmetica speculativa*, ed. P. Ciruelo, Paris 1495; Richard Swineshead, *Calculatoris sublime et prope divinum opus*, ed. J. Martínez Silíceo, Salamanca 1520.

49 For more details, see bibliography in Rice (ed.), *The Prefatory Epistles*.

genealogy of the discipline that begins with Pythagoras and ends with Lefèvre d'Étaples and Josse Clichtove: 'I would say that all of these things are rightly indebted to these men.'[50] The book closes with a poem addressed to Clichtove, praising Silíceo's work as the latest eminent contribution to the same enterprise: 'But no elegance seems in our age, I think, more polished than the art of Silíceo; A brighter flash of lightning consumes the smoke, just as a great river exists from the smallest streams.'[51] The *aemulatio* makes clear just which tradition Silíceo saw himself working within.

Ciruelo's books engaged the work of Lefèvre and his colleagues even more directly. Ciruelo was a talented mathematical author in his own right, with editions and writings ranging across the breadth of late medieval mathematical culture.[52] His first publication was an edition of Bradwardine's theoretical work on proportions (1495), followed three years later by an edition of Pierre d'Ailly's questions on Sacrobosco's *Sphere*, with his own commentary. Already then, however, Ciruelo cited Lefèvre's own commentary on the *Sphere*, and he concluded the work with a dialogue in much the same mode as Lefèvre's dialogues on physics, emphasizing the role of philosophy as the pursuit of truth, and critiquing 'our Parisian philosophers' who tended to ignore the views of other universities.[53] Ciruelo's engagement with the mathematical tradition of Lefèvre's circle intensified when Ciruelo left Paris to join the new wave of humanistic studies associated with Elio Antonio de Nebrija at Salamanca: his signal contributions to curricular reform were new introductions that were either excerpted from or explicitly modelled on Lefèvre, Clichtove, and Bovelles' mathematical works.[54] As a doctor of theology, Ciruelo was well placed to make methodological pronunciations across the disciplines, and indeed one of his most striking works was a defense of astrology (in light of Pico's reverberating critique), which emphasised the closeness between physics and astronomy.[55] At various points in his works, Ciruelo referred to Lefèvre's works as an example of a philo-

---

50 J. Martínez Silíceo, *Liber arithmetices practice Astrologis phisicis et calculatoribus admodum utilis*, Paris 1513, sig. Aii^r: '… quos debito iure omnium horarum viros esse dixerim' (reading *dixerim* for *diverim*).

51 Silíceo, *Liber arithmetices practice*, sig. fiv^r. 'Nulla tamen nostro visa est facundia seclo | Martini solidi cultior arte puto | Candidiora dabit consumpto fulgura fumo | Grande velut minimo flumine flumen adest.'

52 An overview of recent literature is given by T.M.C. Lanuza-Navarro, 'Astrology in Court: The Spanish Inquisition, Authority, and Expertise,' *History of Science* 55 (2017), 187–209, at 195.

53 P. Ciruelo, *Uberrimum sphere mundi commentum intersertis etiam questionibus domini Petri de Aliaco*, Paris 1498, sig. n iiii^v–n vii^r.

54 P. Ciruelo, *Cursus quatuor mathematicarum artium liberalium*, Alcalá 1516.

55 P. Ciruelo, *Apotelesmata astrologiae christianae*, Alcalá 1521.

sophy of mathematics that might show the physical implications of mathematics.[56] But—despite helping make available some of the Calculators—Ciruelo himself did not try to expound the calculator's version of a mathematical natural philosophy.

It makes some sense that Silíceo and Ciruelo, theologians who worked at mathematics within a reformist plan for university teaching, would perhaps overlook the opportunity to integrate calculatory techniques more systematically into the *cursus* of the arts. But we might expect something else from mathematical practitioners, those who made their reputation and money chiefly through teaching, publishing, and advising on mathematical matters. The chief example would be Oronce Fine (1594–1555), who came to define the profile of Parisian mathematics in the sixteenth century.[57] While teaching arts at the Collège de Navarre and the Collège de Maître Gervais, Fine was also active in the print trade, and his work as a mathematician was informed by his work as an editor and designer of mathematical woodcuts, especially for Lefèvre's collaborating printer Henri Estiennes the Elder and his successor Simon de Colines. He edited Silíceo's *Arithmetica* in 1519, and was responsible for repeated editions of Lefèvre's mathematical works through the 1520s and 1530s. Thanks in part to Guillaume Budé's support, in 1532 Fine was made the professor of mathematics in Francis I's new Collège Royal, on the strength of his publications in arithmetic, geometry, cosmography, optics, and his skill in crafting maps, astronomical instruments, and dialls.

Fine's works reveal a deep interest in the relation of mathematics to nature's structures, yet I have not found calculatory techniques anywhere in his *oeuvre*. Angela Axworthy has analysed at length Fine's claims about the value of geometry and arithmetic for explaining nature: geometry 'has taught us the quantities of all bodies, their shapes, movements, bounds, and positions.'[58] Despite arguing strenuously, and over decades, that mathematics was the path to all philosophy, Fine in the main argued that mathematics was best for training the student's mind. Even more than Silíceo and Ciruelo (and like Lefèvre) Fine downplayed the technical, mechanical insights that mathematics might offer philosophy. In his most public statements, he presented the sciences of meas-

56 Oosterhoff, *Making Mathematical Culture*, 161–162, 205, 211–212.

57 The starting point is now A. Axworthy, *Le mathématicien renaissant et son savoir. Le statut des mathématiques selon Oronce Fine*, Paris 2016 (Histoire et philosophie des sciences, 11), and A. Marr (ed.), *The Worlds of Oronce Fine. Mathematics, Instruments and Print in Renaissance France*, Donington 2009.

58 O. Fine, *Epistre exhortative*, Paris 1531, sig. B1^{v}, ed. Axworthy, *Le mathématicien renaissant*, 365–380, at 374.

uring as beneficial, but chiefly to a public man, one engaged in the arts of ruling. Therefore, as part of the liberal arts, mathematical practitioners best defended their place in public by claiming how mathematics would render the *esprits* and *ingenia* of youth ready for noble service. A mistrusted mathematical culture began to bid for greater prestige, but in the context of a Parisian clerical elite that oriented university education increasingly towards the court of Francis I, it found less time for the analytical languages of the Calculators. This first explanation, then, for declining interest in intension and remission of forms, first and last instances, etc., is that adepts like Fine (and to some extent Silíceo and Ciruelo) had a different professional trajectory, needing little from natural philosophy.

A second explanation may be that Paris humanists like Fine had turned to other philosophical linkages between mathematics and physics. A line of analysis might consider the term *physiologia*. Since antiquity, the main meaning of *physiologia* was as an inkhorn Graecism for 'natural philosophy', a study of the properties and changes of natural objects—as Cicero observed, the *ratio naturae* was 'what the Greeks called physiology.'[59] But the term was not widely used until the late sixteenth century, when popular textbooks such as Johannes Magirus' *Physiologia peripatetica* (first published at Frankfurt, 1597) made it stand in for the main subjects of natural philosophy. Thus it showed up in the subtitle of Robert Hooke's famous work, *Micrographia, or, some physiological descriptions of minute bodies made by magnifying glasses* (London, 1665).[60] In these works, mathematics often featured as a preliminary study for natural philosophy. The dedicatory letter to Magirus proclaims the value of 'mathematics and physiology, namely natural science.'[61] Whereas mathematics measures natural bodies, physiology studies their motions.

This doublet makes more sense in light of the rehabilitation of the term *physiologia* one century earlier. Giorgio Valla's encyclopaedic *De expetendis et fugiendis rebus*, published posthumously by a student in 1501, tried to refound the entire cycle of education on a Platonic account of mathematics. Valla had

59 Cic. *Nat. de.* 1.8.20.

60 Cf. G. Vossius, *De theologia gentili et physiologia Christiana*, Amsterdam 1641; M.F. Wendelin, *Contemplationum physicarum sectio I. qua physiologia generalis, De principiis et affectionibus Corporibus naturalis*, Cambridge 1648; W. Charleton, *Physiologia Epicuro-Gassendo-Charltoniana*, London 1654; T. Hobbes, *Decameron physiologicum, or, Ten Dialogues of Natural Philosophy*, London 1678.

61 Conrad Nebenius, dedicatory letter to J. Magirus, *Physiologia peripatetica*, Frankfurt am Main 1619, 6: '... Mathesis, aut Physiologia seu scientia naturalis est.' The word *mathesis* clearly refers to mathematics, since the next three paragraphs are a paean on the dignity and use of this discipline, *circa corporum dimensiones occupata paedia est*.

helped to circulate and translate late antique Platonist texts such as Proclus' *Commentary on the First Book of Euclid*—extracts were silently incorporated into the body of *De expetendis*.[62] As a whole, the two large folio volumes approximate a progression from theory to practice: they begin with the mathematical quadrivium, move through its applications in natural philosophy and medicine, and take up the social disciplines of grammar, logic, and poetics, to their applications in public life, in *oeconomia* and *politica*. For us the important thing is the placement of natural philosophy: it comes under the name *physiologia*, popular among contemporary Greek scholars such as Ficino and Poliziano. *Physiologia* follows directly after astronomy. Valla presents this astronomy (or astrology) as a study of heavenly structures that chiefly has use in medicine. Therefore the four books on astrology map the configuration of the heavenly macrocosm onto the structure of the human microcosm. Indeed, Book II of Valla's *Astrologia* is simply twenty-five pages of tables correlating body parts with longitudes and latitudes of heavenly bodies.[63] This account also deeply marks Valla's four books on *physiologia*. The order of nature and its movements in fact are inescapably the working out of divine 'metaphysical seeds' (*semina metaphysices*) that Valla finds strewn throughout the frame of nature. The relation of macrocosm and microcosm explains why Valla begins this natural philosophy with the human soul and *daemones*, before considering elemental principles.[64]

There is no need for calculatory technique in such an account. Natural change is explained not by a succession of different qualities, sensed through an Aristotelian process of abstraction. For the Aristotelian, mathematics merely models the experience of change. For Valla, mathematics carries the mind into the eternal seeds (*semina*) that exert power over physical form. Sometimes described as a *spiritus*, this force is itself actively rational—therefore it should be explained by reasoning about principles, not by a posteriori reasoning about experience. As a result, we might suppose that Valla's vision of philosophy, though interested in the mathematical principles of natural philo-

62 G. Valla, *De expetendis et fugiendis rebus*, Venice 1501. For context see A.A. Raschieri, 'Giorgio Valla, Editor and Translator of Ancient Scientific Texts,' in: P. Olmos (ed.), *Greek Science in the Long Run: Essays on the Greek Scientific Tradition (4th c. BCE–17th c. CE)*, Cambridge 2012, 127–149; P.L. Rose, 'Bartolomeo Zamberti's Funeral Oration for the Humanist Encyclopedist Giorgio Valla,' in C.H. Clough, *Cultural Aspects of the Italian Renaissance: Essays in Honour of Paul Oskar Kristeller*, Manchester 1976, 299–310. A foundational resource is J.L. Heiberg, *Beiträge zur Geschichte Georg Vallas und seiner Bibliothek*, Leipzig 1896.

63 Valla, *De expetendis et fugiendis rebus*, sig. dd 2r–ee 6r.

64 Valla, *De expetendis et fugiendis rebus*, at sig. ii 8r, and sig. hh 4r.

sophy, depends on a very different concept of 'principle' than most Aristotelian accounts.

Valla's account had close, attentive readers in early sixteenth-century Paris, particularly among those reflecting on the power of mathematics in Oronce Fine's circle. It is clear that Fine himself was influenced by this account of the disciplines, for his arguments about geometry closely follow Proclus—only occasionally rewording the translation of Valla in *De expetendis et fugiendis rebus*.[65] Although Fine may not have been sharply aware of this divergence from the Aristotelian doctrine of abstraction, his account of intuition of mathematical principles was influential among followers at the Collège Royal such as Antoine Mizauld, and eventually Peter Ramus.[66]

Fine's student Jean Fernel is a particularly interesting corroborating case. As a physician teaching at Paris, Fernel has long been understood as one of the most influential and subtle medical theorists of the sixteenth century, a Galenist counterpoint to Vesalius, whose *Physiologia* is usually cited as the first use of the term as a domain of medical study.[67] But before he devoted himself to medicine, Fernel's first career choice had been mathematics, and he worked closely with Fine in the 1520s. Some have seen Fernel's move away from mathematics as a kind of intellectual conversion.[68] But in fact the title of the *Physiologia* hints at ways that Fernel's approach to mathematics in the 1520s continued to animate his work. The work itself followed a conception of the human body's relation to seeds and principles—what he famously called *spiritus*. The sources of such pneumatic ideas are surely many: Fernel himself cited Galen, ancient Stoics, Aristotle's own claims about *entelechia*, and various representatives of the Platonic tradition.[69] Fernel's own decision to call this a *physiologia* surely also

65 Axworthy, *Le mathématicien renaissant*, 60–64, 101–104.

66 On Ramus' intuitive mathematics, see R. Goulding, *Defending Hypatia: Ramus, Savile, and the Renaissance Rediscovery of Mathematical History*, New York 2010 (Archimedes, New Studies in the History of Science and Technology, 25). In forthcoming work, Goulding also address Ramus' efforts to apply such intuition to natural philosophy.

67 For some support of this view, see A. Cunningham, 'The Pen and the Sword: Recovering the Disciplinary Identity of Physiology and Anatomy before 1800: I: Old Physiology—the Pen,' *Studies in History and Philosophy of Biological and Biomedical Sciences* 33/4 (2002), 631–665. An argument for more continuity is V. Nutton, 'Physiologia from Galen to Jacob Bording,' in M. Horstmanshoff, H. King and C. Zittel (eds.), *Blood, Sweat and Tears: The Changing Concepts of Physiology from Antiquity into Early Modern Europe*, Leiden 2012 (Intersections, 25), 27–40.

68 C. Sherrington, *The Endeavour of Jean Fernel, With a List of the Editions of his Writings*, London 1946; J. Henry, '"Mathematics Made No Contribution to the Public Weal": Why Jean Fernel (1497–1558) Became a Physician,' *Centaurus* 53 (2011), 193–220.

69 See the notes to J.M. Forrester and J. Henry (eds.), *The 'Physiologia' of Jean Fernel (1567)*,

reflects his wish to use the classicizing vocabulary of Pseudo-Galen.[70] But the architectonic claim that this discipline was a *physiologia* becomes more intelligible in light of Valla's physiology, where the structure of the cosmos is linked to the human body's functions, so that reasoning about planetary powers informs medical remedies. Likewise, Fernel named *physiologia* as that part of medicine that lay between a study of the body's anatomy and the study of remedies. Valla presented reasoning from one to the other as a deductive process—it is mathematical or scientific reasoning, not merely prudential craft. Likewise, Fernel explained that physiology proceeded not by abstraction from anatomical experience, but on demonstrative reasoning:

> So, since the human body is already divided up by anatomy into the parts open to sense, the next stage must be to move across to those that are learned by thought alone, and to track down further the elements that contribute to each part, and their elemental mixture, their temperament, the powers and faculties that lie hidden in them, and the spirit and heat by which they are maintained. When these matters have been discovered and grasped by analysis, then it will become clear from the sequence of their composition what the efficient causes of everything are, what humors are generated by them, what the functions of individual parts are, and what the natural management of everything is. In this way a comprehensive physiology will be put together, that establishes the natural study of man by the force of demonstration.[71]

Philadephia, PA, 2003 (Transactions of the American Philosophical Society, 93/1). For some relevant framework of body-earth-cosmos analogies, see L. Taub, 'Physiological Analogies and Metaphors in Explanations of the Earth and the Cosmos,' in M. Horstmanshoff, H. King and C. Zittel (eds.), *Blood, Sweat and Tears: The Changing Concepts of Physiology from Antiquity into Early Modern Europe*, Leiden 2012 (Intersections, 25), 41–61. On Fernel in the context of Lefèvre's interests, see J.J. Bono, 'Reform and the Languages of Renaissance Theoretical Medicine: Harvey versus Fernel,' *Journal of the History of Biology* 23/3 (1990), 341–387.

70 Nutton, 'Physiologia,' 29–30.

71 Forrester and Henry (eds.), *The 'Physiologia'*, 180–182: 'Quum igitur humanum corpus anatome in partes sensu conspicuas dissolutum iam sit, ab his deinceps ad ea transeundum, quae cogitatione sola discuntur, et altiùs investigandum ex quibus elementis pars vnaquaeque condita sit, et quae sit elementorum permistio, quae temperatio, quae vires ac facultates lateant in illis, et quo spiritu quóve calore serventor. Quum analysi haec inuenta perceptáque fuerint, compositionis dein ordine conspicuum fiet, quae sint effectrices omnium causae, qui ab his gignantur humores, quae partium singularum functiones sint, et quae naturalis omnium administratio. Sic universa contrahetur physiologia, quae naturalem de homine contemplationem demonstrationis vi constituit' (English translation 181–183).

A serious evaluation of Fernel's potential debt to Valla must await another time. But there is at least a family resemblance in their insistance on physiology as a study of combinations of elements upheld by inborn powers, built around the analogy of macrocosm and microcosm. And for the purposes of this chapter, the implications in either case for the study of natural philosophy are clear: Valla and Fernel represent a *mathesis* in which 'natural principles' meant something different than they meant to the Calculators.[72] In the previous generation, Lefèvre and Ciruelo had been familiar with such principles—Ciruelo described them as belonging to the 'magicians', and offered a sidelong reference to Lefèvre as an example.[73] But they remained responsible teachers of the Aristotelian cursus, and therefore recognised an important place for calculatory techniques. As practitioners of other *métiers*, Fine and Fernel adopted another architectonic account of mathematics in which such analytical languages seemed unnecessary.

Conclusion

Why did these calculatory techniques lose the attractions that had made them such a popular language of analysis in the fourteenth century, in John Murdoch's words? One line of argument has depended on the rhetoric that set apart humanism and scholasticism. This is less convincing on closer examination. Recent studies of these cultures have been alert to the fact that those usually identified as either scholastics or humanists in fact depended on the same institutions, training, career paths, and indeed skills.[74] Moreover, some 'scholastics' such as Gerson railed at *anglicanes subtilitates*, while some 'humanists' like Lefèvre wrote introductions for them. The division of humanism and scholasticism often says more about historians than the history.

72 This leads to questions that have long absorbed historians of later natural philosophy: J. Henry, 'Occult Qualities and the Experimental Philosophy: Active Principles in Pre-Newtonian Matter Theory,' *History of Science*, 24 (1986), 335–381.

73 Oosterhoff, *Making Mathematical Culture*, 211–212.

74 E.g. D.A. Lines, 'Humanism and the Italian Universities,' in C.S. Celenza and K. Gouwens (eds.), *Humanism and Creativity in the Renaissance. Essays in Honor of Ronald G. Witt*, Leiden 2006 (Brill's Studies in Intellectual History, 136), 327–346; D. Hobbins, *Authorship and Publicity Before Print: Jean Gerson and the Transformation of Late Medieval Learning*, Philadelphia, PA, 2009 (Middle Ages Series); A. Broadie, 'John Mair's *Dialogus de Materia Theologo Tractanda*. Introduction, Text and Translation,' in A.A. MacDonald, Z.R.W.M. von Martels and J.R. Veenstra (eds.), *Christian Humanism: Essays in Honour of Arjo Vanderjagt*, Leiden 2009 (Studies in Medieval and Reformation Traditions, 142), 419–430.

Instead, I have argued that the rise of the printed textbook helps us to better explain the decline of calculatory techniques as a genre. After the 1490s (in Paris, but likely elsewhere too), students increasingly encountered natural philosophy in printed textbooks. It would be interesting to find (though hard to prove) that students preferred to buy a cheap book rather than engage in the live disputation that helped train students in the skills necessary for the Calculators' project. Certainly, what counted as peripatetic physics was constrained by what those textbooks contained—and most of them did not include calculatory techniques. Between the introduction of printed textbooks in Paris in 1492 and 1511, there was only one introductory account of Aristotle's natural philosophy that also included the basics on intension and remission of forms, maxima and minima, first and last instances, and so on, and that was Lefèvre's *Totius Aristotelis philosophiae naturalis paraphrases* (1492). And it—adopting a line of university reform that is traceable back to Gerson—was hardly a straightforward example of how to deploy such techniques in university disputations. It seems that an unintended consequence of the two decades it took to come to terms with printed textbooks was that calculatory analytical languages dropped out of regular teaching.

By looking at mathematics in the generations after Lefèvre, I have suggested, we can see two further reasons for the declining interest. First, what we might call the 'professional' trajectory of mathematical interests developed apart from natural philosophy, among those who claimed special expertise in mathematics. Already Ciruelo and Silíceo, picking up Lefèvre's project, did not feel responsible for linking their reform of mathematical teaching to the natural philosophical curriculum. And clearly, for the *lecteur royal* Oronce Fine (and, we might extrapolate, the young Jean Fernel), the identity of the mathematical practitioner—although bolstered by claims of what mathematics might do for philosophers—did not entail publishing textbooks on natural philosophy or calculatory techniques. Even within the university, the growing profile of the mathematical practitioner was linked to astronomy (especially via astrology's place in medicine), mapping, and increasingly machinery, but not natural philosophy per se.[75]

75 On the mathematical practitioner, see R.S. Westman, 'Humanism and Scientific Roles in the Sixteenth Century,' in F. Krafft and R. Schmitz (eds.), *Humanismus und Naturwissenschaften*, Boppard am Rhein 1980 (Beiträge zur Humanismusforschung, 6), 83–100; M. Biagioli, 'The Social Status of Italian Mathematicians,' *History of Science* 27 (1989), 41–95; I. Hantsche, *Der 'Mathematicus': Zur Entwicklung und Bedeutung einer neuen Berufsgruppe in der Zeit Gerhard Mercators*, Bochum 1996 (Duisburger Mercator-Studien, 4). For a later period, L.B. Cormack, S.A. Walton and J.A. Schuster (eds.), *Mathematical Practitioners and the Transformation of Natural Knowledge in Early Modern Europe*, Cham, Switzerland, 2017 (Studies in History and Philosophy of Science, 45).

Finally, the philosophical foundations of mathematical practice had subtly shifted in this new generation, with implications for the Calculators' tradition. Efforts to reimagine the entire cycle of arts on a mathematical foundation, as Giorgio Valla had done a little earlier in Venice, focused on mathematical principles as Platonic ideas, and were less interested in harmonizing with Aristotelian accounts of abstraction. As a result, the analytical languages of the Calculators seemed unnecessary. Put crudely: it was unnecessary because the deductive model held out the promise that—eventually—one might deduce mathematical causes. What need then for merely approximate models?

Acknowledgements

My thanks to Daniel Di Liscia for sharing many kind suggestions and material over the past few years, and for kindly enabling my remote participation in the conference leading to this volume, even when I could not be there in person. I also owe thanks to both Daniel and Edith Sylla for thoughtful and generous suggestions on an earlier draft of this chapter.

CHAPTER 13

Some Aspects of the 'Rules' of *motus difformis* in Angelo da Fossambruno's Commentary on Heytesbury's *De tribus praedicamentis*

Fabio Seller

1 Introduction

In spite of his works' widespread reputation, we have scanty biographical information on Angelo's life—his name has been variously spelled: we find *Forosempronii, Foresimfraneo, Fossinfronte, Fosseimbruno, Fosumbronio, Fossambruno*; here I will simply refer to him as Angelo da Fossambruno.[1] We know that he lived in the second half of the fourteenth century and died at the beginning of the fifteenth, and that he was a master in the universities of Bologna—where he taught logic and natural philosophy in the 1390s—and Padua, as *artium doctor*, where he probably gave lectures on natural philosophy. It should be noticed, as Curtis Wilson has pointed out, that 'it was on the Continent rather than in England that the *Regulae* [by William Heytesbury] enjoyed its greatest and longest period of popularity.'[2] By the end of the fifteenth century Heytesbury's *Regule* and the *Sophismata* had begun to enjoy in Italy a vogue which was to last for a century.

1 Cf. G. Federici Vescovini, 'Il commento di Angelo di Fossombrone al *De tribus praedicamentis* di Guglielmo Heytesbury,' in A. Maierù (ed.), *English Logic in Italy in the 14th and 15th Centuries: Acts of the 5th European Symposium on Medieval Logic and Semantics, Rome, 10–14 November 1980*, Naples 1982 (History of Logic, 1), 359–374. We know very little about Angelo's life and work. The last date of his career is 1402, when he was teaching natural philosophy at Padua. Cf. A. Gloria, *Monumenti dell'Università di Padova (1318–1405)*, Padua 1888 (2 vols.), 1: 495–496. Marshall Clagett placed emphasis on the fact that in Angelo's *De reactione* the name 'Calculator' is clearly mentioned, maybe for the first time in Italy, to designate Richard Swineshead. Cf. M. Clagett, *The Science of Mechanics in the Middle Ages*, Madison, WI, 1959 and 1961 (University of Wisconsin Publications in Medieval Science, 4), 648–649.

2 Cf. C. Wilson, *William Heytesbury: Medieval Logic and the Rise of Mathematical Physics*, Madison, WI, 2nd ed., 1960 (University of Wisconsin Publications in Medieval Science, 3), 25–26; W.J. Courtenay, 'The Early Stages in the Introduction of Oxford Logic into Italy,' in A. Maierù (ed.), *English Logic in Italy in the 14th and 15th Centuries: Acts of the 5th European Symposium on Medieval Logic and Semantics, Rome, 10–14 November 1980*, Naples 1982 (History of Logic, 1), 12–32; D. Buzzetti, 'Linguaggio e ontologia nei commenti di autore Bolognese

 | DOI:10.1163/9789004512054_015

The work I am concerned with here is Angelo's commentary on Heytesbury's *Regule solvendi sophismata*, entitled *Recollectae super Hentisberum de tribus praedicamentis* (or *Expositio de tribus predicamentis*), a text perhaps connected to the teaching along the lines of Heytesbury's book, given at the university of Bologna.[3] Long considered a work by Blasius of Parma (according to Lynn Thorndike, while Paul Oskar Kristeller attributed it to Angelo),[4] this commentary was published in Pavia by Antonio Carcano in 1482 with a different title: *Angelus Forsempronıensis super tractatu de motu locali*. A partial edition appeared in a collection of works, printed in Venice in 1494, and containing, in addition to Heytesbury's *Regule* and *Sophismata*, some commentaries on them, attributed to Gaetano da Thiene, Magister Messinus, Bernardus Tornius, and Simon de Lendenaria.[5] Furthermore, Angelo's commentary also circulated as a separate work, as testified by some manuscripts and editions. In this chapter I will refer to the Venetian edition, in which Angelo's commentary, under the title of *De motu locali*, can be read on folios 64ra–73rb. There is no modern edition of the text, so far.

Angelo's commentary deals with the first two subchapters of Heytesbury's *De tribus predicamentis*, but some arguments are drawn from other thinkers, such as Richard Swineshead.[6] It displays an inner structure, unfolding its con-

al *De tribus predicamentis*,' in D. Buzzetti, M. Ferriani and A. Tabarroni (eds.), *L'insegnamento della logica a Bologna nel XIV secolo*, Bologna 1992 (Studi e memorie. Nuova serie, 8), 579–604.

3 For a list of manuscripts and printed editions of *De tribus predicamentis* see Federici Vescovini, 'Il commento,' 68–71. There are other works attributed to Angelo: the *De reactione* (Cf. Venezia, Biblioteca Nazionale Marciana, Ms. lat. VI. 160 [2816], 248ra–252ra); the *De inductione formarum* or *De maxima seu minima materia* (a commentary on the chapter of Heytesbury's *Regule* dealing with problems *de maximo et minimo*); the *De insolubilibus* (in the following manuscripts: Venezia, Biblioteca Nazionale Marciana, Ms. lat. VI. 155 [3377], 19ra–31vb; Venezia, Biblioteca Nazionale Marciana, Ms. lat. VI. 30 [2547], 63ra–71va; Città del Vaticano, Biblioteca Apostolica Vaticana, Ms. Vat. lat. 2130, 161ra–170va; Klagenfurt, Universitätsbibliothek, Ms. Pap. 168, 61^{r}–67^{v}). As Federici Vescovini, 'Il commento,' 66–67, points out, the *Expositio de tribus predicamentis* and the *De inductione formarum* were merged in the earlier copies, while in the fifteenth century Gaetano da Thiene separated them.

4 Cf. L. Thorndike, *A History of Magic and Experimental Science*, New York 1923–1958 (8 vols.), 4: 654; P.O. Kristeller, *Iter Italicum. A Finding List of Uncatalogued or Incompletely Catalogued Humanistic Manuscripts of the Renaissance in Italian and Other Libraries*, London 1967–1997 (7 vols.), 2: 459.

5 For a description of this edition, see the Appendix to Edith Sylla's contribution to the present volume. There had been a previous edition of Heytesbury's *Regule*, printed at Pavia in 1481, but the Venetian 1494 edition was particularly widespread. Cf. Wilson, *William Heytesbury*, 27.

6 Wilson, *William Heytesbury*, 116.

tent from easier to more complex issues: after some definitions and clarifications about the topics to be discussed, the second part of the work is devoted to the analysis of the interaction between motive power, resistance, and velocity, which has its own *regule* and raises some antinomies (*sophismata*).

The work is primarily didactic or connected to teaching. Nearly all of its parts present the same structure: after a general introduction of the main subject and the language used, there is a detailed investigation of one or more specific cases, together with related objections and solutions. These are in turn supported by numerical examples, intended to clarify the problem. The plan of the work follows very closely the Heytesbury's *Regule*, which suggests a connection with the teaching, maybe in the way of an introductory tool for the reading of Heytesbury's work.[7] Following the English *magister*, Angelo adopts a nominalist approach, by applying logic tools in the description of phenomena that imply different types of change, where an accidental form is intensified or diminished. At the same time, he draws the crucial distinction between the real physical world, which consists of substances and qualities, and their logical/mathematical representation, conceived as just 'imaginary', that is, distinguishable only in reason, though it should not entail logical contradiction. Heytesbury's text first and foremost sets out some *regulae* one needs to conform to in order to give a logical description of phenomena of change (the primary stress is marked upon local motion). This exposition is followed by a section devoted to four sophisms on the intention and remission of motion, which represent an instructive example of how to avoid some contradictions to be solved by a consistent use of the exhibited principles. As will be explained, Angelo tries to provide a clear explanation of Heytesbury's rules and examples, closely following his arguments. The first part of Angelo's work deals with some clarifying statements about motion and its unequivocal definition, expressed in terms of changing: in the strict sense, local motion (the only form of change this analysis concerns with) is seen as the progressive acquisition of essential parts, towards the aim the moving thing tends to. After a brief analysis of the 'dynamical' meaning of velocity, that is of the ratio of moving potency to resistance, Angelo introduces the main subject of his work, concerning the analysis of velocity with respect to its effects (a 'kinematic' perspective). The core of the treatise focuses on local motion and its different types (uniform and difform),

7 Cf. J.A. Weisheipl, 'Ockham and some Mertonians,' *Mediaeval Studies* 30 (1968), 163–213, at 197, repr. in J.I. Catto and T.A.R. Evans (eds.), *The History of the University of Oxford*, 2: *Late Medieval Oxford*, Oxford 1992, 607–658.

with special attention to the acquisition and loss of latitude of velocity, that is intention and remission of motion, which obey to precise rules. By the combination of intention and remission of motion, some sophisms arise: their solution is the last issue of Angelo's work.

2 Motion as *mutatio* and *aptitudo*

At the beginning of his work, Angelo points out the different types of change, in relation to the Aristotelian categories. Broadly speaking, *mutatio*, defined as the actuality of a potentiality as such, occurs in each of the Aristotelian categories. It must be noted, however, that, according to Aristotle, these changes take place *per accidens*, except when they refer to the categories of substance, quality, quantity, and place, where *mutatio* is *per se*.[8] In the other categories, motion first occurs in substance and then consequently in the other category. For instance, in the category of relation such as that of being a father, the ending of being a father first occurs in substance (when the son dies) and only secondarily in the relation (being a father).[9]

Angelo specifies that in a proper sense (*proprie*), motion is defined as occurring *partibiliter*, insofar as it denotes the acquisition of parts one after the other, a process that cannot concern substance. Substance acquires being all at once and nothing else of the same substance can be acquired thereafter. Hence, motion in a proper sense is found in the categories of quantity, quality, and place. And yet, it must be acknowledged that our understanding of velocity in these three categories depends on the twofold way we consider it. As for the cause, or *a priori*, it is evident that it can be identified with the dominance of the power in the mover over the moved; that is to say, velocity in all those categories can be known through one and the same method, that *propter quid*. Instead, in terms of effects, or *a posteriori*, it is not possible to find a general

8 Angelo da Fossambruno, *Super tractatu* De tribus predicamentis ⟨*Hentisberi*⟩. In William Heytesbury, *Regule solvendi sophismata*, Venice 1494 (henceforth: *De motu*), 64ra: 'In omni predicamento potest esse mutatio … Unde in relatione potest esse mutatio de patre in non patrem, quia potest responsio variari per interrogationem istius predicamenti de substantia aliqua. Et similiter accidit in aliis predicamentis. In omnibus enim preter quam in quatuor, scilicet, substantia, quantitate, qualitate et ubi, fit mutatio per accidens solum, et non per se, quia si in aliquo talium fiat variatio in responsione ad interrogationem istius predicamenti, hoc non per se erit, sed propter variationem factam in illo …'

9 Angelo da Fossambruno, *De motu*, 64ra: 'Ut gratia exempli dicetur Sortes pater, deinde non amplius pater. Ista mutatio primo et per se erit propter corruptionem filii, que mutatio primo in substantia erit, et ex consequenti in relatione.'

rule describing different effects of velocity. Sometimes there are finite velocities where there is no resistance.[10]

Angelo adopts an apparently contradictory attitude, widely shared among most of his contemporaries: on one hand, he ostensibly adopts the non-Aristotelian procedure of quantifying qualities. On the other hand, he firmly reacts against every attempt to erode Aristotle's definition of velocity as a result of the relation between motive power and passive resistance. This becomes sufficiently clear in the first question: this Aristotelian position had been questioned by some scholars (Angelo never explicitly mentions them by name). There are reasons for arguing that velocity does not always depend on the proportion of power to resistance, as in the case where the resistance is zero—which concerns celestial motions caused by intellectual movers, whose action is completely voluntary with no *fatigatio*.[11]

An interesting remark clarifies the complexity of this point: if a given cause, in spite of its great inner power, voluntarily moves its mobile body at a slow speed, while a second weaker power in turn moves its own body in such a way that the velocities of the first and that of the second are equal, then the respective proportions of power and motion differ. This leads to the conclusion that the same velocity can be produced by different ratios.[12]

---

10 Angelo da Fossambruno, *De motu*, 64rb: 'De unoquoque igitur talium motuum dicendum est penes quid uniuscuiusque istorum velocitas attendi habet et cognosci. Cognosci autem habet velocitas motus sicut etiam quecunque res nostro cognitioni subijcibilis duplici modo cognoscendi, scilicet a priori et a posteriori, vel penes causam vel effectum, et hi modi duplici processui correspondent, scilicet quia et propter quid. In omnibus igitur his motibus a priori aut penes causam velocitas eodem methodo cognosci potest, quia in quolibet illorum motuum causa est dominium virtutis moventis supra motum. Non autem una communis regula dari potest de velocitate penes effectum.' On the distinction between 'according to the effects' and 'according to the causes', see D.A. Di Liscia, 'Velocidad *quo ad effectus* y velocidad *quo ad causas*: La tradición de los calculadores y la metodología aristotélica,' in D.A. Di Liscia, E. Kessler and C. Methuen (eds.), *Method and Order in Renaissance Philosophy of Nature. The Aristotle Commentary Tradition*, Aldershot 1997, 143–176, part. 167, n. 54.

11 Angelo da Fossambruno, *De motu*, 64rb: 'Et quia de velocitate penes causam multe reperiuntur reprobate false opiniones, ideo presupponit totum quod alibi debet investigari. Dicitur autem hic quod velocitas motuum non potest attendi penes proportionem agentis ad resistentiam. Alicubi enim magna et finita est velocitas, nulla est resistentia. Non resistit enim orbis motori eius, quoniam intelligentia movens simplici velocitate movet et non cum fatigatione et resistentia.'

12 Angelo da Fossambruno, *De motu*, 64rb: 'Similiter et potentia Sortis quia voluntaria est, maiorem potest in moto velocitatem et maiorem ipsa manente continue equalis potentie et moto continue equalis resistentie. Ipsa tamen alia tunc et alia aptitudine movebit, maior namque velocitas cum aliis consequentibus quam velocitate motus inferre pos-

Angelo is here referring to the vexed question of inconsistency between Aristotle's theory of motion and the assumption that a heavenly intelligence, as a cause of motion, does not encounter any resistance. The formula Angelo resorts to is a widespread thesis in late medieval cosmology. It is incorporated in Aquinas' works and conceives the relationship between the intelligence and the celestial sphere moved by it in terms of *aptitudo* of the former vis-à-vis the latter. Accordingly, Angelo speaks of the proportion of velocities as a proportion of aptitudes, thus excluding resistance.[13] In this case, variation of velocities is due merely to free will to move faster or slower. In a wider sense, the cause producing voluntary velocity should be conceived as *aptitudo moventis*.[14] The emphasis placed on motions produced by voluntary efficient causes serves to show the essential difference between voluntary and natural kinds of motions: the former depend on the determination of an animated being, and because of that it is not likely for them to be subjected to the same general rule; only motions of natural, inanimate bodies may be measured by their most rapid velocity, that is the velocity of its most rapidly moved point.[15] This last clarification is of the highest importance, because it ensures the possibility of denominating the motion of a body as a whole, regardless of how other parts of the body can vary.

---

sunt, alia aptitudinem denotant; similiter etiam et in omnibus aliis moventibus naturaliter diversa applicatio diversam dicit aptitudinem.'

13 Angelo da Fossambruno, *De motu*, 64rb: '… licet enim intelligentia non moveat cum resistentia circa suum orbem, movet tamen cum quadam aptitudine ipsum motum; movet enim voluntarie; quanta est voluntas ad movendum, tanta est aptitudo in ea, quia movet quantum vult et alia intelligentia tardius movet quia alia voluntate movet, quia non tantum vult movere.' Cf. Thomas Aquinas, *Quaestio disputata de spiritualibus creaturis*, ed. J. Cos, Rome 2000 (Opera omnia, 24/2), 66. See also Averroes, *In Physicam* VII, comm. 35, Venice 1562, repr. Frankfurt am Main 1962 (Aristotelis opera cum Averrois commentariis, 4), 335. Cf. J. Celeyrette, 'Le mouvement selon la cause chez Messino da Codronchi et Angelo di Fossambruno,' in J. Biard and S. Rommevaux (eds.), *Mathématiques et théorie du mouvement, XIVe–XVIe siècles*, Villeneuve d'Ascq 2008 (Histoire des sciences), 15–29, part. 22–26.

14 Angelo da Fossambruno, *De motu*, 64rb.

15 Angelo da Fossambruno, *De motu*, 64rb: 'Item etiam magnitudo rei mote ratione cuius aliis paribus maius spatium pertransitur, maiorem dicit aptitudinem. Et universaliter, ut dicantur omnia que concurrunt ex parte efficientie motus, constituunt aptitudinem moventis; opposita vero constituunt aptitudinem modicam eiusdem. Et si dicatur quod omnia ista similiter possunt comprehendi per hoc nomen *proportio*, et tunc velocitas attenditur penes proportionem, dicitur quod bene potest sic accipi, sed omnino non adhuc esset illa regula generalis in omnibus motibus, scilicet, voluntariis et naturalibus, sed in naturalibus solum, ubi fit actio solum secundum ultimum sui posse.' Here Angelo adopts a position not dissimilar from that held by Heytesbury. Cf. Wilson, *William Heytesbury*, 117, 120.

3 Uniform vs. Difform Motion

Turning to the effects of velocity, Angelo is aware of the possibility of different kinds of motion that are to be analysed *seriatim*, as it is not possible to establish a single universal rule covering all types of motions. Therefore, in order to account for each kind of motion, it is necessary to examine the different cases in detail, that is *descendere ad particularia*.[16]

First of all, and most basically, Angelo focuses on the search for a *regula in motu speciali*; but this step needs a preliminary clarification concerning what is to be regarded as *motus localis*. This kind of motion is prior to any other, both because of its nature and nobility (this is clear from the fact that local motion pertains to the most noble body, that of the celestial spheres), and because every motion in the sublunar region (even those of augmentation and alteration) depends on that of the celestial spheres.[17] Moreover, any local motion has its own essential and accidental characteristics, the former being in the properties *sursum* and *deorsum*; the latter leading to the distinction between uniform, non-uniform, linear, oblique motion etc.[18] A third class of characteristics pertaining to all motions consists of the main attributes of 'fast' and 'slow'.[19]

The first kind of motion to be defined is the simplest kind of motion, that of uniform local motion, conceived as one in which 'in equal parts of time equal spaces may be traversed' (*motus localis uniformis est quo in partibus equalibus temporis equalia spatia pertransirentur*).[20] Yet, this definition is far from comprehensive: one could note, for example, that it does not apply under the assumption of the existence of a void or under the hypothesis of a varying space or body (e.g., becoming more or less dense during the motion).[21] In the

16 Angelo da Fossambruno, *De motu*, 64rb: '… non potest una universalis regula dari que satisfaceret in omnibus … oportet tum descendere ad particularia …'

17 Angelo da Fossambruno, *De motu*, 64va: 'Dicendum est ergo de motu locali primo. Est igitur motus localis aliis prior natura et nobilitate, quoniam per se priori et nobiliori corpori competit ut celo. Et etiam quia localis motus est aliorum motuum causa. Motus enim celi est causa motus augmenti et alterationis in istis inferioribus.'

18 Angelo da Fossambruno, *De motu*, 64va: 'Insuper quantum ad essentiales differentias distinguitur motus localis in sursum et deorsum, unde dicit Commentator in libro *Celi*, quantum ad accidentales differentias quod multipliciter potest distingui, scilicet in uniformem, difformem, rectum, obliquum, et sic de aliis.'

19 Angelo da Fossambruno, *De motu*, 64va: 'Et iterum ultra differentias essentiales proprias per alias essentiales communes omnibus motibus indifferenter competentes, scilicet velox et tardum.'

20 Angelo da Fossambruno, *De motu*, 64va.

21 Angelo da Fossambruno, *De motu*, 64va: 'Contra quem sic arguitur primo: si aliquis motus localis uniformis fieret in vacuo, si esset, non poterit tunc diffinitio sibi competere, quia

first case, there will not be any space at all, so that it would be nonsensical to speak of equal parts of space to be traversed in equal parts of time. In the second, the space itself will be subject to change of length, and it would be impossible to determine equal spaces. Angelo has no hesitation in claiming that this definition needs some further clarification on the constancy of both space and body.[22]

On the other hand, the *motus difformis* is due to a variation of the space traversed or the time required. Furthermore, it is all too evident that difformity in local motion does not correspond to the definition of uniformly difform motion, because a proportion between the parts of the time elapsed and those of the space traversed cannot always be precisely established.[23]

Difformity and uniformity could also be considered with respect to the parts of the mobile. If the velocity of the parts of a moving subject is the same simultaneously (as in most cases), the resulting motion is *uniformis quoad partes subiecti*. However, this is not the case of a rotating wheel—even when its points move at a constant speed—because not all its parts have the same velocity, since the parts lying on the circumference move faster than the others.[24]

---

nullum spatium pertransiretur ... Item contra idem ponatur quod mobile augeatur in quantitate aut diminuatur, ut posito gratia exempli quod mobile, utputa *a*, sit bipedalis quantitatis, et ponatur quod *c* motu uniformi moveatur *a* mobile per medietatem unius hore super spatio *b*, aliquam eius partem in pertranseundo ipso *a* non variato, ita quod gratia exempli pertransibit unum pedalem *c* spatii. Et ponatur quod in alia medietate eiusdem *a* mobile condensabitur aut fiet in ipso albatio partium in extremo quod est versus terminum ad quem motus, ita quod propter illam minorationem mobile valde minus de spatio pertransibit in secunda medietate quam in prima nulla facta in potentia minoratione. Et tamen motus est uniformis continue. Ergo diffinitio mala.'

22 Angelo da Fossambruno, *De motu*, 64va: 'Respondetur quod diffinitio est male data. Et ideo non est tenenda nisi sit intellecta sic: motus localis uniformis est quo non intenso neque remisso quibuscumque partibus equalibus temporis assignatis per quod durat adequate iste motus gratia sue velocitatis, apta nata sunt spatia pertransiri equalia.'

23 Angelo da Fossambruno, *De motu*, 64va: 'Difformitas autem que accidit ex spatii minoratione, non est gratia velocitatis talis motus, sed per accidens et gratia alterius ... Motus localis difformis ⟨est⟩ quodam modo opposito motui uniformi difformiter. Est enim motus localis difformis quo intenso vel remisso quibuscumque partibus temporis equalibus assignatis per quod adequate durat motus ille, gratia velocitatis eius, spatia non sunt apta nata pertransiri equalia. Et hoc contingit dubitare, quia aut per spatium in partibus equalibus temporis spatia pertransirentur inequalia aut apta nata sunt pertransiri, aut quia in partibus inequalibus temporis spatia pertransirentur equalia, et multis modis etiam variari contingit.'

24 Angelo da Fossambruno, *De motu*, 64va: 'Et aliquis potest esse motus uniformis quoad tempus et etiam quoad partes subiecti: ut si unum mobile moveatur non variatum intrinsecus per condensationem aut rarefactionem semper eodem gradu motus non intenso, nec remisso motu simpliciter recto moveatur uniformiter quoad tempus ... Movetur etiam

Some perplexities are expressed about motion in a void. In fact, in so far as it implies the absence of any space, the void represents a condition under which previous definitions of motion are repealed. If we admit the void, Angelo argues, any body placed in it will immediately extend to fill up all empty space, according to the principle stating that *natura abhorret vacuum*. This explanation is not accepted by Angelo, and his refusal is an interesting point of the chapter, as it shows the very aim of this kind of inquiry. To deal with the interaction between body and void is a physical problem (it is *de rebus*), while what it is actually about is solving a puzzle, that is, what logically results from the assumption of some conditions, without claiming to give an account of the way things behave.[25] Angelo seems to adopt here an approach *secundum imaginationem*, which has been defined as conceptualistic: his analysis appears to be characterised as a semantic one, rather than claiming to display an accurate description of the physical world.[26] Angelo's response supports this reading. After a condensed overview of traditional arguments against the existence of

---

uniformiter quoad partem subiecti moti, quia omnes partes omnino velocitate consimili moventur. Moveri enim uniformiter quoad partes subiecti est omnes partes eque velociter moveri. Aliquid enim contingit uniformiter moveri quoad tempus et difformiter quoad partes subiecti, sicut rota mota circulariter ab eadem potentia continue equali nixu, ita quod circunferentia ipsius continue equaliter movetur. Tunc patet quod totum moveatur difformiter quoad partes subiecti, quia velocius moventur partes versus circunferentiam quam versus centrum.'

25 Angelo da Fossambruno, *De motu*, 64vb: 'Gratia dictorum ad evidentiam dicendorum est dubitandum utrum si daretur vacuum esset motus localis posito mixto, in eo aut aliquo alio mobili. Ad quod dicitur a quibusdam quod non esset motus aliquo modo in eo. Et moveri possunt hi duplici ratione: primo quia in vacuo non esset spatium aliquod. Et si sic, igitur penes nihil posset velocitas illius attendi. Secundo, quia dato vacuo et corpore posito in eo … illud corpus subito rarefieret ad replendum illud vacuum, quod tota natura abhorret … Ista responsio stare non potest … Non enim queritur nunc quid fieret de corpore quod poneretur in vacuo an mutaretur an moveretur … Sed illud quod nunc queritur est an illo corpore conservato et interim intrinsece non se aliter habente in illo vacuo possit esse motus localis …'

26 Cf. J.E. Murdoch, 'The Development of a Critical Temper. New Approaches and Modes of Analysis in Fourteenth-Century Philosophy,' *Medieval and Renaissance Studies* 7 (1978), 51–79; J.E. Murdoch, 'Propositional Analysis in Fourteenth-Century Natural Philosophy: A Case Study,' *Synthese* 40 (1979), 117–146; J.E. Murdoch, '*Scientia mediantibus vocibus*: Metalinguistic Analysis in Late Medieval Natural Philosophy,' in A. Speer (ed.), *Sprache und Erkenntnis im Mittelalter*, 1, Berlin 1981 (Miscellanea Mediaevalia, 13), 73–106. For more details on this question, cf. J.M.M.H. Thijssen, 'Buridan on Mathematics', *Vivarium* 23, 1 (1985), 55–78; E.D. Sylla, 'Mathematical physics and imagination in the work of the Oxford Calculators: Roger Swineshead's *On Natural Motions*', in E. Grant and J.E. Murdoch (eds.), *Mathematics and Its Applications to Science and Natural Philosophy in the Middle Ages. Essays in Honor of Marshall Clagett*, Cambridge 1987, 69–101.

a void, the author points out that there are three ways to approach this issue. The first considers motion in a void as similar to other local motions, because even if there is no space, velocity can indifferently result by a consideration of real or imaginary space (*spatium verum vel imaginarium*). In a second way, this motion is regarded as belonging to the same species of local motion even if it is not strictly speaking a local motion. Angelo supports a third conception: every motion in a void is a local motion. Astronomers, after all, have never denied this motion in the sky, even when they have questioned the void.[27]

Against the first solution, one could argue that the very fact that this imaginary space is traversed brings about a real space, so that the same statement would be proper to two kinds of space, one opposite the other. The space traversed cannot be a real space—properly speaking, there is no space at all in the void: this means that the space considered is just an imaginary one, and the traversing itself is to be thought as problematic. The heart of the reasoning appears to be the basically ambiguous status of an imaginary space. This clearly emerges if we envision two moving things, in such a way that the first (a) is moving in a plenum and the second (b) in a void, under the assumption that b moves hundredfold faster than a. Therefore, in one hour a will traverse one foot, while b will traverse 100 feet. Does this entail that the velocity of b is greater than that of a, considering that in a void there is no space? Even if we acknowledge this way, nonetheless it is allowed 'to imagine' that a single foot traversed by a be hundredfold the 'space' of b. Here we have a clear absurdity, that a mobile body can traverse a real space equivalent to an imaginary one traversed by another body with a greater speed. The core of this argument is the clear remark that *de facto* there is no space at all in a void.[28]

---

27 Angelo da Fossambruno, *De motu*, 64vb–65ra: 'Ad illud dubium sunt tres modi respondendi; unus dicit quod est motus localis, quamvis non sit ibi spatium, et quod velocitas in motu intenditur penes spatium verum vel imaginarium ... Secundus modus dicit quod non est motus localis, sed est idem specie cum motu locali. Tertius modus quem puto verum dicit quod est motus localis. Et restat postea respondere ad motiva ... Deinde dicitur quod iste motus attenditur penes spatium imaginatum pertransitum. Ex quo sequitur quod in tali motu aliquod est spatium imaginatum pertransitum ... Et philosophi et astrologi nunquam dixerunt celum moveri alio motu quam motu locali. Sed secundum istum modum solum restat respondere ad rationes que fiebant contra.'

28 Angelo da Fossambruno, *De motu*, 64vb–65ra: 'Contra primum modum respondendi arguitur primo, quia peccat contra veritatem sermonis. Dicit enim quod in vacuo nullum est spatium et similiter ultra ultimam speram quamvis improprie dicitur ultra vel extra. Deinde dicitur quod iste motus attenditur penes spatium imaginatum pertransitum. Ex quo sequitur quod in tali motu aliquod est spatium imaginatum pertransitum. Et si sic, cum omne spatium imaginatum quod sit pertransitum sit aliquod spatium verum, tunc est aliquod spatium pertransitum, cuius oppositum dicit. Item, si penes spatium verum vel

As for the second solution, Angelo argues that speaking of a motion of the same species as local motion, but not local motion, is an inherent contradiction. In fact, the words *motus localis* stand for the essential species of motion, while any distinction is allowed only between two things of the same species but with accidental disparity.[29]

In support of this statement, Angelo notes that in the syntagm *motus localis*, the word *localis* does not stand for something to be designated by its immediate meaning (it is not a *nomen actuale actualiter supponens*). It is quite true that motion is defined by means of the space traversed over some time, and therefore every discussion on motion entails a notion of space as a divisible extension. Nevertheless, an analysis of the statement 'to move is intrinsically to stand in different ways in relation to place' (*moveri est aliter intrinsece se habere localiter*) reveals that the adverb *localiter* does not refer to the place *in se* (something that is *in loco*); rather its precise function is to specify the kind of motion intended by the verb *moveri*—and as such it stands for 'local' motion.[30]

imaginatum attenditur velocitas talis motus localis, aut igitur attenditur penes spatium imaginatum verum aut penes illud quod imaginatur tantum esse spatium pertransitum: non primum, quia illud non esset verum in vacuo; igitur secundum. Sed contra: tunc sequitur illa conclusio, quod *a* mobile in centuplo movetur tardius quam *b*, et *b* movetur in centuplo tardius quam *a*; conclusio implicat; que sic deducitur: moveatur *a* in rei veritate in centuplo tardius quam *b*, et hoc in pleno et *b* in vacuo; pertranseat enim *a* gratia exempli unum pedale in hora et *b* centum pedalia in eadem hora … Tunc patet quod responsio prima fuit nulla, quia tunc nihil est dicere quod in vacuo attenditur penes spatium imaginatum pertransitum in tanto tempore vel in tanto, quia ibi nullum est spatium verum vel imaginatum.'

29 Angelo da Fossambruno, *De motu*, 65ra: 'Contra secundum modum arguitur, quia notum est cum ille sit motus verus et successivus, aut ergo est motus ad quantitatem aut ad qualitatem aut localis; non primum neque secundum potest dici; igitur relinquitur tertium. Consequentia est clara per ea que dicta sunt in principio, quia tantum tres sunt motus veri et proprii. Item cum quilibet motus sit factus in aliquo predicamento, queritur in quo predicamento est iste motus, et non potest dici in alio quam in ubi; et sic ergo est motus localis. Item talis responsio implicat in verbis, quia sequitur: iste motus est idem specie cum motu locali, ergo est motus localis.'

30 Angelo da Fossambruno, *De motu*, 65ra: 'Ad primum enim negatur consequentia: iste non est motus in loco, igitur non est motus localis. Unde hoc nomen *locale* non est nomen actuale actualiter supponens pro omni quod importat … Ad secundum negatur consequentia: nullum spatium per illum motum est pertransitum, igitur non est motus localis. Nam dicitur, ut infra concludetur immediate, quod velocitas in motu locali attendenda est penes spatium quod per eundem motum est aptum natum pertransiri in tanto tempore vel in tanto et non penes illud quod est pertransitum absolute. Ad tertium dicitur negando quod moveri est aliter intrinsece se habere secundum locum, sed est aliter intrinsece se habere localiter: dicitur quod illud quod movetur in vacuo aliter se habet intrinsece localiter. Et non valet tunc consequentia: igitur movetur in loco, quia hoc adverbium *localiter*

The bulk of Angelo's treatise is mainly devoted to the study of non-uniform, or difform, motion. From the very beginning, the author specifies that a movement can be totally difform (*totus difformis*), when none of its parts moves uniformly, even though it may well be that one (or more) of its parts could move uniformly.[31] It is not the same as the *motus uniformiter difformis*, defined on the presupposition that its parts all move coherently, with the same velocity, although the velocity itself can change. At any rate, because of their intrinsic heterogeneity, the two types of motion need to be analysed in very different ways. An adequate description of difform motion cannot be provided by solely focusing on the relation between the entire space traversed and the whole time it requires, because this relation could vary at any instant. Needless to say, here Angelo is referring to the puzzling concept of instantaneous motion. As Angelo points out, at first glance there seems to be an almost unsurmountable hurdle, because the analysis of difform motion requires a consideration of velocity *punctualiter*: if something moves difformly, an adequate analysis of its motion should focus on an accurate evaluation of velocity at a single point of its path. But a point is by definition 'that which has no part': it is in no way extended, and has no dimensions. It would therefore be impossible to conceive velocity as a ratio between space and time, for the simple reason that no space is traversed in a single instant.[32]

---

non supponit pro loco, sed determinat hoc verbum *habere* ad standum pro specie motus localis, sicut dicendo *hoc movetur localiter*, si *localiter* determinat hoc verbum *movetur* ad standum pro specie motus localis.'

31 Angelo da Fossambruno, *De motu*, 66va: 'Sequitur nunc dicere de motu difformi penes quid in eo attenditur velocius moveri vel tardius. Ubi nota quod motus potest esse totus difformis, ita quod nulla eius pars sit uniformis. Et potest esse talis quod aliqua pars ipsius sit uniformis.'

32 Angelo da Fossambruno, *De motu*, 66^{va-vb}: 'In motu secundum se totum difformi non est attendenda velocitas penes spatium quod pertransitur in toto tempore durationis istius motus, sed in quocunque instanti istius temporis mensurantis istum motum est sic vel sic velociter moveri, istius mobilis attendendum penes spatium quod in tanto tempore pertransiretur si in tanto tempore vel in tanto mobile moveretur uniformiter illo gradu quo in illo instanti movetur. Gratia exempli: movebitur Sortes in tota hora ventura latitudine motus uniformiter difformi, ita quod per nullum tempus adequatum movebitur Sortes aliquo determinato gradu punctualiter. Tunc dicitur quod neque debet sumi quod Sortes tantum vel tantum velociter moveatur in tota hora, sed pro quolibet instanti erit attendenda velocitas qua pro tunc Sortes diceretur sic vel sic moveri; ut puta dato aliquo instanti in quo Sortes movebitur quatuor. Et ponatur pro constanti quod gradu uniformi ut quatuor in hora transiretur unum pedale, gratia exempli. Tunc dicitur quod Sortes pro illo instanti movebitur velociter ut quatuor; que velocitas sive velociter moveri, ut prius dicitur, non attenditur penes spatium quod pertransiretur a Sorte in tota hora toto motu mediante, nec etiam penes spatium quod pertransiretur in illo instanti in quo movetur velociter ut quatuor, quia in illo instanti nullum spatium pertransitur.'

The solution of the puzzle is based on the concept of uniformity. To say that a moving body in a precise moment has its degree of velocity, means that we are considering this body *as if* it moved uniformly for a length of time at the same speed. It is, therefore, a sort of *reductio* of time to the instant taken into account.[33] And yet, this reasoning does not seem to apply to non-uniform motion, even if it is a *uniformiter difformis motus*. Angelo gives us—as he often does—an illuminating example. Let us conceive two men, Socrates and Plato, both moving for one hour from zero degree to 8 degrees of velocity, but with different ratios of respective times (and spaces). Socrates moves for the first quarter of the hour increasing his velocity from *non-gradus* to 6, and for the remaining time up to 8. Conversely, Plato picks up velocity from *non-gradus* to 6 for three quarters of the hour, and from 6 to 8 over the last quarter. It is quite easy to figure out that, notwithstanding their equal velocities (in the sense of instantaneous velocities), Socrates will traverse in that hour—conceived as a whole (*cathegorematice*)—more space than Plato.[34] Put otherwise, difformity of motion, *quoad spatium petransitum*, cannot be treated the way uniform motion is treated, because of its own variety. But there is another way to calculate the space traversed, in the case of uniformly difform motion, which leads to the result of a medium degree (this is the well-known mean speed rule, which Angelo here just states without proving it).[35]

And yet, this way of considering difform motion can reveal itself to be somewhat inconsistent. Suppose, for instance, that a body *b* moves for the first half hour from zero degree to 4, and for the second half in the opposite way, from 4 to zero degree, while a second body *a* moves with a uniform velocity of 4. Then, when the mean speed rule is applied, *b* will traverse as much space as if it moved with a velocity of 2. As a consequence, *a* will traverse twice as much

33 Angelo da Fossambruno, *De motu*, 66vb: 'Sed velocitas illa attenditur penes spatium quod pertransitur ab eodem mobili vel alio si moveretur in hora vel alio tempore illo motu uniformiter quo in illo instanti Sortes movetur. Et quia si aliquid in tali tempore, ut gratia exempli in hora, tali gradu ipsum uniformiter pertransiret unum pedalem, quod pertransire uniformiter in hora est moveri velociter ut quatuor, ideo dicitur illud mobile in illo instanti moveri velocitate ut quatuor; et sic pro quocunque alio instanti est attendendum.'

34 Angelo da Fossambruno, *De motu*, 66vb: '… si Sortes moveatur in prima quarta hore latitudine uniformiter difformi a non gradu ad sex, eam uniformiter acquirendo et in tribus aliis quartis moveatur latitudine uniformiter difformi a sex ad octo, eam etiam acquirendo uniformiter, Plato vero acquirat omnino totam istam latitudinem econtrario, scilicet, in tribus primis quartis a non gradu ad sex et in ultima a sex ad octo, tunc patet quod Sortes et Plato in tota hora cathegorematice sumpta precise eadem vel omnino consimili latitudine motus movebuntur in eodem tempore, scilicet in illa hora, et tamen in eadem valde maius spatium pertransibit Sortes quam Plato, ut patet intuenti.'

35 Angelo da Fossambruno, *De motu*, 66vb.

space as *b*, although each of them has the same instantaneous velocity the other has.[36]

Another peculiarity of non-uniform motion, as regarded in its entirety, is that all the single motions it is composed of are not always evident in the final calculation made using the mean speed rule. So, if a uniform motion occurs as a component of the entire motion (a condition called by Angelo *motus non secundum se totum difformis*), and the uniform section is at mean its velocity, it does not affect the final result obtained by the mean speed rule, as if it were quite indifferent.[37]

Other problems that are to be broached about non-uniform motions imply a consideration of the end points of a *latitudo*. Clearly, Angelo is dealing with the topic of *incipit et desinit*, which features in chapter 4 of Heytesbury's *Regule*. But I will not dwell on this topic here.

4 *Intensio* of Motion

Before analysing intension and remission as applied to local motion, Angelo calls attention to a terminological clarification about the meaning of the terms involved, which implies the clarification of the reciprocal connection between logical and physical levels. In fact, *intensio* can be understood by its denominative meaning—by which, within a semantic context, a certain motion can be *formaliter* called intense for such-and-such determinations, and whose oppos-

36 Angelo da Fossambruno, *De motu*, 67[ra]: 'Et pro secunda ponatur quod *b* in prima medietate hore future moveatur uniformiter intendendo motum suum a non gradu usque ad gradum ut quatuor inclusive, et in secunda remittat motum suum ad non gradum; *a* vero continue moveatur velocitate ut quatuor. Tunc per regulam infra ostendendam de correspondentia gradus medii in difformibus, in dicta hora tantum spatium lineale pertransibit *b* quantum si moveretur continue velocitate ut duo. Et sic sequitur quod duplum spatium pertransibit *a* ad illud quod pertransibit *b*. Et quod tamen precise equevelociter in eadem hora movebitur *a* sicut *b* patet, quia aliqualiter velociter movebitur *b* et etiam *a* et non velocius movebitur *a* in illa hora quam *b* …'

37 Angelo da Fossambruno, *De motu*, 67[ra]: 'Item in motu non secundum se totum difformi attendi potest velocitas pro aliquo adequato tempore, ut posito quod *a* in prima tertia hore intendat motum suum a non gradu ad quatuor, in secunda continuet motum suum uniformem ut quatuor, et in tertia intendat a quatuor ad 8. Tunc velociter moveri ipsius pro secunda tertia hore adequate attenditur penes lineam descriptam a puncto eius velocissime moto, aut qui indivisibiliter velocius movetur quam aliquis punctus in ea, et ut supra de motu uniformi, totius autem motus penes nihil attendi poterit. Et similiter captis partibus difformibus istius motus observande et regula aut habita.'

ite is the privative meaning of *remissio*.[38] In another sense, *intensio* (*remissio*) refers to a change, a process through which something is acquired (lost). In accordance with this concept, *intensio* and *remissio* refer, respectively, to an *acquisitivus* or a *deperditivus* motion.[39] When he deals with *intension* and *remissio*, Angelo declares to refer to the latter sense.

It is worth calling attention to the distinction between motion and the intension of motion, and consequently between the velocity of intension and the velocity of motion. For instance, in simple uniform motion there is no intension, nor velocity of intension of motion. This subtle distinction can be exemplified through the effects themselves: the effect of the velocity of motion is the traversed space, while in the case of the velocity of intension of motion it is the latitude of velocity acquired. The same explanation—but in terms of loss of velocity—can be provided for the *remissio*.[40]

An immediate consequence of this distinction is that two bodies can move in one hour with the same speed of intension (i.e., of acceleration), and yet they may not have the same speed of local motion. This point is illustrated by an example: a body moving with a uniformly increasing speed from 4 to 8 will acquire 4 degrees of the latitude of velocity, just equal to that obtained by another body moving uniformly from rest to a velocity of 4. This does not imply that their motions are totally equal, since their mean velocities are quite different:[41]

---

38 On this issue, cf. S.J. Livesey, 'The Oxford Calculatores, Quantification of Qualities, and Aristotle's Prohibition of Metabasis,' *Vivarium* 24 (1986), 50–69, at 61–66.

39 Angelo da Fossambruno, *De motu*, 67va: 'Dicto de motu locali tam uniformi quam difformi. Restat dicere de intensione et remissione eisdem motus localis. Pro quo notandum est primo quod *intensio* aliquando accipitur pro denominatione qua formaliter motus dicitur intensus, et cui privative opponitur remissio. Aliquando accipitur pro mutatione qua aliquis motus fit intensior; unde intensio sic est motus quidam quo aliquid acquiritur; sic pertransitio est motus quo aliud pertransitur vel est in pertransiri, cui opponitur remissio, prout est motus alicuius deperditivus. Et de illo secundo modo est tractandum principaliter in sequentibus.'

40 Angelo da Fossambruno, *De motu*, 67va: 'Sciendum est etiam pro intellectu sequentium quod differunt motus et intensio motus, capiendo *intensionem* ut dictum est; differunt etiam velocitas motus et velocitas intensionis motus. Motus et intensio motus differunt, quia quandoque est motus ubi non est intensio motus, ut est in motu uniformi, ubi nullus motus intenditur. Differunt autem velocitas motus et velocitas intensionis motus, quia sicut patet, alicubi est velocitas motus, scilicet in motu uniformi, ubi tamen non est velocitas intensionis motus, quia scilicet nullus motus intenditur.'

41 Angelo da Fossambruno, *De motu*, 67va: 'Si *a* moveatur, gratia exempli, ut quatuor, et in hora futura uniformiter intendat motum suum ad octo, et *b* quiescat et a quiete intendat motum uniformiter usque ad quatuor in eadem hora. Tunc enim eque velociter

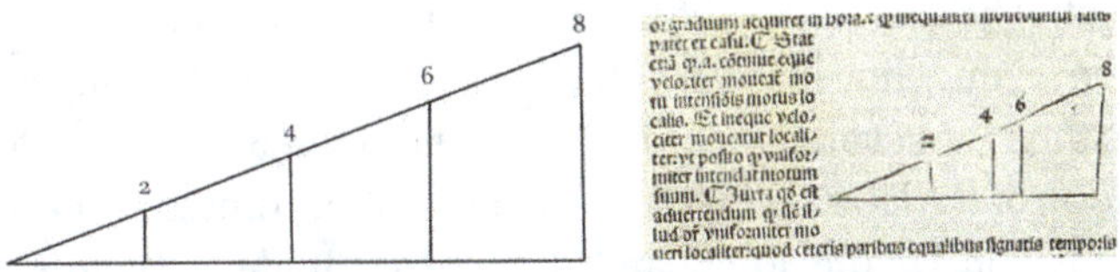

FIGURE 13.1 In local uniform motion, the increasing motion acquires equal latitudes of velocity. Right: Illustration printed in the edition Venice 1494, 67va (Wolfenbüttel, Herzog August Bibliothek, 126.4 Quod. 2° [2]. Reproduction with kind permission of the library). Left: reconstruction of the representation correcting the lacking of proportionality for the latitude with six degrees

With regard to intension, this motion has a constant speed of increase, but from a 'local' point of view (*localiter*), its velocity varies uniformly. It is just on the basis of this distinction that one can provide a definition of local uniform motion as 'traversing equal parts of space in equal parts of time.' It is the same to say that a uniformly increasing/decreasing motion acquires/loses equal latitudes of velocity.[42] Let us consider a movement for one hour, which increases for the first quarter of the time from *non-gradus* to 2, for the second from 2 to 4, for the third from 4 to 6, and so on: when this constancy vanishes, the motion is no longer uniformly difform but difformly difform. In respect of these terminological refinements, it results that a uniformly difform latitude of motion corresponds to a uniform intension of motion or rate of increase of velocity (*intensio motus*), and vice-versa.[43]

With these definitions in mind, it is possible to address the question of infinite motion. Angelo is here questioning whether an increasing motion could acquire an infinite latitude over a finite time interval. The impossibility of this case could be proved by a simple argument: acquiring an infinite latitude implies that one terminus of it is at an infinite degree. Angelo seems to argue

---

ambo movebuntur motus intensionis, quia equales latitudines motus in equali tempore acquirent, quia utrunque latitudinem quatuor graduum acquiret in hora; et quod inequaliter movebuntur satis patet ex casu'.

42 Angelo da Fossambruno, *De motu*, 67va: 'Difformiter vero moveri motu intensionis, vel difformiter intendi, dicitur econverso quod in partibus equalibus temporis latitudines motus acquirit inequales vel inequalibus ⟨partibus temporis⟩ equales; utroque namque modo dicitur difformiter intendi. Et similiter dicitur de difformiter remitti ...'

43 Angelo da Fossambruno, *De motu*, 67vb: 'Et iuxta hoc etiam imaginandum est quod latitudini motus uniformiter difformi est correspondens latitudo uniformis intensionis motus et econtrario. Est enim ibi uniformis latitudo intensionis et uniformiter difformis motus.'

that this latitude cannot be acquired in the first half of the time, because it must only be gained in the second time interval. If this latitude were 'reached' in first part of the motion, then the second one would be superfluous. But that would involve the acquisition of the latitude in just the first half of the time, so contradicting the presupposition. On the other hand, if the two latitudes—acquired, respectively, in the first and the second half of the time—are not equal (there is a difference between finite and infinite), that means that *motus totalis* is not a uniform one.[44]

Nevertheless, dynamically speaking, the increase of motion depends on the proportion between agent and patient, and nothing prevents us from thinking of a proportion of greater inequality, where the force is infinite (or the resistance is zero), involving an infinite latitude. Or, better still, we are allowed to conceive such an infinite latitude, *secundum imaginationem*.[45] It is noteworthy that, in dealing with this point, Angelo seems to suggest that an infinite proportion is at least thinkable, because no proportion is great enough to prevent us from thinking of a greater one. This is the precise meaning of a 'syncategorematic' infinite, although Angelo does not use this term expressly. Here, the problem is addressed in terms of the proportion between motive power and resistance. Therefore, since every motion with a counteracting resistance can be represented in terms of proportions, the greater is the motive power, the greater/larger is the latitude gained in a time interval (with a constant resistance), and we could increase it so as to achieve an infinite latitude.[46] The

---

44 Angelo da Fossambruno, *De motu*, 67vb: 'Ex diffinitione data de uniformiter intendi sequitur quod latitudo totalis motus, cum ipsa sit infinita saltem secundum unum extremum, non possit uniformiter acquiri ab aliquo mobili in aliquo tempore finito. Hoc palam sequitur, quoniam si potest uniformiter acquiri, sit hoc in hora. Et tunc sequitur quod precise tantum de ea acquiritur in secunda medietate hore sicut in prima. Aut ergo in prima acquiritur infinitum de ipsa aut solum finitum. Non infinitum, quia in tota hora precise ponitur debere acquiri infinitum sive latitudinem infinitam solum; ergo infinitum acquiretur in secunda medietate, et per consequens non tantum acquiretur in prima sicut in secunda. Et si sic, igitur non uniformiter acquiretur. Quod erat probandum.'

45 On this issue, see J.E. Murdoch, 'From Social into Intellectual Factors: An Aspect of the Unitary Character of Late Medieval Learning,' in J.E. Murdoch and E.D. Sylla (eds.), *The Cultural Context of Medieval Learning*, Dordrecht 1975 (Boston Studies in the Philosophy of Science, 26), 271–348, esp. 280–281 and 291–292, 297; E. Jung, 'Mathematics and the *secundum imaginationem* Procedure in Richard Kilvington,' *Przegląd Tomistyczny* 22 (2016), 109–120.

46 Angelo da Fossambruno, *De motu*, 67vb: 'Constat autem latitudinem motus esse infinitam versus extremum intensius. Non enim repugnat motum quemlibet augeri et intendi secundum proportionem adductam agentis supra passum. Cum ergo latitudo proportionis maioris inequalitatis imaginata quam non repugnat acquiri in infinitum est infinita, ergo latitudo motus est infinita vel saltem infinitam imaginemur.'

author deems it useful to clarify that this infinity is conceived as a motion with no limit or stop (*status*). It is an imaginary concept, which cannot be grounded upon some conditions of real things, where primary qualities arise from elementary forms in their highest intensions and are *naturaliter finitae*.[47]

The objection is primarily based on the simple consideration that the acquisition of latitudes requires a certain amount of time, and hence an infinite time should be required in order to acquire infinite latitudes. The possibility of such a motion is deemed untenable, also by noting that if it were linear, then the space traversed should be infinite, which is an absurdity. Nor could it be circular, because the fastest circular motion, that is peculiar to the skies, has proved to be finite.[48]

But nevertheless, there are some arguments *pro infinito*, as is the case of a mixed body composed of fire and water, with their respective powers, in a proportion of equality. This case is clearly taken form *On generation and corruption* (I.10, 328a28–32), where Aristotle lays out his theory of mixture. Angelo argues that when fire heats water, assimilating it by parts, the proportion of the powers will be greater and greater, and the resulting motion will be infinite—a categorematic infinity—starting from *non-gradus*.[49] Another example is drawn from natural considerations that presuppose *horror vacui*. Let us envision two contiguous bodies immersed in air and let us imagine that they begin to separate. It follows that the space in between will be suddenly filled up with air (some parts of air will move while the total amount of air in the universe will

47 Angelo da Fossambruno, *De motu*, 67vb: 'Cum ergo ibi non sit status, nec igitur in latitudine motus status erit versus extremum intensius, sed in extremo remissiori terminatur, scilicet in non gradu motus, patet ergo latitudinem talem motus acquirendam imaginari esse infinitam, quod de aliis latitudinibus realium non contingit ...'

48 Angelo da Fossambruno, *De motu*, 67vb: 'Sed dubitatur utrum talis latitudo motus in casu possibili de facto naturali possit totaliter acquiri in tempore finito, et tunc aut uniformiter aut difformiter. Arguitur quod non, quia acquirendo aliquas latitudines motus certum tempus requiritur; ergo ad acquirendum infinitam infinitum requiritur. Item si hoc datur in motu recto, tunc necessario per eum pertransibitur spatium infinitum, quod tamen non datur. Et non potest esse in motu circulari, quia non datur motus circularis velocior motu celi, quem tamen esse finitum probatum est notum.'

49 Angelo da Fossambruno, *De motu*, 67vb: 'In oppositum arguitur: ponatur quod *a* ignis sit in potentia equalis *b* aqua, ita quod inter ipsa sit proportio equalitatis; et ponatur quod propter aliquid calidum adiuvans incipiat *b* aqua diminui in potentia, ita quod incipiat *a* ignis aliquam partem aque assimilare successive, quousque totam sibi assimilaverit ipso *a* nunquam aliquo modo variato. Tunc motus quo *b* alterabitur ab *a* erit infinitus cathegorematice et incipiet a non gradu, quia proportio incipiet acquiri a non gradu et augmentabitur in infinitum; ergo motus incipiet a non gradu et intendetur in infinitum, cum motus insequatur proportionem. Et sic sequitur intentum in casu mere naturali.'

be at rest). This kind of motion occurs instantaneously, so that we are obliged to conclude that its velocity must be infinite.[50]

To solve these puzzles, Angelo remarks that every mobile, uniformly increasing or decreasing its motion from zero degree (or from a definite degree) to some (other) degree, should be regarded as moving with two different motions, namely a uniformly difform local motion and the motion of intension of local motion. Through the former what is acquired is the path traversed, while through the latter a latitude is gained, which is nonetheless acquired uniformly.[51] Thus, looking back at a previous case, if a mobile body starting from *non-gradus* uniformly increases its motion up to a certain degree for one hour, and for another hour it conversely decreases from that acquired degree to *non-gradus*, the spaces traversed over those two hours will be equal. In terms of motion of intension, the latitude of velocity uniformly acquired will correspond to that uniformly lost. In this case, the mean speed rule is still valid, and it should be applied to latitudes rather than to the first and last points—that both are zero degree.[52] This passage should not be misunderstood: the author points out that it would be misleading to think of this kind of uniformly difform latitude as corresponding to its mean degree for its intensity (*intensive*).

---

50 Angelo da Fossambruno, *De motu*, 67vb: 'Consimiliter probatur et in motu locali, posito quod sint aliqua corpora plana immediata secundum superficiem latam et incipiant sublevari; itaque unum istorum equedurantis incipiat levari in aere ab alio etc. Tunc capto toto aere de mundo arguitur sic. Immediate post hoc aliqua pars aeris erit mota usque ad centrum propter non dari vacuum in medio; ergo immediate post hoc iste aer erit motus ad centrum; patet consequentia, quia si aliqua pars quantitativa alicuius movetur, illud dicitur moveri et non requiritur ipsum secundum se totum, et quodlibet sui simpliciter moveri, ut notum est.'

51 Angelo da Fossambruno, *De motu*, 68ra: 'Ad secundum dicitur quod in motu recto potest hoc esse. Et cum infertur "ergo tali motu erit pertransitum spatium infinitum," dicitur negando consequentiam, quia per nullum tempus adequatum durabit ille motus adequate infinitus, sed per infinitum modicum tempus. Licet enim totalis latitudo motus non possit uniformiter in tempore finito acquiri, quoniam ipsa versus extremum intensius infinita existit, quelibet tamen latitudo motus incipiens a gradu finito versus extremum intensius et terminata ad non gradum potest in tempore finito acquiri vel deperdi.'

52 Angelo da Fossambruno, *De motu*, 68ra: 'Sciendum est tamen quod mobile dum motum suum uniformiter intendit vel remittit ab aliquo gradu ad non gradum seu certum gradum vel econverso, duobus motibus distinctis intelligitur moveri, ut dicebatur superius, scilicet motu locali uniformiter difformi et motu intensionis motus localis, qui motus, scilicet intensionis, est uniformis. Per motum primum sibi acquiritur spatium pertransitum, per secundum latitudo motus uniformiter difformis, que tamen uniformiter acquiritur … Iuxta quod hoc est commune principium in illa materia, quod omnis latitudo motus uniformiter difformis, sive ab aliquo gradu incipiat sive a non gradu, dummodo ad aliquem certum gradum terminetur et uniformiter acquiratur seu deperdatur, correspondet gradui suo medio.'

It should be rather known in the light of its effects (*effective*), consisting of the space traversed by a mobile body, *as if* it moved with its mean degree. Secondly, this rule is only applicable under the condition that motion has a finite extreme point, since an infinite extension has no middle point (*nullius infiniti datur medium*). Moreover, at the mean degree the distance from both end-points is to be equal, but nothing is as far from a finite point as from an infinite one.[53] The problem arises when a latitude is not uniformly acquired, as in the foregoing cases, because they could not be addressed by the application of the mean speed rule, insofar as they are difformly—rather than uniformly—acquiring latitudes.[54]

Having expressed these reservations, Angelo can indeed set out a most general principle, expressed in terms of proportions: in every given uniformly difform latitude, starting from a maximum finite degree and ending with *non-gradus*, its mean degree precisely corresponds to the *subduplum* (or half) of the most intense degree of the latitude.[55] Furthermore, an additional corollary states that the mean degree of a latitude whose extremes are both different from zero degree is greater than half the maximum degree in the latitude.[56]

It is easy to see that, as a consequence of the above-mentioned principles, a mobile moving from zero degree to a maximum degree will traverse twice half as much as if it moved with a uniform motion, whose degree was the maximum of the first motion. Thus, supposing that Socrates moves from *non-gradus* to 8 degrees, following the first principle he will traverse as much space as if he moved uniformly with the mean-degree, that is 4; and 4 is half the maximum 8.[57]

---

53 Angelo da Fossambruno, *De motu*, 68ra: 'Secundo notandum est quod notanter addi debet "dummodo ad gradum finitum terminetur in extremo intensiori," quoniam si ad gradum infinitum terminaretur, nullum haberet gradum medium cui corresponderet, quia nullius infiniti datur medium. Et etiam quia medium equaliter distat ab utroque extremorum; implicat autem aliquid equedistare a finito et infinito.'

54 Angelo da Fossambruno, *De motu*, 68ra.

55 Angelo da Fossambruno, *De motu*, 68rb: 'Aliud etiam ponitur principium pro hac materia communiter tentum, quod cuiuslibet latitudinis uniformiter difformis incipientis ab aliquo gradu in extremo intensiori et terminantis ad non gradum in alio extremo, gradus medius est precise subduplus ad gradum istam latitudinem terminantem in extremo intensiori.'

56 Angelo da Fossambruno, *De motu*, 68rb: 'Aliud etiam ponitur circa idem, quod cuiuslibet latitudinis uniformiter difformis terminantis ad certos gradus in utroque extremorum gradus medius est maior quam subduplus ad gradum istam latitudinem terminantem in extremo intensiori.'

57 Angelo da Fossambruno, *De motu*, 68rb.

5 *Intensio*, *remissio*, and Their *regulae*

From a dynamic point of view, there is a double way in which motion may become faster (*dupliciter potest motus intendi*), according to whether it is power or passive resistance that increases or decreases: an acceleration (*intensio*) can result from a growing force with a constant resistance, or from a steady force with a decreasing resistance, because in both cases the proportion between motive force and resistance becomes greater. Obviously, deceleration (*remissio*) can be caused by a weakened force or a greater resistance, *aliis paribus*.[58] Angelo remarks, once again, that in local motion the velocity of motion corresponds to the space traversed in equal lengths of time, while the velocity of intension results from the increase/decrease of the latitude of motion, which corresponds to the velocity acquired/lost in a given interval of time.[59]

In light of these considerations, Angelo goes on to discuss two sets of rules (*regulae*). The first set comprises four rules and concerns simple cases, involving the existence of a resistance. It is worth mentioning the close correspondence of some of Angelo's *regulae* with those discussed by Swineshead in his treatise on local motion in his *Liber calculationum*. In fact, these rules spread in different forms during the fourteenth century.[60]

The first rule states that if some power uniformly increases in degree in a given time interval, while the associated passive resistance remains constant, the velocity of motion will increase more and more slowly. This means that the proportion of the power to resistance grows more and more slowly, according to Bradwardine's rule, and according to Swineshead's rule 19 of *tractatus* XIV of his *Liber calculationum*.[61] The rule is illustrated with an example: if we divide

58 Angelo da Fossambruno, *De motu*, 69vb: 'Pro secundo notandum est etiam quod dupliciter potest motus intendi, uno modo augmentata potentia, resistentia vero non variata cum aliis paribus. Alio modo potentia non variata et resistentia diminuta. Utroque enim istorum modorum contingit proportionem ex qua habet intendi motus augeri. Similiter etiam contingit motum remitti uno modo potentia diminuta et non variata resistentia, alio modo potentia manente et maiorata resistentia continue, alia paria supponendo.'

59 Angelo da Fossambruno, *De motu*, 69vb: 'Item etiam sciendum est quod, sicut in motu locali velocitas attenditur penes spatium quod mobili acquiretur in tanto tempore vel in tanto, ita velocitas intensionis motus attenditur penes latitudinem motus in tanto tempore vel in tanto acquisitam. Et velocitas diminutionis motus attenditur penes maximam latitudinem motus in tanto vel in tanto tempore deperditam.'

60 Cf. Celeyrette, 'Le mouvement,' 26; E.D. Sylla, *The Oxford Calculators and the Mathematics of Motion, 1320–1350. Physics and Measurement by Latitudes*, Ph.D. dissertation, Harvard University, 1970 (repr. with preface and errata, New York 1991).

61 Angelo da Fossambruno, *De motu*, 69vb: 'Sit igitur hec prima regula: si aliqua potentia per certum tempus eius resistentia penitus non variata, idest non aucta nec diminuta in

the whole time during which an intension of motion is produced (from 2 to 8) into three, in the first third the power will be at 4 degrees, in the second at 6 degrees, and in the third at 8 degrees. But, as a result of the resistance, the proportions 4 : 2, 6 : 4, 8 : 6 show a progressive decrease of the increment of the power: the first proportion is a double (*dupla*), the others are respectively a sesquialternate (*sesquialtera*) and a sesquitertian (*sesquitertia*).[62]

In the second rule, Angelo states that for a power to increase its motion uniformly against constant resistance, it must increase its velocity. In the proportion between this power and the resistance, it is the first that grows, implying a continuous proportional increment of the velocity.[63]

The comparison between two different powers and resistance is established in the third rule: if these powers equally increase uniformly, and the resistance remains unchanged, the motion of the weaker power will increase more quickly than that of the stronger. This is so, because the weaker power gains a greater proportion. The whole argument is clearly built up on the assumption of proportion, between degrees of powers and between these results and

gradu, uniformiter maioretur in gradu eius, post continue per talem maiorationem uniformem tardius et tardius illa potentia motum suum intendet.' Cf. Richard Swineshead, *Liber Calculationum*, Venice 1520, 43vb: 'Si potentia crescat uniformiter respectu resistentie non variate, tardius et tardius intendetur motus.' Cf. Thomas Bradwardine, *Tractatus de proportionibus velocitatum in motibus*, ed. H.L. Crosby, Jr., *Thomas of Bradwardine. His* Tractatus de proportionibus. *Its Significance for the Development of Mathematical Physics*, Madison, WI, 1955 (University of Wisconsin Publications in Medieval Science, 2), 112–117.

62 Angelo da Fossambruno, *De motu*, 69vb–70ra: 'Sit *a* potentia ut duo, *b* resistentia ut unum, et ponatur quod uniformiter maioretur illa potentia adequate in hora ad octo resistentia non aucta neque diminuta. Et arguitur quod tardius et tardius intendetur motus, quia tardius et tardius acquiretur proportio; ergo tardius et tardius intendetur motus ... Quia dividatur gratia exempli hora in qua debet fieri huiusmodi intensio motus in tres tertias, et sequitur quod, cum uniformis erit maioratio potentie, quoniam in prima tertia illa potentia erit ut quatuor, et per consequens acquiret unam proportionem duplam; in secunda vero tertia erit ut sex, in qua per consequens acquiretur proportio sexquialtera, scilicet, a sex ad quatuor; in tertia vero tertia acquiretur proportio sexquitertia, scilicet, ab octo ad sex; cum igitur minor fit proportio sexquitertia quam sexquialtera, et minor sexquialtera quam dupla, et iste acquirentur in partibus equalibus temporis, igitur sequitur quod continue tardius et tardius acquiret proportionem.'

63 Angelo da Fossambruno, *De motu*, 70ra: 'Secunda regula sit illa, ad hoc quod aliqua potentia eius resistentia manente ipsa aucta uniformiter motum suum intendat, oportet quod ipsa continue velocius et velocius maioretur. Illa patet sic, quia proportionem equaliter augendo respectu maioris et respectu minoris, maior excessus acquiritur maiori quam minori. Cum ergo potentiam continue oporteat fieri maiorem et maiorem, si debet continue motus intendi per eius augmentum solum, sequitur quod ad continue acquirendum equaliter proportionem precise continue maiorem et maiorem excessum acquirent, et si sic, igitur continue velocius et velocius debet illam potentiam maiorari.'

the resistance.[64] In the example suggested in the text, given two powers whose starting degrees are 4 and 2 respectively (the resistance is supposed to be 1), their uniform increase leads, in one hour, the first to 8 and the second to 6. By reasoning on the two halves of the whole time, the first power will acquire the proportion 6 : 4 (6 being the mean degree) during the first half, while in the remaining time the proportion acquired will be 8 : 6. A similar reasoning, applied to the second power, will result in the proportions of 4 : 2 and 6 : 4: by comparing the results, the first power increases by a sesquialternate and then by a sesquitertian, while the second one goes up by a double and then by a sesquialternate. In both the cases, the change results to be greater in the first half than in the second; but the second power proportionally acquires more than the first one, that is, its velocity results greater than that of the former.[65]

The fourth rule is a sort of corollary derived from the third one. It is quite evident that for two unequal powers to increase with the same velocity, the one with the greater proportion should raise more quickly.[66]

The aforementioned four rules are an appropriate model to use when taking into consideration other combinations, in which resistance is subject to an increase, while the power remains without changes (*stante potentia*): these two cases are argued also for *remissio*.

---

64 Angelo da Fossambruno, *De motu*, 70[ra]: 'Tertia regula: si due potentie inequales per certum tempus uniformiter et equaliter continue maiorantur respectu eiusdem resistentie non variate, velocius continue intendet supra illam resistentiam potentia minor quam maior motum suum. Illa sequitur, quia per equale acquisitum plus proportionabiliter crescit minus quam maius; ergo cum ille potentie equaliter acquirant, *d* gradu potentie continue maiorem proportionem supra se acquirit minor quam maior; sed omnem proportionem quam aliqua illarum acquirit supra se, acquirit supra suam resistentia; ergo etc.'

65 Angelo da Fossambruno, *De motu*, 70[ra]: 'Item probatur in exemplo: sit *a* potentia ut quatuor, *b* potentia ut duo et *c* resistentia ut 1; augeantur in hora ille potentie uniformiter et eque velociter, ita quod *a* fiat ut octo, *b* ut sex. Et sequitur quod in prima medietate hore *a* acquiret supra *c* resistentiam ultra proportionem habitam in principio solum proportionem sexquialteram, scilicet a sex ad quatuor, quia in medio instanti *a* erit ut sex; *b* vero in prima medietate acquiret proportionem duplam, scilicet a quatuor ad duo, quia in medio instanti *b* erit ut quatuor; in secunda vero medietate acquiret *a* (*b ed.*) proportionem ab octo ad sex, que est sexquitertia; *b* vero acquiret proportionem a sex ad quatuor, que est sexquialtera. Ex quo sequitur intentum.'

66 Angelo da Fossambruno, *De motu*, 70[ra]: 'Ex hac sequitur regula quarta quod ad hoc quod due potentie inequales respectu resistentie non variate eque velociter per earum augmentationes motus earum intendant, oportet quod in ea proportione maior earum velocius alia maioretur, in qua proportione est alia maior. Patet, quia continue hoc stante iste potentie equaliter et eque proportionabiliter maiorabuntur; ergo etc.'

Angelo turns his attention in comparing *intensio* and *remissio*: each of them, in fact, could be subject to a comparison to itself or to the other one, in such a way to give rise to four combinations: *intensio cum remissione, remissio cum intensione, remissio cum remissione, intensio cum intensione.* Obviously, if these two terms are intended as qualities which *formaliter* denominate a specific mode of being, they cannot be mutually compared, because it would be improper to confront positive with negative qualities. Nonetheless, the comparison is made possible if *intensio* and *remissio* are thought *pro motibus*, by which some latitude is acquired or lost. Closely following Heytesbury's text, Angelo devotes the last part of his treatise to the analysis of these cases, which gives birth to a lot of contradictory conclusions or *sophismata*.[67] Here we shall consider only the first and the second sophisms.

The first sophism compares two uniform motions in the same time interval, one increasing from a given degree of intension (*c*) to its double, the other decreasing from the same degree *c* to zero: it can be argued that, in the same time interval, the two mobiles move/do not move with equal velocities. Supposing that the first motion (*a*) starts moving at 4 degree and uniformly accelerates to its double, 8 degree; a second motion, *d*, begins its uniform deceleration at 4 degree and decelerates to zero degree; then *d* should successively remit its velocity to its *subduplum*, then *subquadruplum*, *suboctuplum* and go on, in an infinite series. In order to equate the rate of the motion *d* with that of the motion *a*, this one should intend its velocity to the double, quadruple, eightfold, *in infinitum*; but this last sentence is impossible, because the motion *a* can only reach 8 degree.[68] The solution suggested by Angelo takes into account

67 Angelo da Fossambruno, *De motu*, 70vb: 'Circa comparationem intensionis motus cum remissione eiusdem vel alterius motus, vel remissionis cum intensione, vel intensionis cum intensione, vel remissionis cum remissione, multa sophismata in illa materia fieri contingit. Comparatur enim intensio cum intensione, ut cum dicitur velocius intenditur *a* motus quam motus *b*. Comparatur autem intensio cum remissione ut dicendo velocius intenditur *a* motus quam remittatur *b*. Et nota quod capiendo *intensionem* et *remissionem* ut sint qualitates formaliter denominantes aliquid esse intensum et remissum, pro quibus hec nomina "intensio" et "remissio" supponunt positive et privative, differentes non sunt ille ullo modo invicem comparabiles, ut alibi clare ostenditur. Sed accipiendo ut supra pro motibus quibus aliquid sic denominatur intensum vel remissum, quorum unus sit latitudinis motus acquisitivus et alter deperditivus, bene sunt comparabiles intensio et remissio. Quamvis enim magnitudo unius non possit dici maior quam parvitas alterius, tamen bene dici potest unum maiorari velocius quam reliquum minoretur et quod velocior est maioratio alterius.'. Cf. William Heytesbury, *Regulae solvendi sophismata*, Venice, Bonetus Locatellus, 1494, 41vb–44rb; Wilson, *William Heytesbury*, 126–128.

68 Angelo da Fossambruno, *De motu*, 70vb: 'Primum sophisma circa hoc est posito, quod *a* mobile incipiat motum suum uniformiter intendere a *c* gradu, qui sit ut quatuor, usque ad

the difference between equal velocities and equal proportionality: the comparison of the two motions must be set in the terms of acquired and lost latitude of motion. Under this assumption, the latitude gained by motion *a* equals the latitude lost by motion *d*. The objection that *intensio* and *remissio* cannot be compared (as mentioned above) is overcome by the clarification that the whole argument can be related to the sole latitude of intention, gained or lost.[69]

In the second sophism there are two assumptions: (1) that two mobile bodies move with equal increasing uniform velocity to a given degree (*c*); and (2) that beyond this degree, the velocity of one body (*a*) is twice that of the other (*b*). From these assumptions, one is led to two opposite conclusions, namely that the first body (*a*) will move and will not move twice as rapidly as the second (*b*). Here is how Angelo formulates this sophism:

> Aliud sophisma est posito quod *a* et *b* mobilia nunc moveantur eque velociter, ut puta *c* gradu, et intendat uterque motum suum uniformiter a *c* gradu; *a* tamen continue postea in duplo velocius quam *b*. Tunc contingit in hoc probari duo contradictoria, scilicet quod aliquando *a* movebitur in duplo velocius quam tunc movebitur *b*, et tamen *a* nunquam movebitur in duplo velocius quam pro nunc movebitur *b*.[70]

The first conclusion is argued in this way: supposing a *d* degree four times greater than *c*, toward which *a* and *b* tend, it is well known that the distances *ad* and *bd* are equal. Yet, *a* increases its motion twice as quickly as *b*. That implies that once the *d* degree has been attained by *a*, *b* will be at a degree twice that

---

gradum duplum in hora; et sit gradus duplus ad *c*, *d*. Et sit aliud mobile *b*, quod in eadem hora remittat motum suum uniformiter ab ipso *c* gradu usque ad non gradum; quo stante, arguuntur hec duo contradictoria, scilicet, quod *a* in illa hora eque velociter motum suum intendet sicut *b* remittet. Et tamen non eque velociter intendet *a* motum suum sicut *b* remittet'. Cf. Wilson, *William Heytesbury*, 126.

69 Angelo da Fossambruno, *De motu*, 71[ra]: 'Ad sophisma respondetur faciliter scito prius quod velocitas intensionis vel remissionis motus attenditur penes latitudinem motus acquisitam vel deperditam absolute … notata est prius differentia inter terminos istos, scilicet, eque velociter et eque proportionabiliter … Tunc, enim, dicitur concedendo quod precise eque velociter *a* intendetur sicut *b* remittetur et econtrario, quia tanta latitudinem intensionis motus acquiret *a*, quanta deperdet *b* et econtrario, quia inter quemcunque gradum et suum duplum est tanta latitudo sicut inter ipsum et non gradum …; et negatur antecedens, quia illo modo intensio et remissio non sunt comparabiles ut dictum est; negatur consequentia, quia velocitas remissionis non attenditur penes latitudinem remissionis acquisita, sed penes latitudinem intensionis deperditam.'

70 Angelo da Fossambruno, *De motu*, 71[ra].

of *c*, because four times *c* is equivalent to two times the double of *c*.[71] On the other hand, Angelo also proves the other part of the contradiction, that is the impossibility that *a* moves with a velocity twice that of *b*. In fact, if we suppose that *a* begins to move with a velocity twice that of *b*, the rate of *intension* of *a* will remain double that of *b*, because a sum of equal proportions gives equal proportions as result. But in this case, in every instant *a* will have a velocity greater than the previous case, and then the velocity in the first case cannot be the same—that is, the double—of the second one.[72]

The first argument is not valid on the basis of reaffirming that the velocity of *a* will never be twice that of *b*.[73] It is a universal truth that when two bodies are at the same distance from a fixed degree, the body moving with a velocity double that of the other one, is the faster body.[74] In order to offer a clear explanation, Angelo illustrates this point with an example. Let us suppose that two bodies, *a* and *b*, begin to move uniformly from 2 degree to 8 degree, and that the intension of *a* is twice that of *b*. Then, in one hour *a* will gain the terminus twice as rapidly as *b*. That amounts to saying that when *a* reaches 8 degree, it

---

71 Angelo da Fossambruno, *De motu*, 71ra: 'Prima pars probatur, quia signetur aliquis gradus in quadruplo intensior *c*, ultra quem gradum *a* et *b* intendent motus suos, et sit iste gradus *d*. Tunc arguitur: *a* et *b* nunc precise equaliter distant a *d*, et continue post hoc in duplo velocius accedit ad illum gradum *a* quam *b*, quia *a* in duplo velocius intendat motum suum quam *b*; ergo in duplo velocius deveniet *a* ad gradum quadruplum ad *c* quam *b*; et per consequens, quando *a* habebit gradum quadruplum ad *c*, habebit precise *b* gradum duplum ad *c*; et si sic, igitur pro tunc movebitur *a* in duplo velocius quam pro tunc movebitur *b*. Patet consequentia, quia gradus quadruplus ad *c* est duplus ad gradum duplum ad idem *c*.'

72 Angelo da Fossambruno, *De motu*, 71ra: 'Alia pars contradictionis probatur, scilicet, quod *a* nunquam in tali casu movebitur in duplo velocius pro tunc quam movebitur *b*, quia toto illo casu stante propter hoc quod sit *a* in principio duplus ad *b*, tunc continue movebitur *a* in duplo velocius quam pro tunc moveretur *b*; sed *a*, isto casu stante, continue moveretur velocius quam pro tunc movebitur et *b* precise eque velociter; igitur tunc nunquam *a* movebitur in duplo velocius quam pro tunc movebitur *b*. Consequentia est satis nota et bona sicut patet intuenti. Et maior etiam, quia si *a* motus nunc sit duplus ad *b*, et sic est duplus ad *b*, ita continue in duplo velocius eo intendatur, semper *a* motus erit duplus ad *b*, quoniam si eque proportionabilibus continue eque proportionabilia addantur continue, eque proportionabilia remanebunt.'

73 Angelo da Fossambruno, *De motu*, 71^{ra-rb}: 'Ad hoc respondetur concedendo quod nunquam *a* motus erit duplus ad *b* motum vel nunquam movebitur *a* in duplo velocius *b* …'

74 Angelo da Fossambruno, *De motu*, 71rb: 'Et ad secundum pro alia parte conceditur quod, dato gradu quadruplo ad *c* et quocunque alio ab illis ambobus acquirendo, quod in duplo citius erit *a* sub illo gradu quam *b*, quia hoc generaliter verum est quod a quocunque termino aliqua duo equaliter distant, ad quem terminum unum illorum in duplo alio velocius movetur aliis paribus in duplo citius erit illud quod velocius movetur sub illo termino quam aliud.'

will acquire 6 degrees, while *b* will be at 5 degree, because it gains 3 degrees (half 6 degrees). To reach the remaining three degrees, *b* will spend another hour. Which means that *a* is twice as fast as *b* and at every degree of *b* corresponds a double degree acquired by *a*.[75] Obviously, this does not mean that the degree of *b* is just half that of *a*: in the given example, when *a* has 8 degrees, the degree of *b* is 5, not 4. What is relevant here is the amount of degrees gained by the faster body, and it is half this amount which the other body has acquired.[76] Angelo focuses on the fact that the mean degree between *c* and its quadruple turns out to be something more than the double of *c*.[77]

In order to clarify this point, Angelo offers an additional example, with an infinite increase of velocity (*per infinitam latitudinem*), proving once again that the excess of the motion of *a* will never be twice that of *b*.[78] What is assumed here is that the two motions increase non-uniformly, so far as to acquire every degree of velocity imaginable (*usque quod utrumque ipsorum habuerit omnem gradum velocitatis imaginabilem*). This would entail the admission of an infinite degree of motion (*scilicet, gradus infinitus motus*) leading to some contradictions. If the body *a* repeatedly acquires a proportional part that is twice that acquired by *b*, the ratio between the respective velocities of the two bodies will

75 Angelo da Fossambruno, *De motu*, 71[rb]: 'Sint ergo nunc *a* et *b* in gradu ut duo. Et debeant ambo acquirere gradum ut octo, ad quem *a* in duplo velocius intendatur quam *b*. Si igitur gratia exempli in hora *a* attingat istum gradum ut octo uniformiter acquirendo sex gradus supra eos quos habet *b*, tunc quod in duplo tardius intendetur in eadem hora attinget gradum ut quinque, quia acquirere solum debet tres gradus postquam in duplo tardius debet intendi; et cum utrunque intendetur, sequitur quod *b* alios tres gradus acquiret in alia hora sequenti precise. Et sic solum in alia hora attinget istum gradum ut octo, et cum *a* acquiret illum in prima hora, sequitur quod in duplo citius erit *a* sub illo gradu quam *b*. Et sic de omni gradu inferiori *c* est arguendum. Igitur verum est illud, scilicet quod quocunque omni gradu dato ab ambobus acquirendo, in duplo citius erit *a* sub illo gradu quam *b*.'

76 Angelo da Fossambruno, *De motu*, 71[rb]: 'Tunc igitur ad argumentum: cum ex illo antecedente concluditur quod quando *a* erit sub gradu quadruplo ad *c*, erit *b* precise sub gradu duplo, negatur consequentia; immo tunc erit *b* sub gradu maiori quam duplo, scilicet sub gradu ut quinque, ut patet calculanti.'

77 Angelo da Fossambruno, *De motu*, 71[rb]: 'Causa autem quare consequentia non valet est quia gradus medius inter *c* gradum et gradum quadruplum ad *c* non est gradus duplus ad *c*, sed maior quam duplus ad *c*. In instanti enim in quo *a* erit sub gradu ut octo, *b*, quod in duplo tardius intenditur, erit pro tunc sub gradu medio inter *c* et *d*, quia *b* tunc precise acquisivit medietatem totius distantie inter *c* et 8; medium autem gradum inter *c* et 8 est maior *c* magis quam in duplo.'

78 Angelo da Fossambruno, *De motu*, 71[rb]: 'Et iuxta hoc conceditur illa conclusio quod per infinitam latitudinem *a* motus excedet *b* motum, et tamen nunquam erit *a* motus duplus ad *b* motum …'

never be that of 2 : 1, that is to say, the degree acquired by *a* will never be equal to the double of the degree gained by *b*. In fact, by adding repeatedly the same rate of gained velocities to *a* and *b* – 4 in the first case, 2 in the second—there will be two series, posited that the starting value be 2: the first is 6, 10, 14, and so on; the second is 4, 6, 8, and so forth. It is easy to see that, notwithstanding the assumption that the velocity, in any part of time, of *a* is greater than that of *b*, none of the values belonging to the first series is twice that of the values in the second series, and yet each value of the first series will never be the double of the corresponding value of the second series. Angelo feels authorized to extend this sequence *ad infinitum*, stating that this kind of exceeding values of instantaneous velocity will continue *per infinitum*.[79] As Curtis Wilson has noticed, what is at stake here shows that Angelo draws upon Swineshead, rather than Heytesbury's work. In particular, it is argued that the conception of infinite degree is not consistently posed in Heytesbury's examples, because his very definition of uniformly increasing/decreasing motion is not given so as to cover all possible cases.[80] Infinite, in other words, cannot be a point to which the velocities of two bodies could be compared at any instant, because this relation proves to be indeterminate.

6 Concluding Remarks

We know that in 1496 a Statute of the Faculty of Arts of Padua established Heytesbury's *Regule* as a standard text for the teaching of logic, a ratification of the wide diffusion of that text in Italian universities. The location of Angelo's work in the Venice edition would suggest its relevance as a useful instrument in introducing and clarifying some complex passages of the chapters about motion. It has often been remarked that Angelo's aim was to provide students

79 Angelo da Fossambruno, *De motu*, 71[rb]: 'Nunc sint *a*, *b* equales, gratia exempli, uterque ut duo, et intendantur ambo ad gradum infinitum in illa hora secundum partes proportionales, *a* tamen continue in duplo velocius intendatur quam *b*, scilicet in quacunque parte proportionali acquirat enim *a* in quacunque parte proportionali quattuor gradus et *b* duos. Tunc patet quod *a* in fine prime partis proportionalis excedit *b* per duos gradus, quia tunc erit *a* ut sex et *b* ut quattuor; in fine secunde partis proportionalis per quattuor gradus *a* excedit *b* ... Et sic ulterius reperietur quod continue magis et magis excedet *a* *b*; per aliquantam ergo latitudinem *a* in illa hora excedet *b*, aliquando per duplam et per triplam, et sic in infinitum aliquando excedet; ergo etc. Et nunquam *a* erit duplus ad *b*. Probatur ut prius est argutum omnino. Ex quo etiam concedi potest adhuc in simili quod per infinitum *a* excedet *b*, et tamen nunquam *a* bis continebit *b*.'

80 Cf. Wilson, *William Heytesbury*, 195, n. 12.

with an easy-to-handle textbook on this issue, and this is the reason why illustrative examples are so prevalent in the text. Certainly, we will not find as high a level of sophistication in it as we can read in other authors' works (for instance, in Swineshead).

In so far as it consists in an abridgement of academic classes, Angelo's commentary was intended to provide students with basic concepts and a distinctive vocabulary to apply to natural philosophy. It is worth recalling that Angelo's teachings spread in a milieu imbued with Aristotelianism, that he taught and wrote for students who were familiar with Aristotelian texts, and in several passages apparently tried to reconcile this tradition with that of the *Calculatores*. When Pomponazzi, in the first decades of the sixteenth century, lectures against the theories of the *proportionistae*, he is essentially challenging Swineshead's theses, but it is also true that some of his remarks are centred on basic concepts of motion, latitude, remission and intension. In broad terms, one can say that the widespread diffusion of these issues in northern Italian universities is also undoubtedly due to familiarity with Oxonian and Parisian literature about *calculationes* transmitted, at a first level, by works such as that of Heytesbury and Angelo. The latter's *De velocitate motus* represents one—and perhaps not the least influential—medium through which the essential elements of these novelties entered the Italian cultural context.

CHAPTER 14

Leibniz and the Calculators

Edith D. Sylla

1 Introduction*

The calculatory movement, which began in Oxford in the first half of the fourteenth century, spread quickly to Paris. Soon calculatory approaches appeared elsewhere on the Continent. By the time of printed editions, France, Italy, Portugal, Spain, and Eastern Europe hosted most calculatory activity. In the seventeenth century, Gottfried Wilhelm Leibniz proposed to reprint Richard Swineshead's *Liber calculationum* as part of a treasury of philosophy. What Leibniz had to say about 'Suisset, popularly known as the Calculator' (i.e., Richard Swineshead) and his group casts light on the impact, spread, and decline of the Calculators' tradition. Meanwhile Leibniz's own work in a sense carried on the Calculators' programme to develop new mathematical sciences. This chapter explores these developments.

In approximately the second quarter of the fourteenth century there existed at Oxford a group of bachelors and masters of arts who, in recent decades, have been referred to as the 'Oxford Calculators.'[1] Thomas Bradwardine's *Tractatus de proportionibus velocitatum in motibus* was the founding document of this group.[2] In his work, Bradwardine promoted the application of mathematics to natural philosophy, and he posited a rule connecting the ratios of forces to resistances causing motion, on the one hand, and the velocities they cause,

* This article uses the following abbreviations: A = G.W. Leibniz, *Sämtliche Schriften und Briefe* (Akademie Ausgabe), Berlin 1923– (various series and volumes); GM = *Leibnizens Mathematische Schriften*, ed. C.I. Gerhardt, 7 vols., 1848–1863, repr. Hildesheim 1962; Couturat = *Opuscules et fragments inédits de Leibniz*, ed. L. Couturat, Paris 1903.

1 My doctoral dissertation, *The Oxford Calculators and the Mathematics of Motion, 1320–1350. Physics and Measurement by Latitudes*, Ph.D. dissertation, Harvard University, 1970 (repr. with preface and errata, New York 1991), is an early example of usage of the term. Still earlier, Ernest A. Moody referred to 'these fourteenth-century "calculators" of Merton College,' 'the Oxford calculators,' and 'the Oxford school.' See E.A. Moody, 'Laws of Motion in Medieval Physics,' *Scientific Monthly* 72 (1951), 18–23; reprinted in E.A. Moody, *Studies in Medieval Philosophy, Science and Logic. Collected Papers 1933–1969*, Berkeley, CA, 1975, 189–201.

2 For a different view, see Elżbieta Jung's contribution in this volume. I remain unconvinced that Kilvington preceded Bradwardine in the works at issue here. Despite Jung's assertions, the datings of Kilvington's works (and even their authorship) are not established with any

 | DOI:10.1163/9789004512054_016

on the other. Not long after Bradwardine's *Tractatus de proportionibus*, William Heytesbury completed his *Regule solvendi sophismata*, which extended Bradwardine's position on the measures of motion to the three categories of local motion, alteration, and augmentation/diminution. In his *Regule* Heytesbury proved what has come to be called the mean (or better middle) degree theorem of the measure of uniformly difform motion with respect to effect. Then Richard Swineshead composed the crowning work of this movement, the *Liber calculationum*, and John Dumbleton worked on but did not complete his large *Summa logicae et philosophiae naturalis*. Other related works, some discussed in other chapters in this volume, emanated from Oxford during the same period.

Since the Oxford Calculators were not so labelled in their lifetimes, who exactly should be called an Oxford Calculator remains an open question. Moreover, what criteria can be used to define the Calculators?[3] Because Richard Kilvington's *Sophismata* is closely related to Heytesbury's *Regule solvendi sophismata* and *Sophismata*, Kilvington may be considered an Oxford Calculator. Since there is no evidence that Kilvington was ever a fellow of Merton College, we no longer call the Oxford Calculators 'the Merton School,' as once was the case.

In this chapter, I use the extant papers of Gottfried Wilhelm Leibniz as sources of evidence about the Calculators. Leibniz repeatedly wrote about Swineshead and looked for a copy of his *Liber calculationum* after other evidence of Calculators' work had largely disappeared. Leibniz's perspective makes the use of mathematics a key characteristic of the Calculators. Ideally, according to Leibniz, when one makes use of mathematics, one not only quantifies, but one also writes in a demonstrative format (with analogies to Euclid's *Elements*), which, regrettably, many scholastics failed to do because they were wedded to scholastic disputations. Writing a treatise (*tractatus*) in a demonstrative format rather than as a report of a disputation appears to have increased the chances that a work would have an impact and lasting influence.

For Leibniz, Swineshead/Suisset was the leader (*princeps*) of the Calculators sect (*sectatores*) who had no equal and few followers. Prominent among the followers he did have were Thomas Bradwardine and Nicole Oresme. Oresme wrote a work modelled on Bradwardine's *Tractatus de proportionibus veloci-*

---

degree of certainty. It is possible that Kilvington (or the author of works some attribute to Kilvington) had Bradwardine's work on his desk when he wrote, contra Jung. Since the Calculators shared their insights with one another (and with us), our disagreement here is something of a tempest in a teapot.

3 Mark Thakkar's chapter in this volume deals insightfully with these questions.

tatum in motibus. He also wrote a longer work *De configurationibus qualitatum et motuum*. Should we also consider Oresme a Calculator? If so, then what about Albert of Saxony and other Parisians? The title of the 2018 Munich conference, 'Quantifying Aristotle: The Impact, Spread, and Decline of the *Calculatores* Tradition,' which led to this book, implies that there was a Calculators tradition while leaving open how to define it. By not specifying 'Oxford Calculators,' it admits the possibility that not all Calculators were located at Oxford. In fact, at least one of the individuals often considered among the early Oxford Calculators, namely Walter Burley, did some of his calculatory work in Toulouse and Paris rather than in Oxford, where he had studied in the Arts faculty before moving to Paris to study theology.

There is another tacit implication of the conference title, namely that the Calculators were Aristotelians, or perhaps scholastic and/or nominalistic Aristotelians. What does this mean for the characterization of the '*Calculatores* Tradition'? Are we going to say that what makes one a 'Calculator' is being termed a 'Calculator' in late medieval or early modern sources? Or are we going to say, for instance, that to be a Calculator an author must emphasize and use mathematics? Or something else? In this chapter, I propose that the Calculators were Richard Swineshead (according to Leibniz 'Suisset, commonly known as the "Calculator"') and his *sectatores*, especially Thomas Bradwardine and Nicole Oresme, as well as some others.

Pierre Duhem, who was the first to study extensively the new physical science of the fourteenth century, had a mistaken idea of the relative chronology of developments in Paris and Oxford.[4] Whereas the innovations of Oxford starting with Bradwardine preceded the new science of latitudes of forms of Nicole Oresme, Duhem had the idea that Oresme's work preceded Heytesbury's. Anneliese Maier, who had a better grasp of the chronology, in her *An der Grenze von Scholastik und Naturwissenschaft*, while putting the Oxford thinkers earlier than the Parisian, nevertheless tucked 'Die Calculationes des 14. Jahrhunderts,' most of which were done by Oxford Mertonians, into Part III, of *An der Grenze*—'Die Mathematik der Formlatituden'—where it was paired with 'Die Wissenschaft von den Formlatituden' and followed by 2. 'Oresmes Methode der graphischen Darstellung', and then 3. 'Die Nachwirkung der Oresme'schen Lehre.'[5] Thus, Maier seemed to treat Oresme's graphical latitudes of forms as a new branch of *Calculationes*.[6]

4 See above all his *Études sur Léonard de Vinci*, 3: *Les précurseurs parisiens de Galilée*, Paris 1913.

5 A. Maier, *An der Grenze von Scholastik und Naturwissenschaft* (Studien zur Naturphilosophie der Spätscholastik, 3), 2nd ed., Rome 1952 (Storia e letteratura, 52), 257–384.

6 For new research concerning this branch see D.A. Di Liscia, 'The "*Latitudines breves*" and

To those who examine fourteenth-century physics from a mathematical perspective, it is clear that Oxford authors made use of arithmetic or numbers for quantification, which would fit with the label '*calculationes*', while Oresme made use of lines or geometry. In proving what came to be called the mean or middle degree theorem for uniformly difform motion, the question arose how the area of a triangle, of which one side represents time and the perpendiculars to that side represent (instantaneous) velocities, represents distance traversed. A line, as a one-dimensional extension, can represent time, and another line can represent the latitude of velocity or a given velocity. But how does an area represent a distance? Thus, quantifying Aristotle and defining concepts so that they became quantifiable was more complicated than one might at first expect.

If Anneliese Maier was right when she said that what we are calling the Calculators Tradition began with Bradwardine's *On the Proportions of Velocities in Motions*, and if it was also insightful on her part to treat Oresme's work on the latitudes of forms to be a new branch of fourteenth-century calculatory science, what does this imply about the nature of the Calculators tradition? For one thing, it means that while the tradition began with Bradwardine's *Tractatus de proportionibus velocitatum in motibus* (and we think that Bradwardine was a realist or Platonist in mathematics), it almost immediately became a theory or scientific discipline in which the magnitudes could, by the Parisians and then by the Oxonians, be understood in a nominalistic way—as concepts or things of reason (*entia rationis*) or imagination. For instance, the latitude of velocity could be considered an imagined or invented concept, which is understood to be infinite and to have no maximum degree. This occurs, for instance, in Heytesbury's *Regule solvendi sophismata*.

Later, as will be discussed below, it seems to have been the Parisian branch of the Calculators that carried on the Calculators' work until it ultimately spread to Spain and Portugal in the early sixteenth century. Of course, the evidence we have for the Calculators tradition is not complete or continuous. There is a large gap in Oxford evidence between, say, 1350 and 1390. A good part of this gap may be explained not by a lack of activity, but by the burning of Wyclif's papers after his death.[7] There are other known instances where documents at

Late Medieval University Teaching,' *SCIAMVS* 17 (2016), 55–120; D.A. Di Liscia, 'Biagio Pelacani da Parma's Geometrisation of Latitudes and the Problems of the Mean Degree Theorem', in J. Biard and A. Robert (eds.), *La philosophie de Blaise de Parme. Physique, psychologie, éthique*, Florence 2019 (Micrologus Library, 96), 129–152; and D.A. Di Liscia, *Eine Wiener* Expositio *zum* Tractatus de latitudinibus formarum. *Edition mit Einführung und Kommentar*, Vienna (forthcoming). For the graphical latitude of forms within the realm of metaphysics, see Di Liscia's contribution in this volume.

7 On Wyclif as a Calculator, see Mark Thakkar's chapter in this volume.

Oxford were removed or destroyed.[8] Still, if there was a European-wide calculatory tradition, did it have different leading centers at different times?

Not only is the spread of the Calculators movement sometimes obscured because records of activity were lost, but the impact of the movement is sometimes hidden because it is merged with scholastic Aristotelianism. Sometimes, as in Dumbleton's *Summa logicae et philosophiae naturalis*, passages of material like that of the Calculators are interspersed with more standard Aristotelian natural philosophy, such as questions on Aristotle's *Physics* or *De generatione et corruptione*. Therefore, just as Anneliese Maier understood Oresme's configurations of forms as a variant of *calculationes*, so *calculationes* merge with Aristotelian natural philosophy.

The following chapter has two main parts. In the first part, I report on Leibniz's extended efforts to recover the *Liber calculationum* and the tradition to which it gave rise. By Leibniz's time the Calculators tradition had fallen into obscurity or sunken out of view within wider Aristotelianism, but Leibniz thought that 'Suisset [Swineshead], commonly known as the "Calculator"' nevertheless should be remembered and honored for being the first to advocate a mathematical approach to natural science. I have collected a great number of passages in which Leibniz wrote about Suisset/Swineshead, but for reasons of space I include only a few of them. In any case, Leibniz often repeated himself in correspondence to various recipients. In the second main part, then, I investigate some ways in which Leibniz's own work carried forward or extended the work of the fourteenth-century *Calculatores*, even when Leibniz did not label his own work 'calculatory'.[9]

2 Historiographical Interlude

Before I turn to the two main parts of my chapter, it will be worth reviewing further the historiographical context within which the label '*Calculatores*' came into use by twentieth-century historians of science. Elsewhere in this volume, Mark Thakkar credits Anneliese Maier with introducing the label, but then he specifies that it was actually Ernest Moody who first used 'Oxford calculators'

8 Many scholastic manuscripts at Oxford were destroyed at the time of Edward VI. See F.M. Powicke, *The Medieval Books of Merton College*, Oxford 1931, 26.

9 Leibniz was not only passively influenced by the Calculators, but rather may have been the agent of his own reception of calculatory positions. Perhaps it is significant that he did refer to his own work as 'calculus'.

in 1951.[10] As I have already indicated, Anneliese Maier, at least in the second edition of *An der Grenze*, finds calculatory work both in Oxford and in Paris.[11] Whether or not the commonalities between Oxford and Paris in this regard were in Daniel Di Liscia's mind when he chose in the title of the conference to refer only to '*Calculatores*' and not to 'Oxford *Calculatores*', I assume he had a motive in choosing to begin with 'Quantifying Aristotle.' In the classic narrative of the Scientific Revolution of the seventeenth century stemming from Alexandre Koyré, quantifying goes together with an affinity to Platonism rather than Aristotelianism. Did the *Calculatores* quantify in a way that we can considerer Aristotelian rather than Platonic? I think that is the case.

In the work done in the second half of the twentieth century on the relation of medieval physics to early modern mechanics, an ongoing issue was whether there was continuity or discontinuity between the two endeavours. The individuals whose work could be connected or not were very frequently Galileo, on the modern side, and the *Calculatores* (including the Parisians) on the medieval side. Marshall Clagett begins the introduction to his *Science of Mechanics in the Middle Ages* (1959) by saying:

> The key role of physics in general and mechanics in particular in the development of modern science has often been recognized, for it was in the mechanical area of early modern physics that the first thoroughgoing application was made of the mathematical and experimental techniques so crucial to the growth of modern science ... It is an obvious fact to the historian of science that the physical concepts of a Galileo or a Descartes, or even a Newton, radical as they may seem, were conditioned in many ways by the ancient and medieval learning that survived into the early modern period. Thus anyone who is honestly interested in the enormously complex *historical* process of the formation of modern science must examine in detail the germinal concepts of the preceding periods. Such an examination will reveal the elements of continuity (and thereby also of novelty) in the new science. This examination will give some insight into how proto scientific theory was criticized and emended until it was no longer a cogent whole. It will also show how the very points of criticism of the older system became points of departure for the new. It will show,

10 Moody, 'Laws of Motion.' Moody does not capitalize 'calculators', so he likely was thinking of calculators not particularly identified as part of a group, but in distinction from Parisians.

11 It is now possible to see the first edition, mostly destroyed in World War II, on the internet with aids allowing searches.

> in short, how medieval mechanics—largely Aristotelian with some traces of Archimedean character—was continually modified to the point where it was seriously undermined, thus requiring a new mechanical system—and it was the Galilean-Newtonian system of the seventeenth century that fulfilled that requirement.[12]

Clagett dedicated *The Science of Mechanics in the Middle Ages* 'To Alexandre Koyré, Anneliese Maier, and Ernest Moody *Historians of Early Mechanics*.' It was Koyré who, in the mid-twentieth century, made popular the concept of 'The Scientific Revolution of the Seventeenth Century,' and it was Anneliese Maier who made the most thoroughgoing study of the predecessors of Galileo in the fourteenth century (*Die Vorläufer Galileis im 14. Jahrhundert*) within her multivolume *Studies on the Natural Philosophy of the Late Scholastics* (*Studien zur Naturphilosophie der Spätscholastik*). Both Koyré and Maier questioned Pierre Duhem's thesis that the origin of modern science could be traced to the Condemnations of 1277 by Étienne Tempier, the Bishop of Paris.

By the 1960s, the groundwork was set for further scholarly study of primary sources to determine with better evidence what actually happened when the Scholastic Aristotelianism of the Middle Ages gave way to early modern science. In the United States, under the general editorship of Marshall Clagett, fifteen volumes were published under the heading 'University of Wisconsin Publications in Medieval Science'. Most of the volumes contained primary source texts, English translations, and commentaries. Particularly relevant to 'Quantifying Aristotle: The Impact, Spread, and Decline of the *Calculatores* Tradition' were the volumes on Bradwardine's *Tractatus de proportionibus*, William Heytesbury's *Regule solvendi sophismata*, Clagett's own volume on *The Science of Mechanics in the Middle Ages*, Edward Grant's volume on Nicole Oresme's *Tractatus de proportionibus proportionum*, Clagett's volume on Oresme's *De configurationibus qualitatum et motuum*, and I.E. Drabkin and Stillman Drake's volume on Galileo's *On Motion and On Mechanics*. Not in the medieval series, but obviously related to it was Stillman Drake's translation of Galileo's *Two New Sciences Including Centers of Gravity and Force of Percussion*.

Did this series of volumes published by the University of Wisconsin Press help to reveal the historical steps by which the late medieval quantification of Aristotelianism was followed by early modern mechanics? Or how was the work described in these volumes related to 'The Scientific Revolution of

---

12 M. Clagett, *The Science of Mechanics in the Middle Ages*, Madison, WI, 1959 and 1961 (University of Wisconsin Publications in Medieval Science, 4), xix.

the Seventeenth Century'? On the one hand, it was probably not the case that, as Clagett wrote in the introduction to his *Science of Mechanics in the Middle Ages*, 'protoscientific theory was criticized and emended until it was no longer a cogent whole' or that 'medieval mechanics—largely Aristotelian with some traces of Archimedean character—was continually modified to the point where it was seriously undermined, thus requiring a new mechanical system—and it was the Galilean-Newtonian system of the seventeenth century that fulfilled that requirement.'[13] First of all, for the most part, what the *Calculatores* were doing was not mechanics, and, secondly, I think that what they were doing was not modified until it was no longer a cogent whole. Rather, one might suggest that some *Calculatores* were continuing, albeit with gaps, to work on their projects which included theological as well as natural issues,[14] while later generations chose not to continue along those lines but rather took up different problems.

If the question is asked whether there was continuity or discontinuity in the development of physics from medieval to early modern times, the balance of evidence so far uncovered seems to be on the side of discontinuity. John Murdoch, a Ph.D. student of Clagett, who as part of his 1957 Wisconsin Ph.D. dissertation edited Thomas Bradwardine's *De continuo*, concluded that there was not a continuous development between the fourteenth-century *Calculatores* and Galileo. As William Newman wrote in an *éloge* of Murdoch:

> Despite his investment in the history of medieval science, John was perfectly forthright in his rejection of the claims that Pierre Duhem and his followers made for the significance of medieval natural philosophy as a forerunner of the mechanics of Galileo and hence for the emergence of modern science more generally. What John wanted was to carve out a place for the history of science in the Middle Ages seen on its own terms, and this resulted in his emphasis on the discontinuity between medieval and early modern science. As he put it in one of many passages where he made his views on the subject crystal clear, 'seen properly and completely, what one was about in the fourteenth century not only did not lead to Galileo and the like, but was not even pointing in that direction.'[15]

13 Clagett, *The Science of Mechanics*, xix.

14 See M. Thakkar, 'Mathematics in Fourteenth-Century Theology,' in E. Robson and J. Stedall (eds.), *The Oxford Handbook of the History of Mathematics*. Oxford 2009, 619–638.

15 W.R. Newman, 'Eloge: John Murdoch, 10 May 1927–16 September 2010,' *Isis* 102/103 (2011), 529–532, quoting J.E. Murdoch, 'Pierre Duhem and the History of Late Medieval Science and Philosophy in the Latin West,' in R. Imbach and A. Maierù (eds.), *Gli studi di filosofia*

Between the 1950s and the 1990s, the overall conception of the history of science between the Middle Ages and the seventeenth century evolved, spurred in part by the reception of Thomas Kuhn's *Structure of Scientific Revolutions* (1962). By 1991, it was no longer taken for granted that there is one true science and that it began to be discovered in the seventeenth century. When historians first began to study the *Calculatores* so-named, it was taken for granted that the application of mathematics was an essential part of modern physics and, if the *Calculatores* were of interest, it was because, like Galileo, they applied mathematics to physics. From the volumes in the Wisconsin Publications in Medieval Science, it is clear that Clagett oriented the series towards evidence of the influence (or lack of evidence) of the medieval Latin authors on Galileo, even to the extent of publishing Galileo's *On Motion* and *On Mechanics* in the medieval series.[16]

Later, in the *Research Report 2002–2003* of the Max Planck Institute for the History of Science in Berlin there is an essay by Jürgen Renn in which he says:

> For the traditional history of science, scientific progress was ultimately a kind of relay race of individual great minds, who passed the baton of their ingenious ideas across the abyss of time. Historians writing the history of science understood as a chronicle of success thus ask the classical who, what, when, and where—questions that are better suited to investigating professional sports than science ... Indeed, how do we evaluate the achievement of Aristotle, the first natural philosopher to lay down rules for natural motions of heavy and light bodies, when these rules are based on a fundamental differentiation between kinds of bodies that is no longer held today? ... Almost inevitably, the traditional history of science must suppress the incompatibility of past conceptions of the world with those of today.[17]

---

medievale fra Otto e Novecento. Contributo a un bilancio storiografico, Rome 1991 (Storia e letteratura, 179), 253–302, at 279.

16 In *Mechanics in Sixteenth-Century Italy*, Drake explains the inclusion of Galileo's *On Motion* and *On Mechanics* in the medieval series: 'Its purpose was to supply the student of the history of medieval science with a convenient means of judging its role in the transition to the physical science of the modern era.' See S. Drake and I.E. Drabkin (eds.), *Mechanics in Sixteenth-Century Italy: Selections from Tartaglia, Benedetti, Guido Ubaldo, and Galileo*, Madison, WI, 1969 (University of Wisconsin Publications in Medieval Science, 13), ix.

17 J. Renn, 'Towards an Historical Epistemology. Ten Years Max Planck Institute for the History of Science, 1994–2004,' *Research Report 2002–2003* of the Max Planck Institute for the

Renn's implication in part is that the historian should look at a wider range of evidence, so that the history of modern science does not appear to be the history of one great scientific hero passing the baton to the next. I propose, then, to study the work of Leibniz, who had a generous and encompassing view of what past and concurrent philosophical or scientific work might be worthy of study and recognition. Although Leibniz rarely appears in 'relay runner' histories of the Scientific Revolution (unless because of his interactions with Isaac Newton and Newton's supporters over absolute space or the calculus), he certainly deserves attention and recognition for his work in physics and mathematics. His point of view is significant. I am not suggesting that Leibniz can be our sole gatekeeper, as Galileo has sometimes been in the past, but rather that the advance of science does not depend upon a single route of transmission. Progress occurs through many concurrent and divergent routes in the interaction of which chance plays a significant role. Here the point is not to assert that one or more Calculator contributed to the advance of science one or more step forward which then fed into the new science of the Scientific Revolution. Instead, I consider Leibniz as a historical witness and look at his writing for clues as to what the *Calculatores* were about and what reception their work had. One possible connection of the *Calculatores* and Leibniz is that the work of the *Calculatores* may have helped to set the stage for Leibniz's invention of the infinitesimal calculus as a new branch of mathematics. This goes together with the observation that, like Leibniz, most of the *Calculatores* took a nominalistic approach to mathematics, one that denies the real existence of mathematical infinitesimals such as points, lines, and surfaces. More comprehensively, the *Calculatores* treat all mathematical entities as imaginary or things of reason (*entia rationis*). That Leibniz shared with the *Calculatores* an affinity for nominalistic mathematics means that he could as a mathematician propose new mathematical entities.

This chapter expands upon an incomplete project I started more than twenty years ago under the rubric 'Leibniz Looking Back', the premise of which was that the history of late medieval and early modern science need not take Galileo's or Newton's physics as the end point toward which medieval physics, if it was valuable, should have been heading. What difference would it make if other possible end points were considered, such as eighteenth-century physics, which largely utilized the Leibnizian calculus? The *Calculatores* began applying mathematics to physical problems in the early fourteenth century and their

---

History of Science in Berlin, URL = https://www.mpiwg-berlin.mpg.de/sites/default/files/2017-09/rr_02_03_complete.pdf, 31–32.

approach seems to have had followers as late as the beginning of the sixteenth century at least. According to Leibniz, Suisset, commonly called 'Calculator' was the leader (*princeps*) of an effort to apply mathematics to philosophy. What more of interest could we find if we look at the *Calculatores* from the perspective of Leibniz? One narrative is to say that Leibniz re-energized the Calculators' efforts to apply mathematics to physics, thereby extending the calculatorial tradition. Another possibility is to suggest that some of the mathematical physics done in the seventeenth century shows the influence of the Calculators' nominalistic-Aristotelian approach to mathematization, rather than the approach of self-professed Platonists.

3 Leibniz and the Application of Mathematics to Aristotelian Natural Philosophy

Unlike many of the reformers or *novatores* of seventeenth-century science, Gottfried Wilhelm Leibniz adopted a conciliatory attitude toward Aristotelian science. In a letter of April 1669 to his teacher Jakob Thomasius (also an eclectic Aristotelian), Leibniz argued that there was less difference between Aristotelian and reformed science than usually supposed—nearly all the doctrines of the Aristotelians could be explained by mind, space, matter, and motion. According to Leibniz, the reformers could be divided into three groups: the stolid such as Paracelsus, Van Helmont, and others, who rejected Aristotle straight out; the audacious, like Descartes, who had little care for older thinkers and treated them with contempt, rendering suspect even their better meditations; and third the true way, which treated Aristotle as a great man and recognized the truth of what he said in most respects.[18]

18 Leibniz to Jakob Thomasius, April 1669; A II, vol. 1, N. 11, 21–22: 'Aristotelicam philosophiam reformatae conciliasse mihi satis videor, verum reformatae ipsius philosophiae ineuitabilis euentus breviter attingendus est ... Orta luce, philosophia reformata triplex est: alia stolida, qualis Paracelsi, Helmontii, aliorumque, Aristotelem prorsus reiicientium; alia audax, quae exigua veterum cura, immo contemtu eorum palam habito, bonas etiam meditationes suas suspectas reduunt, talis Cartesii; alia vera, quibus Aristoteles vir magnus, et in plerisque verus cognoscitur. Nunc conciliata iam cum Aristotle philosophia reformata, restat, ipsius per se veritas ostendatur, prorsus quemadmodum religio Christiana, tum ex ratione et historia, tum ex scriptura sacra probari potest. Probandum autem est, nulla dari entia in mundo: praeter Mentem, Spatium, Materiam, Motum ... Nunc ostendamus, nullis rebus ad explicanda mundi phaenomena et causas eorum possibiles redendas opus esse, imo nec alias res esse posse ... Vitium enim Hypotheseos est non necessaria assumere. Iam vero explicari omnes totius Mundi res ex his solis posse, satis lec-

Despite what he wrote to Thomasius, Leibniz nevertheless thought that new reformed sciences had been introduced in the seventeenth century by the likes of Galileo, Kepler, Gilbert, and Harvey. He drafted histories of various fields of endeavor to which he may well have hoped to contribute. Perhaps Leibniz wanted to situate himself within intellectual traditions because he lacked frequent face-to-face contact with other intellectuals, such as was available within universities, or within the Royal Society of London or the Académie Royale in Paris.[19] We know that Leibniz was interested in discipline or institution building, as in his work to found the Berlin Academy of Sciences. He conducted an enormous correspondence, keeping in touch with correspondents throughout Europe. In describing the birth of Leibniz's differential calculus, Marc Parmentier writes of Leibniz's mathematical optimism and of his new conception of the history of mathematics in which the development of mathematical disciplines would be a collaborative project.[20]

Chronologically, the present chapter belongs to the third period mentioned in the title of the 2018 Munich conference, that labeled 'decline.' Yes, the calculatory tradition was not often apparent by the time of Leibniz. But was the current state and future of physical science any different from what it might have been if the *Calculatores* had never existed? In praising Swineshead for introducing mathematics into natural philosophy and in proposing to reprint his book, the *Liber calculationum*, Leibniz asserted that Suisset, commonly called 'Calculator' could be considered the *princeps* (first or leader) in Europe in taking a mathematical approach to philosophy. Given that Swineshead compiled his book about 1350, whereas Bradwardine completed his *Tractatus de proportionibus velocitatum in motibus* in 1328, we might consider Swineshead, not the first, but the last of the initial group of Oxford *Calculatores*. In order to give *princeps* a fitting translation, we might suppose that Leibniz meant that Swineshead was the leader, not necessarily the first chronologically to attempt to introduce mathematics into philosophy. In his long and frustrated search for a copy of the *Calculationes* to be used as a model for his proposed new edition or printing, Leibniz provides evidence that, despite having been printed several times in the fifteenth and sixteenth centuries, by Leibniz's time the book had

tio recentiorum philosophorum docet, et ex his patet, quae paulo ante posui, Quum possibilitatem conciliationis Aristotelicae ostenderem.'

19 Cf. A. Ramati, 'Harmony at a Distance: Leibniz's Scientific Academies,' *Isis* 87 (1996), 430–452.

20 G.W. Leibniz, *La naissance du calcul différentiel: 26 articles des 'Acta Eruditorum'*, ed. and transl. M. Parmentier, Paris 1989, 13–51, at 42.

fallen into obscurity.[21] But although this chapter belongs to the last period of the narrative of the Munich conference, that covering the decline of the *Calculatores*, we could suppose that while researching the history of the Calculator and his followers, Leibniz in effect adopted the *Calculatores* as his own ancestors, so that he, in some way, continued their programme.

Leibniz somehow learned about Swineshead relatively early in life. In correspondence, he mentioned Julius Caesar Scaliger's *Exotericarum exercitationum liber quintus decimus: De subtilitate, ad Hieronymum Cardanum* for its praise of Swineshead. Repeatedly beginning in the 1670s, Leibniz wrote historical sketches of various fields of science or philosophy in which he mentions Swineshead (called 'Johannes Suisset' or 'Calculator' by Leibniz). He argued that the *Liber calculationum*, together with other important books published since printing began (not all scientific by any means), should be collected and reprinted. Meanwhile, Leibniz tried without success to see for himself the *Liber calculationum*. Finally, while on a trip through Italy in late 1689, he found out that a copy of the book existed in the monastery of San Marco in Florence, but apparently did not have time to study it. This likely was the Padua edition of 1477.[22] Leibniz wrote to Antonio Alberti in January 1690 reporting his success.[23] Meanwhile, he asked Rudolf Christian von Bodenhausen, living in Florence and currently helping him in the revisions of his *Phoranomus*, to go to San Marco to read Swineshead more carefully.[24] Bodenhausen went as asked, but found

21 Early editions include: Padua ca. 1477; Pavia 1498; Venice 1520; Salamanca 1520 and 1524.

22 A. Robinet, *G.W. Leibniz: Iter Italicum (Mars 1689–Mars 1690). La dynamique de la République des Lettres. Nombreux textes inédits*, Florence 1988 (Accademia Toscana di Scienze e Lettere 'La Colombaria', 90), 286, n. 72: 'Ces fragments comportent une triple précision: "livre imprimé", "vers la fin du 15ème siècle", "venant de la bibliothèque de San Marco". L'exemplaire de Suiseth, *Calculationum liber*, qui répond à ces trois conditions est l'incunable de la BNF, n. 2166 du *Catalogo della edizioni del sec. XV nella magliabechiana di Firenze* de G. Molini: exemplaire sans date, mais vers 1480, sans éditeur, mais aux soins de Johannes de Cipro, sans lieu, mais avec ce final: "Deo gratias, Padue". L'exemplaire 2167, ainsi que l'exemplaire de la Marucelliana, Rari, A 321 (5), relèvent de l'édition de 1498, bien datée, des soins de Jean Tolletini, et de l'éditeur de Pavie Francesco Girardengo; le premier est de provenance imprécise, le second vient de la bibliothèque de al SS. Annunziata.'

23 According to Robinet, *G.W. Leibniz*, 64, 'Ant. Alberti' was a pseudonym, taking initials from A. Arnauld, who was an 'agent d'Arnauld', probably in reality Amable de Tourreil.

24 According to the notes I took in Hannover in 1996, I found that Bodenhausen wrote to Leibniz January 28, 1690, that the book consisted of a war of words (*logomachia scholastica*). In March 1690, Leibniz encouraged Bodenhausen to look for Alberti in Rome. Alberti was interested in Leibniz's *Dynamica* and asked when it would be published. Leibniz to Bodenhausen, March 18, 1690, 18–19, responds: '... pone ut duas partes circa dynamica simplicia seu a rebus abstractis; et dynamica concreta ...' Leibniz continues with a table of contents of the proposed work: 'Cap. 3 de gradibus velocitatis in motu varie difformi. Cap. 4

the Gothic print slow going and reported nothing of the content back to Leibniz.[25] Leibniz then urged Bodenhausen to try to obtain the book itself from San Marco, either by purchasing it outright, or by trading some other book or books possibly of greater interest to the monks for it.[26] By the time of Bodenhausen's death in 1698, this effort had apparently come to naught.

Never daunted, Leibniz tried again in 1701, asking Pinsson des Rolles, a Parliamentary Advocate in Paris, if he could obtain a copy there.[27] Pinsson wrote back, including a copy of the table of contents of the Venice 1520 edition that he had located in the Bibliotheque du Roy.[28] Ultimately, the time-consuming expedient was chosen of making a handwritten copy of the printed Venice

… cum multi problemata circa difformium. De quantitate motus seu impetus. De quantitate translationis in effectu motus formali …' In general, in letters 1–54 of the Leibniz-Bodenhausen correspondence there appears to be no mention of Swineshead—that is up through June 12/22, 1691. But then Leibniz to Bodenhausen, 10/20 Aug., 1691. A III, 5–6, 3, 45 says: 'solte M h. H. wohl ein mahl ein paar tage anwenden Können die fundamenta des Joh. Suisset dicti Calculatoris zu examiniren? … Der Suisset ist in Bibliotheca San Marci.' Leibniz's wide range of interest is illustrated when he goes on to talk about his solution of Galileo's catenary problem, 'medicinam spirituum, medicinasque corporales, circa causam accelerationis gravium, novi impetus imprimere,' motion of a ball on a horizontal plane, 'major libertas eligendi …' The way Swineshead is mentioned again on 155ʳ makes it seem that the Calculator is a recurring topic.

25 Bodenhausen to Leibniz, Dec. 1, 1691 A III, 5–6, #8, 55: 'Swissetum habe ich auch eine Stunde in der Bibliothek zu San Marco gelesen, ist aber wegen die alten drucks v. abbreviaturen so zwar leicht, sehr beschwerlich zu lesen, v. hat mehr zeit vonnöthen. Ich wil sehen, ob ich solchen vor geld oder vertauschung anderer bücher bekommen kan.'

26 A III, 5–6, Leibniz to Bodenhausen, 15/25 February, 1692, 'Wenn Suisset, durch Kauff oder Tausch von den Patribus S. Marci zu haben, wäre es guth; Den ob er schon gedruckt, so ist er noch rar.' Cf. Robinet, *G.W. Leibniz*, 286: 'Et aussi ff. 106–107: Leibniz n'a pas obtenu la copie qu'il en attendait de Florence.' This is in the margin sideways '… Buch der Suisset mich ganz au enym waret.' My thanks to Herbert Breger, in the Leibniz archive, for this oral paraphrase of Leibniz's hard to read script: 'don't forget about the book of Swineshead. I would be happy if the monks would sell it.' Sheets are dated Wolfenbüttel, 30 May 1694.

27 Leibniz to Pinsson, August 29, 1701, saying he is looking for a copy of Swineshead: 'Je desire d'obtenir une copie des ouvrages de Suisset, pour les faire entrer dans un recueil Κειμηλίων φιλοσοφικῶν que je medite; avec le livre philosophique de Ratramne que le R.P. Dom Mabillon m'a envoyé, et autres choses semblables plus modernes' (G.W. Leibniz, A I [allg. pol. u. hist. Briefwechsel], vol. 20, eds. M.-L. Babin, G. van den Heuvel and R. Widmaier, Berlin 2006, letter N° 290 [end of September 1701, but dated in 'Bronsvic 29 Aoust 1701']). Available in https://www.gwlb.de/Leibniz/Leibnizarchiv/Veroeffentlichungen/I20B.pdf, 493. Cf. also GM 4, 1859, 95–96; and 5, 1858, 350.

28 In 1996, with the help of the library staff, I was able to locate this very copy in the Bibliothèque nationale de France with its red seal indicating the era of its entry into the library. There was no evidence in the book itself of its use as the basis for Pinsson's copying, nor were there marginal notes by any reader.

1520 edition. This handwritten copy remains to this day in the Niedersächsische Landesbibliothek, an enormous book of over a thousand pages in single columns of beautiful penmanship, with no proof that it was ever read or used by Leibniz.[29]

Given that Leibniz first saw Swineshead's *Liber calculationum* in Florence in late 1689 when he had just composed his *Phoranomus seu De potentia et legibus naturae* (never published in his lifetime because never finished) and just before he began his *Dynamica de potentia*, it is possible that in 1689 he might have picked up some relevant concepts or vocabulary from Swineshead, who was concerned with similar issues. Michel Fichant, Alberto Guillermo Ranea, and

29 Hannover, Niedersächsische Landesbibliothek, Ms. IV.9. Miscellanea 615. 'Copie de Suiseth tirée sur l'original de la bibliotheque du Roy contenant 73 Rolles et demy d'Impression gothique chaque page a deux Colonnes, au dos duquel est escrit R. Suiseth Calculator vulgó dictus et cotté R. 219. Stamped Bibliotheca Regia Hannoverana'; 1119 pages, folio. On title page, hand note: 'Dicitur de Suinisete, alia, sed corrupté (nam varii scribitur et saepius malé Calculator vulgo, quia quasdam [?] calculationes astronomicas edidit) mentio eius habetur apud Baleum in opere de Illus. Anglico Script. pag. 456 uti sic scribit Rogerus Suuinsete. Refertur etiam inter recentiores Joannis Suissetus cognomine Calculator ... Godofredus lib. Proverb, pag. 65 valde senex fleurisse quod ante a se inventa non capevet de se [?]. Scripsit ante Godefredum anno 1555. Cujus et doctissimi et subtillissimi philosophi meminit Julius Caesar Scaliger. Exercit in Cardanum 324 et 340. Calculator Subtilissimi Richardi Suiseth Anglici Calculationes noviter emendatae et revisae. Questio in super de reactione juxta Aristotelis sententiam et commentatoris. Fertur de eo [Victor Trincavello] ... opus subtilissimum de artibus liberalibus de generatione sonorum, de calore solis, etc. vide tractatum editum Venetiis 1514. 1. Victor Trincavellus venetus physicus Christophoro Arretino inter professores medicae facultatis nostra aetate facile Principi.' S. Continues on to page 3 and table of contents, dedication. The *Liber calculationum* ends on page 1035, followed on p. 1037 by the *Questio de reactione. Victoris Trincavelli veneti Questio de reactione juxta doctrinam Aristotelis et Averroes Commen. Inc.*: 'Circa tertium physicorum librum maxime in tex. com. 8. questio admodum difficulis solita est moveri: quae 1° etiam de generatione circa t. com. 53 et 54 a modernis solent perscrutari: quia talis est. Nunquid omne agens materiale dum agit in passum repatiatur ab ipso passo ita ut fugere non possit quin passum etiam aliquo modo in ipsum non agat. Res difficilis valde habens ad utrumque partem magnos fautores et fortes rationes.' At the end on p. 1119: 'Explicit questio de reactione edita ab Excell. Doctore Domino Victore Trinchavello Veneto, nec non Calculationum liber magistri Raymundi Suiseth. Noviter impressa Venetiis ere ac sollerti cura haeredum .q. Domini Octaviani Scoti civis Modoetiensis ac sociorum. Anno a dominica incarnatione Millesimo quingentesimo vigesimo decimo Kal. Aprilis. Ad laudem eius qui maior est omni laude.' On p. 1119 there is a note in French by the scribe that seems to indicate second thoughts about how he has transcribed the Latin: 'Nota que le mot de quum veut dire quando et le mot d'unum veut dire verum tamen.' The 1952 catalog of the Bibliotheque Nationale in Paris, in vol. 180, lists the Venice 1520 edition as copied. In fol. à 2 col. car. goth. 74 ff. fig. [Res. R. 144]. Everything corresponds but the shelf mark.

François Duchesneau have each suggested that it may have had an effect.[30] They quote such passages from Leibniz's *Dynamica* as 'the intensity of the same action is the quantity of velocity.'[31] Another passage is: 'The perfection or the degree of reality in things, and particularly in motion can be estimated using two factors (*suivant deux raisons*) that is by extension, which is here the magnitude of the place or space changed, and by the intensity, which is here the quickness or the velocity of the change or motion.'[32]

Leibniz remained in Florence for only a month, during which time he had many other activities on his calendar. He could have spent at most a few hours or a day or two looking at the *Liber calculationum*, and he probably spent much less. If he had made notes on the *Liber calculationum*, we might expect he would have saved them, as he saved so much else. Have historians checked thoroughly whether Leibniz did not use some of the same concepts and vocabulary before seeing the *Liber calculationum* in Florence?

Here I want to introduce some evidence that Leibniz wrote in calculatory terms even when he did not mention 'Calculator'. Under the heading 'Quelques semaines à Florence (22 nov.–22 déc. 1689),' André Robinet discusses Leibniz's proposed revisions of his first, 1684 article on the calculus. According to Robinet, Leibniz took an exemplar of this 1684 article introducing the calculus and changed the title to *Specimen pro analysi infinitorum seu singulare calculi genus pro aestimatione incrementorum ac decrementorum continuorum in rebus atque inde nascente inventione Tangentium, Maximorumque ac mi-*

30 F. Duchesneau, *La dynamique de Leibniz*, Paris 1994 (Mathesis); F. Duchesneau, 'Leibniz's Theoretical Shift in the *Phoranomus* and *Dynamica de Potentia*,' *Perspectives on Science* 6 (1998), 77–109; M. Fichant, 'Neue Einblicke in Leibniz's Reform seiner Dynamik (1678),' *Studia Leibnitiana* 22 (1990), 38–68; A.G. Ranea, 'The *a priori* Method and the *actio* Concept Revised: Dynamics and Metaphysics in an Unpublished Controversy between Leibniz and Denis Papin,' *Studia Leibnitiana* 21 (1989), 42–68; A.G. Ranea, 'Calculationes, Galileo, and the Misfortune of Leibniz's A Priori Demonstration,' in I. Marchlewitz (ed.), *Leibniz. Tradition und Aktualität. v. Internationaler Leibniz-Kongreß unter der Schirmherrschaft des Niedersächsischen Ministerpräsidenten Dr. Ernst Albrecht.* Hannover, 14. bis 19. November 1988, Vorträge II. Teil, Hannover 1988–1989, 785–793.

31 From Leibniz's *Dynamica*: 'Diffusio actionis in motu vel actionis extensio est quantitas effectus formalis in motu. Intensio ejusdem actionis est quantitas velocitatis, qua factus est effectus seu qua materia per longitudinem translata est ... Actiones motuum formales sunt in ratione composita diffusionum [= extensionum] et intensionum' (from Duchesneau, *la dynamique de Leibniz*, 186).

32 Leibniz à Papin: 'La perfection ou le degré de réalité dans les choses, et particulièrement dans le mouvement, se peut estimer suivant deux raisons, savoir par l'extension, qui est ici la grandeur du lieu ou l'espace changé, et par l'intension qui est ici la promptitude ou la vitesse du changement ou mouvement' (from Ranea, 'The *a priori* Method,' 56, n. 68).

nimorum, viaeque facillimae qua natura effectus suos praestare solet. Then he added a paragraph:

> The incremental or differential calculus applies wherever there is a change of place. All change in nature is continuous or by degrees and not jumps (*per gradus et non per saltum*) and the transfer occurs by inassignable or indefinitely small quantities the infinite multitude of which constitutes common quantities. All of these are best represented by figures. No kind of increase or decrease can be imagined (*fingi*) which is not most commodiously represented by figures ordered to vary in the same proportions.[33]

Then he returned to the original text. Thus, in Leibniz's proposed revision of his calculus paper concepts and vocabulary of the *Calculatores* have been integrated into Leibniz's shared practice and vocabulary without mention of their connection to Swineshead and his followers. On the other hand, Leibniz elsewhere repeatedly uses intension and remission of forms to characterize what the Calculators did. May we not then consider that elements of the *Calculatores* tradition have a continued life in the work of Leibniz on the calculus?

After his trip to Florence, are there later occasions on which Leibniz might have gained increased knowledge of the work of the *Calculatores*? Since the copy made in Paris of the Venice 1520 edition of the *Liber calculationum* did not reach Leibniz until the early 1700s, might the early 1700s have been a time when its precise content had a significant impact on Leibniz's mathematical physics? These are really moot points, I think, because Leibniz said *before 1689* that he had read work of some of Swineshead's followers (*sectatores*), if not Swineshead himself, and he reported being positively impressed by what he had read. Although Leibniz praises Swineshead for being the *princeps* of an effort to apply mathematics in physics, if he had known more about the

33 'Calculus incrementalis vel differentialis locum habet ubicunque mutationi locus est. Omnis autem mutatio in natura continua est, ac per gradus fit non per saltum, adeoque transitus fiunt per quantitates inassignabiles seu indefinite parvas quarum infinita multitudine opus est ad quantitates communes constituendas. Quamvis autem variatio per minima in omnibus rebus quantitatis capacibus intelligi possit, et in remissione atque intensione formarum, gradibusque qualitatum conspicua sit maxime tamen apparet in gradibus velocitatum; elementisque temporum et spatiorum quemadmodum ex phorometricis nostris intelligi potest; sed omnium optime repraesentatur per figuras, nullum enim crescendi descrescendique genus fingi potest, quod non commodissime figuris exhibeatur, ordinatas habentibus, quae eadem proportione varientur. Sit ergo Axis AX ... etc.' (from Robinet, *G.W. Leibniz*, 265–266).

Calculatores, he might have paid more attention to chronology. For someone interested in the insertion of mathematics into physics, probably Thomas Bradwardine and Richard Swineshead would be the two leading Oxford figures to consider, but Leibniz shows no knowledge or interest concerning development among the Oxford *Calculatores* between 1320 and 1350. This is probably because Leibniz's more detailed knowledge of what was involved in the *Calculatores* movement came from works by Swineshead's *sectatores* that made no distinction of chronology. But before we try to track down what Leibniz may have seen of the work of the Calculator's followers, we should first look at the work of Julius Caesar Scaliger, because Leibniz cites Scaliger as his source, despite the fact that, as I argue, there is little of physical consequence in what Scaliger says about the Calculator.

4 What Leibniz Could Have Learned about Swineshead from Julius Caesar Scaliger

Julius Caesar Scaliger's *Exotericarum exercitationum liber quintus decimus: De subtilitate, ad Hieronymum Cardanum* was obviously one source of Leibniz's interest in the Calculator.[34] Leibniz mentions that Scaliger had praised Suisset, but when one looks at the places where Scaliger mentions Suisset, it becomes apparent that this could not have been the only way that Leibniz knew of Swineshead, because Scaliger says very little that is concrete.

Scaliger's first mention of Swineshead is in section 324 of the *Exotericarum exercitationum*, titled 'Sapientum census' or census of the wise. Scaliger's goal appears to have been to make a short list of the wise, starting with Aristotle. His list begins with Aristotle, John Duns Scotus, Richard Swineshead (whom he calls *Ioannes Suisset calculator*), Archimedes, and Euclid.[35] Then in section 340,

34 According to C.H. Lüthy, 'An Aristotelian Watchdog as Avant-Garde Physicist: Julius Caesar Scaliger,' *The Monist* 84 (2001), 542–561, Scaliger's *Exercitationes* was a truly extraordinary success, going through eleven editions between 1557 and 1665. At the time, according to Lüthy, there was probably no Aristotelian with a higher reputation than Scaliger, even though there were many un-Aristotelian aspects to his philosophy.

35 J.C. Scaliger, *Exotericarum exercitationum liber quintus decimus: De subtilitate, ad Hieronymum Cardanum*, Frankfurt 1607, 1028: 'Qui Aristoteli fabrum antetulisti, non munus ipsis artibus erudito: & Johannis Duns Scotus qui fuit lima veritatis: & Ioanni Suisset calculatori, qui pene modum excessit ingenii humani. Qui Ockan praeteriisti: cuius ingenium ingenia omnia vetera subvertit; nova ad invictas insanias, ob incomprehensibiles subtilitates fabricavit, atque conformavit. Qui post Archimedes statueris Euclidem, quasi lumen qost [! post] lanternam: videris turbine raptus, atque tempestate ingenii tui: quod tibi calcar addidit, non tu ei frenum impostuisti.'

Scaliger's topic is 'De auditu ad sapientiam per autores scientarum.' He begins by considering the order in which studies are taken up, whether the easiest works should be studied first, or the reverse. Then, coming to the Calculator, he tells a story that, when he was old, Suisset's tears flowed as he looked at what he had written when he was young.[36]

These two mentions of Suisset or the Calculator reveal very little about the nature and status of the *Liber calculationum* as it was originally composed at Oxford in the mid-fourteenth century. It was a technically difficult book, but it was also intended for the use of undergraduates in the Arts Faculty, whose minds were to be trained through disputation. The image of Swineshead as an old man crying when he looked at what he had written in his prime could be explained as the result of Swineshead's sorrow at his failing mental powers, but might also have been tears of gratitude in the realization that he had been so brilliant in earlier life. Old age need not bring senility, Scaliger says: the law of nature does not take away life unless it gives a better life. Scaliger also says that he, himself, would have owed Swineshead more if his own mental weakness (*imbecillitatem*) had allowed.

Scaliger may have been taught the calculatory approach when he was a student at Padua—where he claimed as teachers Agostino Nifo, Pietro Pomponazzi, and Niccolò Leoniceno, among others.[37] The fact that memory of the Calculator reached Scaliger and subsequently Leibniz has been thought to be due primarily to continuing interest in the work in Italy. Around the mid-fifteenth century, Nicoletto Vernia had gone to Pavia to study the *Calculationes*, of which at some point he obtained a copy (now Biblioteca Vittorio Emanuele, Ms. 250).[38] Most of the extant manuscripts are in Italy, where most of the early editions were published. In Italy, fifteenth-century scholars were mainly inter-

36 Scaliger, *Exotericarum exercitationum*, 1068: 'Vel cum Dialectices ignaros illico ad Calculatorem deducis rudes adhuc animos initiatorum. Quem virum tam acute scripsisse videmus: ut a paucis plene ac pro meritis intelligatur. Tam docte sibi ille visus est: ut cum senex legeret, quae iuvenis in literas retulisset: ubertim flevisse dicitur. Sive, quod equidem puto, quia quasi recenter fruerentur veteri diligentia sive quod paulo iniquius alii sunt interpretati, quia quae olim scripsisset in aetate florenti in efforto minus assequi posset intelligendo. Dignus profecto, quem neque senium seuem [! senum] faceret, neque naturae lex vita privaret: nisi meliorem vitam apparasset. Haec libentius scribo: quia illi viro maximo multa debeo: plura debiturus, si per ingenii mei imbecillitatem licuisset. Non enim leviter in eius campo nos exercuimus.'

37 However, according to Lüthy, 'An Aristotelian Watchdog,' nn. 34–35, examination of documents has not borne out Scaliger's claims concerning his Paduan education. Scaliger received his M.A. at Padua in 1519.

38 See J.E. Murdoch and E.D. Sylla, 'Swineshead (Swyneshed, Suicet, etc.) Richard,' in C.C. Gillispie (ed.), *Dictionary of Scientific Biography*, 13, New York 1976, 184–213, at 209.

ested in Treatise I of the *Liber calculationum*, on the intension and remission of forms, and in Treatise VII, on reaction, the two treatises that Pietro Pomponazzi commented on. Perhaps it was because of the importance of medicine in Italian universities that interest went in these directions.

Other aspects of the *Liber calculationum* also drew the Italians' attention. They tried to clarify for themselves the conception of ratios and of operations on ratios (especially 'compounding' ratios) used throughout the *Liber calculationum*, and especially when it came to treating dynamics in accordance with Thomas Bradwardine's approach to the ratios of velocities in motion in relation to ratios of forces and resistances. In his *Probatio cuiusdam sententie calculatoris de motu locali*, Giovanni Marliani paid attention to measures of non-uniform motions (as in the so-called 'mean speed theorem').[39] It must be noted, however, that some of the Italian commentators, for instance Pietro Pomponazzi himself, cared little about the mathematical aspects of Swineshead's work.[40]

But before we too quickly assume that Scaliger's knowledge of Suisset came from other Italian sources such as Pomponazzi, it should not be overlooked that Scaliger's mention of Swineshead is in a work whose title ends with the phrase 'On subtlety, to [*ad*] Hieronymus Cardanus.' In fact Scaliger certainly picked up 'Suisset Calculator' in his list of the wise from an earlier list compiled by Girolamo Cardano of the most subtle thinkers of all times. Cardano's list begins:

> In serious studies outstanding men are found in more abundance, and I have chosen twelve of them, leaving to each his own judgment. Archimedes is the first—not just on account of the works of his now published, but on account of the pieces of engineering he used repeatedly to destroy the Roman forces, as Plutarch testifies. In his life of Marcus Marcellus he describes marvellous inventions of Archimedes, and we mention others just as remarkable, on the evidence of Galen, not so much the first author on both subjects, as one beyond imitation. Nor the man who disdained Greeklings (Cicero usually uses this name for all this sort of person) prais-

39 See Murdoch and Sylla, 'Swineshead,' 210.

40 See, e.g., S. Caroti, 'Pomponazzi e la *reactio*: note sulla fortuna del pensiero oxoniense e parigino nella filosofia italiana del Rinascimento,' in G. Federici Vescovini (ed.), *Filosofia e scienza classica, arabo-latina medievale e l'età moderna*, Turnhout 1999 (Textes et études du Moyen Âge, 11), 255–288. Comparing Alvarus Thomas' work to that of Pietro Pomponazzi, I would judge that the Italian comprehension of calculatory work fell well below that in Spain and Portugal.

ing his talent, and searching for his tomb amid the ruins and brambles of the town of Syracuse [this site is still listed in tourist guides].

Second comes Aristotle the Stagirite, tutor of Alexander the Great of Macedonia, who did marvellous work on natural and divine things and on dialectic; he investigated the life, habits and anatomy of animals with wonderful flair. On individual fields of study he wrote what met with approval, yet over the course of so many centuries no notable mistake could be found in his writings.

On similar criteria, the third place goes to Euclid, to Scotus, and to Ioannes Suisset (whom ordinary people call the Calculator) and with equal expectation; but Euclid should get preference through seniority and usefulness. He has two outstanding strengths: the unshaken stability of the doctrines of the book of the *Elements*, and a perfection so complete that you could not rightly compare any other work to that one. It follows that the only people who can distinguish truth from falsehood in difficult problems are those who have Euclid for their friend. He was descended from Megara, and he wrote much else that still survives, of equal subtlety, but much less useful.[41]

Next comes Ioannes Scotus: that was his homeland, and his doctrine earned him an equal distinction everywhere as a subtle teacher.[42] Ioannes Suisset (already mentioned) was an inhabitant of the same island, with the sobriquet 'Calculator', all of posterity has striven to solve one single problem of his, which is contrary to experience about mutual action. People say that when he was old, he wept at not understanding his discoveries when reading them. I think this makes it certain that, as I wrote in the book *De animi immortalitate*, the Barbarians are not our inferiors in talent, since under a foggy sky and divided by a whole sphere, Britain has sent forth two men of such distinguished talent.

The sixth place is owed to Apollonius of Perga who was almost of an age with Archimedes ...[43]

So Scaliger's immediate source of information on the Calculator was almost certainly Cardano.

41 This confuses the Euclid who wrote the *Elements of Geometry* with a different Euclid of Megara.

42 This is John Duns Scotus. The proximity of this John to Suisset may explain why Suisset's first name came to be given as John.

43 G. Cardano, *De subtilitate libri XXI*, Basel 1560, 816 (my translation).

What was Cardano's source of information? Cardano wrote a large mathematical work *De proportionibus*, which is excellent but does not have obvious links to works *De proportionibus* of the *Calculatores* (e.g., by Bradwardine, Albert of Saxony, or Oresme). It is curious that Cardano links Suisset to John Duns Scotus, and notes that both were 'barbarians' from the British Isles. One source suggests that the Italians called all people north of the Alps barbarians. But the link of the *Calculatores* to Britain seems to go beyond that generalization. Of course, the humanist critics of the *Calculatores* made fun of their British names (e.g., 'Dumbleton').

Edward Grant makes a strong argument that Propositions 47 to 52 of Cardano's *Opus novum de proportionibus* (Basel 1570) are based directly or indirectly on propositions of Nicole Oresme's *De commensurabilitate vel incommensurabilitate motuum celi*.[44] Cardano himself says that the principle behind his propositions comes from a book by Campanus of Novara:

> The basis for these propositions was furnished by Campanus of Novara, the expositor of Euclid, in a certain unpublished [*non edito*] booklet, which came to me through the diligence of my father Fazio.[45]

But Grant argues:

> Although specific evidence is lacking to demonstrate convincingly Cardano's dependence on Oresme, the striking agreement between them lends reasonable credence to the conjecture that Oresme—not Campanus … was the source of Cardano's propositions.[46]

Grant suggests that in the booklet given to Cardano by his father, the author of the work may have been misidentified. I suggest that there may have been one or more intermediaries between Oresme and Cardano.

In a later footnote, Grant gives another example of the discussion of incommensurable magnitudes, one that comes from Blasius of Parma's *Questions on the Treatise of Proportions of Thomas Bradwardine*, in which Blasius argues:

44 E. Grant, *Nicole Oresme and the Kinematics of Circular Motion.* Tractatus de commensurabilitate vel incommensurabilitate motuum celi, Madison, WI, 1971 (University of Wisconsin Publications in Medieval Science, 15), 124–161.

45 Grant, *Nicole Oresme and the Kinematics*, 158, n. 186, 'Et harum propositionum principium est traditum a Campano Novariensi, Euclidis expositore, in quodam libello non edito qui diligentia patris mei Facii ad me pervenit.'

46 Grant, *Nicole Oresme and the Kinematics*, 159. Grant leaves open the possibility of an intermediary between Oresme and Cardano.

> There is some local rectilinear motion to which [another] rectilinear motion cannot be proportional. Let two distances (*spatia*) be taken that are related to each other as the diagonal and the side of a square, and I assume that two mobiles are moved over these distances (*spatia*) with motions (*motibus*) that are proportional to these distances. Then this is proved: just as a distance is to a distance in length, so is a motion to a motion in velocity (*motum in velocitate*) comparable by a rational ratio ... Where indeed you maintain that the motion is the mobile itself (*ubi vero tu tenens quod motus est ipsum mobile*), you could say that these two local motions are not mutually comparable, neither according to the common notion of proportion nor according to another. Let this be declared (*declaratur*): that a man and an ass are taken and moved through an hour traversing a distance of one foot. But a man and an ass are not mutually comparable according to any ratio; but a man and an ass made these two motions, therefore, etc.[47]

Although this quotation from Blasius of Parma (d. 1416) is taken out of context, Grant having interpolated it with the goal of providing context for Oresme's *De commensurabilitate*, I include it because, whether or not Blasius would maintain it himself, he uses the phrase 'where indeed you maintain that the motion is the mobile itself.' To identify motion with the mobile is something very much in the spirit of William of Ockham, something that one finds in John Buridan and likewise in the work of *Calculatores* attempting to quantify Aristotle.[48]

So Leibniz's praise of Swineshead as the leader in introducing mathematics into philosophy may have come directly from Swineshead's appearance on Scaliger's list, but we do not know all the likely links starting from Swineshead. There must be some other source or sources of Leibniz's characterization of

47 Grant, *Nicole Oresme and the Kinematics*, 159–160, n. 190.

48 See John Buridan, *Quaestiones super octo libros Physicorum Aristotelis (secundum ultimam lecturam)*, III, q. 7 (*Utrum motus localis sit res distincta a loco et ab eo quod localiter movetur*), eds. M. Streijger and P.J.J.M. Bakker, Guide to the Text by E.D. Sylla, Leiden 2016 (Medieval and Early Modern Philosophy and Science, 27), 75^{9-12}: 'Antiquiores non dubitaverunt de ista quaestione, sed concorditer concesserunt motum localem esse aliam rem a mobili et loco. Sed iam posteriores moderni propter rationes praedictas posuerunt quod motus non sit alia res a mobili.' In Richard Kilvington's question *Utrum continuum sit divisibile in infinitum*, the argument is made that if an extended quality or substance can be considered an extensive quantity without a separate quantitative form, then it follows that an ass and a man are of the very same species. See R. Podkoński, 'Richardus Kilvington, *Utrum continuum sit divisibile in infinitum*,' *Mediaevalia Philosophica Polonorum* 36 (2007), 123–175, 226.

what Swineshead actually did and the source or sources likely include his followers (*sectatores*). Can we find evidence of who these followers were or what printed works Leibniz probably read? What did Leibniz write about the work of Swineshead?

Aside from the efforts, mentioned above, to discern whether the concepts of Leibniz's dynamics reflected Leibniz's chance to look at the *Liber calculationum* in Florence in November and December 1689, I know of no systematic effort to look for changes to Leibniz's conception of the Calculator and his followers over time.[49] What he says in letters about the Calculator can usually be dated by the date on the letter, but what he wrote in drafts of documents is typically not exactly dated. Over recent decades those preparing the *Akademie-Ausgabe* of Leibniz's work attempted to improve dating, but many dates are still conjectural and with margins of error. As a result, I have not spent much time trying to investigate when Leibniz might have received new information or modified his understanding of the Calculator's work. We find Leibniz saying similar things about the Calculator as early as 1670 and as late as 1710.

We know that Leibniz mentions the Calculator in a letter of 1670 to Gottlieb Spitzel concerning the importance of keeping track of history saying, for instance, we should not show ingratitude towards the illustrious, whoever offers their meditations to us which are ingenious even if not true, whoever initiates a new way of speaking or starts a new sect.[50] After receiving the hand-

49 Basil Lourié in a short article says that Leibniz adopted his high view of the Calculator before he saw his work: 'Leibniz's admiration toward Swineshead is a phenomenon whose value is somewhat independent from the historical personality of Swineshead as a scholar' (B. Lourié, 'Intensio: a Story of a Term from Swineshead to Leibniz,' *Studia Humana* 1 [2012], 59–65, at 59). 'According to Leibniz's impression which was already formed as early as in 1682, Swineshead must be placed alongside with Aristotle!' (60). 'When, in December 1689, Leibniz eventually found Swineshead's incunabula in Florence, he was very glad and, of course, did not change his opinion' (60).

50 G.W. Leibniz to Gottlieb Spitzel, 7/17 April 1670: '… At nostro tempore tot praeclarae meditationes, tot egregia inventa, tot vastissimi ac prorsus stupendi labores publicè exstant, ut indigner saepenumero ingratitudini seculi nostri erga suos Ilustratores. Esset igitur optandum magnopere, ut extaret aliquis qui meritissimos de re literaria viros inde usque ab inventa Typographia recenseret ordine temporum; annotaretque illud maximè: quid quisque potissimum in Rempublicam contulerit. Ita fiet ut multi licet doctissimi sint fortasse initio praetereundi (neque enim simul omnes dari possunt), qui tamen nominare non possunt, quid potissimum iis erudito debeat … quicunque proprias nobis meditationes suas, easque si non veras, saltem ingeniosas reliquêre; qui novum aliquod dicendi genus, novam sectam orsi sunt, quibus in medicina ac mathesi, sed et artibus mechanicis utilia reperta debentur, qui historias non aliunde descriptas, sed earum rerum quibus interfuêre ipsi, earumque memorabilium scripsêre: breviter, illi, qui si non vixissent, caeteris rebus in eo qui secutus est statu manentibus, aliquid Res literaria detrimenti passa

written copy from Paris, Leibniz still treated the *Liber calculationum* primarily as a *Cimelion philosophicon* and not as a source for his own efforts to advance physics. When Zacharius Konrad von Uffenbach visited Leibniz in 1710, Leibniz still listed the *Liber calculationum* among the *cimelia* he hoped to republish.[51]

On the other hand, Leibniz thought that the state of mathematical physics in his day was different from what it would have been if the *Calculatores* had never lived. In the late seventeenth century, Leibniz began to write his own history of algebra—he thought John Wallis' history was incomplete. In a manuscript to be found in Hannover, called *De ortu, progressu et natura Algebrae, nonnullisque aliorum et propriis circa eam inventis* (on the internet listed under 1685–1686, that is perhaps immediately after Wallis first published his history), Leibniz wrote that some scholastics, especially the English, produced some subtle calculations concerning intensions and remissions of forms and forces or motions, which he was amazed to find completely neglected so that not even the most celebrated Wallis, even though he was English, remembers them in his history of algebra. Their leader, Leibniz writes, was John Suisset called the Calculator, to whom may be added Thomas Bradwardine, Nicole Oresme, and others.[52] Notably, Leibniz here mentions Bradwardine and Oresme as well as Swineshead. He also mentions Arabic authors and the transmission of algebra and commercial arithmetic from the Arabs to Europe about the time of Frederick II. In this case, we are not talking about mathematical physics in particular, but rather about algebra. That at this early date Leibniz lists Oresme as well as Brad-

esset … quia refert primum prope eorum qui Mathesin ad Physicam applicarunt fuisse Ioh. Suisset Calculatorem ideo Scholasticis appellatum, primos circuli pseudo-quadratores fuisse Bovillum et orontium; et ut ad Theologica transeam, primum omni praesentiam carnis Christi docuisse Fabrum Stapulensem' [continues on with a long list]. A II.1, Berlin 2006, 66, available in https://www.uni-muenster.de/Leibniz/DatenII1/II1_B.pdf.

51 Cf. M. Palumbo *Leibniz e la* Res Bibliothecaria. *Bibliografie*, historiae literariae *e cataloghi nella biblioteca privata leibniziana*, Rome 1993 (Il bibliotecario), 59, n. 23, and E.J. Aiton, *Leibniz. A Biography*, Bristol 1985, 302.

52 GM VII, 210: 'Scholastici quidam maxime Angli moliti sunt singulares quosdam calculos admodum subtiles circa intensiones et remissiones qualitatum et formarum, viresque ac motus, quos miror plane fuisse neglectos, ut ne quidem celeberrimus Wallisius licet ipse Anglus in sua Algebraicarum rerum historia eorum meminerit, cum tamen subsit aliquid solidi et specimen praebetur quasi Metaphysicae cuiusdam Mathematicae. Princeps eorum fuit Johannes Suisset dictus Calculator, cui addendi Thomas Bradwardinus, Nicolaus Orem, et alii. Arabum porro Algebra circa Friderici II tempora videtur et ad Europaeos pervenisse maximeque ab Italis primum fuisse exculta, quibus et inventio Computisticae Mercatoriae debetur. Regiomontanum nostrum intellexisse algebram et in Geometria quoque exercuisse intelligi potest …'

wardine as members of the Calculator's sect indicates to me that from Leibniz's perspective the Calculator's sect included at least some of the French as well as the English, not to mention other nationalities. We also know that the English did some of their calculatory work when they were studying theology in Paris (e.g. Walter Burley and John Dumbleton).

At about this same time, in a document called *Elementa rationis*, dated April to October 1686, Leibniz wrote that although he had not seen the work of the Calculator, he had seen work of some of his followers (*quorundam eius sectatorum scripta*). To trace the fate of the work of the *Calculatores* it would be valuable to identify what texts Leibniz might have seen that would lead to this report. See the Appendix for a survey of the books Leibniz might have seen, concluding that Leibniz's report likely points to Alvarus Thomas' *Liber de triplici motu* (Paris 1509). Alvarus Thomas' book, together with supporting materials, is now available in an open access edition by Stefan Paul Trzeciok.[53] I don't think the availability of Trzeciok's edition of the *Liber de triplici motu* on the internet is the sole cause of my new respect for the work, but it was helpful. Before Trzeciok, William A. Wallace had done a great amount of non-electronic work on the Iberian and other later *Calculatores* and some of what Wallace had assembled is available from the History of Science collection within the Widener Library at Harvard University. Beyond Alvarus Thomas, other figures of interest are John Mair, Gaspar Lax, John Dullaert of Ghent, Juan de Celaya, Pedro Ciruelo, Fernando de Encinas, Antonio Coronel, Domingo de Soto etc., on some of whom Wallace wrote articles.[54]

Although I have written about Alvarus Thomas before (using Wallace's long-term loan of his Xerox bound copy of the *Liber de triplici motu*), until now I did not realize that Alvarus was likely a key character in the transmission of

---

53 S.P. Trzeciok, *Alvarus Thomas und sein* Liber de triplici motu, 1: *Naturphilosophie an der Pariser Artistenfakultät*; 2: *Bearbeiteter Text und Faksimile*, Berlin 2016, URL = https://edition-open-sources.org/media/sources/7/Sources7.pdf. The title page of the volume used for the facsimile edition, the volume conserved in the Bayerische Staatsbibliothek (shelfmark: Res/2 Phys. sp. 30), has the comment 'Liber rarissimus', the stamp 'Bibliotheca Regia Monacensis' and—interestingly enough—the handwritten name of the previous owner, 'Kenelme Digby'. The same hand has also added 'Collegii Paris. Soc. Jesu'. It is known that Digby, for his part, mentions Swineshead when discussing the notion of *reactio* in his *Two Treatises*. See J.L. Russell, 'Action and Reaction before Newton,' *The British Journal for the History of Science* 9 (1976), 25–38, esp. 32. I am grateful to Daniel Di Liscia for providing this interesting information.

54 See W.A. Wallace, 'The Calculatores in the Sixteenth Century,' in Id., *Prelude to Galileo: Essays on Medieval and Sixteenth-Century Sources of Galileo's Thought*, Dordrecht 1981 (Boston Studies in the Philosophy of Science, 62), 78–90.

the work of the *Calculatores*. What do we know about him? Marshall Clagett writes quite a lot about his context in *The Science of Mechanics in the Middle Ages*:

> The sixteenth century: France, Spain, and Portugal. Concurrent with the publication activity in Italy, the various currents of medieval natural philosophy had a marked revival at the University of Paris in the beginning of the sixteenth century … The principal center of the new interest in the older mechanics was the Collège of Montaigu. For example the masters of Montaigu were responsible for the publication of the crucial works of John Buridan and Albert of Saxony. The Scotch master George Lockert published a collection of works in 1516, which included Albert of Saxony's *Questions* on the *Physics*, the *De Caelo*, and the *De generatione* … (653).
>
> Among the many foreigners at Paris at the turn of the sixteenth century, no group is more interesting than that of the Spaniards and the Portuguese … (655).
>
> The Spanish and Portuguese masters at Paris were, of course, not only acquainted with the dynamics of Buridan, but also with the kinematics of Merton College, particularly as summarized in the works of Heytesbury and Swineshead. We have already mentioned Coronel's knowledge of the uniform acceleration theorem. Celaya is also acquainted with it. Furthermore, Master Alvarus Thomas of Lisbon wrote in 1509 what amounted to a commentary on (and paraphrase of) Swineshead's *Calculationes* and Heytesbury's *Tria praedicamenta de motu*. This work, borrowing its structure from Heytesbury's work, was called *Liber de triplici motu*. Thomas not only used the English works, but he was conversant with the Italian commentaries and paraphrases. He, then, has at his command the whole medieval mechanical tradition. In addition to giving the Merton acceleration theorem, Thomas discussed numerous species of infinite series of the kinds used by Heytesbury and Swineshead; and Thomas, like his contemporary John Major, who discussed the infinite in the context of medieval theories of the infinite, must not be left out of any account of the growing application of concepts of the infinitesimal to problems of motion … The Parisian revival of Nominalism with its emphasis on the English logical subtleties evoked sharp and uninformed criticism from humanists, just as it had earlier in Italy (657–658).

So in the later stages of calculatory work, there was transmission from Paris to scholars in the Iberian peninsula and thence, decades later, to evidence that would have been available to Leibniz, who was searching for the important contributors of the past to the advancement of mathematics and science.[55] Historians of science at the Max Planck Institute for the History of Science have noticed that Thomas Harriot, sometimes called the 'English Galileo', was also familiar with Alvarus Thomas' work.[56]

It appears that Leibniz found in his searches the book, which was the single most important contribution to the transmission, but, even more, to the advance, of the effort to apply mathematics to what was Aristotelian natural philosophy. If the project behind the 2018 Munich conference and the present book was to trace the impact, spread, and decline of the effort to quantify Aristotle that started in the fourteenth century with the so-called 'Oxford Calculators', Alvarus Thomas' *Liber de triplici motu* embodies the core of what that movement had become by the early sixteenth century. Where it was strongest was in Paris (including all the foreigners in Paris) and in the Iberian Peninsula.

5 What Does the *Liber de triplici motu* Contain of the Calculators' Work up to That Point?

First, the *Liber de triplici motu* contains a long section on proportions that explains everything behind Thomas Bradwardine's dynamics or rule for the measurement of motion *tanquam penes causam*. Alvarus Thomas is much clearer on the meaning of Bradwardine's rule than Bradwardine himself. By making the mathematics of compounding ratios clear, he distinguishes what is mathematics and what is physics in Bradwardine. In Part III, Alvarus Thomas provides an overview of many of the various things that had been said in previous centuries about the measures of motion *tanquam penes causam* and *tanquam penes effectum* in the categories of local motion, motion of increase and decrease or rarification and condensation, and motion of alteration. The work does include the usual scholastic questions and pro and con about various possible answers, but it also includes more of what one might expect from a Renaissance writer. A wider range of authors from antiquity is quoted, and also Augustine, Thomas Aquinas, John Duns Scotus, and more. In his edition Stefan Paul Trzeciok supplies a list of Alvarus Thomas' citations, helpfully including in

55 According to Lourié, 'Intensio,' 59, Leibniz had access to the publications of Spanish Jesuit thinkers of the seventeenth century who were, for instance, teaching about possible worlds.

56 See M. Schemmel, *The English Galileo. Thomas Harriot's Work on Motion as an Example of Preclassical Mechanics*, 1: *Interpretation*, Dordrecht 2008 (Boston Studies in the Philosophy of Science, 268).

the background of the search function various versions and spellings of names (for instance 'Peter of Abano' in the list of names leads to 'Conciliator' in the text). Among the citations that probably shaped Leibniz's conceptions of the *Calculatores* were, for instance: 'Respondeo ponendo quandam propositionem quam ponit doctissimus proportionum indagator Magister Nicholaus horen [Oresme].'[57]

In his letters advocating the republication of Swineshead's *Liber calculationum* and in his recommendations for reading older works, Leibniz urged that past contributors to the advance of human knowledge should not be neglected, especially if they had in some way left their mark on the course of history (or history of science). What past authors had proposed might not be true, but it was nevertheless ingenious (*si non veras, saltem ingeniosas*). Science can be advanced in many ways—by observation, experience or experiment, by elaborating disciplines (adopting definitions, suppositions, and showing what conclusions follow from these principles) or by designing new mathematical or logical methods or approaches.

Bradwardine's rule or law did not become part of early modern science in part, at least, because the accepted understanding of Euclid's *Geometry* with regard to ratios no longer included the approach to compounding or addition of ratios that Bradwardine's rule presupposed.[58] Nevertheless because of Bradwardine's advocacy of mathematics and, in particular, of the classical Euclidian approach to adding and subtracting ratios, many more university masters and students spent time working with mathematics than would have been the case if Bradwardine's text had not been so widely adopted.[59] Work of other *Calculatores* on the limits of physical entities—time, distance, force, infinity and continuity, and so forth—developed analytical capacity beyond what it would have been without the *Calculatores*.

I add here that something similar happened in logic, i.e. that the type of logic that the *Calculatores* based their work on was in a later period superseded by different styles of logic. Indeed, a key factor in the fate of the conceptions of the *Calculatores* resulted from developments in the reigning forms of logic. When

---

57 A. Thomas, *Liber de triplici motu proportionibus annexis magistri Alvari Thome Ulixbonensis Suisseth calculationes ex parte declarans*, Paris 1509, q6^{vb} [= Trzeciok, *Alvarus Thomas*, 335^{b}].

58 E.D. Sylla, 'The Origin and Fate of Thomas Bradwardine's *De proportionibus velocitatum in motibus* in Relation to the History of Mathematics,' in W.R. Laird and S. Roux (eds.), *Mechanics and Natural Philosophy before the Scientific Revolution*, Dordrecht 2008 (Boston Studies in the Philosophy of Science, 254), 67–119.

59 Sometimes the *De proportionibus* used was that of Albert of Saxony rather than that of Bradwardine.

Bradwardine initiated the *Calculatores*' emphasis on mathematics, the reigning conception of logic was that of Peter of Spain and other protagonists of the *logica modernorum*, which treated propositions as bearers of truth or falsity, and evaluated that truth or falsity by the way that the subjects and predicates of propositions stood for (had 'supposition' for) entities in the world or in human minds. Not all words in propositions had supposition for entities in the mind, in written form, or in the external world. Some so-called syncategorematic words did not have supposition for things but rather affected the supposition of other terms in the proposition.

When the *Calculatores* attempted to quantify physics, they developed a vocabulary that would work within this conception of the meaning and truth or falsity of propositions. This often meant that scholars understood that common ways of speaking or writing were not exact but shorthand ways of expression. So if someone aimed to quantify what was meant if one said 'Socrates begins to run,' they asked how exactly 'begins' (*incipit*) should be interpreted. The agreed upon answer to this question was to say that motions have no first instants (whereas by stipulation beginning was linked to an instant), so the proposition requires exposition into two propositions. The first proposition is 'Socrates is not now running' and the second proposition is 'before any instant after now, Socrates will have been running'. There is no intrinsic first instant of his motion.

How the *Calculatores* quantified substances, qualities, and motions depended on their logical assumptions about the ways in which scientific propositions could be verified or falsified by the way they were mapped onto the world. Some relevant concepts, for instance 'time', did not correspond in a simple way to any things (*res*, i.e. substances) in the world. Aristotle had defined time as the measure of motion with respect to before and after. To give numbers to motions, it was said, requires a human mind or soul. The motions they thought of were commonly celestial motions. If a unit of time was a month, then time would be measured by the motion of the moon, and there would be no months as units of time if humans had not chosen to count time by the motions of the moon.

In the first half of the fourteenth century, the *logica modernorum* had many varieties depending on views about the categories of existence. The categories considered most often in physics were substances, quantities, and qualities. If Socrates is standing and then sits down, is there any new thing such as a form of sitting that comes into existence? Might one conceive of different labels such as 'modes of being' (*modi rerum*), that are something but not substances, forms, nor matter? If everything inside the orb of the moon (earth, water, air, fire, and mixed bodies) were annihilated, would there be any extension left behind?

What does this have to do with the impact, spread, and decline of quantifying Aristotle? As far as initial impact is concerned, the initial rapid spread of calculatory work from Oxford to Paris was conditioned by the existing influence of Ockham's ideas (nominalism or terminism) in Paris which meant that there were presumed to be fewer forms in each substance than realists believed. In particular the idea of a quantitative form was frequently rejected. Without belabouring historians' disagreements about nominalism or terminism, what can be agreed concerning the spread of the calculatory tradition? One thing that is too widespread to be ignored is that growing numbers of authors began to say that, other than human souls, there are no real indivisibles—no real points, lines, or planes. Geometrical indivisibles are things of reason (*entia rationis*) or imagination.[60]

When Leibniz read documents by the *Calculatores* he was exposed to a nominalistic interpretation of the statements of mathematical physics. Historians of logic or philosophy in recent decades have been considering Leibniz's claim that there was no danger arising from his use of infinitesimals in his calculus because they were syncategorematic and hence fictional. Historians are aware that Heytesbury used the phrase 'syncategorematic infinite', but this has not been fully investigated. If Leibniz studied Alvarus Thomas' *Liber de triplici motu*, he almost certainly read in Alvarus Thomas' treatment of the middle degree theorem that the infinities involved were syncategorematic. The first big step in saying that something is syncategorematically infinite is to recognize that there is not one thing that is infinite, but things that are infinitely many in relation to each other.

If a calculator says that a continuous line is infinitely divisible, that does not mean that it can be divided and can have been completely divided into an infinite number of parts. What it means is that no matter how many times it has been divided, it can be divided more times (*non tantum quin maius*). There is a certain resemblance between the use of the concept of the syncategorematic infinite and the later use of limit proofs using epsilon and delta. In each case, there is a first value chosen and then a second value chosen whose range is determined by the first choice. In the case of syncategorematic infinity, one is not proving that there is a limit, but that there is not a limit. Of course, it is a problem to specify (without using the concept of set, which they had not

60 E.D. Sylla, 'Averroes and Fourteenth-Century Theories of Alteration. *Minima Naturalia* and the Distinction between Mathematics and Physics,' in: P.J.J.M. Bakker (ed.), *Averroes' Natural Philosophy and Its Reception in the Latin West*, Leuven 2015 (Ancient and medieval philosophy, series 1, 50), 141–192.

defined) what the collection is, from which choices are to be made. In analyzing the proposition 'infinite are the finite' (*infinita sunt finita*) presumably the choices are made from the *supposita* of the word 'finita'.

In sum if Leibniz read Alvarus Thomas' *Liber de triplici motu*, then he very likely read Alvarus Thomas' treatment of the middle degree theorem, where he would have seen Alvarus Thomas' use of the syncategorematic infinite. This is plausible because Leibniz once reported that when he was young and reading his father's books, he could read and enjoy the writings of late scholastics such as Suárez in the way that other people read novels.[61]

I have published elsewhere a summary of Alvarus Thomas' treatment of the middle degree theorem.[62] Alvarus begins with the question:

> Whether all motion uniformly difform with respect to time should be measured (*mensurari habet*) by the middle degree and all motion difformly difform with respect to time by means of reduction to uniformity or by the commensuration of the denomination to produce the denomination by which the mobile will be denominated to move.[63]

He gives principal arguments that the answer to the question is no:

> The first principal argument concerns motion that is uniformly difform. It is argued that motion that is uniformly difform is not measurable by its middle degree because, if so, it would follow that everything that is moved uniformly difformly in some time from zero degree up to some certain degree (that is, for instance, from zero up to twelve) would be moved

61 K. Müller and G. Krönert, *Leben und Werk von Gottfried Wilhelm Leibniz. Eine Chronik*, Frankfurt am Main 1969 (Veröffentlichungen des Leibniz-Archivs, 2), 6, concerning early 1661: 'Inzwischen lebte ich in dem Zabarella, dem Rubius und dem Fonseca und anderen Scholastikern mit keinem geringeren Vergnügen als vorher in den Historikern und brachte es so weit daran, dass ich den Suarez mit ebenso grosser Leichtigkeit las, wie man die Milesischen Geschichten, die man heute Romane nennt, liest.'

62 For my previous summary and outline in Latin of Alvarus Thomas' argument, see E.D. Sylla, 'Alvarus Thomas and the Role of Logic and Calculations in Sixteenth-Century Natural Philosophy,' in S. Caroti (ed.), *Studies in Medieval Natural Philosophy*, Florence 1989 (Biblioteca di Nuncius, 1), 257–298, at 270–298.

63 Thomas, *Liber de triplici motu*, $o3^{vb}$ [= Trzeciok, *Alvarus Thomas*, 283^{b}]: 'Utrum omnis motus uniformiter difformis quo ad tempus mensurari habet penes gradum medium et omnis difformiter difformis quo ad tempus penes reductionem ad uniformitatem sive penes commensurationem denominationis qua denominatione denominat mobile moveri.'

> twice as slowly as a mobile moved continuously with degree 12; but the consequent is false; therefore that from which it follows.[64]

His reply to the question and to the principal arguments makes use of the syncategorematic infinite. I summarize it as follows:[65]

> Against the middle speed theorem it had been argued that if when a mobile is moved uniformly difformly from zero degree to some maximum it has a motion equivalent to a uniform motion at its middle degree, then at the middle instant of the first half of its motion, its equivalent uniform velocity will be one fourth of its final velocity. At the middle instant of the first quarter of its motion, its equivalent uniform velocity will be one eighth of its final velocity. At the middle instant of the first eighth of the motion, its equivalent uniform velocity will be one sixteenth of its final velocity, and so forth *ad infinitum*. So it follows that infinitely slowly (in the syncategorematic sense) a mobile will begin to be moved in some proportional part of its motion. The response is that this is all possible if one is talking not about categorematic infinites, but rather about syncategorematic infinites. To the proof Alvarus admits the case, and to the argument he concedes the antecedent taking the word '*infinita*' in the major premise and in the minor premise syncategorematically. And he denies the consequence. From which it follows that in the case posited, any one of those mobiles immediately after this infinitely slowly will be moved, and nevertheless immediately after this infinitely fast will be moved some one of them.

I imagine that, if Leibniz read this section of the disputation in the *Liber de triplici motu* and tried to follow the argument, and if he then got to the ending brandishing the word 'syncategorematic', then if the word was not already

64 Thomas, *Liber de triplici motu*, 04[ra] [= Trzeciok, *Alvarus Thomas*, 285[a]]: 'Arguitur quod motus uniformiter difformis velocitas non est gradu illius medio commensuranda, quia sequeretur quod omne quod movetur in aliquo tempore uniformiter a non gradu usque ad certum gradum, id est a non gradu usque ad duodecimum, moveretur in duplo tardius quam mobile motum per idem tempus gradu duodecimo continuo; sed consequens est falsum; igitur illud ex quo sequiter.'

65 For the Latin text underlying this summary, see Sylla, 'Alvarus Thomas,' Appendix A, 282–298. What may seem a strange word order found in Alvarus' text and the last sentence of this summary is introduced in order to distinguish between categorematic and syncategorematic uses of the word 'infinite'.

known to him, he would likely have searched for what this might signify. Then when Leibniz was pushed to defend his infinitely small differentials and he said that they are syncategorematically infinitely small, he might have drawn on the concept of syncategorematic infinity, even though he had not worked out exactly how a strict argument could be made for syncategorematically infinitely small quantities. It would, however, involve many very small quantities of varying sizes in relation to each other and not single absolutely infinitely small quantities even if those are supposed to vary, as historians writing about Leibniz's use of the syncategorematic sense of infinitesimals assert.[66] Elsewhere in the seventeenth century, Jacob Bernoulli, one of the first to take up the challenge of Leibniz's papers on the calculus, made the argument that what came to be called his weak law of large numbers showed that however certain (not including absolute certainty) you wanted to be about some invisible random situation you could reach that degree of certainty if you made a sufficient number of tests.

In Leibniz's *Nachlass* there are several drafts of a letter that Leibniz sent to Pinsson des Rolles in 1701. The different parts of this draft seem independent and yet their juxtaposition is striking:

> À M. Pinsson Bronsvic 29 Aoust 1701
>
> Monsieur
>
> J' espere que M. l' Abbé Nicase trouvera bon que je vous donne occasion de lire, ce que je luy ecris, pour ne pas repeter les même choses. Les Naudaeana et Patiniana me seront tres agreables …
>
> Je desire d' obtenir une copie des ouvrages de Suisset, pour les faire entrer dans un recueil Cemilia philosophicon [Greek letters] que je medite; avec le livre philosophique de Ratramne que le R.P. Dom Mabillon m' a envoyé, et autres choses semblables plus modernes.
>
> Un des journaux de Trevouz contient quelque methode de M. Jacques Bernoulli et y mêle des reflexions sur le calcul des differences, où j' ay tant de part. L'auteur de ces reflexions semble trouver le chemin par l' infini et l' infiniti de l' infini non pas assés seur et trop eloigné de la methode des anciens, mais il aura la bonté de considerer, que si les decouvertes sont considerables, la nouveauté de la methode en releve plus tost la

66 See e.g. R.T.W. Arthur, 'Leibniz's syncategorematic infinitesimals,' *Archive for History of Exact Sciences*, 67/5 (2013), 553–593. It was not a new idea to use the distinction between categorematic and syncategorematic senses in relation to infinites.

beauté. Mais à l'egard de la seureté du chemin le livre de M. le Marquis de L'Hospital luy pourra donner satisfaction. J'adjouteray même à ce que cet illustre Mathematicien en a dit qu'on n'a pas besoin de prendre l'infiniti icy à la rigueur, mais seulement comme lors qu'on dit dans l'optique que les rayons du soliel viennent d'un point infiniment eloigné et ainsi sont estimés paralleles. Et quand il y a plusieurs degrés d'infini ou infiniment petit, c'est comme le globe de la terre est estimé un point a l'egard de la distance des fixes, et une boule que nous manions est encore un point en comparaison au semidiametre du globe de la terre, de sorte que la distance des fixes est comme un infini de l'infini par rapport au diametre de la boule. Car au lieu de l'infini ou de l'infiniment petit, on prend des quantites aussi grandes aussi et petites qu'il faut pour que l'erreur soit moindre que l'erreur donnée.[67] De sorte qu'on ne differe du style d'Archimede que dans les expressions qui sont plus directes dans nostre Methode, et plus conformes à l'art d'inventer.

Je n'ay pas encor le livre posthume de M. Nicole pour *la grace universelle qu'il combattoit pendant sa vie* ...[68]

A copy of this letter was put into an envelope with a letter going to Pierre Varignon in Paris, which included a discussion of how to argue for the validity Leibniz's infinitesimals.

Obviously, in the note transcribed above what Leibniz writes about wanting works of Suisset is not integrated with what he says next about infinitesimals. Nevertheless, it is clear that he thought about the Calculator and that he thought about how to deal with the infinitely small in close proximity. His use of the word 'syncategorematic' to describe his infinitesimals is evidence of his knowing about the widespread earlier uses of the syncategorematic sense of infinity, as in Alvarus Thomas' *Liber de triplici motu* as already described.

In Leibniz's unpublished papers, there is so much interesting material that it cannot all be combined into one coherent view. At one stage, at least, Leibniz proposed a view with many levels, at the deepest level monads, but, more

---

67 In a given case, a person could choose the maximum error that he would accept.

68 Jacob Bernoulli and his brother John were among the first to take up Leibniz's work on the infinitesimal calculus. In his proof of what came to be called the law of large numbers, Jacob Bernoulli made a similar argument. To investigate the value of an unknown statistical expectation by experiment, one can find the result with a smaller and smaller range of uncertainty by making more and more trials where there is no limit to the exactness that can be achieved by taking more trials. See J. Bernoulli, *The Art of Conjecturing, Together with Letter to a Friend on Sets in Court Tennis*, transl. E.D. Sylla, Baltimore 2006.

superficially or observationally, phenomena and then conceptions, mathematics, and so forth. Continuity might be a property of phenomena, but not of monads.[69]

The *logica modernorum*, which makes so much use of the concept of the supposition of terms in propositions was, in Leibniz's time, being supplanted by other approaches to logic such as that of Petrus Ramus and of those after Ramus, but Leibniz was still familiar with the logical use of supposition such as assumed by the fourteenth-century *Calculatores* and this had an impact on his physics. Basil Lourié conveniently epitomizes some of the features that tie together the *Calculatores* and Leibniz:

> In his earlier period, Leibniz knew the word *intensio* only in the sense of the late Scholasticism, where it was a physical term (roughly with the same meaning as the modern 'intensity') forming a pair with its antonym *remissio*. The fourteenth-century scholastic debate concerning 'intensions' and 'remissions' of forms was about physics. Leibniz, however, was thinking about physics in terms of semantics.[70]

Lourié goes on to say that Leibniz was influenced by the Port Royal Logic (1662), with its distinction between 'extension' and 'comprehension'. But in the inherited logic of supposition, the supposition of terms has a connection to the reference of terms, pointing to individual things or aggregates of things in the external world. According to Lourié, Leibniz always considered his approach to be that of Swineshead and he always explained his intensional calculi in extensional terms as well (calling such an extensional approach the 'scholastic one').

In *New Essays on Human Understanding*, Theo, representing Leibniz, says:

> For when I say *Every man is an animal* I mean that all men are included amongst all animals; but at the same time I mean that the idea of animal

69 See J.E. McGuire, '"Labyrinthus Continui": Leibniz on Substance, Activity, and Matter,' in P.K. Machamer and R.G. Turnbull (eds.), *Motion and Time, Space and Matter: Interrelations in the History of Philosophy and Science*, Columbus, OH, 1976, 290–326. See also more recently C.R. Palmerino, '*Geschichte des Kontinuumproblems* or Notes on Fromondus' Labyrinthus? On the True Nature of LH XXXVII, IV, 57r–58v,' *The Leibniz Review* 26 (2016), 63–98. To consider what education or reading might have influenced Leibniz, one place to begin would be in M.R. Antognazza's, *Leibniz. An Intellectual Biography*, Cambridge 2009.

70 Lourié, 'Intensio', 60. Here I emphasize Lourié's comment that Leibniz was thinking about physics in terms of semantics. I overlook the fact that the argument here concerns Leibniz's modal logic.

> is included in the idea of man. 'Animal' comprises more individuals than 'man' does, but 'man' comprises more ideas or more attributes: one has more instances, the other more degrees of reality; one has the greater extension, the other the greater intension.[71]

In *Elementa calculi* (1679), Leibniz writes: in the Schools, they speak differently, not looking at concepts (*notiones*) but at examples subject to universal notions.[72]

If Alvarus Thomas' *Liber de triplici motu* represents a high point in the tradition of the *Calculatores*, and if Leibniz, as an unintended consequence of his interest in including the *Liber calculationum* in a series of treasures of philosophy, took up some of the ideas of the *Calculatores* via Alvarus Thomas into his own work, then I think Leibniz thereby demonstrated the value of honoring past contributions to the advance of knowledge, even those that have been superseded by later work.

What are the morals of this story? One is that Leibniz in his own mathematical work probably benefitted from the past work of the *Calculatores* even though they had fallen out of explicit notice for the most part. One way that the *Calculatores* would have left the world in a different place than it would have been in had they never existed is that they exemplified an attitude toward applied mathematics that made it a progressive activity where new knowledge may be expected. This, of course, would stem partly from the early alliance between the ideas of Bradwardine and the position of Nicole Oresme and the other Parisians, which included nominalism or terminism in mathematics.

Let us suppose that Leibniz benefitted by all the efforts he made to learn from the writings of past scientists. Nevertheless, another lesson that might be learned is that science may stall or regress not so much through the fault of earlier scientists, but rather through the lack of attention or even failings of later generations. By the seventeenth century, the Calculator was remembered as too smart or subtle for his own good. In his 1623 book *La Doctrine curieuse des beaux esprits de ce temps, or prétendus tels: contenant plusieurs maximes pernicieuses à la religion, à l'Estat et aux bonnes moeurs, combattue et renversée*, P. François Garassus wrote that it is a mistake to diverge from the common road and go off on one's own. In his youth, Suisset was considered the marvel of his

71 G.W. Leibniz, *New Essays on Human Understanding*, Bk. 4, ch. xvii, eds. and transl. P. Remnant and J. Bennett, Cambridge 1981, 8.

72 '… in scholis aliter loquuntur, non notiones spectando, sed exempla notionibus universalibus subjecta.'

century, but in old age he could not understand what he himself had written—thus explaining why he cried when he read his own work:

> Je dis que la plus grande folie qui soit au monde, c'est de s'escarter du grand chemin … Il peut arriver une fois que suivant les sentiers on avance chemin et gaigne-t'on pays: mais pour une fois que la chose succede heureusement, il arrive cent fois qu'on se trouve pris et enfermé de quelque bras de reviere, au reste par une levée esgarée dans une campagne hors de cognoissance, et souvent il faut revenir sur ses pas et aller rencontrer le grand chemin à la course. Outre il en est tout de mesme des esprits escartez in quelque matiere que ce soit, en matiere des sciences nous avons un exemple en la personne de Calculator: en matiere de religion, nous en avons deux en la personne de Berengarius et de Pierre Abailard. Pour Jean Suysset, autrement nommé Calculator, ce fut en sa jeunesse un esprit si perocant, qu'il estoit estimé comme une merveille de son siecle, mais pource qu'il se laissa aller à suivre ses extravagances, et s'escarta du grand chemin en toutes ses opinions, comme il vint sur ses vieux ans, il se fit apporter les oeuvres de sa jeunesse, et en presence de ses disciples se prit à pleurer à chaudes larmes, et comme on luy demandast la raison de ses pleurs: helas, dit-il, mes enfans, c'est pource que je n'entends pas maintenant ce que j'ay faict et composé en ma jeunesse. S'il ne fût point allé par des extravagances, et qu'il eust suivy le chemin battu et les opinions communément soustenuës en l'schole, il n'eut pas receu cet affront en la caducité [decrepitude] de son âge.[73]

I propose that Leibniz in a sense carried forward the *Calculatores*' programme. Some of the aspects of his own work, without naming the *Calculatores*, show a kinship to the *Calculatores* insofar as Leibniz makes use of a nominalistic conception of mathematics. This, among other things encourages a belief that innovation is possible in mathematical sciences, both by representing new information about the world and by designing new mathematical sub-disciplines. There is much evidence that Leibniz had a high opinion of nominalism. John Murdoch quotes Leibniz as saying that the nominalists were '*omnium inter scholasticas profundissima*.'[74] To strengthen my argument, let me look further at the links between Leibniz's calculus and the *Calculatores*.

---

73 P.F. Garassus, *La Doctrine curieuse des beaux esprits de ce temps, or prétendus tells: contenant plusieurs maximes pernicieuses a la religion à l'Estat et aux moeurs, combattue et renversee*, Livre 1, Section 4, Paris 1624, 27–30.

74 J.E. Murdoch, 'A Skewed View: The Achievement of Late Medieval Science and Philosophy as seen from the Renaissance,' in S.E. Young (ed.), *Crossing Boundaries at Medieval Uni-*

6 Leibniz, the Calculators and the Calculus

Heidi Ishiguro argues for Leibniz's nominalism. When she does this, she calls upon much recent philosophy of language such as that of Bertrand Russell which is similar but not exactly the same as that of the fourteenth century logic of the *Calculatores*. If, however, one learns how to translate twentieth-century language into fourteenth-century language and seventeenth-century language, then the connections between Leibniz's positions and those of the *Calculatores* become more apparent.

In *Leibniz's Philosophy of Logic and Language*, Heidi Ishiguro quotes many statements by Leibniz that are consistent with the *Calculatores*' conception of syncategorematic infinities. She says, for instance:

> This brings me to the fourth Leibnizian thesis about attributes and relations, which I will call the nominalist thesis. It is the thesis that our notion of relation is an abstraction from relational facts concerning things or the order and connection of things (just as our notion of colour is an abstraction from coloured things). Relations are abstract entities made by abstraction.[75]

This statement seems to have the connotations of a translation from a Thomist or realist Aristotelian text because it uses the word 'abstract' which is Thomistic or realistic vocabulary with the implication that relations and quantities correspond to forms that exist in the external world, which the mind grasps or separates, leaving aside any underlying substance or body, thereby abstracting them (compare this to the suggestion, as is made in one place, that impetus or momentum is increased in a body in the way that more salt is dissolved in water). The resulting product may be the same or very similar: in either case, one has a mental or an imaginary something, but in the case of a nominalist the being of reason or imagination has been constructed by the thinker rather than 'abstracted' or mentally separated from external entities.

I do not have all the Latin texts that Ishiguro translates, but I think the translation 'abstract quantity' is misleading if in the case of the *Calculatores* it is taken to refer always to something that is part of a body or substance, but is being mentally separated from the physical features of the body. I also think the

---

versities, Leiden 2010, (Education and Society in the Middle Ages and Renaissance, 36), 175–188, at 188.

75 H. Ishiguro, *Leibniz's Philosophy of Logic and Language*, 2nd ed., Cambridge 1991, 138.

translation 'ideal' may be misleading, if it is understood to refer to a realist conception of something that is supposed to exist other than normal Aristotelian things. Ishiguro writes:

> But although relations are not fictitious, they are mental constructions all the same and can be given a contextual definition; sentences referring to them can be reduced to sentences in which individual constants only refer to substances.[76]

I agree with this if 'fictitious' is understood to mean something like a unicorn that is posited to exist externally, but does not. On the other hand, relations may be said to be 'fictitious' in the sense that they have a mental component.

In addition, the idea of a 'contextual definition' borrowed from Bertrand Russell does not perfectly fit Leibniz, because what is varied is not the definition, but the extension of the supposition or reference of the term. For the medieval authors—and I am suggesting that it echoes in Leibniz—it is the propositions that carry truth or falsity and the truth or falsity of propositions depends upon how the words in propositions have supposition for things outside the proposition. Many common propositions use words that do not have supposition for substances or qualities, and so they must be expounded into two or more propositions in which the categorematic terms do have supposition for substances or things. So 'infinite' in a syncategorematic sense does not have a so-called contextual definition but rather an extensional reference to substances (whether real or imaginary). Ishiguro writes:

> What really exist as basic constituents of the world are individual substances for Leibniz. In the objective world we perceive, i.e., in the world of *bene fundata*, we identify and refer to corporeal substances, i.e. individual things that we learn to identify and re-identify in the spatial-temporal world. All other things we refer to when talking of the world, whether relations or qualities, are only entities made by our abstraction from the fact that things or the phenomena we perceive have these properties. According to Leibniz, the nominalists (following Ockham) have 'deduced the rule that everything in the world can be explained without any reference to universals and real forms.' Leibniz agrees that 'nothing is truer than this opinion,' and from very early on to the end of his career he did not change his position. 'For concrete things are really things; abstractions

76 Ishiguro, *Leibniz's Philosophy*, 139.

> are not things but modes of things. Modes are usually nothing but the relation of a thing to the understanding' (1670), and again: 'Every accident is a kind of abstraction; only substance is concrete' (1712). Thus when we refer to relations, as when, according to Leibniz, we refer to places or to ratios, we are referring to entia rationis. (He thought that 'numbers, unities, fractions' also depended on the mind for their existence and thus had 'the nature of relations' [1706]). This holds of every relation, and not only of certain relations which have been arbitrarily determined by human thought. For although Leibniz distinguished between relations which held between things because of the regularity of nature and relations which come about by convention or agreement, like that which holds between an army commander and his men, it is not this conventional feature which he is talking about when he repeatedly asserts that relations are beings of reason. He seems rather to be saying that, although we can quantify over or refer to relations, what we are referring to are not entities which are the basic constituents of the world in the manner that individual substances are.[77]

What I want to add to Ishiguro's analysis is that, like relations, mathematical entities for the nominalistic *Calculatores* are things of reason. By the same token, they are imaginary and fictional. Insofar as Leibniz is in some way influenced by the nominalists/*Calculatores*, infinitesimals would have been for him always things of reason and, more than that, as syncategorematic, they would not have had reference to individual things, but rather to many things related to each other in line with Leibniz's design of the calculus. Given this conception of mathematics, the nominalist has room for creativity in inventing a mathematical vocabulary or notation, while the realist is presumably limited to entities, forms or attributes that really exist in the external world.

Anneliese Maier writes about the *Calculatores* in a chapter entitled 'Der Funktionsbegriff in der Physik des 14. Jahrhunderts' of the second edition of her *Die Vorläufer Galileis im 14. Jahrhundert* (1966). I know that many historians think that functions as now defined had not been conceived of in the fourteenth century. Maier, however, thinks that confusion has arisen because of differing definitions of the term 'function'. I think that there is an affinity between the *Calculatores* and Leibniz in the sense that the quantity of one physical entity can be a function of a complex mathematical expression, as

77 Ishiguro, *Leibniz's Philosophy*, 139–140. The author provides footnotes to the dated passages and others; these are not included here.

you might say that velocity is a function of distance and time insofar as, given proper measures, velocity equals distance divided by time.

Given the idea that scientific truths are the properties of propositions, the quantifications that the *Calculatores* might make would be not of substances, for instance, but of motions or some other newly defined variables such as 'action'. Bradwardine's rule, if proper notation existed, would be a function. In one case, Leibniz considers the measures of force as a cause and force as effect to say that there is conservation of force properly defined before and after a collision (this is related to the controversy whether mv is conserved or mv^2 is conserved). Odd as it may seem, most quantification by the *Calculatores* was not of entities or properties that had previously been familiar, but rather of newly defined concepts—an obvious case is 'velocity'.

The work of William of Ockham had a strong influence on the fourteenth-century *Calculatores* both at Oxford and at Paris. In 1644, concerning the development of a library, Gabriel Naudé wrote that one should include all the first and principal authors old and new, in the best editions and with commentaries. Among the authors he listed were Thomas Aquinas, Ockham, Burley, Giles of Rome, Albert the Great, and many others. But then he said: what should be said about the fact that the works of Richard Suisset, who was listed by Scaliger and Cardano as one of the greatest personages ever, can now not be found, not even in the most famous libraries? And what does it signify that the followers of Ockham, the leader of the nominalists (*prince des nominaux*), are forever deprived of seeing his works?[78]

I trust that the present volume will contribute to a fairer appreciation of the impact, spread, and decline of the *Calculatores*' contribution to the application of mathematics in physical science. It does not necessarily follow that when *Calculatores* ceased to be cited under that label, their contribution to the development of science had come to an end. Though he never reprinted the *Liber calculationum*, Leibniz nevertheless, through his own efforts, brought it about that the earlier existence of the *Calculatores* continued to have an influence on later thought.

---

78 G. Naudé, *Advis pour dresser une bibliothèque*, Paris 1644, Chap. 4, *De quelle qualité et condition ils doivent estre* ... [38–39]: 'la premiere regle que l'on y doit observer est de fournir premierement une bibliotheque de tous les premiers et principaux autheurs vieux et modernes, choisis des meilleures editions, en corps ou in parcelles, et accompagnez de leurs plus doctes et meilleurs interpretes et commentateurs qui se trouvent en chaque faculté, sans oublier celles qui sont le moins communes, et par consequent plus curieuses ... de Sainct Thomas, Occham, Durand, Pierre Lombart, Henry de Gand, Alexandre de Ales, Gilles de Rome, Albert Le Grand, Auriolus, Burlée, Capreolus, Major, Vasquez, Suarez, pour la scholastique; des cours civil et canon ... [86] Quoy doncques sera-il dit que Scaliger et

Appendix: Leibniz's Report of Seeing Some of the Works of the Calculator's *sectatores* and a Survey of the Printed Books He Might Have Seen

1 *A Leibniz Reference to the Work of Swineshead and His* sectatores

'... Hinc Physicae Scientiae Perfectio (praeter experimenta) sine controversia in eo consistit, ut reducatur ad Geometriam; detectis, quoad ejus fieri potest naturae mechanismis, qui à partium figuris motibusque pendent; ipsa autem rursus Geometria cum nonnihil adhuc perplexa sit, neque enim omnes figurarum habitudines lineis in charta deductis commodè exprimi possunt, reducta est ad calculum quendam sive numerorum aestimationem, qua efficitur, ut characteribus numericis ac literis alphabeti, numeros indeterminatos designantibus, variè combinatis ipsae figurae corporum mirabili ratione exprimantur, quod vulgò Calculum Speciosam à characteribus sive speciebus rerum vocant ...' [Leibniz's reasoning here is interesting in that he is thinking about figures and geometry, in particular, as a way of introducing mathematics. After talking about the jurisconsults, he goes on:]

'Reperiuntur aliqua hujus generis etiam apud Scholasticos superiorum maxime seculorum, ne quem sua merita laude defraudemus. Fuit enim aliquis Johannes Suisset, dictus Calculator, qui circa motus et qualitatum intensiones in media metaphysicorum regione Mathematicum sine exemplo agere coepit. Hunc mihi videre non contigit. Vidi tamen quorundam ejus sectatorum scripta, unde agnosco, si ingenio eorum hominum ac bonae voluntati, lumen Mathematicorum quod nunc accensum est, accessisset, potuisse ab illis nostros labores praeveniri. Sed istam bonam frugem spinae innumerabiles in eodem agro enatae oppresserunt, certè occultarunt. Nam alioqui tantum abest ut ego Theologis philosophisque scholasticis, quemadmodum ab ignaris solet, detraham, ut contra subtilitatem eorum admirer, agnoscamque lubens plurima apud eos extare solida et praeclara, demonstrationum capacia, quae ab horrida illa obscuritate purgata magno cum fructu in laetius solum transferri et quemadmodum sylvestres arbores cultura mitescere queant.

Cardan, les deux plus grands personnages du dernier siecle s'accordent en un seul poinct, que concerne les loüanges de Richard Suisset, autrement nommé Calculator, qui vivoit il n'y a que trois cens ans, pour le mettre au rang des dix plus grands esprits qui ayent jamais esté, sans que nous puissons trouver ses oeuvres dans toutes les plus fameuses bibliotheques? Et qu'elle apparence y at-il que les sectateurs d'Occham prince des nominaux soient eternellement privez de voir ses oeuvres, aussi bien que tous les philosophes celles de ce grand et renommé Avicenne?'

Restauratis bonis literis, ac restituta dicendi ratione, quod superioris maximè seculi beneficium est, id potissimum nostro servaverant fata, ut quod dixi Lumen Matheseos post tanti temporis Eclipsin rursus effulgeret, detectis atque auctis Archimedeis per indivisibilia et infinita inveniendi artificiis, quae Metaphysicam Geometrarum appellare possis, et quae si quid judico plerisque aliis veterum praeter Archimedem fuêre ignota; productaque simul Analysi illa calculi speciosi, à veteribus partim studiosè occultata, partim non satis explorata; quam nos Vietae debemus, qua tota Geometria ad Arithmeticam singularum reducitur. Sed accessit aliud majus, nempe initia quaedam Physicae ad Geometriam revocandae, quorum specimina primi omnium Galilaeus et Keplerus et Gilbertus dedêre, quibus Harvaeum mechanica circulationis lege reserata meritò jungas. Horum inventis inter se copulatis atque in unum systema adornatis praeclara addidit Cartesius, cui si diuturnior vita contigisset, haud dubie ille nobis aliquando veritates dedisset solidas ...' [Couturat, 335–341, quoted passages on 335–336 and 340–341; A VI, vol. 4A, #162, 714–721].

2 *Where Might Leibniz Have Seen Some of the Work of Swineshead's* sectatores*?*

Although I don't know when (other than before October 1686) Leibniz might have read some of the work of Swineshead's followers (*sectatores*), I think it was most likely in a printed book, although in one case he refers to manuscripts.

A. One possibility is that he saw the Venice 1505 compilation printed by Octavianus Scotus per Bonetum Locatellum, namely:

Bassanus Politus, *Questio de modalibus*
Bassanus Politus, *Tractatus proportionum introductorius ad calculationes Suisset*
Thomas Bradwardine, *Tractatus proportionum*
Nicole Oresme, *Tractatus proportionum Nicholai Oren*
Nicole Oresme, *Tractatus de latitudinibus formarum*
Blasius of Parma, *Tractatus de latitudinibus formarum*
Anonymous, *Tractatus de sex inconvenientibus*
Venetiis: Mandato et sumptibus heredum Octaviani Scoti per Bonetum Locatellum. Kalendis 1 Septembris 1505.

Some of the parts of this volume, together with related works were bound together in varying combinations. The Columbia University copy of this volume also contains (as does the copy of the 1505 work scanned by the Münchener Digitalisierungszentrum after a separate title page):

Giovanni da Casale, *Questio velocitate motus alterationis*
Blasius of Parma, *De tactu corporum durorum*

Whether alone or with such additions, it is not likely that this was the compilation that Leibniz saw, because, oddly in a work entitled *Tractatus proportionum introductorius ad calculationes Suisset*, Bassanus Politus rejects the conception of compounding or adding proportions that is key to Bradwardine's rules and hence to Swineshead's treatise 14 on local motion.[79]

B. Another possibility is that Leibniz saw a collection of works, many by authors connected in some way to the Calculators, which is contained in a volume printed in Venice in 1494:

William Heytesbury, *Regule solvendi sophismata*, 4^{v}–53^{r}
Gaetano da Thiene, *Declaratio* [= *Recollectae*] *super tractatu Hentisberi regularum* (scattered within Heytesbury's text, 7^{r}–52^{r})
Messinus, *Questio cum expositione totius tractatus Hentisberi de tribus predicamentis*, 52va–62vb
Gaetano da Thiene, *Complementum expositionis Messini de tribus predicamentis*, 2vb–64ra
Angelo da Fossambruno, *Super tractatu* De tribus predicamentis ⟨*Hentisberi*⟩, 64ra–73rb
Bernardus Tornius, *Quedam annotata in capitulum de moto locali Hentisberi*, 73va–77va
William Heytesbury, *Sophismata*, 77va–170vb
Anonymous, *Probationes conclusionum*, 188va–203vb

While these 1494 works would generally be included in the Calculatory tradition, they do not appear on the surface to be written by followers of Swineshead, and the works are more logical than physical or mathematical.

C. Third, there is a work by the Portuguese author Alvarus Thomas, which was widely distributed and which was favourable to the *Calculatores*. This seems almost certainly to be the printed book that Leibniz examined:

Alvarus Thomas, *Liber de triplici motu proportionibus annexis magistri Alvari Thome Ulixbonensis Suisseth calculationes ex parte declarans*, Paris:

79 See Sylla, 'The Origin and Fate,' 67–119.

Poinset le Preux, 1509. 141 folios.

Part I, *De proportionibus*, 2^r–8^v

Part II, *De proportionalitatibus et accidentiis*, 9^r–29^r

Part III, *Liber de triplici motu huius operis tertia pars*

De motu quo ad causam

Tractatus secundus of Part III: *de motu locali quo ad effectum*

Tractatus tertius of Part III: *de motu rarefactionis et condensationis*, 88^r

Tractatus quartus of Part III: *de motu alterationis*, 110^r; *de intensione et remissione*, 123^r

This work cites many other authors by name and gives the impression of an active field of study in many centres. I conclude that the book by Swineshead's *sectatores* that Leibniz reported seeing and being positively impressed by before he saw the *Liber calculationum* itself was almost certainly a copy of Alvarus Thomas' *Liber de triplici motu* (Paris 1509).

Acknowledgements

For their considerable help in revising and completing this essay, I wish to thank Paul J.J.M. Bakker, Daniel A. Di Liscia, and Richard Sylla.

Manuscripts

Basel, Universitätsbibliothek, Ms. F. II. 9
Basel, Universitätsbibliothek, Ms. F. VIII. 7
Berlin, Staatsbibliothek Preußischer Kulturbesitz, Ms. lat. fol. 206
Bologna, Biblioteca comunale dell'Archiginnasio, Ms. 985
Brugge, Stedelijke Openbare Bibliotheek, Ms. 503
Brugge, Stedelijke Openbare Bibliotheek, Ms. 188
Bologna, Biblioteca dell' Archiginnasio, Ms. 985.
Bruxelles, Bibliothèque Royale de Belgique, Ms. 1631 (21.856)
Cambridge, Gonville and Caius College Library, Ms. 499/268
Cambridge, Peterhouse Library, Ms. 195
Cesena, Biblioteca Malatestiana, Ms. Plut. VIII, sin. 2
Erfurt, Universitätsbibliothek, Dep. Erf., CA F. 105
Erfurt, Universitätsbibliothek, Dep. Erf., CA O. 74
Erfurt, Universitätsbibliothek, Dep. Erf., CA Q. 135
Erfurt, Universitätsbibliothek, Dep. Erf., CA Q. 277
Erfurt, Universitätsbibliothek, Dep. Erf., CA Q. 325
Erfurt, Universitätsbibliothek, Dep. Erf., CA Q. 387
Firenze, Biblioteca Nationale Centrale, Cod. Conv. Soppr. B. VI. 1681
Genova, Biblioteca Civica Berio, Ms. C.F. 53
Hannover, Niedersächsische Landesbibliothek, Ms. IV, 615
Klagenfurt, Universitätsbibliothek, Ms. Pap. 168
København, Kongelige Bibliotek, Ms. Ny kgl. saml. 190 4°
København, Kongelige Bibliotek, Ms. Thott 583
Kraków, Biblioteka Jagiellońska, Ms. 648
Kraków, Biblioteka Jagiellońska, Ms. 696
Kraków, Biblioteka Jagiellońska, Ms. 704
Leipzig, Universitätsbibliothek, Ms. 1348
Leipzig, Universitätsbibliothek, Ms. 1360
Leipzig, Universitätsbibliothek, Ms. 1423
London, British Library, Harley, Ms. 3243
Lüneburg, Ratsbücherei, Ms. Theol. 2° 99
München, Bayerische Staatsbibliothek, Clm 331
München, Bayerische Staatsbibliothek, Clm 4377
München, Bayerische Staatsbibliothek, Clm 4384
München, Bayerische Staatsbibliothek, Clm 4391
München, Bayerische Staatsbibliothek, Clm 16.520
München, Bayerische Staatsbibliothek, Clm 26.838

München, Bayerische Staatsbibliothek, Clm 26.889
München, Bayerische Staatsbibliothek, Clm 28.395
Oxford, Bodleian Library, Ms. Canon. Misc. 177
Oxford, Bodleian Library, Ms. Canon. Misc. 226
Oxford, Merton College Library, Ms. 306
Oxford, New College Library, Ms. 154
Padova, Biblioteca Universitaria, Ms. 1123
Paris, Bibliothèque Nationale de France, Ms. lat. 6558
Paris, Bibliothèque Nationale de France, Ms. lat. 6559
Paris, Bibliothèque Nationale de France, Ms. lat. 6752
Paris, Bibliothèque Nationale de France, Ms. lat. 14.576
Paris, Bibliothèque Nationale de France, Ms. lat. 15.561
Paris, Bibliothèque Nationale de France, Ms. lat. 16.401
Perugia, Biblioteca Comunale Augusta, Ms. 1062
Praha, Národní Knihovna České Republiky, Ms. III. B. 10
Praha, Národní Knihovna České Republiky, Ms. VIII. G. 19
Sankt Florian, Stiftsbibliothek, Ms. XI. 627
Sankt Gallen, Stiftsbibliothek, Ms. 840
Sevilla, Biblioteca Capitular y Colombina, Ms. 7-6-12
Sevilla, Biblioteca Capitular y Colombina, Ms. 7-7-13
Sevilla, Biblioteca Capitular y Colombina, Ms. 7-7-29
Tortosa, Biblioteca de la Catedral y del Cabildo de la Sanctísima Iglesia Catedral, Ms. 186
Troyes, Médiathèque de l'Agglomération Troyenne, Ms. 1477
Vaticano (Città del), Biblioteca Apostolica Vaticana, Ms. Chigi E. IV. 120
Vaticano (Città del), Biblioteca Apostolica Vaticana, Ms. Chigi E. VI. 197
Vaticano (Città del), Biblioteca Apostolica Vaticana, Ms. Chigi F. IV. 66
Vaticano, Città del), Biblioteca Apostolica Vaticana, Ms. Vat. lat. 955
Vaticano (Città del), Biblioteca Apostolica Vaticana, Ms. Vat. lat. 1108
Vaticano (Città del), Biblioteca Apostolica Vaticana, Ms. Vat. lat. 1110
Vaticano (Città del), Biblioteca Apostolica Vaticana, Ms. Vat. lat. 2130
Vaticano (Città del), Biblioteca Apostolica Vaticana, Ms. Vat. lat. 2148
Vaticano (Città del), Biblioteca Apostolica Vaticana, Ms. Vat. lat. 2159
Vaticano (Città del), Biblioteca Apostolica Vaticana, Ms. Vat. lat. 2225
Vaticano (Città del), Biblioteca Apostolica Vaticana, Ms. Vat. lat. 4353
Vaticano (Città del), Biblioteca Apostolica Vaticana, Ms. Vat. lat. 4429
Vaticano (Città del), Biblioteca Apostolica Vaticana, Ms. Vat. lat. 4545
Venezia, Biblioteca Nazionale Marciana, Ms. lat. VI. 30 (2547)
Venezia, Biblioteca Nazionale Marciana, Ms. lat. VI. 72 (2810)
Venezia, Biblioteca Nazionale Marciana, Ms. lat. VI. 155 (3377)

Venezia, Biblioteca Nazionale Marciana, Ms. lat. VI. 160 (2816)
Venezia, Biblioteca Nazionale Marciana, Ms. lat. VI. 173 (2675)
Wien, Österreichische Nationalbibliothek, Cod. 4953
Worcester, Cathedral and Chapter Library, Ms. F.35
Wrocław, Biblioteka Uniwersytecka, Ms. IV. F. 198
Wrocław, Biblioteka Uniwersytecka, Ms. IV. Q. 3

Bibliography

In the following bibliography and Index, ancient and medieval authors are listed by their first names (e.g., Clement of Alexandria, Blasius of Parma, etc.). Post-medieval authors are listed by their last names (e.g., Descartes, Leibniz, etc.). Medieval authors are generally mentioned under their English names, except for lesser known authors and authors whose foreign names are currently used in English scholarly literature.

1 Primary Sources

1.1 *Manuscripts*

Angelo da Fossambruno, *De insolubilibus*, Venezia, Biblioteca Nazionale Marciana, Ms. lat. VI. 155 (3377), 19ra–31vb; Ms. lat. VI. 30 (2547), 63ra–71va; Città del Vaticano, Biblioteca Apostolica Vaticana, Ms. Vat. lat. 2130, 1061ra–1070va; Klagenfurt, Universitätsbibliothek, Ms. Pap. 168, 61^{r}–67^{v}.

Angelo da Fossambruno, *De reactione*, Venezia, Biblioteca Nazionale Marciana, Ms. lat. VI. 160 (2816), 248ra–252ra.

Anonymous, *Tractatus bonus*, Erfurt, Universitätsbibliothek, Dep. Erf., CA Q. 325, 43^{r}–45^{r}; København, Kongelige Bibliotek, Ms. Ny kgl. saml. 190 4°, 167^{r}–173^{v} (*Utrum superficies triangularis sit uniformiter difformis incipiens a non gradu*).

Anonymous, *Tractatus de sex inconvenientibus*, Paris, Bibliothèque Nationale de France, Ms. lat. 6559, 1ra–41va; Praha, Národní Knihovna České Republiky, Ms. VIII. G. 19, 25^{r}–46^{v}.

Blasius of Parma, *Questiones super octo libros Physicorum*, Città del Vaticano, Biblioteca Apostolica Vaticana, Ms. Vat. lat. 2159, 1^{r}–98^{v}.

Blasius of Parma, *Questiones super tractatum de latitudinibus formarum*, Oxford, Bodleian Library, Ms. Canon. Misc. 177, 86^{vb}(109vb)–89^{va}(113va).

Giovanni Marliani, *Questio de proportione motuum in velocitate*, Città del Vaticano, Biblioteca Apostolica Vaticana, Ms. Chigi E. VI. 197, 102^{v}–126^{r}.

Jacques Legrand, *Compendium utriusque philosophie*, Genova, Biblioteca Civica Berio, Ms. C.F. 53, 2^{r}–235^{v}; Paris, Bibliothèque Nationale de France, Ms. lat. 6752, 4^{r}–236^{r}.

James of Naples, *Tractatus de perfectione specierum*, Bruxelles, Bibliothèque Royale de Belgique, Ms. 1631 (21.856), 68^{r}–75^{v}; Erfurt, Universitätsbibliothek, Dep. Erf., CA Q. 387, 79^{r}–85^{v}; København, Kongelige Bibliotek, Ms. Ny kgl. saml. 190 4°, 179^{r}–197^{r}; Lüneburg, Ratsbücherei, Ms. Theol. 2° 99, 282ra–286vb; München, Bayerische Staatsbibliothek, Clm 26.889, 206^{r}–212^{v}; Clm 4377, 143^{r}–148^{v}; Clm 26.838, 76^{v}–85^{v}, 118^{r}–124^{v}; Oxford, Bodleian Library, Ms. Canon. Misc. 177, 232^{r}–234^{v}; Città del Vaticano, Biblioteca Apostolica Vaticana, Ms. Chigi F. IV. 66, 6^{v}–12^{v}; Ms. Vat. lat. 4545, 32^{r}–48^{r}.

John (Rucherat) of Wesel (or Nicholas of Amsterdam?), *Disputata sive Exercicium super Libris Methaphysice Aristotelis*, Basel, Universitätsbibliothek, Ms. F. VIII. 7, 1^{r}–261^{v} (attr. to Johannes de Wesalia); München, Bayerische Staatsbibliothek, Clm 28.395, 1ra–103ra; Sankt Gallen, Stiftsbibliothek, Ms. 840, 1^{r}–205^{r}.

John Aurifaber, *Quaestiones super libros Physicorum*, Leipzig, Universitätsbibliothek, Ms. 1423, 1^{r}–220^{r}.

John of Holland, *De velocitate motuum*, Città del Vaticano, Biblioteca Apostolica Vaticana, Ms. Vat. lat. 1108, 144^{r}–162^{r}.

John of Holland, *Sophismata*, Leipzig, Universitätsbibliothek, Ms. 1360, 207ra–282rb; Kraków, Biblioteka Jagiellońska, Ms. 704, 35ra–39ra.

John Wyclif, *Dubia super materia librorum Physicorum*, Venezia, Biblioteca Nazionale Marciana, Ms. lat. VI. 173 (2675), 1ra–58vb.

Paul of Venice, *Summa philosophiae naturalis magistri Pauli Veneti noviter recognita … restituta*, Basel, Universitätsbibliothek, Ms. F. II. 9, 98^{r}–134^{v}.

Richard Kilvington, *Expositio super primum librum Physicorum*, Città del Vaticano, Biblioteca Apostolica Vaticana Ms. Vat. lat. 4353, 125^{r}–129^{v}.

Richard Kilvington, *Quaestiones super De generatione et corruptione*, Brugge, Stedelijke Openbare Bibliotheek, Ms. 503, 20vb–50vb.

Richard Kilvington, *Quaestiones super libros Sententiarum*, Città del Vaticano, Biblioteca Apostolica Vaticana, Ms. Vat. lat. 4353, 1^{r}–58^{r}; Bologna, Biblioteca comunale dell'Archiginnasio, Ms. 985, 1ra–52va.

Richard Kilvington, *Quaestiones super Physicam*, Venezia, Biblioteca Nazionale Marciana, Ms. lat. VI. 72 (2810), 81ra–112rb, 168ra–169va; Sevilla, Biblioteca Capitular y Colombina, Ms. 7-7-13, 27ra–50vb.

Richard Swineshead, *Liber calculationum*, Worcester, Cathedral and Chapter Library, Ms. F. 35, 3^{r}, 27^{r}–65^{v}, 70–75^{v} (students' notebook with various fragments); Hannover, Niedersächsische Landesbibliothek, Ms. IV. 9, Miscellanea 615, 1–1119.

Richard Swineshead, *Opusculum de motu*, Cambridge, Gonville and Caius College Library, Ms. 499/268, 212ra–215rb; Sevilla, Biblioteca Capitular y Colombina, Ms. 7-7-29, 27^{v}–33^{r}; Oxford, Bodleian Library, Ms. Digby 154, 42ra–44ra.

Roger Swineshead, *De motibus naturalibus*, Erfurt, Universitätsbibliothek, Dep. Erf., CA Q. 135, 25va–47ra.

Roger Swineshead, *Sophismata logicalia*, Wrocław, Biblioteka Uniwersytecka, Ms. IV. Q. 3, 241^{r}–249^{r} (incomplete); Padova, Biblioteca Universitaria, Ms. 1123, 42ra–50ra.

Thomas Buckingham, *Quaestiones theologicae*, Oxford, New College Library, Ms. 134, 324^{r}–438^{v}.

Thomas Wylton, *Quaestiones libri Physicorum*, Cesena, Biblioteca Malatestiana, Ms. Plut. VIII, sin. 2, 2^{r}–141^{v}.

1.2 *Printed Sources*

Adam Wodeham, *Lectura secunda in librum primum Sententiarum*, eds. R. Wood and G. Gál, New York 1990 (3 vols.).

Albert of Saxony, *Expositio et quaestiones in Aristotelis Physicam ad Albertum de Saxonia attributae*, ed. B. Patar, Louvain-la-Neuve, Leuven 1999 (3 vols.) (Philosophes médiévaux, 39–41).

Albert of Saxony, *Logik*, Lateinisch–Deutsch, ed. H. Berger, Hamburg 2010 (Philosophische Bibliothek, 611).

Alcuin of York, *Epistolae*, ed. E. Dümmler, *Monumenta Germaniae Historica, Epistolae* IV.2., Berlin 1895.

Alexander of Hales, *Glossa in quatuor libros Sententiarum. 2: In librum secundum*, eds. PP. Collegii S. Bonaventurae, Quaracchi 1952 (Bibliotheca Franciscana scholastica medii aevi, 13).

Alexander of Hales, *Summa theologica*, Liber II, eds. PP. Collegii S. Bonaventurae, Quaracchi 1928.

Almain, J., *Embammata phisicalia*, Paris 1506.

Angelo da Fossambruno, *Super tractatu* De tribus predicamentis ⟨*Hentisberi*⟩. In William Heytesbury, *Regule solvendi sophismata*, Venice 1494, 64ra–73rb.

Anonymous, *Calendar on Entries in the Papal Registers relating to Great Britain and Ireland* (CPL), 2: A.D. 1305–1342, Petition to the Pope, London 1896.

Anonymous, *Probationes conclusionum*. In William Heytesbury, *Regule solvendi sophismata*, Venice 1494, 188va–203vb.

Anonymous, *Tractatus bonus*, ed. M. Clagett, *Nicole Oresme and the Medieval Geometry of Qualities and Motions. A Treatise on the Uniformity and Difformity of Intensities known as* Tractatus de configurationibus qualitatum et motuum, Madison, WI, 1968, 576–621.

Anonymous, *Tractatus de sex inconvenientibus*, Venice 1505, sigs. $e1^{ra}$–$h4^{ra}$; ed. S. Rommevaux-Tani, *De sex inconvenientibus, un traité anonyme de philosophie naturelle au* XIV*e* *siècle*, édition, introduction et étude doctrinale, Paris 2022.

Aristotle, *De generatione et corruptione* (*On Generation and Corruption*), transl. H.H. Joachim, in *The Basic Works of Aristotle*, ed. R. McKeon, New York 2001, 470–531.

Aristotle, *Metaphysica*, transl. Guillelmus de Moerbeka, ed. G. Vuillemin-Diem, Leiden 1995 (Aristoteles latinus, 25/3.2).

Aristotle, *Metaphysics*. A Revised Text with Introduction and Commentary, ed. W.D. Ross, Oxford 1970 (2 vols.).

Aristotle, *Meteorologica*. In *The Works of Aristotle*, ed. W.D. Ross, transl. E.W. Webster, 3, Oxford 1931.

Aristotle, *On coming-to-be and passing-away* (*De generatione et corruptione*). A Revised Text with Introduction and Commentary, ed. H.H. Joachim, Oxford 1952.

Aristotle, *Physica, translatio vetus*, eds. F. Bossier and J. Brams, Leiden 1990 (Aristoteles latinus, 7/1).

Aristotle, *Physics*, transl. E. Hussey, in: *Aristotle, Physics Books III and IV*, Oxford 1983 (Clarendon Aristotle Series, 44).

Aristotle, *The Complete Works of Aristotle*, ed. J. Barnes, Princeton, NJ, 1984 (2 vols.) (Bollingen series, 71).

Augustine, *De ordine*, ed. W.M. Green, Turnhout 1970 (Corpus Christianorum Series Latina, 29), 87–137.

Augustine, *Retractationes*, ed. A. Mutzenbecher, Turnhout 1984 (Corpus Christianorum Series Latina, 57).

Averroes, *Commentum magnum super libro De celo et mundo Aristotelis*, eds. F.J. Carmody and R. Arntzen, Leuven 2003 (2 vols.) (Recherches de Théologie et Philosophie médiévales. Bibliotheca, 4).

Averroes, *In Metaphysicam*, Venice 1562, repr. Frankfurt am Main 1962 (Aristotelis opera cum Averrois commentariis, 8).

Averroes, *In Physicam*, Venice 1562, repr. Frankfurt am Main 1962 (Aristotelis opera cum Averrois commentariis, 4).

Benedictus de Victoriis, *Opusculum Benedicti Victorii Bononiensis in Tisberum de sensu composito ac diviso: cum eiusdem collectaneis in suppositiones Pauli Veneti*, Bologna 1504.

Bernardus Tornius, *Quedam annotata in capitulum de moto locali Hentisberi*. In William Heytesbury, *Regule solvendi sophismata*, Venice 1494, 73va–77va.

Bernoulli, J., *The Art of Conjecturing, Together with Letter to a Friend on Sets in Court Tennis*, transl. E.D. Sylla, Baltimore 2006.

Blasius of Parma, Commentary on *Physics VII*, ed. and transl. S. Rommevaux-Tani, URL = https://bdeparme.hypotheses.org/questions-sur-la-physique.

Blasius of Parma, *Quaestio de intensione et remissione formarum*, ed. G. Federici Vescovini, 'La *Quaestio de intensione et remissione formarum* di Biagio Pelacani da Parma,' *Physis* 31 (1994), 433–535 (edition 470–535).

Blasius of Parma, *Questiones super tractatum de latitudinibus formarum*, Padua 1486.

Blasius of Parma, *Questiones super tractatus logice magistri Petri Hispani*, eds. J. Biard and G. Federici Vescovini, Paris 2001 (Textes philosophiques du Moyen Âge, 20).

Boethius, A.M.S., *Institutio arithmetica*, ed. J.-Y. Guillaumin, Boèce, *Institution Arithmétique*, Paris 2002 (Collection des Universités de France).

Brahe, T., *Epistolae astronomicae*, Uraniborg 1596.

Callimachus, *Hymn to Zeus*, ed. and comm. G.R. McLennan, Rome 1977 (Testi e commenti, 2).

Cardano, G., *De subtilitate libri XXI*, Basel 1560.

Cardano, G., *Opus novum de proportionibus*, Basel 1570.

Celaya, J. de, *Expositio in octo libros Physicorum Aristotelis, cum questionibus eiusdem secundum triplicem viam beati Thomae, realium, et nominalium*, Paris 1517.

Charleton, W., *Physiologia Epicuro-Gassendo-Charltoniana*, London 1654.

Ciruelo, P., *Apotelesmata astrologiae christianae*, Alcalá 1521.

Ciruelo, P., *Cursus quatuor mathematicarum artium liberalium*, Alcalá 1516.

Ciruelo, P., *Uberrimum sphere mundi commentum intersertis etiam questionibus domini Petri de Aliaco*, Paris 1498.

Clement of Alexandria, *Stromateis*, Books 1–3, transl. J. Ferguson, Washington, DC, 1991 (The Fathers of the Church, 85).

Coronel, L., *Physice perscrutationes*, Paris 1511.

Descartes, R., *Le monde ou traité de la lumière*. In C. Adam and P. Tannery (eds.), *Oeuvres de Descartes*, Paris 1996, 11: 4–118.

Euclid (Adelard of Bath), *Elementa*, ed. H.L.L. Busard, *The First Latin Translation of Euclid's* Elements *Commonly Ascribed to Adelard of Bath*, Toronto, ON, 1983 (Studies and Texts, 64).

Euclid (Campanus), *Elementa*, ed. H.L.L. Busard, *Campanus of Novara and Euclid's Elements*, Wiesbaden 2005 (Boethius. Texte und Abhandlungen zur Geschichte der Mathematik und der Naturwissenschaft, 51/1).

Euclid (Johannes de Tinemue), *Elementa*, ed. H.L.L. Busard, *Johannes de Tinemue's redaction of Euclid's* Elements *the so-called Adelard III version*, Stuttgart 2001 (Boethius. Texte und Abhandlungen zur Geschichte der Mathematik und der Naturwissenschaft, 45).

Euclid (Robert of Chester), *Elementa*, eds. H.L.L. Busard and M. Folkerts, *Robert of Chester's (?) Redaction of Euclid's* Elements, *the so-called Adelard II Version*, Basel 1992 (Science Networks-Historical Studies, 9).

Euclid, *Elementa*, ed. H.L.L. Busard, *The Mediaeval Latin Translation of Euclid's Elements Made Directly from the Greek*, Wiesbaden 1987 (Boethius. Texte und Abhandlungen zur Geschichte der Mathematik und der Naturwissenschaft, 15).

Fine, O., *Epistre exhortative*, Paris 1531.

Francis of Marchia, *Quaestiones super Metaphysicam*, ed. N. Mariani, Grottaferrata 2012 (Spicilegium Bonaventurianum, 37).

Gaetano da Thiene, *Complementum expositionis Messini de tribus predicamentis*. In William Heytesbury, *Regule solvendi sophismata*, Venice 1494, 2vb–64ra.

Gaetano da Thiene, *Declaratio* [= *Recollectae*] *super tractatu Hentisberi regularum*. In William Heytesbury, *Regule solvendi sophismata*, Venice 1494 (scattered within Heytesbury's text), 7ra–52rb.

Galilei, G., *Opere*, 1, ed. A. Favaro, Florence 1890.

Garassus, P.F., *La Doctrine curieuse des beaux esprits de ce temps, or prétendus tells: contenant plusieurs maximes pernicieuses a la religion à l'Estat et aux moeurs, combattue et renversee*, Paris 1624.

George of Brussels, *Cursus quaestionum super philosophiam Aristotelis*, ed. Thomas Bricot, Paris 1491.

George of Trebizond, *Dialectica*, ed. J. Lefèvre d'Étaples, Paris 1508.

Giles of Rome, *Commentaria in octo libros Phisicorum Aristotelis*, Venice 1502, repr. Frankfurt am Main 1968.

Giovanni da Casale, *Questio velocitate motus alterationis* (*Utrum in moventibus ad qualitatem illud semper velocius moveatur quod in equali tempore acquirit maiorem latitudinem*), Venice 1505.

Giovanni Marliani, *Questio de proportione motuum in velocitate*, Pavia 1482.

Henry of Harclay, *Ordinary Questions*, ed. M. Henninger, Oxford 2008 (2 vols.) (Auctores Britannici Medii Aevi, 17–18).

Hobbes, T., *Decameron physiologicum, or, Ten Dialogues of Natural Philosophy*, London 1678.

Hugolino of Orvieto, *Commentarius in quattuor libros Sententiarum*, 1, ed. W. Eckermann, Würzburg 1980 (Cassiciacum, Supplement, 8).

Hugolino of Orvieto, *Commentarius in quattuor libros Sententiarum*, 2, ed. W. Eckermann, Würzburg 1984 (Cassiciacum, Supplement, 9).

Hugolino of Orvieto, *Quaestiones super quattuor libros Physicorum*, ed. W. Eckermann, *Der Physikkommentar Hugolins von Orvieto OESA. Ein Beitrag zur Erkenntnislehre des spätmittelalterlichen Augustinismus*, Berlin 1972 (Spätmittelalter und Reformation, 5).

Hugolino of Orvieto, *Tractatus de perfectione specierum*, ed. F. Corvino, 'Ugolini de Urbe Veteri Tractatus de perfectione specierum,' *ACME Annali della Facoltà die Filosofia e Lettere dell'Università Statale di Milano* 8 (1955), 119–204.

Jacobus de Sancto Martino (?), *Tractatus de latitudinibus formarum*, ed. T.M. Smith, *A Critical Text and Commentary upon* De latitudinibus formarum, Ph.D. Dissertation, University of Wisconsin, Madison, WI, 1954.

John Buridan, *Quaestiones super libris quattuor De caelo et mundo*, ed. E.A. Moody, Cambridge, MA, 1942.

John Buridan, *Quaestiones super octo libros Physicorum Aristotelis* (*secundum ultimam lecturam*), Libri III–IV, eds. M. Streijger and P.J.J.M. Bakker, Guide to the Text by E.D. Sylla, Leiden 2016 (Medieval and Early Modern Philosophy and Science, 27).

John Buridan, *Questiones Elencorum*, eds. R. van der Lecq and H.A.G. Braakhuis, Nijmegen 1994 (Artistarium, 9).

John Duns Scotus, *Lectura in librum primum Sententiarum. Distinctiones 8–45*, eds. C. Balić et al., Vatican City 1966 (Opera omnia, 17).

John Mair, *Octo libri physicorum cum naturali philosophia atque metaphysica Johannis Maioris hadigtonani theologi Parisiensis*, Paris 1526.

John of Holland, *Four Tracts on Logic* (*Suppositiones, Fallaciae, Obligationes, Insolubilia*), ed. E.P. Bos, Nijmegen 1985 (Artistarium, 5).

John of Ripa, *Lectura super primum Sententiarum: prologi quaestiones I & II*, ed. A. Combes, Paris 1961 (Textes philosophiques du Moyen Âge, 8).

John Pecham, *Perspectiva communis* ed. and transl. D.C. Lindberg, *John Pecham and the Science of Optics*, Madison, WI, 1970 (University of Wisconsin Publications in Medieval Science, 14).

John Wyclif, *De civili dominio*, vol. 3, ed. J. Loserth, London 1903.

John Wyclif, *De eucharistia tractatus maior*, ed. J. Loserth, London 1892.

John Wyclif, *Sermones*, vol. 3, ed. J. Loserth, London 1889.

John Wyclif, *Tractatus de benedicta incarnacione*, ed. E. Harris, London 1886.

John Wyclif, *Tractatus de ecclesia*, ed. J. Loserth, London 1886.

John Wyclif, *Tractatus de Logica*, ed. M. Dziewicki, London 1893–1899 (3 vols.).

John Wyclif, *Tractatus de statu innocencie*, eds. J. Loserth and F. Matthew, London 1922.

John Wyclif, *Tractatus de universalibus*, ed. I. Mueller, Oxford 1985.

John Wyclif, *Trialogus*, ed. G. Lechler, Oxford 1869.

Jordanus de Nemore, *Elementa super demonstrationem ponderum*, ed. E.A. Moody. In M. Clagett and E.A. Moody, *The Medieval Science of Weights (Scientia de ponderibus): Treatises Ascribed to Euclid, Archimedes, Thabat Ibn Qurra, Jordanus de Nemore and Blasius of Parma*, Madison, WI, 1952 (University of Wisconsin Publications in Medieval Science, 1), 119–141.

Jordanus de Nemore, *Liber de ponderibus*, ed. E.A. Moody. In M. Clagett and E.A. Moody, *The Medieval Science of Weights (Scientia de ponderibus): Treatises Ascribed to Euclid, Archimedes, Thabat Ibn Qurra, Jordanus de Nemore and Blasius of Parma*, Madison, WI, 1952 (University of Wisconsin Publications in Medieval Science, 1), 143–165.

Kant, I., *Über das Mißlingen aller philosophischen Versuche in der Theodizee* (1791), ed. W. Weischedel, Frankfurt am Main 1974 (Werke, 11), 103–124.

Kästner, A.G., *Geschichte der Mathematik*, 1: *Geschichte der Künste und Wissenschaften seit der Widerherstellung derselben bis an das Ende des achtzehnten Jahrhunderts. Erster Zeitraum; bis zum Ende des sechszehnten Jahrhunderts*, Göttingen 1796.

Kepler, J., *Ad Vitellionem paralipomena, quibus astronomiae pars optica traditur*, Frankfurt am Main 1604.

Lawrence of Lindores, *Questiones super libris Physicorum*, ed. (partial) T. Dewender, *Das Problem des Unendlichen im ausgehenden 14. Jahrhundert. Eine Studie mit Textedition zum Physikkommentar des Lorenz von Lindores*, Amsterdam 1999 (Bochumer Studien zur Philosophie, 36).

Lefèvre d'Étaples, J. (ed.), *Egregii patris et clari theologi Ricardi [Sancti Victoris] De superdivina Trinitate theologicum opus hexade librorum distinctum. Commentarius artificio analytico*, Paris 1510.

Lefèvre d'Étaples, J. (ed.), *Georgii Trapezontii dialectica*, Paris 1508.

Lefèvre d'Étaples, J., and J. Clichtove, *Artificialis introductio per modum Epitomatis in decem libros Ethicorum Aristotelis adiectis elucidata [Jodoci Clichtovei] commentariis*, Paris 1502.

Lefèvre d'Étaples, J., and J. Clichtove, *Totius philosophiae naturalis paraphrases, adiecto commentario [Clichtovei]*, Paris 1502.

Lefèvre d'Étaples, J., *Dialogus difficilium physicalium introductorius*, Paris 1492.

Lefèvre d'Étaples, J., *Introductio in metaphysicorum libros Aristotelis*, ed. J. Clichtove, Paris 1494.

Lefèvre d'Étaples, J., *Totius Aristotelis philosophiae naturalis paraphrases*, Paris 1492.

Leibniz, G.W., *Die Theodizee*, ed. H. Herring, Frankfurt am Main 1965 (Philosophische Schriften, 2.1).

Leibniz, G.W., *La naissance du calcul différentiel: 26 articles des 'Acta Eruditorum'*, ed. and transl. M. Parmentier, Paris 1989.

[Leibniz, G.W.], *Leibnizens Mathematische Schriften*, ed. C.I. Gerhardt, Berlin 1848–1863, repr. Hildesheim 1962 (7 vols.).

Leibniz, G.W., *New Essays on Human Understanding*, eds. and transl. P. Remnant and J. Bennett, Cambridge 1981.

[Leibniz, G.W.], *Opuscules et fragments inédits de Leibniz*, ed. L. Couturat, Paris 1903.

Leibniz, G.W., *Sämtliche Schriften und Briefe*, Berlin 1923–.

Magirus, J., *Physiologia peripatetica*, Frankfurt am Main 1619.

Martínez Silíceo, J., *Liber arithmetices practice Astrologis phisicis et calculatoribus admodum utilis*, Paris 1513.

Messinus, *Questio cum expositione totius tractatus Hentisberi de tribus predicamentis*. In William Heytesbury, *Regule solvendi sophismata*, Venice 1494, 52va–62vb.

More, T., 'Letter to Martin Dorp,' ed. D. Kinney, New Haven, CT, 1986 (The Complete Works of St. Thomas More, 15).

Naudé, G., *Advis pour dresser une bibliothèque*, Paris 1644.

Nicole Oresme, *Écrits métaphysiques et théologiques. La résompte de 1362*, ed. A. Boureau, Paris 2021 (2 vols.) (Bibliothèque scolastique, 17).

Nicole Oresme, *Le livre du ciel et du monde*, eds. A.D. Menut and A.J. Denomy, Madison, WI, 1968 (University of Wisconsin Publications in Medieval Science, 11).

Nicole Oresme, *Questiones super geometriam Euclidis*, ed. H.L.L. Busard, Stuttgart 2010 (Boethius, 57).

Nicole Oresme, *Questiones super Physicam (Books I–VII)*, eds. S. Caroti, J. Celeyrette, S. Kirschner and E. Mazet, Leiden 2013 (Studien und Texte zur Geistesgeschichte des Mittelalters, 112).

Nicole Oresme, *Tractatus de commensurabilitate vel incommensurabilitate motuum celi*, ed. E. Grant, *Nicole Oresme and the Kinematics of Circular Motion.* Tractatus de commensurabilitate vel incommensurabilitate motuum celi, Madison, WI, 1971 (University of Wisconsin Publications in Medieval Science, 15).

Nicole Oresme, *Tractatus de configurationibus qualitatum et motuum*, ed. M. Clagett, *Nicole Oresme and the Medieval Geometry of Qualities and Motions. A Treatise on the Uniformity and Difformity of Intensities Known as* Tractatus de configurationibus qualitatum et motuum, Madison, WI, 1968 (University of Wisconsin Publications in Medieval Science, 12).

Nicole Oresme, *Tractatus De proportionibus proportionum* and *Ad Pauca respicientes*, ed. E. Grant, Madison, WI, 1966 (University of Wisconsin Publications in Medieval Science, 9).

Nifo, A., *Augstini Niphi ... Dialectica ludicra tyrunculis atque veteranis utillima peripateticis consona: inuioribus sophisticantibus contraria*, Venice 1521.

Nifo, A., *Dilucidarium Augustini Niphi Suessani Philosophi solertissimi metaphysicarum disputationum in Aristotelis Decem et quattuor Libros Metaphysicorum ...*, Venice 1559.

Oliverius Senensis, *Tractatus rationalis* (= *Opus tripartitum*), Siena 1491.

Paul of Venice, *Lectura super libros Metaphysicorum Aristotelis*, ed. G. Galluzzo, *The Medieval Reception of Book Zeta of Aristotle's Metaphysics*, Leiden 2012 (2 vols.) (Studien und Texte zur Geistesgeschichte des Mittelalters, 110).

Paul of Venice, *Summa philosophiae naturalis magistri Pauli Veneti noviter recognita ... restituta*, Venice 1503, repr. Hildesheim 1974.

Peter of Candia, *In Sententias*, eds. C. Schabel and S.F. Brown, URL = http://candia.ucy.ac.cy/index.htm.

Pomponazzi, P., *Tutti i trattati peripatetici*, eds. F.P. Raimondi and J.M. García Valverde, Milan 2013.

Richard Kilvington, *Quaestiones de motu* (Polish translation). In E. Jung, *Arystoteles na nowo odczytany. Ryszarda Kilvingtona 'Kwestie o ruchu'*, Łódź 2014, 107–316.

Richard Kilvington, *Quaestiones super libros Ethicorum*, ed. M. Michałowska, Leiden 2016 (Studien und Texte zur Geistesgeschichte des Mittelalters, 121).

Richard Kilvington, *Sophismata*, eds. N. Kretzmann and B.E. Kretzmann, Oxford 1990 (Auctores Britannici Medii Aevi, 12).

[Richard Kilvington], *The Sophismata of Richard Kilvington*, transl. N. Kretzmann and B.E. Kretzmann, Cambridge 1990.

Richard Swineshead, *Calculatoris sublime et prope divinum opus*, ed. J. Martínez Silíceo, Salamanca 1520.

Richard Swineshead, *Liber calculationum*, Padua ca. 1477; Pavia 1498; Venice 1520; Salamanca 1520, 1524.

Robert Holcot, *Questions on Future Contingents*, eds. P.A. Streveler and K.H. Tachau, *Robert Holcot. Seeing the Future Clearly. Questions on Future Contingents*, Toronto, ON, 1995 (Studies and Texts, 119).

Roger Bacon, *Liber secundus Communium naturalium Fratris Rogeri De celestibus*, ed. R. Steele, Oxford 1913 (Opera hactenus inedita, 4).

Roger Bacon, *Perspectiva*, ed. D.C. Lindberg, *Roger Bacon and the Origins of* Perspectiva *in the Middle Ages*: a Critical Edition and English Translation of Bacon's *Perspectiva* with Introduction and Notes, Oxford 1996.

Scaliger, J.C., *Exotericarum exercitationum liber quintus decimus: De subtilitate, ad Hieronymum Cardanum*, Paris 1557; Frankfurt 1607.

Tartaret, P., *Expositio totius philosophiae nec non metaphysicae Aristotelis*, Paris 1495.

Tartaret, P., *Tractatus de intensione, rarefactione et condensatione formarum utilis ad totam phisicam intelligendam extractus a Gregorio de Herimino et aliis doctoribus*, Paris 1493.

Themo Judaeus, *Question sur le mouvement de la lune* (1350), ed. H. Hugonnard-Roche, *L'oeuvre astronomique de Thémon Juif, Maître parisien du XIVe siècle*, Geneva 1973.

Thomas Aquinas, *Commentaria in octo libros Physicorum Aristotelis*, eds. Fratres Praedicatores, Rome 1884 (Opera omnia, 2).

Thomas Aquinas, *In octo libros Physicorum Aristotelis Expositio*, ed. P.M. Maggiòlo, Turin 1954.

Thomas Aquinas, *Quaestio disputata de spiritualibus creaturis*, ed. J. Cos, Rome 2000 (Opera omnia, 24/2).

Thomas Aquinas, *Quaestiones disputatae de anima*, ed. B.-C. Bazán, Rome 1996 (Opera omnia, 24/1).

Thomas Aquinas, *Scriptum super libros Sententiarum*, eds. P. Mandonnet and M.F. Moos, Paris 1929–1947 (4 vols.).

Thomas Bradwardine, *Arithmetica speculativa*, ed. P. Ciruelo, Paris 1495.

Thomas Bradwardine, *De causa Dei contra Pelagium et de virtute causarum ad suos Mertonenses libri tres*, ed. H. Savile, London 1618.

Thomas Bradwardine, *De causa Dei contra Pelagium*, ed. and transl. (partial) E.A. Lukács, Göttingen 2013.

Thomas Bradwardine, *De futuris contingentibus*, ed. J.-F. Genest, 'Le *De futuris contingentibus* de Thomas Bradwardine,' *Recherches Augustiniennes et Patristiques*, 14 (1979), 249–336.

Thomas Bradwardine, *De memoria artificiali adquirenda*, ed. M. Carruthers, 'Thomas Bradwardine, *De memoria artificiali adquirenda*,' *The Journal of Medieval Latin* 2 (1992), 25–43.

Thomas Bradwardine, *Geometria speculativa*, ed. and transl. A.G. Molland, Stuttgart 1989 (Boethius. Texte und Abhandlungen zur Geschichte der exakten Wissenschaften, 18)

Thomas Bradwardine, *Insolubilia*, ed. and transl. S. Read, Leuven 2010 (Dallas Medieval Texts and Translations, 10).

Thomas Bradwardine, *Tractatus de continuo*, ed. J.E. Murdoch, *Geometry and the Continuum in the Fourteenth Century: A Philosophical Analysis of Thomas Bradwardine's* Tractatus de Continuo, unpublished Ph.D. dissertation, University of Wisconsin, 1957.

Thomas Bradwardine, *Tractatus de proportionibus velocitatum in motibus*, ed. H.L. Crosby, Jr., *Thomas of Bradwardine. His* Tractatus de proportionibus. *Its Significance for the Development of Mathematical Physics*, Madison, WI, 1955 (University of Wisconsin Publications in Medieval Science, 2).

Thomas Bradwardine, *Traité des rapports entre les rapidités dans les mouvements*, transl. S. Rommevaux, Paris 2010 (Sagesses médiévales).

Thomas Bricot and George of Brussels, *Textus abbreviatus in cursum totius physices et metaphysicorum Aristotelis*, Lyon 1486.

Thomas Bricot, *Textus abbreviatus in cursum totius logices Aristotelis*, Paris 1489.

Thomas Buckingham, *Quaestiones in quatuor libros Sententiarum*, Paris 1505.

Thomas Wylton, *Quodlibet*, eds. L.O. Nielsen and C. Trifogli, *On the Intellectual Soul*, transl. G. Trimble, Oxford 2010 (Auctores Britannici Medii Aevi, 19).

Thomas, A., *Liber de triplici motu proportionibus annexis magistri Alvari Thome Ulixbonensis Suisseth calculationes ex parte declarans*, Paris 1509.

Valla, G., *De expetendis et fugiendis rebus*, Venice 1501.

Vossius, G., *De theologia gentili et physiologia Christiana*, Amsterdam 1641.

Walter Burley, *De primo et ultimo instanti*, eds. C. Shapiro and H. Shapiro, 'De primo et ultimo instanti des Walter Burley,' *Archiv für Geschichte der Philosophie* 47 (1965), 159–173.

Wendelin, M.F., *Contemplationum physicarum sectio I. qua physiologia generalis, De principiis et affectionibus Corporibus naturalis*, Cambridge 1648.

William de Chambre, *Continuatio Historiae Dunelmensis*, ed. J. Raine, Newcastle 1839.

William Heytesbury, 'The Compounded and Divided Senses', in N. Kretzmann and E. Stump (eds. and transl.), *The Cambridge Translations of Medieval Philosophical Texts*, 1: *Logic and Philosophy of Language*, Cambridge 1988, 413–434.

William Heytesbury, *On 'Insoluble' Sentences: Chapter One of his Rules for Solving Sophisms*, transl. P.V. Spade, Toronto, ON, 1979 (Mediaeval Sources in Translation, 21).

William Heytesbury, *Regule solvendi sophismata*, Venice 1494, 4^{v}–53^{r}.

William Heytesbury, *Sophismata*. In William Heytesbury, *Regule solvendi sophismata*, Venice 1494, 77va–170vb.

William of Ockham, *Expositio in libros Physicorum Aristotelis. Libri IV–VIII*, eds. R. Wood, R. Green, G. Gál, J. Giermek, F. Kelley, G. Leibold and G. Etzkorn, St. Bonaventure, NY, 1985 (Opera philosophica, 5).

William of Ockham, *Quaestiones in libros Physicorum Aristotelis*, ed. S. Brown, St. Bonaventure, NY, 1984 (Opera philosophica, 6), 395–813.

2 Secondary Sources

Aiton, E.J., *Leibniz. A Biography*, Bristol 1985.

Allott, S., *Alcuin of York: His Life and Letters*, York 1974.

Amerini, F., and G. Galluzzo (eds.), *A Companion to the Latin Medieval Commentaries on Aristotle's Metaphysics*, Leiden 2014 (Brill's companions to the Christian Tradition, 43).

Antognazza, M.R., *Leibniz. An Intellectual Biography*, Cambridge 2009.

Arthur, R.T.W., 'Leibniz's syncategorematic infinitesimals,' *Archive for History of Exact Sciences*, 67/5 (2013), 553–593.

Axworthy, A., *Le mathématicien renaissant et son savoir. Le statut des mathématiques selon Oronce Fine*, Paris 2016 (Histoire et philosophie des sciences, 11).

Bakker, P.J.J.M., 'Fifteenth-Century Parisian Commentaries on Aristotle's *Metaphysics*,' in F. Amerini and G. Galluzzo (eds.), *A Companion to the Latin Medieval Commentaries on Aristotle's Metaphysics*, Leiden 2014 (Brill's companions to the Christian Tradition, 43), 575–630.

Bakker, P.J.J.M., 'Natural Philosophy and Metaphysics in Late Fifteenth-Century Paris. I: The Commentaries on Aristotle by Johannes Hennon,' *Bulletin de philosophie médiévale* 47 (2005), 125–155.

Bakker, P.J.J.M., 'Natural Philosophy and Metaphysics in Late Fifteenth-Century Paris. II: The Commentaries on Aristotle by Johannes le Damoisiau,' *Bulletin de philosophie médiévale* 48 (2006), 209–228.

Bakker, P.J.J.M., 'Natural Philosophy and Metaphysics in Late Fifteenth-Century Paris. III: The Commentaries on Aristotle by Johannes de Caulaincourt (alias Johannes de Magistris),' *Bulletin de philosophie médiévale* 49 (2007), 195–237.

Bakker, P.J.J.M., 'Nicholas of Amsterdam on Accidental Being: A Study and Edition of Two Questions from his Commentary on the *Metaphysics*,' *Bochumer Philosophisches Jahrbuch für Antike und Mittelalter* 15 (2012), 131–180.

Bakker, P.J.J.M., and J.M.M.H. Thijssen (eds.), *Mind, Cognition and Representation. The Tradition of Commentaries on Aristotle's* De anima, Aldershot 2007 (Ashgate Studies in Medieval Philosophy).

Bale, J., *Index Britanniae Scriptorum (ca. 1550)*, eds. R. Poole and M. Bateson, Oxford 1902.

Bartha, P., and L. Pasternack (eds.), *Pascal's Wager*, Cambridge 2018 (Classic Philosophical Arguments, 4).

Bartocci B., and S. Masolini, 'Reading Aristotle at the University of Louvain in the Fifteenth Century: A First Survey of Petrus de Rivo's Commentaries on Aristotle (II),' *Bulletin de philosophie médiévale* 56 (2014), 281–383.

Bartocci B., and S. Read, 'John Dumbleton on insolubles: an edition of an Epitome of his solution to insoluble,' *Noctua: La tradizione filosofica dall'antico al moderno*, 2022 (forthcoming).

Bartocci B., S. Masolini and R.L. Friedman, 'Reading Aristotle at the University of Louvain in the Fifteenth Century: A First Survey of Petrus de Rivo's Commentaries on Aristotle (I),' *Bulletin de philosophie médiévale* 55 (2013), 133–175.

Baxter, J.H.S., 'The Philosopher Laurence of Lindores,' *Philosophical Quarterly* 5 (1955), 348–354.

Beltrán, E., 'Jacques Legrand O.E.S.A. Sa vie et son oeuvre,' *Augustiniana* 24 (1974), 132–160 and 387–414.

Beltrán, E., 'Jacques Legrand prédicateur,' *Analecta Augustiniana* 30 (1967), 148–209.

Berger, H., 'Helmold of Zoltwedel (†1441): His Academic Career, Scientific Works, and Philosophical Alignment,' in O. Pavlíček (ed.), *Studying the Arts in Late Medieval Bohemia*, Turnhout 2021 (Studia Artistarum, 48), 175–203.

Biagioli, M., 'The Social Status of Italian Mathematicians,' *History of Science* 27 (1989), 41–95.

Biard, J., 'God as First Principle and Metaphysics as a Science,' in R.L. Friedman and L.O. Nielsen (eds.), *The Medieval Heritage in Early Modern Metaphysics and Modal Theory, 1400–1700*, Dordrecht 2003 (The New Synthese Historical Library, 53), 75–97.

Biard, J., 'Matière, forme, qualités: Blaise de Parme et le statut de la matière,' in T. Suarez-Nani and A. Paravicini Bagliani (eds.), *Materia: nouvelles perspectives de recherche dans la pensée et la culture médiévales (XIIe–XVIe siècles)*, Florence 2017 (Micrologus Library, 83), 191–215.

Biard, J., 'Nature et degrés de perfection chez Blaise de Parme,' in L. Bianchi, O. Grassi, and C. Panti (eds.), *Edizioni, traduzioni e tradizioni filosofiche (secoli XII–XVI): Studi per Pietro B. Rossi*, Cantenaro 2018 (2 vols.), 1: 387–410.

Biard, J., and A. Robert (eds.), *La philosophie de Blaise de Parme. Physique, psychologie, éthique*, Florence 2019 (Micrologus Library, 96).

Biard, J., and J. Celeyrette (eds.), *De la théologie aux mathématique. L'infini au XIVe siècle*, Paris 2005 (Sagesses médiévales).

Bocheński, J.M., *A History of Formal Logic*, transl. I. Thomas, Notre Dame, IN, 1961 (2nd ed., New York 1970).

Bocheński, J.M., *Formale Logik*, Freiburg 1956 (6th ed., Freiburg 2015).

Bodnár, I., 'Facinus de Ast. Opera Philosophica I. *Tractatus de maximo et minimo*,' *Archives d'histoire doctrinale et littéraire du Moyen Âge* 64 (1997), 405–437.

Bodnár, I., 'Facinus de Ast. Opera Philosophica II. Questiones super libros Physicorum,' *Archives d'histoire doctrinale et littéraire du Moyen Âge* 65 (1998), 331–414.

Bono, J.J., 'Reform and the Languages of Renaissance Theoretical Medicine: Harvey versus Fernel,' *Journal of the History of Biology* 23/3 (1990), 341–387.

Borgo, M., 'Latin Medieval Translations of Aristotle's Metaphysics,' in F. Amerini and G. Galluzzo (eds.), *A Companion to the Latin Medieval Commentaries on Aristotle's Metaphysics*, Leiden 2014 (Brill's companions to the Christian Tradition, 43), 19–56.

Bos, E.P. (ed.), *John of Holland, Four Tracts on Logic (Suppositiones, Fallaciae, Obligationes, Insolubilia)*, Nijmegen 1985 (Artistarium, 5).

Bottin, F., 'Gli "insolubilia" nel "curriculum" scolastico tardo-medievale (con l'edizione di un anonimo compendio degli "Insolubilia" di Giovanni Dumbleton),' in C. Giacon (ed.), *Logica, epistemologia, storia della storigrafia*, Padua 1980, 13–32.

Bottin, F., 'L'*Opinio de insolubilibus* di Richard Kilmyngton,' *Rivista critica di storia della filosofia* 28 (1973), 408–421.

Bottin, F., *Le antinomie semantiche nella logica medievale*, Padua 1976.

Bottin, F., 'The Mertonians' Metalinguistic Sciences and the *insolubilia*,' in P.O. Lewry (ed.), *The Rise of British Logic*, Toronto, ON, 1985 (Papers in Mediaeval Studies, 7), 235–248.

Bouton-Touboulic, A.-I., *L'ordre caché. La notion d'ordre chez Saint Augustin*, Paris 2004 (Collection des études augustiniennes. Série Antiquité, 174).

Boyer, C.B., *The Concepts of the Calculus*, New York 1939.

Brînzei, M., 'When Theologians Play Philosopher: a Lost Confrontation in the *Principia* of James of Eltville and His *Socii* on the Perfection of Species and Its Infinite Latitude,' in M. Brînzei and C. Schabel (eds.), *The Cistercian James of Eltville (†1393). Author in Paris and Authority in Vienna*, Turnhout 2018 (Studia Sententiarum, 3), 42–77.

Britton, J.P., *Models and Precision: The Quality of Ptolemy's Observations and Parameters*, New York 1992 (Sources and Studies in the History and Philosophy of Classical Science, 1).

Broadie, A., 'John Mair's *Dialogus de Materia Theologo Tractanda*. Introduction, Text and Translation,' in A.A. MacDonald, Z.R.W.M. von Martels and J.R. Veenstra (eds.), *Christian Humanism: Essays in Honour of Arjo Vanderjagt*, Leiden 2009 (Studies in Medieval and Reformation Traditions, 142), 419–430.

Broadie, A., *The Circle of John Mair. Logic and Logicians in Pre-Reformation Scotland*, Oxford 1985.

Brodrick, G.C., *Memorials of Merton College with Biographical Notices of the Wardens and Fellows*, Oxford 1885.

Bruin, F., and M. Bruin, 'The Equator Ring, Equinoxes, and Atmospheric Refraction,' *Centaurus* 20 (1976), 89–111.

Burns, W.E., '"A Proverb of Versatile Mutability": Proteus and Natural Knowledge in Early Modern Britain,' *The Sixteenth Century Journal* 32 (2001), 969–980.

Burton, D., *Nicole Oresme's* De visione stellarum (*On Seeing the stars*)*: a Critical Edition of Oresme's Treatise on Optics and Atmospheric Refraction*, Leiden 2007 (Medieval and Early Modern Philosophy and Science, 7).

Buzzetti, D., 'Intensione delle forme e strutture linguistiche negli scritti bolognesi di filosofia naturale del secolo XIV,' *Medioevo* 29 (2004), 133–176.

Buzzetti, D., 'Linguaggio e ontologia nei commenti di autore Bolognese al *De tribus predicamentis*,' in D. Buzzetti, M. Ferriani and A. Tabarroni (eds.), *L'insegnamento della logica a Bologna nel XIV secolo*, Bologna 1992 (Studi e memorie. Nuova serie, 8), 579–604.

Caroti, S., 'Da Walter Burley al *Tractatus sex inconvenientium*: La tradizione inglese della discussione medievale *De reactione*,' *Medioevo* 21 (1995), 257–374.

Caroti, S., 'La discussione sull'*intensio et remissio formarum* nelle università italiane (sec. XIV),' in L. Bianchi and C. Crisciani (eds.), *Forme e oggetti della conoscenza nel XIV secolo: studi in ricordo di Maria Elena Reina*, Florence 2014 (Micrologus Library, 61), 415–460.

Caroti, S., 'Nicole Oresme: dalla *questio* alla *glose*. La presenza del dibattito universitario nelle glosse di *Le livre du ciel et du monde*,' in N. Bray and L. Sturlese (eds.), *Filosofia in volgare nel medioevo. Atti del convegno della Società Italiana per lo studio del pensiero medievale*, Louvain-la-Neuve 2003, (Textes et études du Moyen Âge, 21), 155–190.

Caroti, S., 'Pomponazzi e la *reactio*: note sulla fortuna del pensiero oxoniense e parigino nella filosofia italiana del Rinascimento,' in G. Federici Vescovini (ed.), *Filosofia e scienza classica, arabo-latina medievale e l'età moderna*, Turnhout 1999 (Textes et études du Moyen Âge, 11), 255–288.

Carruthers, M., and J.M. Ziolkowski (eds.), *The Medieval Craft of Memory. An Anthology of Texts and Pictures*, Philadelphia, PA, 2002.

Carruthers, M., *The Book of Memory: A Study of Memory in Medieval Culture*, New York 1990.

Catto, J.I., 'Wyclif and Wycliffism at Oxford 1356–1430,' in J.I. Catto and T.A.R. Evans (eds.), *The History of the University of Oxford*, 2: *Late Medieval Oxford*, Oxford 1992, 175–261.

Catto, J.I., and T.A.R. Evans (eds.), *The History of the University of Oxford*, 2: *Late Medieval Oxford*, Oxford 1992.

Celenza, C.S., *The Intellectual World of the Italian Renaissance: Language, Philosophy, and the Search for Meaning*, New York 2018.

Celeyrette, J., 'Le mouvement selon la cause chez Messino da Codronchi et Angelo di Fossambruno,' in J. Biard and S. Rommevaux (eds.), *Mathématiques et théorie du mouvement, XIVe–XVIe siècles*, Villeneuve d'Ascq 2008 (Histoire des sciences), 15–29.

Celeyrette, J., 'L'argumentation mathématique dans la physique d'Oresme,' in C. Grellard (ed.), *Méthodes et statut des sciences à la fin du Moyen-Âge*, Villeneuve d'Ascq 2004 (Histoire des sciences), 201–215.

Cesalli, L., 'Propositions: their Meaning and Truth,' in C. Dutilh Novaes and S. Read (eds.), *The Cambridge Companion to Medieval Logic*, Cambridge 2016, 245–264.

Clagett, M. (ed.), *Nicole Oresme and the Medieval Geometry of Qualities and Motions. A Treatise on the Uniformity and Difformity of Intensities Known as* Tractatus de configurationibus qualitatum et motuum, Madison, WI, 1968 (University of Wisconsin Publications in Medieval Science, 12).

Clagett, M., *Archimedes in the Middle Ages*, 5: *Quasi-Archimedean Geometry in the Thirteenth Century. A Supplementary Volume Comprising the* Liber de motu *of Gerard of Brussels, the* Liber philotegni *of Jordanus de Nemore together with its Longer Version Known as the* Liber de triangulis Iordani, *and an Appended Text of John Dee's* Inventa circa illam coni recti atque rectanguli sectionem quae ab antiquis mathematicis parabola appellabatur, Philadelphia, PA, 1984 (Memoirs of the American Philosophical Society, 157).

Clagett, M., *Giovanni Marliani and Late Medieval Physics*, New York 1941.

Clagett, M., 'Richard Swineshead and Late Medieval Physics,' *Osiris* 9 (1950), 131–161.

Clagett, M., *Studies in Medieval Physics and Mathematics*, London 1979 (Collected Studies Series, 103).

Clagett, M., 'The Impact of Archimedes on Medieval Science,' *Isis* 50 (1959), 419–429.

Clagett, M., 'The *Liber de motu* of Gerard of Brussels and the Origins of Kinematics in the West,' *Osiris* 12 (1956), 73–175.

Clagett, M., *The Science of Mechanics in the Middle Ages*, Madison, WI, 1959 and 1961 (University of Wisconsin Publications in Medieval Science, 4).

Clagett, M., and E.A. Moody, *The Medieval Science of Weights (Scientia de ponderibus): Treatises Ascribed to Euclid, Archimedes, Thabat Ibn Qurra, Jordanus de Nemore and Blasius of Parma*, Madison, WI, 1952 (University of Wisconsin Publications in Medieval Science, 1).

Coleman, J., 'Jean de Ripa, O.F.M. and the Oxford Calculators,' *Medieval Studies* 37 (1975), 130–189.

Combes, A., 'La intensité des formes d'après Jean de Ripa,' *Archives d'histoire doctrinale et littéraire du Moyen Âge* 45 (1970), 17–147.

Combes, A., 'Présentation de Jean de Ripa,' *Archives d'histoire doctrinale et littéraire du Moyen Âge* 23 (1956), 145–242.

Conti, A., 'John Wyclif,' in E.N. Zalta (ed.), *The Stanford Encyclopedia of Philosophy* (Spring 2017 Edition), URL = https://plato.stanford.edu/archives/spr2017/entries/wyclif/.

Conti, A., 'Paul of Venice,' in E.N. Zalta (ed.), *The Stanford Encyclopedia of Philosophy* (Summer 2017 Edition), URL = https://plato.stanford.edu/archives/sum2017/entries/paul-venice/.

Conti, A., 'Paul of Venice's Commentary on the *Metaphysics*,' in F. Amerini and G. Galluzzo (eds.), *A Companion to the Latin Medieval Commentaries on Aristotle's Metaphysics*, Leiden 2014 (Brill's companions to the Christian Tradition, 43), 551–574.

Conti, A., 'Wyclif's Logic and Metaphysics,' in I.C. Levy (ed.), *A Companion to John Wyclif. Late Medieval Theologian*, Leiden 2006 (Brill's Companions to the Christian Tradition, 4), 67–125.

Coope, U., *Time for Aristotle. Physics IV.10–14*, Oxford 2005 (Oxford Aristotle Studies), 125–139.

Cormack, L.B., S.A. Walton and J.A. Schuster (eds.), *Mathematical Practitioners and the Transformation of Natural Knowledge in Early Modern Europe*, Cham, Switzerland, 2017 (Studies in History and Philosophy of Science, 45).

Corsten, S., 'Universities and Early Printing,' *Bibliography and the Study of 15th-Century Civilization* 5 (1987) 83–123.

Corvino, F., 'Il "De perfectione specierum" di Ugolino D'Orvieto,' *ACME Annali della Facoltà die Filosofia e Lettere dell'Università Statale di Milano* 8 (1954), 73–105.

Cotton, J., 'Benedetto Reguardati of Nursia (1398–1469),' *Medical History* 13 (1969), 175–189.

Courtenay, W.J., *Schools and Scholars in Fourteenth-Century England*, Princeton, NJ, 1987.

Courtenay, W.J., 'The Early Stages in the Introduction of Oxford Logic into Italy,' in A. Maierù (ed.), *English Logic in Italy in the 14th and 15th Centuries: Acts of the 5th European Symposium on Medieval Logic and Semantics, Rome, 10–14 November 1980*, Naples 1982 (History of Logic, 1), 12–32.

Courtenay, W.J., 'The Role of English Thought in the Transformation of University Education in the Late Middle Ages,' in J.M. Kittelson and P.J. Transue (eds.), *Rebirth, Reform and Resilience. Universities in Transition, 1300–1700*, Columbus, OH, 1984, 108–115.

Courtenay, W.J., 'The University of Paris at the Time of Jean Buridan and Nicole Oresme,' *Vivarium* 42 (2004), 3–17.

Courtenay, W.J., 'Theology and Theologians from Ockham to Wyclif,' in J.I. Catto and T.A.R. Evans (eds.), *The History of the University of Oxford, 2: Late Medieval Oxford*, Oxford 1992, 1–34.

Crombie, A.C., *Medieval and Early Modern Science*, Vol. II: Science in the Latter Middle Ages and Early Modern Times: XIII–XVII Centuries, New York 1959.

Crosby, H.L., Jr. (ed.), *Thomas of Bradwardine. His* Tractatus de proportionibus. *Its Significance for the Development of Mathematical Physics*, Madison, WI, 1955 (University of Wisconsin Publications in Medieval Science, 2).

Cunningham, A., 'The Pen and the Sword: Recovering the Disciplinary Identity of Physiology and Anatomy before 1800: I: Old Physiology—the Pen,' *Studies in History and Philosophy of Biological and Biomedical Sciences* 33/4 (2002), 631–665.

Cunningham, A., and S. Kusukawa, 'Translators' Introduction,' in A. Cunningham and S. Kusukawa (eds.), *Natural Philosophy Epitomised: Books 8–11 of Gregor Reisch's* Philosophical Pearl *(1503)*, Farnham 2010, ix–lxxiv.

Curtze, M., 'Eine Studienreise. Rechenschaftsbericht über seine mit Unterstützung der Königl. Preußischen Akademie der Wissenschaften zu Berlin in den Monaten August, September und Oktober 1896 unternommenen Forschungen zur Geschichte der Geometrie im Mittelalter,' *Centralblatt für Bibliothekswesen*, XVI/6–7, (1899), 257–306.

Da Valsanzibio, S., 'La "via Scoti" in Gaetano di Thiene,' in C. Bérubé (ed.), *Regnum Hominis et Regnum Dei, 2: La tradizione scotista veneto-padovana*, Rome 1978, 199–205.

Da Valsanzibio, S., *Vita e dottrina di Gaetano di Thiene, filosofo dello Studio di Padova (1387–1465)*, Verona 1948, repr. Padua 1949.

De Bellis, E., *Bibliografia di Agostino Nifo*, Lecce 2005.

De Bellis, E., *Il pensiero logico di Agostino Nifo*, Lecce 1998.

De Bellis, E., *Nicoletto Vernia e Agostino Nifo. Aspetti storiografici e metodologici*, Lecce 2003.

Dewender, T., *Das Problem des Unendlichen im ausgehenden 14. Jahrhundert. Eine Studie mit Textedition zum Physikkommentar des Lorenz von Lindores*, Amsterdam 1999 (Bochumer Studien zur Philosophie, 36).

Di Liscia, D.A., 'Biagio Pelacani da Parma's Geometrisation of Latitudes and the Problems of the Mean Degree Theorem,' in J. Biard and A. Robert (eds.), *La philosophie de Blaise de Parme. Physique, psychologie, éthique*, Florence 2019 (Micrologus Library, 96), 129–152.

Di Liscia, D.A., 'Der Kommentar des Johannes Rucherat de Wesalia zur Aristotelischen Physik: seine Bedeutung und Überlieferung,' *Codices Manuscripti et Impressi* 99/100 (2015), 9–28.

Di Liscia, D.A., 'Die fallenden Körper und die Rätsel des Domingo de Soto,' in M. Segre and E. Knobloch (eds.), *Der ungebändigte Galilei: Beiträge zu einem Symposion*, Stuttgart 2001, 9–22.

Di Liscia, D.A., *Eine Wiener* Expositio *zum* Tractatus de latitudinibus formarum. *Edition mit Einführung und Kommentar*, Vienna (forthcoming).

Di Liscia, D.A., 'Kalkulierte Ethik: Vives und die "Zerstörer" der Moralphilosophie (Le Maistre, Cranston und Almain),' in S. Ebbersmeyer and E. Kessler (eds.), *Ethik: Wissenschaft oder Lebenskunst? Modelle der Normenbegründung von der Antike bis zur frühen Neuzeit*, Münster 2007, 75–105.

Di Liscia, D.A., 'La "latitud de las formas" y la geometrización de la ciencia del movimiento,' *Mediaevalia. Textos e estudos* 36 (2017), 75–114.

Di Liscia, D.A., 'La *conclusio pulchra, mirabilis et bona*: una ingeniosa demostración atribuible a Nicole Oresme,' *Mediaevalia. Textos e estudos* 37 (2018), 139–168.

Di Liscia, D.A., 'The "*Latitudines breves*" and Late Medieval University Teaching,' *SCIAMVS* 17 (2016), 55–120.

Di Liscia, D.A., 'The Questions on the *Metaphysics* attributed to Johannes Rucherat de Wesalia (A Textual Analysis),' *Filosofia. Revista da Faculdade de Letras da Universidade do Porto* 35–36 (2018–2019), 57–75.

Di Liscia, D.A., 'The Subject Matter of Physics and Metaphysics in Jacques Legrand's *Compendium utriusque philosophie*,' *Revista Española de Filosofía Medieval* 24 (2017), 249–265.

Di Liscia, D.A., 'Transmutación y movimiento según el tiempo en Jacques Legrand (*Compendium utriusque philosophie* IV, 1–2),' in C.J. Fernández and M. Pérez Carrasco (eds.), Per philosophica documenta. *Estudios en honor de Francisco Bertelloni*, Buenos Aires 2021, 151–175.

Di Liscia, D.A., 'Velocidad *quo ad effectus* y velocidad *quo ad causas*: La tradición de los calculadores y la metodología aristotélica,' in D.A. Di Liscia, E. Kessler and C. Methuen (eds.), *Method and Order in Renaissance Philosophy of Nature. The Aristotle Commentary Tradition*, Aldershot 1997, 143–176.

Di Liscia, D.A., 'Walter Burley, Paulus Venetus und die Tradition *De Instanti* (mit dem "*Tractatus de instanti*" des Paulus Venetus nach Hs. Florenz, Biblioteca Nazionale

Centrale, II. IV. 553, foll. 68va–69va),' in A. Speer and D. Wirmer (eds.), *Das Sein der Dauer*, Berlin 2008 (Miscellanea Mediaevalia, 34), 130–135.

Di Liscia, D.A., *Zwischen Geometrie und Naturphilosophie: die Entwicklung der Formlatitudenlehre im deutschen Sprachraum*, unpublished Ph.D. dissertation, University of Munich, 2003.

Dillon, J., *The Golden Chain. Studies in the Development of Platonism and Christianity*, Aldershot 1990 (Collected Studies Series, 333).

Dionisotti, C., 'Ermolao Barbaro e la Fortuna di Suiseth,' *Medioevo e Rinascimento: Studi in onore di Bruno Nardi*, Florence 1955, 1: 219–253.

Drake, S., 'Bradwardine's Function, Mediate Denomination and Multiple Continua,' *Physis* 12 (1970), 51–68.

Drake, S., 'Medieval Ratio Theory vs. Compound Medicines in the Origins of Bradwardine's Rule,' *Isis* 64 (1973), 67–77.

Drake, S., and I.E. Drabkin (eds.), *Mechanics in Sixteenth-Century Italy: Selections from Tartaglia, Benedetti, Guido Ubaldo, and Galileo*, Madison, WI, 1969 (University of Wisconsin Publications in Medieval Science, 13).

Duba, W.O., and C. Schabel, 'Remigio, Scotus, Auriol and the Myth of the Two-Year *Sentences* Lecture at Paris,' *Recherches de Théologie et Philosophie Médiévales* 84 (2017), 143–179.

Duchesneau, F., 'Leibniz's Theoretical Shift in the *Phoranomus* and *Dynamica de Potentia*,' *Perspectives on Science* 6 (1998), 77–109.

Duchesneau, F., *La dynamique de Leibniz*, Paris 1994 (Mathesis).

Duhem, P., 'Dominique Soto et la scolastique parisienne (*suite*),' *Bulletin hispanique* 14 (1912), 275–299, repr. in Id., *Études sur Léonard de Vinci*, 3: *Les précurseurs parisiens de Galilée*, Paris 1913, 531–555.

Duhem, P., *Études sur Léonard de Vinci*, 3: *Les précurseurs parisiens de Galilée*, Paris 1913.

Duhem, P., 'La dialectique d'Oxford et la scolastique italienne,' *Bulletin italien* 12 (1912), 6–26, 93–120, 203–223, 289–298, repr. in Id., *Études sur Léonard de Vinci*, 3: *Les précurseurs parisiens de Galilée*, Paris 1913, 405–481.

Duhem, P., 'La tradition de Buridan et la science italienne au XVIe siècle,' *Bulletin italien* 9 (1909), 338–360, repr. in Id., *Études sur Léonard de Vinci*, 3: *Les précurseurs parisiens de Galilée*, Paris 1913, 113–137.

Dutilh Novaes, C., and S. Uckelman, 'Obligationes,' in C. Dutilh Novaes and S. Read (eds.), *The Cambridge Companion to Medieval Logic*, Cambridge 2016, 370–395.

Ebbesen, S., and F. Goubier, *A Catalogue of 13th-Century Sophismata*, I: *Introduction and Indices*; II: *Catalogue of* Sophismata, Paris 2010 (Sic et non).

Ehrle, F., *Der Sentenzenkommentar Peters von Candia des Pisaner Papstes Alexanders V. Ein Beitrag zur Scheidung der Schulen in der Scholastik des vierzehnten Jahrhunderts und zur Geschichte des Wegestreites*, Münster 1925 (Franziskanische Studien, Beiheft 9).

Élie, H., 'Quelques maîtres de l'Université de Paris vers l'an 1500,' *Archives d'histoire doctrinale et littéraire du Moyen Âge* 18 (1950–1951), 193–243.

Emden, A.B., *A Biographical Register of the University of Oxford to A.D. 1500*, Oxford 1957–1959 (3 vols.).

Eneström, G., 'Sur l'auteur d'un traité 'de motu' auquel Bradwardin a fait allusion en 1328,' *Archivio di Storia della Scienza* 2 (1921–1922), 133–136.

Farge, J.K., *Biographical Register of Paris Doctors of Theology, 1500–1536*, Toronto, ON, 1980 (Subsidia mediaevalia, 10).

Federici Vescovini, G., 'Blaise de Parme et l'intelligibilité mathématique,' in J. Biard and A. Robert (eds.), *La philosophie de Blaise de Parme. Physique, psychologie, éthique*, Florence 2019 (Micrologus Library, 96), 107–128.

Federici Vescovini, G., 'Il commento di Angelo di Fossombrone al *De tribus praedicamentis* di Guglielmo Heytesbury,' in A. Maierù (ed.), *English Logic in Italy in the 14th and 15th Centuries: Acts of the 5th European Symposium on Medieval Logic and Semantics, Rome, 10–14 November 1980*, Naples 1982 (History of Logic, 1), 359–374.

Federici Vescovini, G., 'La *Quaestio de intensione et remissione formarum* di Biagio Pelacani da Parma,' *Physis* 31 (1994), 433–535.

Federici Vescovini, G., 'Simone di Castello e la medicina dei "moderni",' in Ead., *Arti e filosofia nel secolo XIV: Studi sulla tradizione aristotelica e i 'moderni'*, Florence 1983, 213–229.

Fernandez Walker, G., 'A New Source of Nicholas of Autrecourt's *Quaestio*: The Anonymous *Tractatus de sex inconvenientibus*,' *Bulletin de philosophie médiévale* 55 (2013), 57–69.

Fichant, M., 'Neue Einblicke in Leibniz's Reform seiner Dynamik (1678),' *Studia Leibnitiana* 22 (1990), 38–68.

Field, H., 'Maudlin's *Truth and Paradox*,' *Philosophy and Phenomenological Research* 73 (2006), 713–720.

Fleming, B., *Thomas de Bradwardine, Oxford Scholar, Royal Servant, and Archbishop of Canterbury*, unpublished Ph.D. dissertation, Leuven University, 1964.

Fletcher, J.M., 'The Faculty of Arts,' in J.I. Catto and T.A.R. Evans (eds.), *The History of the University of Oxford, 2: Late Medieval Oxford*, Oxford 1992, 369–399.

Forrester, J.M., and J. Henry (eds.), *The 'Physiologia' of Jean Fernel (1567)*, Philadephia, PA, 2003 (Transactions of the American Philosophical Society, 93/1).

Fryde, E.B., and J.R.L. Highfield, 'An Oxfordshire Deed of Balliol College,' *Oxoniensia* 20 (1955), 40–45.

Gabriel, A.L., 'Metaphysics in the Curriculum of Studies of the Mediaeval Universities,' in P. Wilpert (ed.), *Die Metaphysik im Mittelalter. Ihr Ursprung und Ihre Bedeutung*, Berlin 1963 (Miscellanea Mediaevalia, 2), 92–102.

Garin, E., *L'età nuova. Richerche di storia della cultura dal XII al XVI secolo*, Naples 1969 (Materiali di filosofia, 11).

Genest, J.-F., 'Le *De futuris contingentibus* de Thomas Bradwardine,' *Recherches Augustiniennes et Patristiques*, 14 (1979), 249–336.

Genest, J.-F., 'Les premiers écrits théologiques de Bradwardine: textes inédits et découvertes récentes,' in G.R. Evans (ed.), *Medieval Commentaries on the* Sentences *of Peter Lombard. 1: Current Research*, Leiden 2002, 395–421.

Genest, J.-F., *Prédétermination et liberté créée à Oxford au XIV*e *siècle. Buckingham contre Bradwardine*, Paris 1992 (Études de philosophie médiévale, 70).

Genest, J.-F., and K.H. Tachau, 'La lecture de Thomas Bradwardine sur les Sentences,' *Archives d'histoire doctrinale et littéraire du Moyen Âge*, 57 (1990), 301–306.

Gilbert, N.W., 'Richard de Bury and the *Quires of Yesterday's Sophisms*,' in E.P. Mahoney (ed.), *Philosophy and Humanism. Renaissance Essays in Honour of Paul Oskar Kristeller*, New York 1976, 229–257.

Gloria, A., *Monumenti dell'Università di Padova (1318–1405)*, Padua 1888 (2 vols.).

Goddu, A., *The Physics of William of Ockham*, Leiden 1984 (Studien und Texte zur Geistesgeschichte des Mittelalters, 16).

Goldstein, B.R., 'Refraction, Twilight, and the Height of the Atmosphere,' *Vistas in Astronomy* 20 (1976), 105–107.

Goulding, R., *Defending Hypatia: Ramus, Savile, and the Renaissance Rediscovery of Mathematical History*, New York 2010 (Archimedes, New Studies in the History of Science and Technology, 25).

Grabmann, M., 'Der Anteil Deutschlands am Aristotelismus des Mittelalters,' in Id., *Mittelalterliches Geistesleben: Abhandlungen zur Geschichte der Scholastik und Mystik*, Munich 1956 (3 vols.), 3: 219–231.

Grabmann, M., *Methoden und Hilfsmittel des Aristotelesstudiums im Mittelalter*, München 1939 (Sitzungsberichte der Bayerischen Akademie der Wissenschaften, Philosophisch-historische Abteilung, 1939/5).

Grant, E., 'Cosmology,' in D.C. Lindberg (ed.), *Science in the Middle Ages*, Chicago 1978 (Chicago History of Science and Medicine), 265–302.

Grant, E., *Much Ado About Nothing. Theories of Space and Vacuum from the Middle Ages to the Scientific Revolution*, Cambridge 1981.

Grant, E., *Nicole Oresme and the Kinematics of Circular Motion.* Tractatus de commensurabilitate vel incommensurabilitate motuum celi, Madison, WI, 1971 (University of Wisconsin Publications in Medieval Science, 15).

Grant, E., *Nicole Oresme*, Tractatus de Proportionibus proportionum *and* Ad Pauca respicientes, Madison, WI, 1966 (University of Wisconsin Publications in Medieval Science, 9).

Grellard, C., 'L'usage des nouveaux langages d'analyse dans la *quaestio* de Nicolas d'Autrécourt. Contribution à la théorie autrécourienne de la connaissance,' in S. Caroti and J. Celeyrette (eds.), Quia inter doctores est magna dissensio. *Les débats de philosophie naturelle à Paris au XIV*e *siècle*, Florence 2004 (Biblioteca di Nuncius, 52), 69–95.

Hájek, A., 'The (In)validity of Pascal's Wager,' in P. Bartha and L. Pasternack (eds.), *Pascal's Wager*, Cambridge 2018 (Classic Philosophical Arguments, 4), 123–147.

Hanke, M., 'Cajetan of Thiene on the Logic of Paradox,' *Studia Neoaristotelica* 13 (2016), 71–95.

Hanke, M., and E. Jung, 'William Heytesbury,' in E.N. Zalta (ed.), *The Stanford Encyclopedia of Philosophy* (Spring 2018 Edition), URL = https://plato.stanford.edu/archives/spr2018/entries/heytesbury/.

Hantsche, I., *Der 'Mathematicus': Zur Entwicklung und Bedeutung einer neuen Berufsgruppe in der Zeit Gerhard Mercators*, Bochum 1996 (Duisburger Mercator-Studien, 4).

Hasse, D.N., 'Influence of Arabic and Islamic Philosophy on the Latin West,' in E.N. Zalta (ed.), *The Stanford Encyclopedia of Philosophy* (Fall 2014 Edition), URL = https://plato.stanford.edu/archives/fall2014/entries/arabic-islamic-influence/.

Heiberg, J.L., *Beiträge zur Geschichte Georg Vallas und seiner Bibliothek*, Leipzig 1896.

Henry, J., '"Mathematics Made No Contribution to the Public Weal": Why Jean Fernel (1497–1558) Became a Physician,' *Centaurus* 53 (2011), 193–220.

Henry, J., 'Occult Qualities and the Experimental Philosophy: Active Principles in Pre-Newtonian Matter Theory,' *History of Science*, 24 (1986), 335–381.

Herold, V., 'The Spiritual Background of the Czech Reformation: Precursors of Jan Hus,' in F. Šmahel and O. Pavlíček (eds.), *A Companion to Jan Hus*, Leiden 2015 (Brill's Companions to the Christian Tradition, 54), 69–95.

Hissette, R., *Enquête sur les 219 articles condamnés à Paris le 7 mars 1277*, Leuven 1977 (Philosophes médiévaux, 22).

Hobbins, D., *Authorship and Publicity Before Print: Jean Gerson and the Transformation of Late Medieval Learning*, Philadelphia, PA, 2009 (Middle Ages Series).

Hugonnard-Roche, H., 'Analyse sémantique et analyse *secundum imaginationem* dans la physique parisienne au XIVe siècle,' in S. Caroti (ed.), *Studies in Medieval Natural Philosophy*, Florence 1989 (Biblioteca di Nuncius, 1), 133–153.

Hussey, E., *Aristotle Physics Books III and IV*, Oxford 1983 (Clarendon Aristotle Series).

Ierodiakonou, K., and S. Roux (eds.), *Thought Experiments in Methodological and Historical Contexts*, Leiden 2011 (Medieval and Early Modern Philosophy and Science, 15).

Iogna-Prat, D., *Order and Exclusion: Cluny and Christendom Face Heresy, Judaism, and Islam (1000–1150)*, Ithaca, NY, 2002.

Ishiguro, H., *Leibniz's Philosophy of Logic and Language*, 2nd ed., Cambridge 1991.

Jones, J., *Balliol College: A History, 1263–1939*, Oxford 1988.

Jung-Palczewska, E., 'Works by Richard Kilvington,' *Archives d'histoire doctrinale et littéraire du Moyen Âge* 67 (2000), 181–223.

Jung, E., *Arystoteles na nowo odczytany. Ryszarda Kilvingtona 'Kwestie o ruchu'*, Łódź 2014.

Jung, E., 'Intension and Remission of Forms,' in H. Lagerlund (ed.), *Encyclopedia of Medieval Philosophy*, Dordrecht 2011, 551–555.

Jung, E., 'Mathematics and the *secundum imaginationem* Procedure in Richard Kilvington,' *Przegląd Tomistyczny* 22 (2016), 109–120.

Jung, E., 'Richard Kilvington on Local Motion,' in P.J.J.M. Bakker (ed.) with the collaboration of E. Faye and C. Grellard, *Chemins de la pensée médiévale. Études offertes à Zénon Kaluza*, Turnhout 2002 (Textes et études du Moyen Âge, 20), 116–128.

Jung, E., 'Richard Kilvington,' in E.N. Zalta (ed.), *The Stanford Encyclopedia of Philosophy* (Winter 2016 Edition), URL = https://plato.stanford.edu/archives/win2016/entries/kilvington/.

Jung, E., and R. Podkoński, 'Richard Kilvington on Continuity,' in C. Grellard and A. Robert (eds.), *Atomism in Late Medieval Philosophy and Theology*, Leiden 2009 (Medieval and Early Modern Philosophy and Science, 9), 65–84.

Jung, E., and R. Podkoński, 'Richard Kilvington on Proportions,' in J. Biard and S. Rommevaux (eds.), *Mathématiques et théorie du mouvement, XIVe–XVIe siècles*, Villeneuve d'Ascq 2008 (Histoire des sciences), 81–101.

Jung, E., and R. Podkoński, *Towards the Modern Theory of Motion. Oxford Calculators and the New Interpretation of Aristotle*, Łódź 2020 (Research on Science and Natural Philosophy, 4).

Kaluza, Z., 'La nature des écrits de Jean de Ripa,' *Traditio* 43 (1987) 257–298.

Kaluza, Z., *Les querelles doctrinales à Paris: nominalistes et réalistes aux confins du XIVe et du XVe siècles*, Bergamo 1988 (Quodlibet, 2).

Kaluza, Z., *Nicolas d'Autrécourt. Ami de la vérité*, Paris 1995 (Histoire littéraire de la France, 42/1).

Keele, R., 'The Logical and Scientific Treatises of John Chilmark (†1396),' *Bulletin de philosophie médiévale* 49 (2007), 119–137.

King, P., 'Scotus on Metaphysics,' in T. Williams (ed.), *The Cambridge Companion to Duns Scotus*, Cambridge 2003, 15–68.

Kirschner, S., 'Oresme on Intension and Remission of Qualities in His Commentary on Aristotle's *Physics*,' *Vivarium* 38 (2000), 255–274.

Kitanov, S., and C. Schabel, 'Thomas Bradwardine's Questions on Grace and Merit from His *Lectura* on the *Sentences* at Oxford, 1332–1333,' *Archives d'histoire doctrinale et littéraire du Moyen Âge* 89 (2022) (forthcoming).

Kok, F.J., 'John Buridan's Commentary on the *Metaphysics*,' in F. Amerini and G. Galluzzo (eds.), *A Companion to the Latin Medieval Commentaries on Aristotle's Metaphysics*, Leiden 2014 (Brill's companions to the Christian Tradition, 43), 495–549.

Korolec, G.B., *Repertorium commentariorum medii aevi in Aristotelem Latinorum quae in Bibliotheca olim Universitatis Pragensis nunc Státní Knihovna ČSR vocata asservantur*, Wrocław 1977.

Kretzmann, N., 'Continua, Indivisibles, and Change in Wyclif's Logic of Scripture,' in A. Kenny (ed.), *Wyclif in His Times*, Oxford 1986, 31–65.

Kretzmann, N., 'Richard Kilvington and the Logic of Instantaneous Speed,' in A. Maierù and A. Paravicini Bagliani (eds.), *Studi sul XIV secolo in memoria di Anneliese Maier*, Rome 1981 (Storia e letteratura, 151), 143–178.

Kretzmann, N., '*Tu scis hoc esse omne quod est hoc*: Richard Kilvington and the Logic of Knowledge,' in N. Kretzmann (ed.), *Meaning and Inference in Medieval Philosophy. Studies in Memory of Jan Pinborg*, Dordrecht 1988 (Synthese Historical Library, 32), 225–245.

Kristeller, P.O., *Iter Italicum. A Finding List of Uncatalogued or Incompletely Catalogued Humanistic Manuscripts of the Renaissance in Italian and Other Libraries*, London 1967–1997 (7 vols.).

Löb, M., 'On a Solution of a Problem of Leon Henkin,' *Journal of Symbolic Logic* 20 (1995), 115–118.

Laird, W.R., 'The School of Merton and the Middle Sciences,' *Bulletin de philosophie médiévale*, 38 (1996), 41–51.

Langer, U., *Perfect Friendship: Studies in Literature and Moral Philosophy from Boccaccio to Corneille*, Geneva 1994 (Histoire des idées et critique littéraire, 331).

Lanuza-Navarro, T.M.C., 'Astrology in Court: The Spanish Inquisition, Authority, and Expertise,' *History of Science* 55 (2017), 187–209.

Larsen, A.E., 'John Wyclif, c. 1331–1384,' in I.C. Levy (ed.), *A Companion to John Wyclif. Late Medieval Theologian*, Leiden 2006 (Brill's Companions to the Christian Tradition, 4), 1–65.

Latham, R.E. et al. (eds.), *Dictionary of Medieval Latin from British Sources*, London 1986 and 1997.

Leff, G., *Bradwardine and the Pelagians*, Cambridge 1957 (Cambridge Studies in Medieval Life and Thought, Series 2, 5).

Lehn, W.H., and S. van der Werf, 'Atmospheric Refraction: A History,' *Applied Optics* 44/27 (2005), 5624–5636.

Leigh, M., *From Polypragmon to Curiosus: Ancient Concepts of Curious and Meddlesome*, Oxford 2013.

Levy, I.C. (ed.), *A Companion to John Wyclif. Late Medieval Theologian*, Leiden 2006 (Brill's Companions to the Christian Tradition, 4).

Lička, L., 'Studying and Discussing Optics at the Prague Faculty of Arts: Optical Topics and Authorities in Prague Quodlibets and John of Borotín's *Quaestio* on Extramission,' in O. Pavlíček (ed.), *Studying the Arts in Late Medieval Bohemia: Production, Reception and Transmission of Knowledge*, Turnhout 2021 (Studia Artistarum, 48), 251–303.

Lindberg, D.C., *A Catalogue of Medieval and Renaissance Optical Manuscripts*, Toronto, ON, 1965 (Subsidia Mediaevalia, 4).

Lindberg, D.C., 'The Cause of Refraction in Medieval Optics,' *The British Journal for the History of Science* 4 (1968), 23–38.

Lindberg, D.C., 'The Genesis of Kepler's Theory of Light: Light Metaphysics from Plotinus to Kepler,' *Osiris* 2 (1986), 4–42.

Lines, D.A., 'Humanism and the Italian Universities,' in C.S. Celenza and K. Gouwens (eds.), *Humanism and Creativity in the Renaissance. Essays in Honor of Ronald G. Witt*, Leiden 2006 (Brill's Studies in Intellectual History, 136), 327–346.

Lines, D.A., 'Lefèvre and French Aristotelianism on the Eve of the Sixteenth Century,' in G. Frank and A. Speer (eds.), *Der Aristotelismus in der Frühen Neuzeit: Kontinuität oder Wiederangeignung?*, Wiesbaden 2007, 273–290.

Lines, D.A., 'Teaching Physics in Louvain and Bologna: Franz Titelmans and Ulysse Aldrovandi,' in E. Campi, S. De Angelis, A.-S. Goeing and A. Grafton (eds.), *Scholarly Knowledge: Textbooks in Early Modern Europe*, Geneva 2008 (Travaux d'humanisme et Renaissance, 447), 183–203.

Livesey, S.J., 'The Oxford Calculatores, Quantification of Qualities, and Aristotle's Prohibition of Metabasis,' *Vivarium* 24 (1986), 50–69.

Lohr, C.H., *Latin Aristotle Commentaries I.1. Medieval Authors A–L*, Florence 2013 (Corpus philosophorum medii aevi. Subsidia, 17).

Lohr, C.H., *Latin Aristotle Commentaries I.2. Medieval Authors M–Z*, Florence 2010 (Corpus philosophorum medii aevi. Subsidia, 18).

Lohr, C.H., 'Metaphysics,' in C.B. Schmitt, Q. Skinner, E. Kessler and J. Kraye (eds.), *The Cambridge History of Renaissance Philosophy*, Cambridge 1992, 537–638.

Lohrmann, D., '*Motus continuus* und *motus perpetuus* in der mittelalterlichen Technik und Physik,' in A. Speer and D. Wirmer (eds.), *Das Sein der Dauer*, Berlin 2008 (Miscellanea Mediaevalia, 34), 224–243.

Longeway, J., *William Heytesbury: On Maxima and Minima. Chapter 5 of* Rules for Solving Sophismata *with an Anonymous Fourteenth-Century Discussion*, Dordrecht 1984 (Synthese Historical Library, 26).

Lourié, B., 'Intensio: a Story of a Term from Swineshead to Leibniz,' *Studia Humana* 1 (2012), 59–65.

Lovejoy, A.O., *The Great Chain of Being. A Study of the History of an Idea*, Cambridge, MA, 1936.

Löwe, C.L., 'Gregory of Rimini on the Intension and Remission of Corporeal Forms,' *Recherches de Théologie et Philosophie Médiévales* 81 (2014), 273–330.

Lukács, E.A., 'Bradwardine and Buckingham on the Extramundane Void,' *Bochumer Philosophisches Jahrbuch für Antike und Mittelalter* 17 (2014), 123–149.

Lukács, E.A., '"Our Stories" and "Your Stories": Narrating the Passion and the Resurrection in Thomas Bradwardine's *De causa Dei*,' *Essays in Medieval Studies* 31 (2015), 183–200.

Lüthy, C.H., 'An Aristotelian Watchdog as Avant-Garde Physicist: Julius Caesar Scaliger,' *The Monist* 84 (2001), 542–561.

Mahoney, E.P., 'Lovejoy and the Hierarchy of Being,' *Journal for the History of Ideas* 48 (1987), 221–230.

Mahoney, E.P., 'Metaphysical Foundations of the Hierarchy of Being According to Some Late-Medieval and Renaissance Philosophers,' in P. Morewedge (ed.), *Philosophies of Existence, Ancient and Medieval*, New York 1982, 165–257.

Maier, A., *An der Grenze von Scholastik und Naturwissenschaft* (Studien zur Naturphilosophie der Spätscholastik, 3), 2nd ed., Rome 1952 (Storia e letteratura, 52).

Maier, A., *Ausgehendes Mittelalter. Gesammelte Aufsätze zur Geistesgeschichte des 14. Jahrhunderts*, 1, Rome 1964 (Storia e letteratura, 97).

Maier, A., 'Das Problem der intensiven Größe,' in Ead., *Zwei Grundprobleme der scholastischen Naturphilosophie. Das Problem der intensiven Größe. Die Impetustheorie* (Studien zur Naturphilosophie der Spätscholastik, 2), 3rd ed., Rome 1968 (Storia e letteratura, 37), 3–109.

Maier, A., 'Der Funktionsbegriff in der Physik des 14. Jahrhunderts,' *Divus Thomas*, 3rd series, 24 (1946), 147–166.

Maier, A., 'Die Mathematik der Formlatituden,' in Ead., *An der Grenze von Scholastik und Naturwissenschaft* (Studien zur Naturphilosophie der Spätscholastik, 3), 2nd ed., Rome 1952 (Storia e letteratura, 52) (1st ed., Essen 1943), 255–384.

Maier, A., 'Die *Quaestio de velocitate* des Johannes von Casale,' *Archivum Franciscanum Historicum*, 53 (1960), 276–306, repr. in Ead., *Ausgehendes Mittelalter. Gesammelte Aufsätze zur Geistesgeschichte des 14. Jahrhunderts*, 1, Rome 1964 (Storia e letteratura, 97), 381–411.

Maier, A., *Die Vorläufer Galileis im 14. Jahrhundert* (Studien zur Naturphilosophie der Spätscholastik, 1), 2nd ed., Rome 1966 (Storia e letteratura, 22).

Maier, A., 'Diskussionen über das aktuell Unendliche in der ersten Hälfte des 14. Jahrhunderts,' *Divus Thomas*, 3rd series, 25 (1947), 147–166, 317–337, repr. in Ead., *Ausgehendes Mittelalter. Gesammelte Aufsätze zur Geistesgeschichte des 14. Jahrhunderts*, 1, Rome 1964 (Storia e letteratura, 97), 41–85.

Maier, A., *Metaphysische Hintergründe der Spätscholastischen Naturphilosophie* (Studien zur Naturphilosophie der Spätscholastik, 4), Rome 1955 (Storia e letteratura, 52).

Maier, A., Review of H.L. Crosby, Jr. (ed.), *Thomas of Bradwardine. His* Tractatus de proportionibus (1955), *Isis* 48 (1957), 84–87.

Maier, A., 'Verschollene Aristoteleskommentare des 14. Jahrhunderts', in *Autour d'Aristote. Recueil d'études de philosophie ancienne et médiévale offert à Monseigneur A. Mansion*, Leuven 1955, 515–541, repr. with addenda in Ead., *Ausgehendes Mittelalter. Gesammelte Aufsätze zur Geistesgeschichte des 14. Jahrhunderts*, 1, Rome 1964, 237–264.

Maier, A., 'Zu Walter Burleys Traktat *De intensione et remissione formarum*,' *Franciscan Studies* 25 (1965), 293–321.

Maierù, A. (ed.), *English Logic in Italy in the 14th and 15th Centuries: Acts of the 5th European Symposium on Medieval Logic and Semantics, Rome, 10–14 November 1980*, Naples 1982 (History of Logic, 1).

Maierù, A., *Terminologia logica della tarda scolastica*, Rome 1972 (Lessico Intellettuale Europeo, 8).

Maierù, A., and E.D. Sylla, 'Daughter of Her Time: Anneliese Maier (1905–1971) and the Study of Fourteenth-Century Philosophy,' in J. Chance (ed.), *Women Medievalists and the Academy*, Madison, WI, 2005, 625–645.

Markowski, M., 'Der Buridanismus an der Krakauer Universität im Mittelalter,' in: O. Pluta (ed.), *Die Philosophie im 14. und 15. Jahrhundert. In memoriam Konstanty Michalski (1879–1947)*, Amsterdam 1988 (Bochumer Studien zur Philosophie, 10), 245–260.

Markowski, M., 'Komentarz Jana Aurifabera z Erfurtu do *Fizyki* Arystotelesa w rękopisie 1423 Bibliotheki Univerwersyteckiej w Lipsku,' *Acta mediaevalia* 12 (1999), 335–343.

Markowski, M., *Repertorium commentariorum medii aevi in Aristotelem Latinorum qui in Bibliotheca Universitatis Lipsiensis asservantur*, Kraków 2012.

Marr, A. (ed.), *The Worlds of Oronce Fine. Mathematics, Instruments and Print in Renaissance France*, Donington 2009.

Martin, G.H., and J.R.L. Highfield, *A History of Merton College*, Oxford 1997.

Mazet, E., 'Pierre Ceffons et Oresme: leur relation revisitée,' in S. Caroti and J. Celeyrette (eds.), Quia inter doctores est magna dissensio. *Les débats de philosophie naturelle à Paris au XIVe siècle*, Florence 2004 (Biblioteca di Nuncius, 52), 175–194.

McGuire, J.E., '"Labyrinthus Continui": Leibniz on Substance, Activity, and Matter,' in P.K. Machamer and R.G. Turnbull (eds.), *Motion and Time, Space and Matter: Interrelations in the History of Philosophy and Science*, Columbus, OH, 1976, 290–326.

McVaugh, M.R., 'Arnald of Villanova and Bradwardine's Law,' *Isis* 58 (1967), 56–64.

Michałowska, M. (ed.), *Richard Kilvington*, Quaestiones super libros Ethicorum, Leiden 2016 (Studien und Texte zur Geistesgeschichte des Mittelalters, 121).

Miethke, J., 'Kirchenstruktur und Staatstheorien im Zeitalter der Scholastik,' in B. Schneidmüller and S. Weinfurter (eds.), *Ordnungskonfigurationen im hohen Mittelalter*, Ostfildern 2006 (Vorträge und Forschungen, 64), 127–151.

Molland, A.G., 'Swineshead, Roger (*d.* 1365?), Natural Philosopher and Benedictine Monk,' in *Oxford Dictionary of National Biography* 53, Oxford 2004, doi:10.1093/ref:odnb/52682.

Molland, A.G., *The* Geometria speculativa *of Thomas Bradwardine. Text with Critical Discussion*, unpublished Ph.D. dissertation, University of Cambridge, 1967.

Molland, A.G., 'The Geometrical Background on the "Merton School",' *British Journal for the History of Science* 4 (1968–1969), 108–125.

Moody, E.A., 'Galileo and Avempace: The Dynamics of the Leaning Tower Experiment,' *Journal of the History of Ideas* 12 (1951), 163–193, 375–422.

Moody, E.A., 'Laws of Motion in Medieval Physics,' *Scientific Monthly* 72 (1951), 18–23.

Moody, E.A., *Studies in Medieval Philosophy, Science and Logic. Collected Papers 1933–1969*, Berkeley, CA, 1975, 189–201.

Moonan, L., 'Lawrence of Lindores on Life in the Living Being,' *Classica et mediaevalia* 27 (1966), 349–374.

Moore, G.E., 'Moore's paradox,' in T. Baldwin (ed.), *G.E. Moore: Selected Writings*, London 1993, 207–212.

Morewedge, P. (ed.), *Philosophies of Existence, Ancient and Medieval*, New York 1982.

Mueller, I. (ed.), *John Wyclif*, Tractatus de universalibus, Oxford 1985.

Müller, K., and G. Krönert, *Leben und Werk von Gottfried Wilhelm Leibniz. Eine Chronik*, Frankfurt am Main 1969 (Veröffentlichungen des Leibniz-Archivs, 2).

Murdoch, J.E., 'A Skewed View: The Achievement of Late Medieval Science and Philosophy as seen from the Renaissance,' in S.E. Young (ed.), *Crossing Boundaries at Medieval Universities*, Leiden 2010, (Education and Society in the Middle Ages and Renaissance, 36), 175–188.

Murdoch, J.E., 'From Social into Intellectual Factors: An Aspect of the Unitary Character of Late Medieval Learning,' in J.E. Murdoch and E.D. Sylla (eds.), *The Cultural Context of Medieval Learning*, Dordrecht 1975 (Boston Studies in the Philosophy of Science, 26), 271–348.

Murdoch, J.E., *Geometry and the Continuum in the Fourteenth Century: A Philosophical Analysis of Thomas Bradwardine's* Tractatus de Continuo, unpublished Ph.D. dissertation, University of Wisconsin, 1957.

Murdoch, J.E., 'Infinity and Continuity,' in N. Kretzmann, A. Kenny, J. Pinborg and E. Stump (eds.), *The Cambridge History of Later Medieval Philosophy. From the Rediscovery of Aristotle to the Disintegration of Scholasticism, 1100–1600*, Cambridge 1982, 564–591.

Murdoch, J.E., 'Mathematics and Sophisms in Late Medieval Natural Philosophy,' in *Les genres littéraires dans les sources théologiques et philosophiques médiévales: définition, critique et exploitation. Actes du colloque international de Louvain-la-Neuve, U.C.L., Institut d'études médiévales*, Louvain-la-Neuve 1982 (Textes, Études, Congrès, 5), 85–100.

Murdoch, J.E., '*Mathesis in philosophiam scholasticam introducta*. The Rise and Developement of the Application of Mathematics in Fourteenth-Century Philosophy and Theology,' in *Arts libéraux et philosophie au Moyen Âge. Actes du quatrième Congrès international de philosophie médiévale, Montréal 1967*, Montreal 1969, 215–254.

Murdoch, J.E., 'Philosophy and the Enterprise of Science in the Later Middle Ages,' in Y. Elkana (ed.), *The Interaction between Science and Philosophy*, Atlantic Highlands, NJ, 1974 (The Van Leer Jerusalem Foundation Series), 51–74 (plus discussion, 91–113).

Murdoch, J.E., 'Pierre Duhem and the History of Late Medieval Science and Philosophy in the Latin West,' in R. Imbach and A. Maierù (eds.), *Gli studi di filosofia medievale fra Otto e Novecento. Contributo a un bilancio storiografico*, Rome 1991 (Storia e letteratura, 179), 253–302.

Murdoch, J.E., 'Propositional Analysis in Fourteenth-Century Natural Philosophy: A Case Study,' *Synthese* 40 (1979), 117–146.

Murdoch, J.E., '*Scientia mediantibus vocibus*: Metalinguistic Analysis in Late Medieval Natural Philosophy,' in A. Speer (ed.), *Sprache und Erkenntnis im Mittelalter*, 1, Berlin 1981 (Miscellanea Mediaevalia, 13), 73–106.

Murdoch, J.E., '*Subtilitates Anglicanae* in Fourteenth-Century Paris: John of Mirecourt and Peter Ceffons,' in M.P. Cosman and B. Chandler (eds.), *Machaut's World: Science and Art in the Fourteenth Century*, New York 1978 (Annals of the New York Academy of Sciences, 314), 51–86.

Murdoch, J.E., 'The Development of a Critical Temper. New Approaches and Modes of Analysis in Fourteenth-Century Philosophy,' *Medieval and Renaissance Studies* 7 (1978), 51–79.

Murdoch, J.E., 'The Involvement of Logic in Late Medieval Natural Philosophy,' in S. Caroti, (ed.), *Studies in Medieval Natural Philosophy*, Florence 1989 (Biblioteca di Nuncius, 1), 3–28.

Murdoch, J.E., 'The Medieval Language of Proportions: Elements of the Interaction with Greek Foundations and the Development of New Mathematical Techniques,' in A.C. Crombie, *Scientific Change. Historical Studies in the Intellectual, Social and Technical Conditions for Scientific Discovery and Technical Invention, from Antiquity to the Present. Symposium on the History of Science, University of Oxford 9–15 July, 1961*, London 1963, 237–271.

Murdoch, J.E., and E.D. Sylla, 'Swineshead (Swyneshed, Suicet, etc.) Richard,' in C.C. Gillispie (ed.), *Dictionary of Scientific Biography*, 13, New York 1976, 184–213.

Murdoch, J.E., and E.D. Sylla, 'The Science of Motion,' in D.C. Lindberg (ed.), *Science in the Middle Ages*, Chicago 1978 (Chicago History of Science and Medicine), 206–264.

Nauta, L.W., 'Latin as a Common Language: The Coherence of Lorenzo Valla's Humanist Program,' *Renaissance Quarterly* 71 (2018), 1–32.

Navarro-Brotons, V., *Disciplinas, saberes y prácticas: Filiosofía natural, matemáticas y astronomía en la sociedad española de la época moderna*, Valencia 2014.

Newman, W.R., 'Eloge: John Murdoch, 10 May 1927–16 September 2010,' *Isis* 102/103 (2011), 529–532.

Nielsen, L.O., and C. Trifogli, 'Thomas Wylton's Questions on Number, the Instant, and Time,' *Documenti e studi sulla tradizione filosofica medievale* 16 (2005), 88–105.

Nutton, V., 'Physiologia from Galen to Jacob Bording,' in M. Horstmanshoff, H. King and C. Zittel (eds.), *Blood, Sweat and Tears: The Changing Concepts of Physiology from Antiquity into Early Modern Europe*, Leiden 2012 (Intersections, 25), 27–40.

Oosterhoff, R.J., 'Lovers in Paratexts: Oronce Fine's Republic of Mathematics,' *Nuncius* 31/3 (2016), 549–583.

Oosterhoff, R.J., *Making Mathematical Culture: University and Print in the Circle of Lefèvre d'Étaples*, Oxford 2018 (Oxford-Warburg Studies).

Oosterhoff, R.J., 'The Dialogue of Ingenuous Students: Early Printed Textbooks at Paris,' in S. Berger and D. Garber (eds.), *Teaching Philosophy in the Seventeenth Century: Text and Image*, Dordrecht 2021, 11–30.

Palmerino, C.R., '*Geschichte des Kontinuumproblems* or Notes on Fromondus' Labyrinthus? On the True Nature of LH XXXVII, IV, 57r–58v,' *The Leibniz Review* 26 (2016), 63–98.

Palumbo M., *Leibniz e la* Res Bibliothecaria. *Bibliografie*, historiae literariae *e cataloghi nella biblioteca privata leibniziana*, Rome 1993 (Il bibliotecario).

Pantin, W.A., *Canterbury College, Oxford*, 4, Oxford 1985 (Oxford Historical Society, New Series, 30).

Pantin, W.A., *The English Church in the Fourteenth Century*, Cambridge 1955.

Pasnau, R., *Metaphysical Themes 1274–1671*, Oxford 2011.

Patera A., and A. Podlaha, *Soupis rukopisů knihovny Metropolitní kapitoly Pražské*, Prague 1910–1922 (2 vols.).

Penn, S., *Truth, Time and Sacred Text: Responses to Medieval Nominalism in John Wyclif's* Summa de Ente *and* De Veritate Sacrae Scripturae, unpublished Ph.D. dissertation, University of York, 1998.

Pesic, P., 'Shapes of Proteus in Renaissance Art,' *Huntington Library Quarterly* 73 (2010), 57–82.

Petrescu, L., 'The Threefold Object of the Scientific Knowledge. Pseudo-Scotus and the Literature on the *Meteorologica* in Fourteenth-Century Paris,' *Franciscan Studies* 72 (2014), 465–502.

Pinborg, J., 'Die Aristoteles-Quaestionen des magister Nicolaus von Amsterdam,' *Classica et mediaevalia* 25 (1964), 244–261.

Pironet, F., 'William Heytesbury and the Treatment of *Insolubilia* in Fourteenth-Century England Followed by a Critical Edition of Three Anonymous Treatises *De insolubilibus* Inspired by Heytesbury,' in S. Rahman, T. Tulenheimo and E. Genot (eds.), *Unity, Truth and the Liar. The Modern Relevance of Medieval Solutions to the Liar Paradox*, Dordrecht 2008 (Logic, epistemology, and the unity of science, 8), 255–333.

Pluta, O., 'Materialism in the Philosophy of Mind. Nicholas of Amsterdam's *Quaestiones De anima*,' in P.J.J.M. Bakker and J.M.M.H. Thijssen (eds.), *Mind, Cognition and Representation. The Tradition of Commentaries on Aristotle's* De anima, Aldershot 2007 (Ashgate Studies in Medieval Philosophy), 109–126.

Pluta, O., 'Nicholas of Amsterdam: Life and Works,' *Bochumer Philosophisches Jahrbuch für Antike und Mittelalter* 16 (2013), 185–265.

Podkoński, R., 'Richard Swineshead's *Liber calculationum* in Italy: Some Remarks on Manuscripts, Editions and Dissemination,' *Recherches de Théologie et Philosophie Médiévales* 80 (2013), 307–361.

Podkoński, R., 'Richard Swineshead's *Liber calculationum* in Italy: the Codex Bibl. Naz.

San Marco lat. VI.226 and its significance,' *Recherches de Théologie et Philosophie Médiévales* 84 (2017), 407–426.

Podkoński, R., *Richard Swineshead's Theory of Motion*, Łódź 2019 (Research on Science and Natural Philosophy, 2).

Podkoński, R., 'Richardus Kilvington, *Utrum continuum sit divisibile in infinitum*,' *Mediaevalia Philosophica Polonorum* 36 (2007), 123–175.

Podkoński, R., *Suisetica inania, Ryszarda Swinesheada spekulatywna nauka o ruchu lokalnym*, Łódź 2018.

Podkoński, R., 'Thomas Bradwardine's Critique of the *Falsigraphus*'s Concept of Actual Infinity,' *Studia Antyczne i Mediewistyczne* 36 (2003), 147–153.

Powicke, F.M., *The Medieval Books of Merton College*, Oxford 1931.

Pozzi, L., *Il Mentitore e il Medioevo. Il dibattito sui paradossi dell'autoriferimento. Scelta di testi, commento, traduzione*, Parma 1987 (Quaderni di filosofia, 10).

Prior, A.N., 'Epimenides the Cretan,' *Journal of Symbolic Logic* 23 (1958), 261–266.

Raimondi, F.P., 'Pomponazzi e la polemica anticalculatoria,' in F.P. Raimondi and J.M. García Valverde (eds.), *Tutti i trattati peripatetici*, Milan 2013, 9–86.

Raimondi, F.P., 'Pomponazzi's Criticism of Swineshead and the Decline of the Calculatory Tradition in Italy,' *Physis*, New series, 37 (2000), 311–358.

Ramati, A., 'Harmony at a Distance: Leibniz's Scientific Academies,' *Isis* 87 (1996), 430–452.

Ranea, A.G., 'Calculationes, Galileo, and the Misfortune of Leibniz's A Priori Demonstration,' in I. Marchlewitz (ed.), *Leibniz. Tradition und Aktualität. V. Internationaler Leibniz-Kongreß unter der Schirmherrschaft des Niedersächsischen Ministerpräsidenten Dr. Ernst Albrecht.* Hannover, 14. bis 19. November 1988, Vorträge II. Teil, Hannover 1988–1989, 785–793.

Ranea, A.G., 'The *a priori* Method and the *actio* Concept Revised: Dynamics and Metaphysics in an Unpublished Controversy between Leibniz and Denis Papin,' *Studia Leibnitiana* 21 (1989), 42–68.

Raschieri, A.A., 'Giorgio Valla, Editor and Translator of Ancient Scientific Texts,' in: P. Olmos (ed.), *Greek Science in the Long Run: Essays on the Greek Scientific Tradition (4th c. BCE–17th c. CE)*, Cambridge 2012, 127–149.

Read, S., 'Bradwardine's Revenge,' in J.C. Beall (ed.), *Revenge of the Liar*, Oxford 2007, 250–261.

Read, S., 'Medieval Theories: Properties of Terms,' in E.N. Zalta (ed.), *The Stanford Encyclopedia of Philosophy* (Spring 2019 Edition), URL = https://plato.stanford.edu/archives/spr2015/entries/medieval-terms/.

Read, S., 'Paradoxes of Signification,' *Vivarium* 54 (2016), 335–355.

Read, S., 'Richard Kilvington and the Theory of Obligations,' *Vivarium* 53 (2015), 391–404.

Read, S., and M. Thakkar, 'Robert Fland—or Elandus Dialecticus?,' *Mediaeval Studies* 78 (2016), 167–180.

Renaudet, A., *Préréforme et humanisme à Paris pendant les premières guerres d'Italie, 1494–1517*, 2nd ed., Paris 1953.

Renn, J., 'Towards an Historical Epistemology. Ten Years Max Planck Institute for the History of Science, 1994–2004,' *Research Report 2002–2003* of the Max Planck Institute for the History of Science in Berlin, URL = https://www.mpiwg-berlin.mpg.de/sites/default/files/2017-09/rr_02_03_complete.pdf.

Rice, E.F., Jr. (ed.), *The Prefatory Epistles of Jacques Lefèvre d'Étaples and Related Texts*, New York 1972.

Rief, J., *Der Ordobegriff des jungen Augustinus*, Paderborn 1962 (Abhandlungen zur Moraltheologie, 2).

Rijk, L.M. de (ed.), *Some 14th Century Tracts on the* Probationes Terminorum (*Martin of Alnwick O.F.M., Richard Billingham, Edward Upton and Others*), Nijmegen 1982 (Artistarium, 3).

Rijk, L.M. de, '*Logica Cantabrigiensis*: A Fifteenth Century Cambridge Manual of Logic,' *Revue Internationale de Philosophie*, 29 (1975), 297–315.

Robert, A., 'Atomism at Oxford after John Wyclif. The Cases of Robert Alyngton and Roger Whelpdale,' in L. Campi and S. Simonetta (eds.), *Before and After Wyclif. Sources and Textual Influences*, Basel 2020 (Textes et Études du Moyen Âge, 97), 57–94.

Robinet, A., *G.W. Leibniz: Iter Italicum (Mars 1689–Mars 1690). La dynamique de la République des Lettres. Nombreux textes inédits*, Florence 1988 (Accademia Toscana di Scienze e Lettere 'La Colombaria', 90).

Robson, J.A., *Wyclif and the Oxford Schools: The Relation of the* Summa de ente *to Scholastic Debates at Oxford in the Late Fourteenth Century*, Cambridge 1961 (Cambridge Studies in Medieval Life and Thought. New series, 8).

Rommevaux-Tani, S., 'La détermination de la rapidité d'augmentation dans le *De sex inconvenientibus*: comparaison avec les développements sur le même sujet de William Heytesbury,' in C. Grellard (ed.), *Miroir de l'amitié. Mélanges offerts à Joël Biard*, Paris 2017 (Études de philosophie médiévale, 106), 153–162.

Rommevaux, S. (transl.), *Thomas Bradwardine: Traité des rapports entre les rapidités dans les mouvements, suivi de Nicole Oresme: Sur les rapports de rapports*, Paris 2010 (Sagesses médiévales).

Rommevaux, S., 'A Treatise on Proportion in the Tradition of Thomas Bradwardine: The *De proportionibus libri duo* (1528) of Jean Fernel,' *Historia Mathematica* 40 (2013), 164–182.

Rommevaux, S., 'L'ajout de conditions dans les règles du mouvement formulées par Blaise de Parme dans la seconde rédaction de ses *Questions sur la Physique*,' in J. Biard and A. Robert (eds.), *La philosophie de Blaise de Parme. Physique, psychologie, éthique*, Florence 2019 (Micrologus Library, 96), 153–174.

Rommevaux, S., *Les nouvelles théories des rapports mathématiques du XIVe au XVIe siècle*, Turnhout 2014 (De diversis artibus, 92).

Rommevaux, S., 'Six inconvénients découlant de la règle du mouvement de Thomas Bradwardine dans un texte anonyme du XIVe siècle,' in M. Malpangotto, V. Jullien and E. Nicolaïdis (eds.), *L'homme au risque de l'infini. Mélanges d'histoire et de philosophie des sciences offerts à Michel Blay*, Turnhout 2013 (De diversis artibus, 93), 35–47.

Rommevaux, S., 'Un auteur anonyme du XIVe siècle, à Oxford, lecteur de Pierre de Maricourt,' *Revue d'histoire des sciences* 67 (2014), 5–33.

Rommevaux, S., 'Un exemple de *Question* mathématique au Moyen Âge,' *Annals of science* 63/4 (2006), 425–445.

Rommevaux, S., P. Vendrix and V. Zara (eds.), *Proportions. Science, musique, peinture et architecture*. Actes du LIe Colloque International d'Études Humanistes, Turnhout 2011 (Études Renaissantes, 6).

Rose, P.L., 'Bartolomeo Zamberti's Funeral Oration for the Humanist Encyclopedist Giorgio Valla,' in C.H. Clough, *Cultural Aspects of the Italian Renaissance: Essays in Honour of Paul Oskar Kristeller*, Manchester 1976, 299–310.

Roth, F., 'Jacques Legrand,' *Augustiniana* 7 (1957), 316–326.

Roudaut, S., 'Hugolinus of Orvieto and the Controversies about the Perfection of Species,' *Augustiniana* 69 (2019), 299–331.

Roudaut, S., *La mesure de l'être. Le problème de la quantification des formes au Moyen Âge (ca. 1250–1370)*, Leiden (forthcoming) (History of Metaphysics, Ancient, Medieval, Modern, 3).

Russell, B., 'On "Insolubilia" and Their Solution by Symbolic Logic (1906),' in B. Russell, *Essays in Analysis*, ed. D. Lackey, London 1973, 190–214.

Russell, B., and A.N. Whitehead, *Principia Mathematica*, 2nd ed., Cambridge 1927.

Russell, J.L., 'Action and Reaction before Newton,' *The British Journal for the History of Science* 9 (1976), 25–38.

Ryle, G., 'Heterologicality,' *Analysis* 11 (1951), 61–69.

Sabra, A.I., 'The Authorship of the *Liber de Crepusculis*, an Eleventh-Century Work on Atmospheric Refraction,' *Isis* 58 (1967), 77–85.

Saville, C., *Holcot and Kilvington: Questions 81 and 82 in Context* (forthcoming).

Sbrozi, M., 'Metodo matematico e pensiero teologico nel *De causa Dei*,' *Studi medievali* 31 (1990–1991), 143–191.

Schabel, C., 'Ockham, the *Principia* of Holcot and Wodeham, and the Myth of the Two-Year *Sentences* Lecture at Oxford,' *Recherches de Théologie et Philosophie Médiévales* 87 (2020), 59–102.

Schabel, C., 'Richard FitzRalph vs. William Skelton, 1331–1332: The Attribution of the *Determinationes* in a Florence Manuscript,' in: M. Dunne and I. Nolan (eds.), *A Companion to Richard FitzRalph* (forthcoming).

Schemmel, M., *The English Galileo. Thomas Harriot's Work on Motion as an Example of Preclassical Mechanics*, 1: *Interpretation*, Dordrecht 2008 (Boston Studies in the Philosophy of Science, 268).

Schumacher, L., *Early Franciscan Theology: Between Authority and Innovation*, Cambridge 2019.

Shapiro, C., and H. Shapiro, 'De primo et ultimo instanti des Walter Burley,' *Archiv für Geschichte der Philosophie* 47 (1965), 159–173.

Shapiro, H., 'Walter Burley and the Intension and Remission of Forms,' *Speculum* 34 (1959), 413–427.

Sherrington, C., *The Endeavour of Jean Fernel, With a List of the Editions of his Writings*, London 1946.

Silagiová, Z., and F. Šmahel, *Catalogi librorum vetustissimi Universitatis Pragensis*, Turnhout 2015 (Corpus Christianorum. Continuatio Mediaevalis, 271).

Šmahel, F., *Die Prager Universität im Mittelalter: Gesammelte Aufsätze—Charles University in the Middle Ages: Selected Studies*, Leiden 2007 (Education and Society in the Middle Ages and Renaissance, 28).

Šmahel, F., *Verzeichnis der Quellen zum Prager Universalienstreit 1348–1500*, Wrocław 1980 (Mediaevalia Philosophica Polonorum, 25).

Smith, A.M., *Alhacen's Theory of Visual Perception. A Critical Edition, with English Translation and Commentary, of the first three Books of Alhacen's* De aspectibus, *the Medieval Latin Version of Ibn al-Haytham's Kitāb al-Manāẓir*, Philadelphia, PA, 2001 (2 vols.) (Transactions of the American Philosophical Society, 91).

Smith, A.M., *Ptolemy's Theory of Visual Perception: An English Translation of the* Optics *with Introduction and Commentary*, Philadelphia, PA, 1996 (Transactions of the American Philosophical Society, 86/2).

Snedegar, K., 'Merton Calculators (*act. c.* 1300–*c.* 1349),' in *Oxford Dictionary of National Biography*, (2006), doi:10.1093/ref:odnb/95034.

Šolcová, A., 'Jan Šindel—pravděpodobný autor matematického modelu pražského orloje,' *Pokroky matematiky, fyziky a astronomie* 54 (2009), 307–317.

Solère, J.-L., 'D'un commentaire l'autre: l'interaction entre philosophie et théologie au Moyen Âge dans le problème de l'intensification des formes,' in M.O. Goulet-Cazé (ed.), *Le commentaire: entre tradition et innovation. Actes du Colloque International de l'Institut des traditions textuelles (Paris et Villejuif, 22–25 septembre 1999)*, Paris 2000 (Bibliothèque d'histoire de la philosophie), 411–424.

Solère, J.-L., 'Plus ou moins: le vocabulaire de la latitude des formes,' in J. Hamesse and C. Steel (eds.), *L'élaboration du vocabulaire philosophique au Moyen Âge. Actes du Colloque International de Louvain-la-Neuve et Leuven, 12–14 septembre 1998*, Turnhout 2000 (Rencontres de philosophie médiévale, 8), 437–488.

Souffrin, P., '*Velocitas totalis*. Enquête sur une pseudodénomination médiévale,' in S. Caroti and P. Souffrin (eds.), *La nouvelle physique du XIVe siècle. Actes du Colloque International (Nice, 3–5 septembre 1993)*, Florence 1997 (Biblioteca di Nuncius, 24), 251–275.

Spade, P.V., *An Anonymous Fourteenth-Century Treatise on* Insolubles: *Text and Study*, unpublished Ph.D. dissertation, University of Toronto, 1969.

Spade, P.V., 'Five Early Theories in the Mediaeval *Insolubilia*-Literature,' *Vivarium* 25 (1987), 24–46.

Spade, P.V., *Lies, Language and Logic in the Late Middle Ages*, London 1988 (Collected Studies Series, 272).

Spade, P.V., 'Richard Brinkley's *De insolubilibus*: A Preliminary Assessment,' *Rivista di storia della filosofia*, 46 (1991), 245–256.

Spade, P.V., 'Roger Swyneshed's *Insolubilia*: Edition and Comments,' *Archives d'histoire doctrinale et littéraire du Moyen Âge* 46 (1979), 177–220.

Spade, P.V., 'The Manuscripts of William Heytesbury's *Regulae solvendi sophismata*: Conclusions, Notes and Descriptions,' *Medioevo* 15 (1989), 271–313.

Spade, P.V., *The Mediaeval Liar: A Catalogue of* Insolubilia-*Literature*, Toronto, ON, 1975 (Subsidia Mediaevalia, 5).

Spade, P.V., and S. Read, 'Insolubles,' in E.N. Zalta (ed.), *The Stanford Encyclopedia of Philosophy* (Fall 2017 Edition), URL = https://plato.stanford.edu/archives/fall2017/entries/insolubles/.

Spunar, P., *Repertorium auctorum Bohemorum provectum idearum post universitatem Pragensem conditam illustrans*, Wrocław 1985 (Studia Copernicana 25).

Stigler, S.M., 'Stigler's Law of Eponymy,' *Transactions of the New York Academy of Sciences* 39 (1980), 147–158.

Swieżawski, S., *Histoire de la philosophie européenne au XVe siècle*, ed. M. Prokopowicz, transl. H. Rollet and M. Prokopowicz, Paris 1990.

Sylla, E.D., 'Alvarus Thomas and the Role of Logic and Calculations in Sixteenth-Century Natural Philosophy,' in S. Caroti (ed.), *Studies in Medieval Natural Philosophy*, Florence 1989 (Biblioteca di Nuncius, 1), 257–298.

Sylla, E.D., 'Averroes and Fourteenth-Century Theories of Alteration. *Minima Naturalia* and the Distinction between Mathematics and Physics,' in: P.J.J.M. Bakker (ed.), *Averroes' Natural Philosophy and Its Reception in the Latin West*, Leuven 2015 (Ancient and medieval philosophy, series I, 50), 141–192.

Sylla, E.D., 'Compounding Ratios: Bradwardine, Oresme, and the First Edition of Newton's *Principia*,' in E. Mendelsohn (ed.), *Transformation and Tradition in the Sciences. Essays in Honour of I. Bernard Cohen*, Cambridge 1984, 11–43.

Sylla, E.D., 'Dumbleton, John,' in C.C. Gillispie, F.L. Holmes, N. Koertge and T. Gale (eds.), *Complete Dictionary of Scientific Biography*, vol. 7, New York 2008, 116–117.

Sylla, E.D., 'Galileo and the Oxford Calculatores: Analytical Languages and the Mean-Speed Theorem for Accelerated Motion,' in W.A. Wallace (ed.), *Reinterpreting Galileo*, Washington, DC, 1986 (Studies in Philosophy and the History of Philosophy, 15), 53–108.

Sylla, E.D., 'Guide to the Text,' in John Buridan, *Quaestiones super octo libros Physicorum Aristotelis (secundum ultimam lecturam)*, Libri III–IV, eds. M. Streijger and P.J.J.M. Bakker, Guide to the Text by E.D. Sylla, Leiden 2016 (Medieval and Early Modern Philosophy and Science, 27), XX–CCXVIII.

Sylla, E.D., 'Infinity and Continuity: Thomas Bradwardine and His Contemporaries,' in S. Shapiro and G. Hellman (eds.), *The History of Continua: Philosophical and Mathematical Perspectives*, Oxford 2021, 49–81.

Sylla, E.D., 'Introduction, Notes, and Translation to Jacob Bernoulli,' in E.D. Sylla (transl.), *J. Bernoulli, The Art of Conjecturing, Together with Letter to a Friend on Sets in Court Tennis*, Baltimore 2006, 1–126.

Sylla, E.D., 'Mathematical Physics and Imagination in the Work of the Oxford Calculators: Roger Swineshead's *On Natural Motions*,' in E. Grant and J.E. Murdoch (eds.), *Mathematics and its Applications to Science and Natural Philosophy in the Middle Ages. Essays in Honour of Marshall Clagett*, Cambridge 1987, 69–101.

Sylla, E.D., 'Mathematics and Physics of First and Last Instants: Walter Burley and William of Ockham,' in F. Goubier and M. Roques (eds.), *The Instant of Change in Medieval Philosophy and Beyond*, Leiden 2018 (Special issue of *Vivarium*, 55 [2017]), 103–129.

Sylla, E.D., 'Mathematics in the *Liber de Triplici Motu* (1509) of Alvarus Thomas of Lisbon,' in L. Saraiva and H. Leitão (eds.), *The Practice of Mathematics in Portugal*, Coimbra 2004 (Acta Universitatis Conimbrigensis), 109–161.

Sylla, E.D., 'Medieval Concepts of the Latitude of Forms: The Oxford Calculators,' *Archives d'histoire doctrinale et littéraire du Moyen Âge* 40 (1973), 223–283.

Sylla, E.D., 'Medieval Quantifications of Qualities: The "Merton School",' *Archive for History of Exact Sciences* 8 (1971–1972), 9–39.

Sylla, E.D., 'Oxford Calculators,' in H. Lagerlund (ed.), *Encyclopedia of Medieval Philosophy*, Dordrecht 2011, 903–908.

Sylla, E.D., Review of S. Rommevaux, *Thomas Bradwardine: Traité des rapports entre les rapidités dans les mouvements*, *Aestimatio* 8 (2011), 23–29.

Sylla, E.D., 'Swyneshed (Swineshead), Roger,' in N. Koertge (ed.), *New Dictionary of Scientific Biography*, 6, Detroit, MI, 2008, 563–564.

Sylla, E.D., 'The Fate of the Oxford Calculatory Tradition,' in C. Wénin (ed.), *L'homme et son univers au Moyen Âge. Actes du septième congrès international de philosophie médiévale (30 août–4 septembre 1982)*, Louvain-la-Neuve 1986 (Philosophes médiévaux, 26), 2: 692–698.

Sylla, E.D., 'The Origin and Fate of Thomas Bradwardine's *De proportionibus velocitatum in motibus* in Relation to the History of Mathematics,' in W.R. Laird and S. Roux (eds.), *Mechanics and Natural Philosophy before the Scientific Revolution*, Dordrecht 2008 (Boston Studies in the Philosophy of Science, 254), 67–119.

Sylla, E.D., 'The Oxford Calculators and Mathematical Physics: John Dumbleton's *Summa logicae et philosophiae naturalis*, Parts II and III,' in S. Unguru (ed.), *Physics, Cosmology, and Astronomy, 1300–1700: Tension and Accomodation*, Dordrecht 1991 (Boston Studies in the Philosophy of Science, 126), 129–161.

Sylla, E.D., *The Oxford Calculators and the Mathematics of Motion, 1320–1350. Physics and*

Measurement by Latitudes, Ph.D. dissertation, Harvard University, 1970 (repr. with preface and errata, New York 1991).

Sylla, E.D., 'The Oxford Calculators in Context,' *Science in Context* 1 (1987), 257–279.

Sylla, E.D., 'The Oxford Calculators,' in N. Kretzmann, A. Kenny, J. Pinborg and E. Stump (eds.), *The Cambridge History of Later Medieval Philosophy. From the Rediscovery of Aristotle to the Disintegration of Scholasticism, 1100–1600*, Cambridge 1982, 540–563.

Sylla, E.D., 'The Oxford Calculators' Middle Degree Theorem in Context,' *Early Science and Medicine* 15 (2010), 338–370.

Sylla, E.D., 'Thomas Bradwardine's *De continuo* and the Structure of Fourteenth-Century Learning,' in E.D. Sylla and M.R. McVaugh (eds.), *Texts and Contexts in Ancient and Medieval Science. Studies on the Occasion of John E. Murdoch's Seventieth Birthday*, Leiden 1997 (Brill's Studies in Intellectual History, 78), 148–186.

Sylla, E.D., 'Transmission of the New Physics of the Fourteenth Century from England to the Continent,' in S. Caroti and P. Souffrin (eds.), *La nouvelle physique du XIVe siècle. Actes du Colloque International (Nice, 3–5 septembre 1993)*, Florence 1997 (Biblioteca di Nuncius, 24), 65–110.

Sylla, E.D., 'Walter Burley's *Tractatus primus*: Evidence Concerning the Relations of Disputations and Written Works,' *Franciscan Studies* 44 (1984) (*William of Ockham [1285–1347] Commemorative Issue*), 1: 257–274.

Tachau, K.H., 'French Theology in the Mid-Fourteenth Century: Vatican Latin 986 and Wrocław, Milich F.64,' *Archives d'histoire doctrinale et littéraire du Moyen Âge* 51 (1984), 41–80.

Tachau, K.H., and P.A. Streveler (eds.), *Robert Holcot. Seeing the Future Clearly. Questions on Future Contingents*, Toronto, ON, 1995 (Studies and Texts, 119).

Taub, L., 'Physiological Analogies and Metaphors in Explanations of the Earth and the Cosmos,' in M. Horstmanshoff, H. King and C. Zittel (eds.), *Blood, Sweat and Tears: The Changing Concepts of Physiology from Antiquity into Early Modern Europe*, Leiden 2012 (Intersections, 25), 41–61.

Thakkar, M., 'Mathematics in Fourteenth-Century Theology,' in E. Robson and J. Stedall (eds.), *The Oxford Handbook of the History of Mathematics*, Oxford 2009, 619–638.

Thakkar, M., 'Wyclif's *Logica* and the *Logica Oxoniensis*,' in L. Campi and S. Simonetta (eds.), *Before and After Wyclif. Sources and Textual Influences*, Basel 2020 (Textes et Études du Moyen Âge, 97), 1–31.

Thijssen, J.M.M.H., 'The Buridan School Reassessed. John Buridan and Albert of Saxony,' *Vivarium* 42 (2004), 18–42.

Thijssen, J.M.M.H., 'The Debate over the Nature of Motion: John Buridan, Nicole Oresme and Albert of Saxony. With an Edition of John Buridan's *Quaestiones super libros Physicorum, secundum ultimam lecturam*, Book III, q. 7,' *Early Science and Medicine* 14 (2009) 186–210.

Thomson, S.H., 'Unnoticed MSS and Works of Wyclif II,' *Journal of Theological Studies* 38 (1937), 139–148.

Thomson, W.R., *The Latin Writings of John Wyclyf. An Annotated Catalog*, Toronto, ON, 1983 (Subsidia Mediaevalia, 14).

Thorndike, L., *A History of Magic and Experimental Science*, New York 1923–1958 (8 vols.).

Thorndike, L., 'An Anonymous Treatise in Six Books on Metaphysics and Natural Philosophy,' *The Philosophical Review* 40 (1931), 317–340.

Toomer, G.J., 'Hipparchus' Empirical Basis for His Lunar Mean Motions: A Historical Footnote to Olaf Pedersen, A Survey of the *Almagest*, 161–164,' *Centaurus* 24 (1980), 97–109.

Trifogli, C., 'Il problema dello statuto ontologico del tempo nelle *Quaestiones super Physicam* di Thomas Wylton e di Giovanni di Jandun,' *Documenti e studi sulla tradizione filosofica medievale* 1 (1990), 491–548.

Trifogli, C., 'Motion and Time,' in A.D. Conti (ed.), *A companion to Walter Burley. Late Medieval Logician and Metaphysician*, Leiden 2013 (Brill's Companions to the Christian Tradition, 41), 267–299.

Trifogli, C., 'The Quodlibet of Thomas Wylton,' in C. Schabel (ed.), *Theological Quodlibeta in the Middle Ages: The Fourteenth Century*, Leiden 2007 (Brill's Companions to the Christian Tradition, 7), 231–266.

Trifogli, C., 'The Reality of Time in the Commentary Tradition on the *Physics*: the Case of Wylton and Burley,' in G. Fioravanti, C. Leonardi and S. Perfetti (eds.), *Il commento filosofico nell'occidente latino (secoli XIII–XV). Actes du Colloque International de Florence-Pise, octobre 2000*, Turnhout 2002 (Rencontres de philosophie médiévale, 10), 233–251.

Trifogli, C., 'Thomas Wylton on the Instant of Time,' in A. Zimmermann (ed.), *Mensch und Natur im Mittelalter*, Berlin 1991 (Miscellanea Mediaevalia, 21/1), 308–318.

Trifogli, C., 'Thomas Wylton's Question *An contingit dare ultimum rei permanentis in esse*,' *Medieval Philosophy and Theology* 4 (1994), 91–141.

Trifogli, C., 'Walter Burley on the *Incipit* and *Desinit* of an Instant of Time,' in F. Goubier and M. Roques (eds.), *The Instant of Change in Medieval Philosophy and Beyond*, Leiden 2018 (Special issue of *Vivarium*, 55 [2017]), 85–102.

Truhlář, J., *Catalogus codicum manu scriptorum latinorum qui in C.R. Bibliotheca publica atque universitatis Pragensis asservantur*, Prague 1905–1906 (2 vols.).

Trzeciok, S.P., *Alvarus Thomas und sein* Liber de triplici motu, 1: *Naturphilosophie an der Pariser Artistenfakultät*; 2: *Bearbeiteter Text und Faksimile*, Berlin 2016, URL = https://edition-open-sources.org/media/sources/7/Sources7.pdf.

Uiblein, P., *Die Universität Wien im Mittelalter. Beiträge und Forschungen*, Vienna 1999 (Schriftenreihe des Universitätsarchivs, Universität Wien, 11).

Ulman, C., *Johann Wessel, ein Vorgänger Luthers. Zur Characteristik der christlichen

Kirche und Theologie in ihrem Uebergang aus dem Mittelalter in die Reformationszeit, Hamburg 1834.

Vasoli, C., *Studi sulla cultura del rinascimento*, Manduria 1968 (Biblioteca de studi moderni, 5).

Verde, A.F., *Lo Studio fiorentino 1473–1503. IV: La vita universitaria*, 1: *Anni scolastici 1473/74–1481/82*, Florence 1985.

Wallace, W.A., 'Domingo de Soto and the Iberian Roots of Galileo's Science,' in K. White (ed.), *Hispanic Philosophy in the Age of Discovery*, Washington, DC, 1997 (Studies in Philosophy and the History of Philosophy, 29), 113–129.

Wallace, W.A., *Prelude to Galileo: Essays on Medieval and Sixteenth-Century Sources of Galileo's Thought*, Dordrecht 1981 (Boston Studies in the Philosophy of Science, 62).

Wallace, W.A., 'The *Calculatores* in Early Sixteenth-Century Physics,' *The British Journal for the History of Science* 4 (1969), 221–232.

Wallace, W.A., 'The Calculatores in the Sixteenth Century,' in Id., *Prelude to Galileo: Essays on Medieval and Sixteenth-Century Sources of Galileo's Thought*, Dordrecht 1981 (Boston Studies in the Philosophy of Science, 62), 78–90.

Wallace, W.A., 'The Enigma of Domingo de Soto: *Uniformiter Difformis* and Falling Bodies in Late Medieval Physics,' *Isis* 59 (1968), 384–401.

Wayne Hellmann, J.A., *Divine and Created Order in Bonaventura's Theology*, St. Bonaventure, NY, 2001 (Theology Series).

Weisheipl, J.A., 'Curriculum of the Faculty of Arts at Oxford in the Early Fourteenth Century,' *Mediaeval Studies* 26 (1964), 143–185.

Weisheipl, J.A., *Early Fourteenth Century Physics of the Merton 'School', with Special Reference to Dumbleton and Heytesbury*, unpublished Ph.D. dissertation, Oxford University, 1956.

Weisheipl, J.A., 'Ockham and some Mertonians,' *Mediaeval Studies* 30 (1968), 163–213, repr. in J.I. Catto and T.A.R. Evans (eds.), *The History of the University of Oxford*, 2: *Late Medieval Oxford*, Oxford 1992, 607–658.

Weisheipl, J.A., 'Repertorium Mertonense,' *Medieval Studies* 31 (1969), 174–224.

Weisheipl, J.A., 'Roger Swyneshed, OSB, Logician, Natural Philosopher, and Theologian,' in *Oxford Studies Presented to Daniel Callus*, Oxford, 1964 (Oxford Historical Society, New Series, 16), 231–252.

Weisheipl, J.A., 'The Interpretation of Aristotle's Physics and the Science of Motion,' in N. Kretzmann, A. Kenny, J. Pinborg and E. Stump (eds.), *The Cambridge History of Later Medieval Philosophy. From the Rediscovery of Aristotle to the Disintegration of Scholasticism, 1100–1600*, Cambridge 1982, 519–536.

Weisheipl, J.A., 'The Place of John Dumbleton in the Merton School,' *Isis* 50 (1959), 439–454.

Westman, R.S., 'Humanism and Scientific Roles in the Sixteenth Century,' in F. Krafft and R. Schmitz (eds.), *Humanismus und Naturwissenschaften*, Boppard am Rhein 1980 (Beiträge zur Humanismusforschung, 6), 83–100.

Wilks Dolnikowski, E., *Thomas Bradwardine. A View of Time and a Vision of Eternity in Fourteenth-Century Thought*, Leiden 1995 (Studies in the History of Christian Thought, 65).

Wilson, C., *William Heytesbury: Medieval Logic and the Rise of Mathematical Physics*, Madison, WI, 2nd ed., 1960 (University of Wisconsin Publications in Medieval Science, 3).

Wind, E., *Pagan Mysteries in the Renaissance. An Exploration of Philosophical and Mystical Sources of Iconography in Renaissance Art*, London 1958.

Włodek, Z., 'Le commentaire de Nicolas d'Amsterdam sur le *De anima* d'Aristote. Introduction—Textes inédits,' *Mediaevalia Philosophica Polonorum* 11 (1963), 23–42.

Wöhler, H.-U., 'Der operationale Konzeptualismus des Nikolaus von Amsterdam vor dem Hintergrund der Universaliendebatte,' *Bochumer Philosophisches Jahrbuch für Antike und Mittelalter* 15 (2012), 26–36.

Wöhler, H.-U., 'Nicolaus de Amsterdam, *Quaestiones circa libros Metaphysicae Aristotelis*, Liber VII, Quaestio 12: *Utrum ponenda sint aliqua universalia realia a singularibus secundum esse distincta*,' *Bochumer Philosophisches Jahrbuch für Antike und Mittelalter* 17 (2014), 221–250.

Wood, W., 'The Wager and Pascal's Theology,' in P. Bartha and L. Pasternack (eds.), *Pascal's Wager*, Cambridge 2018 (Classic Philosophical Arguments, 4), 45–63.

Worstbrock, F.J., 'Adolf von Wien,' in K. Langosch and K. Ruh (eds.), *Die deutsche Literatur des Mittelalters: Verfasserlexikon*, 1 (Aalen—Futerer), Berlin, 2nd ed., 1978, 68–71.

Worstbrock, F.J., 'Ulrich von Wien,' in K. Langosch and K. Ruh (eds.), *Die deutsche Literatur des Mittelalters: Verfasserlexikon*, 10 (Ulr-–Z), Berlin, 2nd ed., 1999, 53–57.

Zanin, F., *L'analisi matematica del movimento e i limiti della fisica tardo-medievale. La ricezione della* perspectiva *e delle* calculationes *alla Facoltà delle Arti di Parigi* (*1340–1350*), Padua 2004 (Subsidia Mediaevalia Patavina, 6).

Zepeda, H., *The First Latin Treatise on Ptolemy's Astronomy: The Almagesti minor* (*c. 1200*), Turnhout 2018 (Ptolemaeus Arabus et Latinus, Texts Volume 1).

Zetterberg, J.P., 'The Mistaking of "the Mathematics" for Magic in Tudor and Stuart England,' *The Sixteenth Century Journal* 11 (1980), 83–97.

Zimmermann, A., *Ontologie oder Metaphysik? Die Diskussion über den Gegenstand der Metaphysik im 13. und 14. Jahrhundert: Texte und Untersuchungen*, Leuven 1998 (Recherches de Théologie et Philosophie médiévales. Bibliotheca, 1).

Index Nominum